TABLE OF CONTENTS

Page

INTRODUCTION -- Y. Pomeranz ix

PART A: Economics and Utilization

1. ECONOMIC AND TECHNICAL CONSTRAINTS IN
   IMPLEMENTING INDUSTRIAL UTILIZATION OF
   WHEAT IN NORTH AMERICA AND EUROPE
   -- D. J. Leuck 1

2. WHEAT: RIGHT FOR BREAD, BUT FOR MUCH
   MORE TOO -- C. W. Wrigley and
   G. J. McMaster 21

3. THE ECONOMICS OF WHEAT FLOUR AS A RAW
   MATERIAL FOR GLUTEN AND STARCH
   MANUFACTURE -- B. W. Morrison 41

4. A NEW AGRICULTURAL SYSTEM IN EUROPE?
   -- L. Munck, F. Rexen, B. Petersen, and
   P. Bjørn Petersen 49

PART B: Uniqueness of Wheat Structure and Quality

5. HOW THE STRUCTURE OF THE WHEAT CARYOPSIS
   SHOULD BE MODIFIED TO INCREASE ITS
   END-USE VALUE -- D. B. Bechtel 71

6. NUTRITIONAL POTENTIAL AND LIMITATIONS OF
   WHEAT -- A. A. Betschart 85

7. CORRELATION BETWEEN VISUAL ASSESSMENT OF
   PRE-HARVEST SPROUTING OF WHEAT AND THE
   α-AMYLASE CONTENT OF GRAIN AND FLOUR
   -- A. de Francisco, L. Munck, and
   J. Ruud-Hansen 103

8. CHALLENGES WITH ALPHA-AMYLASE INHIBITION
   IN SPROUT-DAMAGED WHEAT
   -- U. A. Zawistowska 117

9. IMMUNODIAGNOSTIC APPROACHES TO QUALITY
   AND PROCESS CONTROL IN WHEAT BREEDING
   AND UTILISATION -- J. H. Skerritt 131

10. NOVEL APPROACHES TO GRAIN QUALITY AND
    PROCESS CONTROL -- I. L. Batey,
    C. W. Wrigley, and P. W. Gras 161

11. CHARACTERIZATION OF WHEAT STARCH AND
    GLUTEN AS RELATED TO END-USE PROPERTIES
    -- A.-C. Eliasson 177

PART C: Uniqueness of Wheat Components

12. UNIQUENESS OF WHEAT STARCH
    -- W. R. Morrison 193

13. PROPERTIES OF WHEAT STARCH COMPARED
    TO NORMAL MAIZE STARCH -- Y.-C. Shi and
    P. A. Seib 215

14. MODIFICATION OF WHEAT STARCH
    -- S. A. S. Craig 235

15. MINOR COMPONENTS OF WHEAT STARCH AND
    THEIR TECHNOLOGICAL SIGNIFICANCE
    -- T. Galliard, P. Bowler, and
    P. J. Towersey 251

16. WHEAT GLUTEN IS GOOD NOT ONLY FOR
    BREADMAKING -- W. Bushuk and C. Wadhawan ... 263

17. GLUTENIN STRUCTURE IN RELATION TO WHEAT
    QUALITY -- D. D. Kasarda 277

18. TELLING DIFFERENCES AMONG WHEAT PROTEINS
    MAY MAKE A DIFFERENCE IN MARKETING
    -- J. A. Bietz 303

19. WHEAT LIPIDS ARE UNIQUE
    -- W. R. Morrison 319

20. FUNCTIONAL SIGNIFICANCE OF WHEAT LIPIDS
    -- O. K. Chung 341

21. THE USE OF WHEAT FLOURS IN EXTRUSION
    COOKING -- R. C. E. Guy 369

22. THE TWIN-SCREW EXTRUSION COOKER AS A
    VERSATILE TOOL FOR WHEAT PROCESSING
    -- P. K. Linko 379

23. EXTRUSION COOKING OF WHEAT PRODUCTS
    -- B. van Lengerich, F. Meuser,
    and W. Pfaller 395

PART D: Wheat Processing, Milling, and Fractionation

24. PROCESSING AND UTILIZATION OF
    AIR-CLASSIFIED WHEAT FLOUR FRACTIONS
    -- F. W. Sosulski and D. M. Nowakowski 431

25. A SHORT MILLING PROCESS FOR WHEAT
    -- L. E. Gram, L. Munck, and
    M. P. Andersen 445

26. WHEAT FRACTIONATION AND UTILIZATION
    -- Y. Mälkki, J. Sorvaniemi,
    O. Myllymäki, J. Peuhkuri, and E. Pessa 457

27. A STUDY OF THE FACTORS AFFECTING THE
    SEPARATION OF WHEAT FLOUR INTO STARCH
    AND GLUTEN -- R. J. Hamer, P. L. Weegels,
    J. P. Marseille, and M. Kelfkens 467

28. DEVELOPMENTS IN THE EXTRACTION OF STARCH AND GLUTEN FROM WHEAT FLOUR AND WHEAT KERNELS -- F. Meuser, F. Althoff, and H. Huster 479

29. THE ENGINEERING OF A MODERN WHEAT STARCH PROCESS -- D. J. Barr 501

30. PRODUCTION OF WHEAT STARCH AND GLUTEN: HISTORICAL REVIEW AND DEVELOPMENT INTO A NEW APPROACH -- W. R. M. Zwitserloot 509

31. PROCESS FOR THE INDUSTRIAL PRODUCTION OF WHEAT STARCH FROM WHOLE WHEAT -- W. Kempf and C. Röhrmann 521

32. NEW INDUSTRIAL APPLICATIONS OF CHEMICALLY MODIFIED WHEAT GLUTEN -- W. Kempf, W. Bergthaller, and B. Pelech 541

33. THERMAL MODIFICATION OF GLUTEN AS RELATED TO END-USE PROPERTIES -- J. C. Autran, O. Ait-Mouh, and P. Feillet 563

34. THE INTERACTIONS THAT PRODUCE UNIQUE PRODUCTS FROM WHEAT FLOUR -- R. C. Hoseney 595

35. ADVANCES IN WHEAT PROCESSING AND UTILIZATION IN JAPAN -- S. Nagao 607

36. NEW AND POTENTIAL MARKETS FOR WHEAT STARCH -- W. M. Doane 615

PART E: New Products

37. TRUDEX - A SPRAY DRIED TOTAL WHEAT SUGAR -- N. Wookey 633

38. THE GLUCOTECH PROCESS -- H. W. Doelle and W. J. Wells III........................ 641

39. UTILIZATION OF BY-PRODUCTS OF WHEAT-
    BASED ALCOHOL FERMENTATION -- Y. V. Wu 657

40. THE DEVELOPMENT AND UTILIZATION OF WHEAT
    DISTILLERS' GRAINS WITH SOLUBLES (DDGS)
    AS A FOOD INGREDIENT -- B. A. Rasco 675

41. WHEAT STARCH IN THE FORMULATION OF
    DEGRADABLE PLASTICS -- G. J. L. Griffin 695

WRAP-UP OF CONFERENCE -- Y. Pomeranz 707

SUBJECT INDEX 713

INTRODUCTION

This book contains edited proceedings of the Wheat Industry Utilization Conference, which was held in San Diego, CA on October 7-8, 1988. The meeting was organized by the National Association of Wheat Growers Foundation in cooperation with the Agricultural Research Service of the U.S. Department of Agriculture and the American Association of Cereal Chemists.

From 1962 to 1977, 10 wheat utilization conferences were held. The proceedings of those conferences included 216 papers printed on 1756 pages. They covered many aspects of food-feed- and industrial uses of wheat. Why then this conference? The reasons are that over the last decade much new information has been gained worldwide. That information warrants a review of the state-of-the-art, critical evaluation of what has been accomplished, and projection of most promising future developments. Paraphrasing E. G. Leterman: If you and I have a coin , and we exchange the coins, each of us has still only one coin. If you and I have an idea, and we exchange ideas, each of us has two ideas.

The objectives of the conference were to bring together experts from around the world to explore, report on, discuss, and evaluate the following factors affecting the industrial utilization of wheat:
I. unique physical, chemical, and nutritional properties of wheat and wheat products;
II. non-conventional uses of wheat and wheat products;
III. novel and more economical approaches to wheat processing and utilization; and
IV. new prospects, developments, and projections related to wheat utilization.

The focus of the Conference was Wheat is Unique in Structure, Composition, Processing, End-Use

Properties, and Products. The 43 papers were presented by prominent researchers, engineers, and representatives of the wheat production and processing industries from Australia, Canada, Denmark, Finland, France, Japan, The Netherlands, Scotland, Sweden, the United Kingdom, the United States, and West Germany.

The presentations were arranged in five sessions:
I. The Politics, Economics and Statistics of Wheat Production, Processing, and Industrial Utilization;
II. Uniqueness of Wheat Structure and Quality;
III. Uniqueness of Wheat Components;
IV. Wheat Processing, Milling, and Fractionating; and
V. Wheat Products.

The first session began with a review of the economic and technical constraints on implementing industrial utilization of wheat in North America and Europe. The papers were presented in the context of the underpinning and overriding consideration that economics (and politics) is what makes industrial utilization start, succeed and continue, or fail and end. It was followed by a discussion of some novel European and Australian concepts for building agricultural factories for processing and utilizing the whole wheat plant.

The presentations on wheat structure described how the wheat kernel is put together and whether it is possible by modern tools of genetic engineering to create a better kernel from the standpoint of processing, end-use properties, and utilization. Following a discussion of the nutritional potential and limitations of wheat as a food and feed, was a series of five papers on wheat quality (sprout damage assessment and prevention, novel approaches to process control, and general characterization of major wheat components as related to end-use properties).

The prominent British scientist, J. B. S. Haldane, has been quoted to say: a) that our true knowledge and understanding of biological systems and materials comes from detailed knowledge of their biochemistry, and b) that one page of well substantiated data is by far superior to 500 pages of loose talk. In the context of the conference, biochemistry could be equated with uniqueness of wheat components. Insofar as quality of data is concerned, the speakers were reminded of the statement attributed to Mark Twain that the great thing about science is that one may get such a great return of speculation for such a small investment of facts. Imaginative projections were encouraged, provided they were based on sound evidence. To this end, uniqueness of wheat starch (including its properties, modification, and minor components and their technological significance), wheat proteins, and wheat lipids were reviewed in nine papers. The subsequent three papers in the session dealt with extrusion cooking of wheat and wheat products - as an introduction to the next session on wheat milling, air classification, and wet fractionation. The eight papers concerned some novel and unique engineering and biotechnological approaches and practical experiences. The papers document new processing technologies and demonstrate that things, which were considered only a decade ago in the realm of impossible, have been made a reality (processing of whole wheat flour, low protein flour, or low quality flour); that separation can be made more effective, yields can be increased, effluent formation can be reduced or even prevented, and that quality of the separated products can be improved. It makes you realize that until you succeed you are stubborn, once you succeed you are persistent. In a transition to the last session, were a review on new industrial applications of chemically modified gluten and thermal modification of gluten as related to end-use properties and papers on interactions that produce unique products from wheat flour, advances in wheat processing and utilization in Japan (including a process for production of gluten hydrolyzates), and a survey of new and potential markets for wheat starch (based largely on the experience gained from maize starch).

The last session, on new products, dealt with wheat sugar, efficient production of ethanol, use of wheat gluten and starch as functional ingredients in surimi-based products, utilization of by-products of wheat-based alcohol fermentation as a food ingredient, and, last but not least, wheat starch in the formulation of degradable plastics. It is a new challenging world of novel uses and products - some traditional with a new twist and some esoteric and most unusual.

The challenge and charge to the speakers was to delineate the status of basic research that can be converted into applied utilization technology and to identify and make known areas in which applied utilization technology is ripe and ready to be converted into practical use. I am confident that anyone who attended attentively the conference was more knowledgeable; I like to think that at the end of the conference (or after reading this book) many of us are wiser; and I am hopeful that we found ways to put the knowledge and wisdom to better use in wheat industry utilization. For that we are all indebted to the speakers who prepared the papers included in this book.

V. Smail (previously with the NAWG-Foundation and presently with Biotechnica Agriculture) plowed with me through the first stages of organizing the meeting and J. Rees (NAWG-Foundation) was in charge of the fine local arrangements. Colorado Association of Wheat Growers, E. I. du Pont de Nemours and Co., International Wheat Gluten Association, Kansas Wheat Commission, Minnesota Wheat Research and Promotion Council, Montana Wheat and Barley Committee, Nebraska Wheat Board, North Dakota Wheat Commission, Oklahoma Wheat Commission, and South Dakota Wheat Commission are thanked for the financial support. Mrs. Virginia Hansel, my secretary, is thanked for her uncompromising technical proofing, checking references, retyping many manuscripts, and fine eye and mind in assuring uniformity of style and of appearance. The American Association of Cereal Chemists is thanked for the valuable input into organizing the symposium and for prompt and fine publication of these proceedings.

                      Y. Pomeranz
                      Technical Program Chair

November, 1988
Pullman, WA

1

ECONOMIC AND TECHNICAL CONSTRAINTS IN IMPLEMENTING INDUSTRIAL UTILIZATION OF WHEAT IN NORTH AMERICA AND EUROPE[1]

Dale J. Leuck

United States Department of Agriculture
Economic Research Service
1301 New York Avenue
Washington, D.C. 20005

INTRODUCTION

The separation of the starch and gluten fractions in wheat flour is termed "wheat washing" because the fractions are washed apart with water. The wheat washing industry has grown rapidly in the European Community (EC), but slowly in North America (Table I). Wheat starch increased from only 3 percent of a 3.0 million ton EC starch market in 1974 to 10 percent of a 4.0 million ton market in 1984 (Kempf, 1984). The growth of this industry in the EC was facilitated by technical advances and EC agricultural (economic) policies. Further expansion of wheat washing in the EC will continue to depend

---

[1] Views expressed in this chapter the author's and do not necessarily reflect those of the United States Department of Agriculture, other government agencies, private institutions, or persons.

This Chapter is in the public domain and not copy-rightable. It may be freely reprinted with customary crediting of the source. American Association of Cereal Chemists, 1989.

Table I. Wheat Gluten Production in North America and the EC[a]

Region	1980	1985	1986
	– 1,000 metric tons –		
North America	35	45	47
European Community	25	75	130[b]
Other regions	20	71	76
Total world	80	191	253

[a]Source: (Sosland, 1986) and author's estimates for North America and other regions for 1980.
[b]Capacity.

upon favorable agricultural policies, and could further affect on EC trade in wheat and wheat gluten.

TECHNICAL CHARACTERISTICS OF THE WHEAT WASHING INDUSTRY

The wheat washing industry in both North America and the EC uses similar technologies, and its outputs are used for similar purposes (Pomeranz, 1983; Radley, 1976; Rexen and Munck, 1984; Sosland, 1986).

Three basic wheat washing processes exist, all of which produce gluten, A-starch, B-starch, and other by-products in relatively fixed proportions. The three processes, discussed elsewhere in these proceedings, are: the Martin-process; the "wet" screening process; and the Alfa-Laval process. Flour from high protein (about 13 percent) hard wheat is generally used as the raw material in North America and flour from low protein (between 10 percent and 11 percent) soft wheat is generally used in Europe.

The raw A-starch is generally modified (chemically or otherwise) for food or industrial uses. Wheat starch and corn starch substitute in many uses because they have similar properties, but wheat starch granules are slightly larger and have a lower gelatinization temperature than corn starch granules (Rexen and Munck, 1984). The B-starch and other byproducts are used mainly for livestock feed.

Wheat gluten contains 75 to 80 percent protein and its major use is to provide a desired volume in baked goods (Chamberlain, 1980; Collins and Evans, 1983; Home Grown Cereals Authority, 1985). Wheat gluten, which is properly milled in order to serve this purpose, is known as "vital" gluten. A common "rule of thumb" in Europe is that a mixture of 98 percent low protein (ie. 10 percent) soft wheat flour fortified with 2 percent vital wheat gluten has the same baking qualities as a high protein (11.3 percent) hard wheat flour (Chamberlain, 1980; Collins and Evans, 1983). Devitalized wheat gluten is used as a protein additive in many human and pet foods.

AGRICULTURAL POLICIES IN NORTH AMERICA

The U.S. grain policy entitles farmers who enroll in acreage reduction programs to receive deficiency payments and pledge their crop as security for nonrecourse commodity loans. The deficiency payment per bushel is the difference between a target price and either the market price or loan rate (which ever is higher). If market prices rise above the loan rate, producers may repay their loan with interest and sell their crop. Otherwise, they may forfeit their crop as settlement for the loan.

The U.S. program allows mutual influence of U.S. and world prices at levels above the U.S. loan rate because little grain is placed under loan. However, as declines in world prices pull U.S. prices below the loan rates, the increase of supplies under loan limits the decline in U.S. and world prices. High participation rates in the U.S. program and the influence of U.S. grain exports on world prices have generally restrained both U.S. and world grain prices from falling much below the U.S. loan rate.

Transportation subsidies, input subsidies, and income stabilization payments are the primary tools of Canadian grain policy. These subsidies generally have not prevented Canadian grain prices from varying with world prices. Canada did have a two-tier price system for wheat, but it did not generally drive a significant wedge between Canadian and world prices (Economic Research Service, 1988).

AGRICULTURAL POLICIES IN THE EUROPEAN COMMUNITY

The EC's Common Agricultural Policy (CAP) determines EC grain prices independently of world prices. The CAP raises the prices of imported grain to a <u>minimum import</u> (<u>threshold</u>) level by charging a <u>variable levy</u>, equal to the difference between the threshold price and the lowest c.i.f. offer price, on any import. The CAP also prevents the prices of surplus grain from falling below <u>intervention prices</u> by allowing exporters to bid for export refunds (subsidies) with which to export surplus grain.

The CAP has increased the intervention price for wheat less rapidly than the threshold prices for wheat and corn (Figure 1). The EC has exceeded self sufficiency in low protein soft wheat since the early 1970's, so that soft wheat prices have been supported by the intervention price. The EC has been deficit in both high protein hard wheat and "high quality" corn used in wet milling (about one-half of which has been grown in the EC in recent years), so that their market prices are determined by the threshold prices.

"Incentive" prices facing the wheat washing and the corn wet milling industries are calculated by subtracting "production refunds" from the wheat intervention and corn threshold prices, respectively. A system of production refunds paid to all starch producers out of the common budget operated by the EC Commission was implemented in 1967 for the purpose of compensating EC starch producers for higher EC grain prices, and significantly modified in 1975 (Commission of the European Community, Reg 2727/75).

The decline in the wheat incentive price relative to the corn incentive and wheat threshold prices has encouraged shifts from corn starch to wheat starch and from hard wheat flour to gluten-fortified flour (Figure 1). Variable import levies and export subsidies for starch and wheat gluten keep EC starch and gluten prices above world levels.

EVALUATING THE PAST PROFITABILITY OF THE EC WHEAT WASHING INDUSTRY

The change in the profitability of these

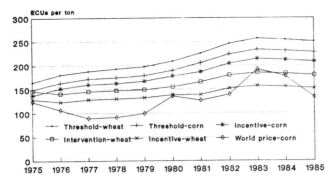

Fig.1.EC policy and incentive prices for corn and wheat, and world corn price.
Source: Eurostat, Agricultural Prices.

industries may be measured by changes in the gross margin, or percentage value added (PVA) to the costs of wheat and corn because these grains represent the major share (roughly two-thirds) of total production costs for both industries. The PVA's are calculated from partial budgets, using data derived from interviews and (limited) published sources.

The proprietary nature of industry data makes it difficult to derive representative technical coefficients. Therefore, these budgets are intended to indicate the changes in technology and price incentives facing both industries between 1980 and 1985. The wheat washing industry is treated as having emerged from using a "lagging" technology assumed to represent the less efficient firms in 1980 and having adopted an "advanced" technology widely available by 1985. The intervention price for wheat and the threshold price for corn represent the costs of these grains.

The technical coefficients characterizing the corn wet milling industry are presented in column 1 of table II. The coefficients representing the wheat washing technology are less well known and have been more variable over time. The technical coefficients indicative of the "lagging" technology in 1980 and those which are indicative of the "advanced"

Table II

Characteristics of the Corn Wet Milling Industry, 1980 and 1985

Item	Technology[a] Metric ton	1980/81[b] Price (ECU)	Value (ECU)	1985/86[c] Price (ECU)	Value (ECU)
Revenues:					
Corn starch	.68	338	230	400	272
Corn oil	.03	619	19	743	22
Gluten feed	.20	177	35	212	42
Germ cake	.04	177	7	212	8
Gluten meal	.05	327	16	392	20
Production refund	–	12	12	19	19
Total revenues	–	–	319	–	383
Cost of corn	1.00	190	190	228	228
Percent value added	–	–	68	–	68

[a] Source: Rexen and Munck, 1984.
[b] Sources: Prices for the corn fractions are from Rexen and Munck (1984), adjusted by author's estimates; the cost of corn (threshold price) and the production refund are from Commission of the European Communities.
[c] Same as for 1980/81, except that prices for the corn fractions other than starch were adjusted by author's estimates.

technology in 1985 are presented in columns 1 and 4, respectively, of table III.

Stable technology and price incentives resulted in an unchanged PVA for the corn wet milling industry. The prices of starch and corn are the major monetary determinants of the PVA. Since corn was the major source of starch during this period, starch and corn prices are estimated to have risen by similar proportions between 1980 and 1985.

Increased profitability of the wheat washing industry occurred because higher extraction rates for A-starch and wheat gluten more than offset a small decline in the wheat gluten price and a small increase in the wheat intervention price. Even without technical advances, however, the PVA for the wheat washing industry would have increased to 103 percent, underscoring the importance of the rising price of starch relative to wheat.

TECHNICAL AND ECONOMIC DEVELOPMENTS AFFECTING THE FUTURE OF WHEAT WASHING IN THE EC

The EC wheat washing industry is already adjusting to the technical and economic constraints likely to affect it for the rest of this century.

Technical Constraints

Technical advances have occurred in the wheat washing industry, and higher flour extraction rates have also allowed greater utilization of the wheat endosperm, which contains starch, gluten, and germ. Laboratory experiments in 1977 showed that it was possible to obtain an 80.8 percent flour milling rate (Kempf, 1977). This milling rate is consistent with an A-starch recovery rate of 54 percent, compared to past industry standards of 45 to 50 percent. Gluten extraction rates are less flexible, but are assumed to increase marginally. Thus, a "superior" technology having higher extraction rates for gluten and starch than the "advanced" technology is assumed available by 1995.

No major technical advances are expected to occur in the corn wet milling industry.

Table III

Characteristics of the Wheat Washing Industry, 1980 and 1985

Item	1980/81			1985/86		
	Lagging Tech.[a] (metric ton)	Price[b] (ECU)	Value (ECU)	Adv. Tech.[c] (Metric Ton)	Price[c] (ECU)	Value (ECU)
Revenues:						
Feed from flour mill	.30	150	45	.25	156	39
A-starch	.42	338	142	.48	400	192
B-starch	.14	172	24	.128	180	23
Feed from wheat wash	.07	150	11	.06	156	9
Wheat gluten	.07	1,283	90	.085	1,200	102
Production refund	–	28	28	–	28	28
Total revenues	–	–	340	–	–	393
Cost of wheat	1.00	175	175	1.00	179	179
Percent value added	–	–	94	–	–	120

[a] Source: Rexen and Munck (1984) and author's estimates.
[b] Prices for A-starch and wheat gluten are from Rexen and Munck (1984); prices for other wheat fractions are author's estimates; the cost of wheat (intervention price) and the production refund are from Commission of the European Communities.
[c] Sources: Kempf, 1977; Godon et al, 1983; House of Lords, Select Committee on the European Communities, 1985; and author's estimates.
[d] Gluten prices are from industry sources; other sources are the same as 1980.

EC Grain and Starch Policies

Increasing budgetary outlays under the CAP have resulted in the wheat intervention remaining at 179.44 European Currency Units (ECUs) since 1985/86, where it is assumed to remain for this analysis.

A new Starch Regime intended to encourage more starch usage by the chemical and biotechnology industries took effect in June, 1986 (Commission of the European Community, Reg 2169/86). Under the old Starch Regime, variable import levies were charged on the starch content of (mainly) food products, but not on the starch content of industrial products. Thus, the biotechnology industry viewed location inside the EC as uneconomic because the prices of its products could be undercut by imports (House of Lords, Select Committee on the European Communities, 1985).

Under the new Starch Regime, variable refunds are computed quarterly and paid <u>only</u> to starch users who make products which are not "protected" by the variable levy system.[2] The refund is computed by multiplying the difference between the corn intervention price and the average corn c.i.f. (import) price over the previous three months by 1.6, a coefficient intended to represent the "theoretical" amount of corn required per ton of starch. This coefficient is also used to calculate the production refund for wheat starch. The refunds under the new Regime were fully phased in by the beginning of the 1987/88 crop year, and those paid under the old system will be phased out at the end of 1988/89.

A regulation to provide production refunds for users of sugar was introduced in 1986 (Commission of the European Communities, Reg 1010/86) in order to maintain a "balance" between starch and sugar used as carbohydrate sources.

---

[2]See (Commission of the European Communities, Reg 1009/86) for a list of "protected" products.

THE FUTURE ECONOMICS WHEAT WASHING IN THE EC

Under the new Starch Regime, both industries lose gross revenues equal to the production refunds they received under the old system. The profitability of the corn wet milling industry declines below its 1980 level (Table IV) while the profitability of the wheat washing industry remains above its 1980 level (Table V). Thus, the wheat washing industry may satisfy increased demand for its outputs at starch and gluten prices of 400 and 1,200 ECUs per ton, respectively.

Adoption of the superior technology may allow the wheat washing industry to remain profitable at lower starch and gluten prices. The production refunds under the new Regime are (theoretically) designed to facilitate increased starch demand without starch prices falling below the 1985 base level. However, no policy mechanisms currently exist to restrict gluten prices from declining. The partial budgeting analysis suggests that the wheat washing industry could remain profitable with gluten prices as low as 1,000 ECUs per ton (see Table V, last column). Even at gluten prices as low as 800 ECUs per ton, the PVA would only fall to about the level which prevailed in 1980 when the industry was beginning to expand.

Starch Demand Under the New Starch Regime

The European Chemical Industry Federation (CEFIC) projected the annual usage of starch and sugar by the chemical and biotechnology industries to increase by 1.35 and 0.4 million tons, respectively, by 1990 if favorable "...changes in market conditions..." were made in 1985 (House of Lords, Select Committee on the European Communities, 1985).

The projections of starch usage made by CEFIC are optimistic for three reasons. Firstly, CEFIC's original projections were considered too optimistic by many experts. Secondly, the full refunds of the new Starch Regime took effect two years after CEFIC's estimates, in mid-1987. Thirdly, increased sugar use could exceed 0.4 million tons, but by only a small

Table IV

Characteristics of the Corn Wet Milling Industry Under the New Starch Regime

Item	Technical Coeff.[a] (metric ton)	Price[b] (ECU)	Value Under the New Starch Regime (ECU)
Revenues:			
Corn starch	.68	400	272
Corn oil	.03	743	22
Gluten feed	.20	212	42
Germ cake	.04	212	8
Gluten meal	.05	392	20
Production refund	—	0	0
Total revenues	—	—	364
Cost of corn	1.00	228	228
Percent value added	—	—	60

Source: See footnotes in Table II.

[a]

Table V

Characteristics of the Wheat Washing Industry Under the New Starch Regime

Item	Price[a] (ECU)	Advanced Technology[a] (metric ton)	Value Under the New Starch Regime 1986 (ECU)	Superior Technology[b] (metric ton)	Value Under The New Starch Regime 1986 (ECU)
Revenues:					
Feed from flour mill	156	.25	39	.19	30
A-starch	400	.48	192	.54	216
B-starch	180	.128	23	.122	23
Feed	156	.06	9	.057	9
Wheat gluten	1,200	.085	102	–	–
	1,000	–	–	.088	88
Production refund	28	–	0	–	0
Total revenues	–	–	365	–	366
Cost of wheat	179	1.00	179	1.00	179
Percent value added	–	–	104	–	104

[a]Source: See footnotes in Table III.
[b]Source: Kempf (1977) and author's estimates

amount because the new regulations allow the production refund for sugar to be changed if it is viewed as disturbing the starch market. For these reasons, it is assumed that 1995 is the year by which a maximum of 1.35 million tons more starch may be used, annually, instead of by 1990.

The Use of Subsidies to Maintain Gluten Prices
in View of a Limited EC Market
For Gluten-Fortified Flour

Further expansion of the wheat washing industry will also increase the production of wheat gluten. However, the expansion of gluten use in fortified flour is limited by the 2.6 million tons of high protein hard wheat imported by the EC in 1985. Assuming a hard wheat milling rate of 0.73, these imports are equivalent to 1.9 million tons of flour. The 2 percent "rule of thumb" inclusion rate for gluten implies the use of 38,000 additional tons of gluten in the 1.9 million tons of fortified flour. However, the starch extraction rate (0.54) and the gluten extraction rate (0.088) associated with the "superior" technology imply about 0.163 tons of gluten produced per ton of starch (0.088/0.54). Thus, annual production of 1.35 million additional tons of starch would also produce about 220,000 more tons of wheat gluten. These figures suggest that wheat gluten prices could fall below the 1,000 ECUs per ton necessary for sustained profitability in the wheat washing industry, leading the EC Commission to search for policy actions to maintain gluten prices.

One policy could be an initial use of export subsidies for gluten to support its use as "vital" gluten in fortified flour on international markets. However, the world vital gluten market was only 191,000 tons in 1985, suggesting that eventually the world market for vital gluten could be satiated.

A policy providing subsidies to direct gluten into the much larger livestock feed market as a protein source may be financially feasible. In 1985, the price of 44 percent protein meal in the EC was 204 ECUs per ton, so that the price for gluten as a protein source would be 348 ECUs (204*0.75/0.44). A maximum export or internal subsidy of 652 ECUs per ton for gluten would be required to bridge the gap between 1,000 ECUs and its 348 ECUs per ton feed value. For the 88 Kg. of gluten produced from a ton of wheat, this would amount to a maximum export subsidy of 57 ECUs per ton. Although average export subsidies per ton of wheat are unpublished, the difference between the intervention price for wheat and the average f.o.b. export price of all EC wheat exports in 1985/86 was 55.4 ECUs (Economic Research Service, 1988). In addition, the EC provides subsidies to partially offset transportation costs and to place wheat into low income markets. Selected review of *Agra Europe* indicates that many shipments in 1985/86 received subsidies in excess of 100 ECUs per ton, which is well in excess of 57 ECUs for the gluten equivalent per ton of wheat.

PAST AND POTENTIAL FUTURE TRADE IMPACTS OF THE WHEAT WASHING INDUSTRY

Past and potential production and trade impacts resulting from the expansion of wheat washing are presented in Table VI. These data indicate the broad magnitudes by which this industry has affected EC markets for gluten, starch, and wheat in the past and by which it may affect these markets in the future, and should not be viewed as "point" estimates.

Data available on the production of wheat gluten are used to compute the approximate production of starch, utilization of soft wheat, and displacement of imported wheat in 1980 and 1985. Total wheat gluten production tripled between 1980 and 1985, the use of gluten in fortified flour nearly quadrupled, and other uses doubled. Starch production more than tripled (due to increased efficiencies in extraction), and the use of EC wheat in washing and fortified flour rose more than threefold. EC wheat exports in 1985 would therefore have been more than 20 percent higher had the wheat washing industry remained at its 1980 level.

The calculations also indicate that imported wheat displaced by gluten-fortified flour rose from 1 million tons in 1980 to 3.8 million tons in 1985.

Table VI
Past and Potential Production and Trade Impacts
Resulting from Expansion of the Wheat
Washing Industry, EC-10

	1,000 Metric Tons		
Item	1980	1985	By 1995
Wheat gluten[a]			
Use in fortified flour	15	55	95[h]
Other EC uses	10	18	45[F]
Exports	0	2	164[i]
Total production	25	75	304[g]
Wheat starch			
Total production	150[e]	514[e]	1,864[F]
Low protein (EC) soft wheat			
Use in washing	357[f]	1,071[f]	3,452[j]
Use in fortified flour[b]	980	3,593	6,272
Total use	1,337	4,664	9,724
Soft wheat exports[c]	15,200	16,100	NA
High protein (imported) hard wheat			
Displaced by gluten-fortified flour[d]	1,027	3,767	6,367[F]
Total imports[c]	4,500	2,600	0[F]
Technical conversion rates			
Wheat gluten	.07	.07	.088
Wheat starch	.42	.48	.54

15

[a] Source: Industry and author's estimates for 1980 and 1985; calculated from projected starch production for 1995, except forecast by author for other uses.
[b] The amount of soft wheat flour per unit of gluten (.98/.02), converted to soft wheat, assuming a 75% soft flour milling rate, (.98/.02)*b/.75.
[c] Source: Eurostat, Crop Production, Statistical Office of the European Communities, selected issues.
[d] The sum of gluten and soft wheat flour in fortified flour, converted to hard wheat, [(0.98/.02)b+b]/0.73, where 0.73 is the hard flour milling rate.
[e] Total gluten production converted to the amount of wheat used in washing and then to starch produced, (a/l)*m.
[f] Total gluten production converted to the amount of wheat washed, (a/l).
[g] Forecasted wheat starch production converted to the amount of wheat washed and then to the amount of gluten produced, (e/m)*l.
[h] The additional gluten used by 1995 in fortified flour is calculated as the maximum amount of hard wheat imports displaced (2,800), multiplied by the the hard flour milling rate (0.73) and the 2% of this flour for which gluten substitutes. This is added to the 55,000 tons of gluten used in 1985.
[i] a-b-c.
[j] Total wheat starch converted to wheat washed (e/m).
[F] Forecast by author
NA = Not available

This would imply a displacement of 2.8 million tons of wheat imports. However, wheat imports declined by only 1.9 million tons, leaving the use of 0.9 million tons of fortified flour unaccounted for. The discrepancy may be explained by lags between production and use, the quality of data used in comparing only 2 base years, and the use of "rules of thumb" to calculate this substitution instead of precisely known parameters. The important point is that growth in wheat washing accounted for a majority of the decline in wheat imports over this period. The estimates in Table VI are consistent with other studies (Home Grown Cereals Authority, 1985).

Potential production and trade impacts by 1995 are determined by the growth in starch production and other uses of gluten, and by the level of high protein hard wheat imports. Starch production is assumed to increase by 1.35 million tons and to be exclusively from wheat. The calculations in Table VI are based on the assumption that export and/or domestic subsidies are provided for gluten and that EC corn prices remain close to the threshold price. Under this scenario, EC wheat imports are assumed to be fully displaced by fortified flour by 1995.

The effects on EC wheat and gluten trade are quite significant under the above assumptions. Production of gluten could rise to over 300,000 tons by 1995. Plant capacity in the EC currently exceeds 140,000 tons. Future increases in the use of gluten-fortified flour in the EC will be constrained by the lack of large amounts of high protein hard wheat imports to be displaced. Although the figures in Table VI should be viewed cautiously given the assumptions and methodology, they suggest that gluten available for exports might exceed 160,000 tons by 1995. This would be three and one-half times the amount of gluten produced in North America and 65 percent of total world production in 1986. It should be noted that any permanent increase in high quality corn produced in the EC, which drives corn prices below the threshold price, could limit these impacts because more corn starch would be produced.

The implications for EC and world wheat and flour markets are also significant. An additional 5 million tons of EC wheat could be used in the EC by 1995. To the extent that "vital" gluten is exported and used to fortify softer wheat flour, the world prices of hard wheat and hard wheat flour may be reduced.

CONCLUSIONS

The CAP has altered EC price incentives in a way that encouraged, and may continue to encourage, the expansion of the wheat washing industry. The future growth of the industry may be limited by: the internal market for starch used in industrial (mainly biotechnology) applications, the absence of domestic or export subsidies provided for wheat gluten, and declines in corn prices below the threshold price.

Continued expansion of the wheat washing industry would continue to affect EC trade in high protein hard wheat, low protein soft wheat, and wheat gluten. An aggressive program to support wheat gluten prices with domestic or export subsidies could enable the EC to become the world's major exporter of wheat gluten and reduce its growth of soft wheat exports. EC gluten exports could rise even if EC gluten prices fall to their value as a feed ingredient. Without subsidies being provided for gluten, however, the wheat washing industry may be unable to expand much beyond its current capacity of about 140,000 tons.

LITERATURE CITED

AGRA EUROPE. 1985. Selected issues. Agra Europe, Ltd., Tunbridge Wells, UK.

COLLINS, T. H. and EVANS, K. 1983. Chorleywood bread process: Loaf volume improvement from gluten addition to flour. FMBRA Bull. Apr:43.

COMMISSION OF THE EUROPEAN COMMUNITIES. Various years. Selected issues. Off. J. Eur. Comm., Brussells, Belgium.

ECONOMIC RESEARCH SERVICE. 1988. Estimates of Producer and Consumer Subsidy Equivalents, Government Intervention in Agriculture,

1982-86. ERS/USDA Rpt. No. AGES880127, Washington DC.
GODON, B., LEBLANC, M. P., and POPINEAU, Y. 1983. A small scale devise for wheat gluten separation. Quality Plant Foods Hum. Nutr. 33:161.
HOME GROWN CEREALS AUTHORITY. 1982. The UK Starch Industry and the EEC Arrangements on Starch. HGCA Marketing Note 17(6), London.
HOUSE OF LORDS, SELECT COMMITTEE ON THE EUROPEAN COMMUNITIES. 1985. EEC Starch Regime. Her Majesty's Stationery Office, London, UK.
KEMPF, W. 1977. Einsatzmoglichkeiten nicht backfahiger Weizensorten in der Starkeindustrie. Starch 29:307.
KEMPF, W. 1984. Recent trends in European Community and West German starch industries. Starch 36:333.
POMERANZ, Y., ed. 1983. Industrial Uses of Cereals. Am. Assoc. Cereal Chem., St. Paul, MN.
RADLEY, J. A. 1976. Industrial Uses of Starch and Its Derivatives. Applied Science Publ., Ltd., London.
REXEN, F. and MUNCK, L. 1984. Cereal Crops For Industrial Use in Europe. EEC and the Carlsberg Research Laboratory.
SOSLAND, N. H. 1986. Production of vital wheat gluten grows dramatically since 1980. World Grain 4(9).

2

WHEAT:
RIGHT FOR BREAD, BUT FOR MUCH MORE TOO

C. W. Wrigley
CSIRO Division of Plant Industry
Wheat Research Unit
P O Box 7, North Ryde (Sydney), NSW 2113,
Australia

G. J. McMaster
Bread Research Institute of Australia
P O Box 7, North Ryde (Sydney), NSW 2113,
Australia

To the Western mind, wheat is used to make leavened bread - no more! Yet the unique properties of wheat suit it to so many more traditional and novel products, as is outlined in this paper.

USES OF AUSTRALIAN WHEAT

Some decades ago, Australian wheat was used for bread and little more. Traditionally, our wheat had been sent back to Europe for bread production in "the old country" (England - "home" for so many English who had migrated to Australia). Since the Second World War, Australia has become much more cosmopolitan. With the many "New Australians" from Europe, the Mediterranean and more recently Asia, has come a wider range of bread types than an English heritage would provide, together with even more diverse uses of wheat and flour. Nevertheless, about half of domestic flour use is for production of pan bread.

In addition, there have been dramatic changes in the markets (and thus uses) for Australian wheat since

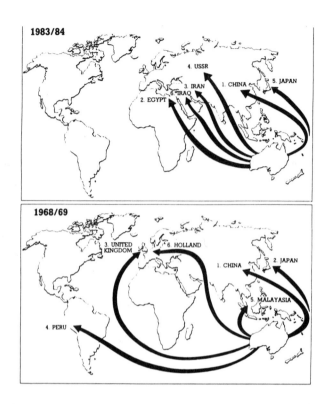

Figure 1. Major destinations for exports of Australian wheat in two specific years. Source: Bulk wheat (1985), page 13, reproduced by permission.

Europe achieved self-sufficiency in wheat and since such markets diminished for Australian wheat. Figure 1 shows part of this picture as a map, showing the major destinations of Australian wheat in years taken from the sixties and the eighties. In recent years, our major markets have been the Middle East, China, USSR, South East Asia and North Asia.

Wheat Production and Future Prospects

Australia's wheat production currently exceeds consumption by about 10:1. Average annual production is 15 million tonnes, of which only about 1.5 million tonnes is utilized by the domestic market; the remainder of the crop is exported. Current production

TABLE I
Australia's Production and Export of Wheat Grain in Recent Years (1 July to 30 June). Source: Australian Bureau of Agricultural and Resource Economics, Canberra

Year	Production (m.tonnes)	Production (A$m)	Export (m.tonnes)
1988/9*	13.2	2,160	10.3
1987/8	12.5	2,030	12.3
1986/7	16.2	2,460	14.9
1985/6	16.2	2,719	16.2
1984/5	18.7	3,203	15.7
1983/4	22.0	3,606	10.6

* Forecast by ABARE

and export figures are given in Table I. Recent decreases in annual production reflect grower reactions to falling world prices, causing diversification into other farming activities. Despite a current firming of prices, there is still a long-term concern about international wheat prices. In particular, the increasing export of wheat to developing countries (who cannot afford to pay for commodities) is becoming a major concern.

A recent forecast by the International Wheat Council indicated that the developing countries may account for almost 80 percent of world grain trade by the year 2000, compared with 50 percent in 1985. The forecast stated, "There is an evident, and possibly widening, gap between the prices which many of the importers can afford to pay for their grain and the returns which producers, even low cost ones, need if they are to remain in business."

Clearly, there is an incentive for wheat exporting countries to re-examine fundamental opportunities that may be provided by novel processing and value addition to wheat as means of increasing

TABLE II
National Flour Mix for Years Ending
June 30, 1987 and 1988
(Based on confidential sales questionnaire
to Australian Flour Mills by the
Bread Research Institute)

User of Flour	1987 k tonnes	%	1988 k tonnes	%
Starch/Gluten Mfrs	270	22	300	24
Other Industrial Users	3	0	6	1
Bread Bakers	541	45	563	45*
Pastrycooks, etc	90	8	93	7*
Biscuit Mfrs	86	7	81	6
Pasta Mfrs	41	3	44	4
Packeted Flour & Mixes	82	7	80	6
Food Mfrs	95	8	97	7
TOTAL (Domestic Uses)	1,208	100	1,264	100
Export of Flour Products	73		86	
GRAND TOTAL	1,281		1,350	

* Estimates as of September 20, 1988

returns from this important primary product. Current thought in Australia is that opportunities for utilizing/processing wheat and adding value to wheat require more attention for the future viability of the economy. Whilst it is conceded at the outset that adding value to wheat past the flour-milling processes is difficult, there are opportunities for doing so. Some of these will be included in this paper.

Domestic Processing of Wheat

The approximately 10% of wheat production that is processed within Australia accounts for about 4% of manufacturing employment. As Tables II and III show, starch/gluten production is a major activity, which

TABLE III
Trends in Flour Use and Export for Australia
(in thousands of tonnes)

Year	Starch/ Gluten	Bread	Export	Total
1979	208	551	92	851
1980	218	570	78	866
1981	229	568	100	897
1982	209	567	102	878
1983	193	570	91	854
1984	232	545	63	841
1985	259	540	61	859
1986	274	523	61	859
1987	270	541	73	884
1988	300	563	86	949

contributes significantly to exports of processed wheat products (Table IV) as well as domestic consumption. Overall, only about 5% of domestically processed wheat products are exported.

An increasingly important item in this category is biscuits (cookies), which are made in Australia in a much wider range of types than is generally available in North America. For many years, breeders and flour millers have given particular attention to the production of specialty soft wheats for the biscuit industry, thus positioning industry well to manufacture and export quality products.

Uses of Australia's Wheat Exports

Loaf bread represents less than one third of the processing overseas of wheat exported from Australia. As Figure 2 shows, Arabic breads and Asian noodles are each of at least equal importance with pan bread. Chinese steamed bread is also significant. The rising importance of these 'exotic' products in the utilization of our wheat has introduced important

TABLE IV
Wheat-derived Products Exported from
Australia (A$m) (Wrigley, 1988)
Source: Australian Bureau of Statistics, Canberra

Product	1983/4	1984/5	1985/6	3 yr mean
Gluten	23.6	28.1	28.1	35%
Flour, mill products	23.6	25.8	20.4	31%
Cakes, biscuits	15.4	18.0	20.9	24%
Breakfast cereals	2.8	3.5	3.9	5%
Starches	2.6	4.5	4.7	5%

changes in the evaluation of quality type in new varieties of wheat and in the assessment of quality for grain receivals at harvest. These products are important to Australia, as many require white wheat of only moderate protein content, the type most readily available in Australia. Arabic breads include flat breads and pocket breads of a range of dimensions. Steamed bread is similar to a dumpling in appearance and may have a filling of meat or vegetable material. Noodles are in three major classes - alkaline (made from a dough of pH from 7.5 to 10.5), salted (made from flour, salt and water) or 'Instant' (cut from a near neutral dough, steamed, fried and wrapped). Alkaline and salted noodles may be sold raw, dried or boiled.

NEW QUALITY GUIDELINES FOR AUSTRALIAN WHEATS

Ensuring Suitable Wheat Types for 'Exotic' Products

During the past 20 years, guest workers sponsored by the Australian Wheat Board have been hosted at the Bread Research Institute from various Asian and Middle East countries in order to collaborate with them in a systematic study of quality evaluation of such products as Arabic flat breads, various types of noodles and Chinese steamed breads and dumplings. An

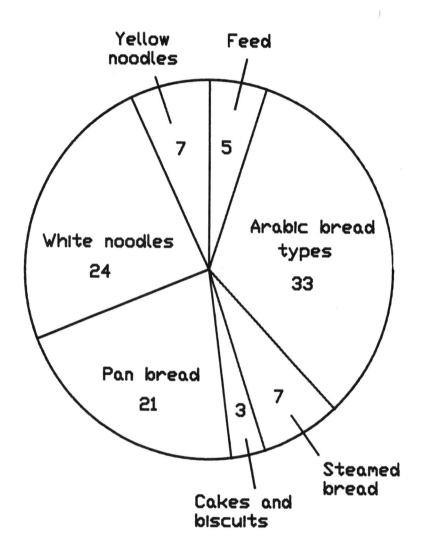

Figure 2. Current uses of wheat exported from Australia, based on current industry estimates (as % of total).

TABLE V
Quality Requirements According to Grade and Use

Grade	Use	Quality Requirement
Prime Hard	Pan bread, gluten Alkaline noodle Blending	(High yield, white flour (Strong, extensible (Moderately hard grain
Hard No. 1	Pan bread Arabic bread	(Hard grain (Extensible dough
Hard ASW	Arabic bread Chapattis Pan bread	(Hard grain (Extensible dough
Soft ASW	Noodles Grocery lines	(Soft, very white flour (High paste viscosity
Soft	Biscuits, cakes	(Low flour water absorption (Weak, white flour

important part of these collaborations has been to devise small-scale tests for each of the many products and variations of processing conditions to permit evaluation of the suitability of existing cultivars, new breeders' lines, particular wheat grades and cargoes of wheat. The results of many of these studies have been published as reports and in the scientific press (Moss, 1971, 1980, 1982, 1985; Moss and Miskelly, 1984; Moss et al, 1986, 1987; Miskelly and Moss, 1985; Miskelly, 1984; Lee et al, 1987; Qarooni et al 1987, 1988). As a result, Australian wheat lines approaching registration are checked for qualities suitable for many products besides pan bread. These products have quite different requirements (Table V). For instance, flat breads are best made from hard wheat flour, whereas noodles require flour milled from a mixture of hard and soft grain, or soft wheat of greater dough strength than traditional soft wheat. Flour colour is important with all noodles, and starch gelatinizing properties are important with some.

These studies on wheat types have since been extended to examine the components of quality and the chemical constituents of the grain that are most critical to product quality. The various exotic products differ in flour-quality requirements from those for leavened bread, but the unique properties of wheat gluten underlie most aspects of these requirements. The interactions between protein content and grain hardness for the various products are set out in Figure 3. More specific guidelines for the flour-quality requirements of Australian wheat classes and their product suitabilities are provided by O'Brien and Blakeney (1985). In parallel with these studies in Australia, we have become extensively involved through the Australian Wheat Board in visits to the countries using Australian wheat for these products. These opportunities to see processing and product at close hand have improved our understanding of the value of wheat beyond its processing into loaf bread. Such experience has further added to Australia's production and marketing advantages in being able to match the quality requirements of specific importing countries. Accordingly, this exposure to the marketplace by people actually performing the research has enabled research to be effectively targeted more specifically to market requirements for Australian wheat and for valid research ideas to be developed.

Ensuring Suitable Grain Receival

Considerable effort is required to ensure that the advantages of correct selection of quality type in breeding are carried through into grain receival by strict varietal selection. For this purpose, varietal-controlling legislation has been enacted in all Australian states; methods of variety identification have been devised and are being further developed (Wrigley et al 1987).

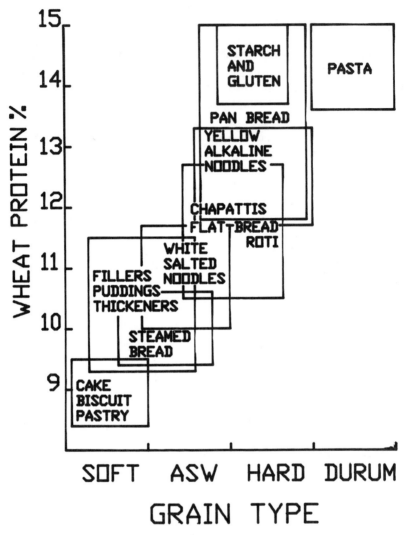

Figure 3. Requirements of products made from wheat in terms of protein content and grain hardness and type (ASW=Australian Standard White). The figure shows that yellow alkaline noodles (Cantonese) are best made from hard wheat with protein content in the range of about 11.7 to 13.3%, and that flat breads (including chapattis and roti) require hard grain of slightly lower protein content. Steamed breads, on the other hand, are suited to medium/soft wheat of about 10% protein.

Improved procedures have also been devised for ensuring the soundness of grain (Ross et al. 1987) and for quickly testing protein content to allow segregation of receival on this basis also (Ronalds and Miskelly 1985). A further item requiring evaluation has been the sensitivity to rain damage for some products such as noodles. The results of this work have been described by Orth and Moss (1987); in further work, we are examining the contributions of specific enzymes of sprouting (Edwards and Ross 1987). Fuller details on these and other improved testing procedures were described in proceedings of this conference by Drs Batey and Skerritt of the CSIRO Wheat Research Unit.

Ensuring Suitable Processing

A further dimension, required to match the small-scale tests for quality described above has involved scaling up the size of product evaluation, thereby permitting more realistic evaluation of processing variables as well as of flour requirements. In recent years, the Bread Research Institute has upgraded its pilot-scale facilities to further improve capabilities in this field.

A pilot flour mill (0.75 tonnes wheat per hour) was commissioned in 1986. This facility is allowing research to be conducted on a commercial scale in the milling technology of Australian wheat, thereby evaluating the impact of various milling techniques on product quality. The pilot bakery is currently being expanded and upgraded to permit evaluation of flours at a commercial level for a range of baked products.

Bakery Premixes and Noodles

In recent decades, flour exports from Australia have progressively decreased. New opportunities are offered, however, by providing the convenience and consistency of flour/ingredient performance of pre-mixes containing flour and other ingredients, suitable for production of many cereal-based foods. Noodles, in their various forms, are also well suited to local production for export. A group of wheat growers in western New South Wales is currently studying the feasibility of an operation that would permit them growing specialty varieties for local processing into noodles, using expertise accumulated by the contacts with noodle-consuming countries as described already. Joint-venture conditions with a company based in the importing country is likely to be the most satisfactory basis for this type of operation. The Australian government is actively encouraging Australian industry to seek out potential export situations for value-added goods manufactured in Australia. The export of wheaten products represents a challenge to Australian food processors - high risks but potentially excellent rewards.

GLUTEN AND STARCH

Gluten Production

As previously mentioned (Tables II and III), gluten production is a major use of wheat flour in Australia. Our production and exports of gluten have remained roughly stable during the past few years (Table VI) (McMaster and Wrigley 1987), although the exports had increased earlier. World-wide, however, there has been a remarkable increase in gluten production/ consumption - from 88,000 tonnes in 1980/1 to 240,000 tonnes in 1987/8 (according to the International Wheat Gluten Association).

Ten years ago, Australia was responsible for about half the world's production of gluten. Since then, many very large plants have been commissioned, particularly in Europe, where production capacity now greatly exceeds consumption, but also in North America, which continues to be an important importer of gluten (Table VII). Overall, world production capacity exceeds consumption, so trading is competitive.

Gluten production focuses attention on the unique property of wheat flour - its dough-forming protein. Even gluten production does not take us far from leavened bread, because the baking industry is the

TABLE VI
Australian Production and Export of Gluten

Year	Production (tonnes)	Export (tonnes)	Proportion exported
1981/2	37,964	22,417	59%
1982/3	35,249	19,716	56%
1983/4	39,613	21,282	54%
1984/5	40,959	24,970	61%
1985/6	40,222	23,277	58%

main user of gluten, as is indicated in the utilization figures for USA (Table VIII). This great increase in gluten use has been largely due to the increased popularity of wholemeal breads (e.g. from 20% of bread consumption 15 years ago in Australia to about 50% at present) and the need to increase protein content to maintain bread quality.

In addition, however, gluten has enjoyed increasing popularity due to its versatility as a food and feed protein in the wide range of applications listed in Table VIII. Future increases might be expected in the use of gluten in processed meats, pasta and cheese analogs, and particularly in aquaculture feeds and seafood analogs.

Modification of gluten offers further possibilities for increasing its value in food processing, since chemical and/or enzyme treatments can provide gluten with new properties such as foam stabilization and emulsification (Batey 1986). Some of our modification studies have also indicated the potential of modified gluten to substitute for 'chemical' additives used in baking (Asp et al. 1986).

Wheat Starch

The dramatic increase in gluten production has necessarily been accompanied by a similar increase in production of wheat starch, thereby inevitably affecting trade in and utilization of the wide range

TABLE VII
Production and Consumption of Wheat Gluten, according to Major Country Groupings in the Year 1987/8. (Figures Provided by the International Wheat Gluten Association)

Region	No. major plants	Production capacity (tonnes)	Consumption (tonnes)
Europe, EEC	28	135,000	100,000
Europe, Non EEC	10	12,000	5,000
North America	5	50,000	75,000
Australia	4	40,000	20,000
Others	16	23,000	40,000
Total	63	260,000	240,000

TABLE VIII
Utilization of Wheat Gluten (by Proportion) in Various Food Products in the United States of America (Source: the International Wheat Gluten Association)

Product	1980/1	1987/8
Baking	77%	65%
Flour fortification	4	10
Pet foods	10	8
Breakfast cereals	7	4
Meat	0	3
Snack foods	1	2
Aquaculture feeds	0	2
Seafood analogs	0	1
Pasta	0	1
Cheese analogs	0	1
Breading/batters	0	1
Other	1	2

of industrial starches. Munck et al (1988) have recently reviewed the wide range of uses of cereal starches (native, modified and hydrolyzed) for food, animal feed, paper, board, textiles, adhesives, pharmaceuticals and various other aspects of the chemical industry. They also pointed to the greater potential of cereal starches to fulfill needs of industry now and in future scenarios, depending on the pricing of alternative feed stocks (particularly petrochemicals). Maize starch will always be a formidable competitor for wheat starch, but the need for gluten (with its unique properties) ensures the continuation of wheat starch as a volume commodity. Any inadequacies in wheat starch for chemical processing must be corrected by breeding (Timmerman, 1988; Kelfkens et al, 1988), since it is important for us to consider wheat starch as a critical raw material in all future applications of starch (e.g. see Koch and Roper, 1988). Glucose and fructose are, of course, already major products from starch. Citric acid has been identified by an Australian company as being a valuable product for processing from wheat starch.

NON-FOOD UTILIZATION

Strawboard and Paper Pulp

Munck (1987, and elsewhere) has pointed out the need to make better use of the whole cereal plant, including the stems and leaves in addition to the grain only. Use of the straw for feed and fertilizer has always been practiced, several Australian initiatives have sought to further increase financial returns by using the wheat plant for the manufacture of strawboard or for paper pulp.

'Harvestboard' is the name given to a building material being developed by Australian Value Added Agricultural Projects (Canowindra, NSW, Australia). It is a composite board, similar to particle boards made from wood chips, but produced from wheat-straw waste.

A feasibility study is currently being carried

out on the location in Western Australia, of a paper-pulp plant to produce 50,000 air-dried tonnes of pulp per year. The Soda Anthraquinone process has been chosen for the purpose. It involves dissolution of the lignin from the straw to permit separation of fibres. Capital costs are estimated to be A$74m, and operating costs to be A$460 per tonne dry product.

The use of straw for this purpose has the advantage over some alternatives (such as bagasse from sugar cane) that it is low in moisture and does not contain pith, making its yield of fibre double that for bagasse. On the other hand, the costs of collection and transport of wheat straw are likely to be important to the success of the project. Siting of the plant is critical for this reason, as well as to obtain access to low-cost energy and water. Such sites have been identified in the coastal wheat-growing areas of Western Australia. CMPS Consultants (Mr Paul, P O Box 361, Camberwell, Vic 3124) have been commissioned to undertake the study, in co-operation with Voest-Alpine of Linz, Austria.

The production of ethanol from cereal grains is probably the most often discussed example of a non-food use of cereals. During the Second World War there was significant production of fuel alcohol from wheat grain in the town of Cowra (New South Wales), but it was only justified as a war-time measure. More recently the oil crisis of last decade sparked renewed interest in fuel ethanol, and we were involved in some further research on the topic (Batey et al., 1982), but the economics of the process are only justifiable with Government intervention via tax concessions.

CONCLUSION

The Australian concept of wheat quality has undergone major changes during recent years. Wheat is uniquely 'right' for bread, but this is a 'blinkered' Western view of wheat use. The unique properties of wheat have been capitalized upon in many more products by many other civilizations to yield a vast array of fascinating foods that have evolved within cultures. In gluten we can extract the essence of wheat's uniqueness, and put it to work in bread and in dozens of other food products.

LITERATURE CITED

ASP, E.A., BATEY, I.L., ERAGER, E., MARSTON, P., and SIMMONDS, D.H. 1986. The effect of enzymically modified gluten on the mixing and baking properties of wheat-flour doughs. Food Technol. Aust. 38:247.

BATEY, I.L. 1986. Enzyme solubilization of wheat gluten. J. Appl. Biochem. 7:423.

BATEY, I.L., GRAS, P.W., MacRITCHIE, F., and SIMMONDS, D.H. 1982. Production of fermentable carbohydrate and by-product protein from cereal grains by wet-milling. Food Technol. Aust. 34:356.

EDWARDS, R.A., and ROSS, A.S. 1987. Enzyme activities of rain damaged grain. Proc. 37th Aust. Cereal Chem. Conf., p. 45. Royal Aust. Chem. Inst., Melbourne.

KELFKENS, M., MARSEILLE, J.P., WEEGELS, P., and HAMER, R.J. 1988. Determination of processing quality of wheat flour for the starch industry. Page 227. In: Cereal Breeding Related to Integrated Cereal Production. M.L. Jorna and L.A.J. Slootmaker, eds. Pudoc, Wageningen, The Netherlands.

KOCH, H., and ROPER, H. 1988. New industrial products from starch. Starch 40:121.

LEE, C-H., GORE, P.J., LEE, H-D., YOO, B.S., and HONG, S-H. 1987. Utilization of Australian wheat for Korean style dried noodle making. J. Cereal Sci. 6:283.

McMASTER, G.J., and WRIGLEY, C.W. 1987. Primary production value added? Is there potential in the Australian wheat industry? Food Technol. Aust. 39:433.

MISKELLY, D.M. 1984. Flour components affecting paste and noodle colour. J. Sci. Food Agric. 35:463.

MISKELLY, D.M., and MOSS, H.J. 1985. Flour quality requirements for Chinese noodle manufacture. J.

Cereal Sci. 3:379.

MOSS, H.J. 1971. Quality of noodles prepared from the flours of some Australian wheats. Aust. J. Exp. Agric. Anim. Husb. 11:243.

MOSS, H.J. 1980. The pasting properties of some wheat starches free of sprout damage. Cereal Res. Commun. 8:297.

MOSS, H.J. 1982. Wheat flour quality for Chinese noodle production. Page 234. In: Proc. of Food Conference 1982. C.Y. Theng, W.L. Kwik, and C.Y. Fong, eds. Singapore Institute of Food Science and Technology.

MOSS, H.J. 1985. Ingredient effect in mechanised noodle manufacture. Page 71. In: Proc. 4th S.I.F.S.T. Symposium: Advances in Food Processing. K.K. Lim, C.Y. Chong, C.N. Gwee and C.K. Choo. Singapore Institute of Food Science and Technology.

MOSS, H.J., and MISKELLY, D.M. 1984. Variation in starch quality in Australian flour. Food Technol. Aust. 36:90.

MOSS, H.J., MISKELLY, D.M., and MOSS, R. 1986. The effect of alkaline conditions on the properties of wheat flour dough and Cantonese-style noodles. J. Cereal Sci. 4:261.

MOSS, R., GORE, P.J., and MURRAY, I.C. 1987. The influence of ingredients and processing variables on the quality and microstructure of Hokkien, Cantonese and instant noodles. Food Microstruct. 6:283.

MUNCK, L. 1987. A new agricultural system for Europe? Trends in Biotechnology 5:1.

MUNCK, L., REXEN, F., and HAASTRUP, L. 1988. Cereal starches within the European community - agricultural production, dry and wet milling and potential use in industry. Starch 40:81.

O'BRIEN, L., and BLAKENEY, A.B. 1985. A census of methodology used in wheat variety development in Australia. Cereal Chemistry Division, Royal Australian Chemical Institute, Melbourne.

ORTH, R.A., and MOSS, H.J. 1987. The sensitivity of various products to sprouted wheat. Page 165. In: 4th Intern. Symp. on Pre-harvest Sprouting in

Cereals. D.J. Mares, ed. Westview Press, Boulder, Co.

QAROONI, J., ORTH, R.A., and WOOTTON, M. 1987. A test baking technique for Arabic bread quality. J. Cereal Sci. 6:69.

QAROONI, J., MOSS, H.J., ORTH, R.A., and WOOTTON, M. 1988. The effect of flour properties on the quality of Arabic bread. J. Cereal Sci. 7:95.

RONALDS, J.A., and MISKELLY, D. 1985. Near infrared reflectance spectroscopy. Chem. Aust. 52:302.

ROSS, A.S., WALKER, C.E., BOOTH, R.I., ORTH, R.A., and WRIGLEY, C.W. 1987. The Rapid Visco Analyzer: A new technique for the estimation of sprout damage. Cereal Foods World 32:827.

TIMMERMAN, H. 1988. Wheat for starch processing. Page 223. In: Cereal Breeding Related to Integrated Cereal Production. M.L. Jorna and L.A.J. Slootmaker, eds. Pudoc, Wageningen, The Netherlands.

WRIGLEY, C.W. 1988. How can chemistry help to increase returns to Australia from our cereal grain exports? Chem. Aust. 55:174.

WRIGLEY, C.W., BATEY, I.L., CAMPBELL, W.P., and SKERRITT, J.H. 1987. Complementing traditional methods of identifying cereal varieties with novel procedures. Seed Sci. Technol. 15:679.

3

THE ECONOMICS OF WHEAT FLOUR AS A RAW MATERIAL FOR GLUTEN AND STARCH MANUFACTURE

B. W. Morrison

Ogilvie Mills Ltd
1 Place Ville Marie, Suite 2100
Montreal, Quebec, Canada H3B 2X2

ABSTRACT

Wheat flour is the conventional raw material used in the production of vital wheat gluten and wheat starch by wet separation processes.

The economics of the process are determined by:
1) cost of wheat in mill,
2) flour extraction rate and value of millfeed by-product,
3) processing costs - milling and wet process,
4) flour quality attributes - protein level, solubles (enzyme activity, germ index, starch damage).

Analyses of sensitivity to these cost and quality factors are presented to illustrate effects on washing margin per unit of flour processed, given fixed returns for wheat gluten and both primary and secondary starches.

INTRODUCTION

Wheat gluten is produced in 28 countries throughout the world. In 1986 total production was reported to be 253,000 metric tonnes (Sosland 1986). In the following year the International Wheat Gluten Association estimated that an additional 20,000 m.t. of production had come on stream.

Eight countries (Australia, United States, France, Holland, West Germany, Belgium, United Kingdom and Canada) accounted for 72% of world production in 1987. Of this group only the U.S.A. and the U.K. remain net importers. They are the largest consumers of wheat gluten utilizing currently some 45% of the world supply.

Wheat starch is the natural co-product of gluten production. Because it is present at 4 to 6 times the level of gluten in the wheat kernel starch value has a major influence on the economics of the industry. The extent of its contribution to the viability of the industry varies. In the European Community there was, until recently, preferential production refunds given to wheat starch producers over corn and potato starch. In Australia the starch industry is based on wheat and as such formed the basis for development of the gluten industry in that area.

In North America wheat starch must compete directly against the usually low priced corn product. For this reason the economics of our industry rely heavily on input flour cost, gluten yields and value added starch products.

FLOUR COSTS

In the following economic exercise several assumptions of both fixed and variable cost factors were employed.

Fixed
Extraction rate: 1.3 tonnes wheat/tonne flour
Milling costs: $25 per tonne of flour produced

Variable
Wheat cost in mill: 3.00, 3.50, 4.00 dollars/bu
Millfeed credit: 50.00, 75.00, 100.00 dollars/tonne

Table I shows the effect on flour cost/tonne due to variable wheat cost. In the current calendar year wheat prices have moved sharply in response to drought stress in North America, and in fact have fluctuated between 2.70 and 4.50 per bu

(Minneapolis). Flour by-product or millfeed values will fluctuate on commodity markets, generally in response to change in coarse grain and soybean costs to the formula feed industry. This has provided some off-set in flour cost as grain prices rose this year, however heavy running time in the mills due to strong flour demand has a depressing effect on millfeed values as supply/demand factors enter the picture.

In Table II the effect of variation in millfeed credit value on flour cost is tabulated at each wheat price level.

TABLE I

Effect of Wheat Price on Flour Cost

	Wheat Price ($/bu)		
	3.00	3.50	4.00
Wheat Cost/T Flour $	143	167	191
Millfeed Credit @ $75/T	(22.5)	(22.5)	(22.5)
Net Wheat Cost/T Flour $	120.5	144.5	168.5
Milling Cost/T Flour $	25	25	25
Bulk Flour Cost/T $	145.5	169.5	193.5

TABLE II

Effect of Millfeed Value on Flour Cost
(Bulk Flour Cost $/T)

Wheat Cost	Millfeed Value ($/T)		
($/bu)	50	75	100
3.00	153	145.50	138
3.50	177	169.50	162
4.00	201	193.50	186

WET PROCESS FACTORS

In order to assess the economic value (washing margin) of the starch and gluten process, due to cost and quality factors of the input flour, the following assumptions were made:

Fixed
Wet process cost/tonne flour washed	$ 130
Product revenue – gluten/tonne	$1323
– A starch/tonne	$ 250
– B starch/tonne	$ 125

A to B starch ratio 4:1

Variable
Protein content of flour
Solubles content of flour

STARCH/GLUTEN YIELD ESTIMATION

Flour type:
 – Moisture, protein content, solubles level

Process Balances   – Moisture loss
 + Solubles loss = level X 1.3
 + Dry Process loss
 = Total losses % (25)

Flour processed minus total losses
 = Dry basis yield (75)

@ 9.5% Moisture = Net overall yield (83%)

GLUTEN YIELD

Total protein as gluten $\frac{13}{.80}$ = 16.2%

Protein loss	(3.25%)
D.B. gluten yield	13.00%
Net gluten yield (7% moisture)	14.00%

	STARCH YIELD
Net overall yield	83%
Net gluten yield	(14%)
Starch yield	69%

WASHING MARGINS

Added to the direct impact of wheat cost and flour by-product values on the profitability of the starch and gluten industry, when calculated on the basis of fixed returns for these co-products, the washing margins are heavily influenced by flour quality and its effect on product yields.

FLOUR PROTEIN

Given the fact that good quality flour will normally yield 2.8 to 3.0 lbs of wet gluten (80% protein d.b.) per lb of flour protein and further given the value of gluten over starch, it is evident that flour protein content has a major influence on total revenues. There is an off-set to these higher margins in that a wheat protein premium is usually involved however not to the extent that lower flour proteins are favoured. Significant benefits are obtained when high protein flour fractions are sourced from standard protein grists through flour stream selection. It is possible to obtain two to three percent extra flour protein by this method, but the feasibility of this practice depends on having commercial outlets for the balance of flour production.

Table III shows that the effect on washing margin of 11% protein flour versus 15% protein flour is $58 per tonne of flour processed regardless of the wheat price, although the combination of $4.00/bu wheat and 11% protein flour results in a negative washing margin of $10.53 per tonne. At the other end of the scale $3.00/bu wheat and 15% protein flour gives a positive washing margin of $95.47.

TABLE III

Effect of Flour Protein on Washing
Margin $/T at Various Wheat Prices

Flour Protein	Wheat Price ($/bu)		
(%)	3.00	3.50	4.00
11.0	37.47	13.47	(10.53)
13.0	67.07	43.07	19.07
15.0	95.47	71.47	47.47

Washing margin established from standard flour cost at extraction rate of 1.3 T wheat/T flour, millfeed credit value of $75/T and process balance and yields as described.

TABLE IV

Effect of Various Solubles Level in 13.0% Protein
Flour on Yields and Washing Margin/T Flour Processed

	Flour Solubles		
	6%	10%	14%
Process Balance (%)			
Total Losses	24.7	29.9	35.1
D.B. Recovery	75.3	70.1	64.9
Yield (9.5 moisture)	83.2	77.5	71.7
Gluten Recovery			
Flour Prot/Gluten Protein	15.9	15.9	15.9
Protein Loss (%)	3.0	4.5	5.9
D.B. Gluten Yield (%)	12.9	11.4	10.0
@ 7% Moisture (%)	14.0	12.3	10.8
Starch Yield (%)	69.2	65.2	60.9
Washing Margin ($/T)			
@ 3.00 Wheat	67.50	36.50	6.50
@ 3.50 Wheat	43.50	12.50	(17.50)
@ 4.00 Wheat	20.00	(11.50)	(41.50)

FLOUR SOLUBLES

An elevated level of soluble material in flour processed for starch and gluten recovery has a serious impact on yields and subsequent profitability of the operation as shown in Table IV. An additional negative economic factor is the effect of higher solubles on the effluent stream.

Flour solubles level is influenced by a) soundness of the parent wheat or degree of enzyme activity due to sprouting and b) the inclusion of wheat germ fractions in the flour. Enzyme activity, as measured on the Hagberg apparatus, shows a strong inverse relationship between Falling Number and solubles level.

The development of the GERM INDEX TEST by the Australians (Stenvert and Murray 1981) which provided quantitative measurement of wheat germ in flour streams, has been of great assistance to the miller in tracking the efficiency of germ removal systems.

Table V shows the direct relationship between Germ Index and solubles in flour, attesting to the presence of strong reducing compounds in wheat germ.

The effect of starch damage in flour used in starch production is two-fold.

First, there will be an increase in secondary or B starch production from flour with high starch damage levels.

Secondly, increased starch damage, due to heavy grinding of endosperm, will produce some solubles due

TABLE V

Germ Index vs Flour Solubles

	Germ Index	Solubles (%)
Top Reduction Flour	.097	7
Bran Duster Flour	.397	9
7th Midds Flour	.724	12
Last Reduction Flour	.885	14

TABLE VI

Starch Damage vs Flour Solubles

	Starch Damage Farrand Units	Solubles (%)
Top Break Flours	10	5.7
Top Reduction Flours	50	7.0
Last Break Flours	23	9.3
Last Reduction Flours	35	14

to amylase degradation of this susceptible, or available, starch during wet processing.

Table VI shows this effect, as well as the strong influence of wheat germ presence, in the comparison of starch damage/solubles relationship of first and last reduction flours.

It would be of significant economic benefit to the wheat starch and gluten process if germ could be extracted completely from the wheat kernel prior to entering the 1st break operation of the milling process. This would eliminate those negative washing margins associated with elevated flour solubles and would permit flour of higher extraction rates to be used in the process.

LITERATURE CITED

SOSLAND, N. 1986. History, technology of wheat gluten. World Grain 4(9):11.

STENVERT, N. J., and MURRAY, L. F. 1981. The milling quality of Australian wheat. Milling 164:30.

4

A NEW AGRICULTURAL SYSTEM IN EUROPE ?

Lars Munck, Finn Rexen,
Bent Petersen and Pernille Bjørn Petersen
Department of Biotechnology
Carlsberg Research Laboratory
10 Gamle Carlsberg Vej
DK-2500 Valby, Copenhagen
Denmark

INTRODUCTION

Recently, the causes for and the financing of the surpluses of some agricultural commodities like grains, butter, milk, meat etc. have been intensively discussed in the European Community, mostly by politicians and townspeople who have completely forgotten that foods and oxygen for driving biological and mechanical machinery as well as industry all originate from plants and agriculture. In the Middle Ages, cereal production was a luxury in Europe because the relation between the fertilization of fields by manure on one side and yield on the other was not clear to most people. The population mainly survived on animal husbandry products from grazing animals supplemented by cereals, the latter yielding only 500-1500 kg/ha.

In Denmark the fertile soils and the relatively sparse population permitted export of cereals to England in the beginning of the 19th century boosting the industrial revolution in that country. In the 1850's the cereal production resources of the great plains in North America, South America and Australia came up as major competitors to Denmark because steamships greatly reduced the transportation costs,

and for the first time creating a world market for grains. The Danish farmers not only survived but also prospered by switching from grain export to export of animal husbandry products, such as bacon to the rapidly growing English population. From the end of the 19th century up to now productivity in agriculture increased enormously due to industrial inputs of fertilizers, pesticides and modern equipment.
This development, taking place also in the U.S., must be considered 'the star-war program' of all ages because of the remarkable spin-off effects which are sought to come from such a programme, indirectly stimulating the formation and activity of a range of non-food industries. Thus, most of the present car manufacturers in Europe were founded by selling tractors and other farm equipment to agriculture. In fact, so many industrial and other business interests have joined the agricultural "bandwagon" that farmers' input is now (in Denmark) just about a quarter of the value of the food products and a few percent of the gross national product. At the same time, the agriculturally associated operations are still in the range of 30% of the entire industrial production.

The world is becoming increasingly smaller, and we must now acknowledge the fact that neither Europe nor the U.S. are isolated, economically or enviromentally. Consequently, we have to reconsider our future and the symbiosis between agriculture and industry to obtain a long-term sustainable production for the good life of all people around the world. Obviously it is not possible to install refrigerators in all Chinese households without risking the precious ozone layer in the stratosphere so important for plant production. Alternatives to freon exist and should be expediently implemented. Moreover, a general change to a recycling technology at a higher degree and based on agricultural products than at present, is mandatory.

AGRI-CULTURE VERSUS AGRI-COMMODITY PRODUCTION IN THE EC

The agricultural crisis in the industrialized

countries has evolved gradually and is due to a stepwise development stimulated by heavy inputs from machines, fertilizers and chemical industries. This development has evolved

<u>from</u> a general deficit in food <u>to</u> overproduction

<u>from</u> utilization of considerable amounts of agricultural raw materials in the non-food industries <u>to</u> almost total replacement of agriculture-derived products by crude oil-derived substitutes

<u>from</u> employment of about 80% of the population in agriculture <u>to</u> the employment of only about 2% in some EC countries.

Within the EC of 12, individual countries and also parts of each country represent different stages of this development. It is, therefore, not advisable to handle the agricultural problems within the EC by a single approach (Munck, 1986a; Munck and Rexen, 1986). Thus, it is neither possible nor rational to solve the problems of the EC under one heading or label such as 'agriculture'. To make progress we must treat the two major aspects of the agri-oriented society within two separate sectors, agri-culture production and agri-commodity production (Fig. 1). We must realize that the suffix, <u>culture</u>, in the word agri-culture really means <u>culture</u> in the original sense, implying that we in Europe, for instance, must take care of our heritage, our delicious high-quality specialty foods, our crafts, our living countryside and its recreation facilities for the urban and rural populations.

In the subsistent farmer community each member of the society could overview the plant and animal production empirically and get an immediate feed-back with regard to its quality. The main effect of the monetary economy is to split the production chains by stimulating specialization. This enhances productivity but impedes empirical information feedback regarding the production history of products and often requires expensive laboratory analyses for quality

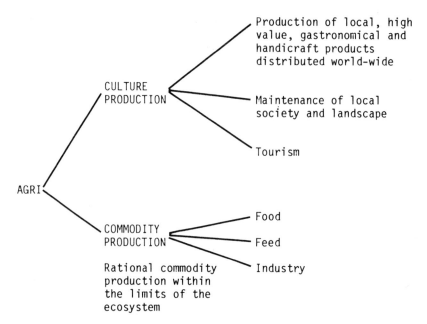

Fig. 1. The fundaments of the agri-sector.

each time a commodity is exchanged for money. One example of re-establishing the empirical link between the consumer and the farmer in a modern society is the Appellation (l'Origine) Contrôllée system introduced on French wines which well could be developed further for other specialities. It is obvious that in the present industrial agriculture, the farmers have lost working opportunities and influences to the other members of the production chain. The number of farmers has also been greatly reduced. The farming communities must, therefore, activate themselves. One option would be to integrate the production chain by development for sale of local specialties in the farm or region or with international distribution as specialty products at a comparatively high price. We have suggested to the Commission (Munck and Rexen, 1986) to support such an Appellation l'Origine Contrôllée system and that the Commission should support not the production but the distribution and control of such products to stimulate the survival of the rich heritage of Europe's fine foods and handicrafts. Because of the small numbers of farmers in most of the old EC countries, even a small percentage conversion of the food market from bulk to Appellation l'Origine Contrôllée would have a considerable impact on the economy of these farms.

The other option which should be pursued, in parallel, is to continue the agri-commodity route emphasizing high yield and efficiency, however, with due consideration to the environment in obtaining a sustainable production. In the following we shall discuss this option emphasizing wheat utilization.

AGRI-COMMODITY PRODUCTION

Present Trends

It is obvious from Table I that there is no general overproduction of agricultural commodities in the EC but a major imbalance. There are surpluses in cereals and sugar while the earlier surplus of animal products now is more or less controlled by the changed EC politics. It is striking, however, that there

TABLE I
Self-sufficiency (% of Consumption)
in the EC

Food/Commodity	1978-80 EC 10	1986 EC 12	
Milk	112	100	
Meat	102	102	
Sugar	124	125	
Cereals	107	112	
Maize	66	83	
Vegetable oil raw materials	25	48	(1985)
Fodder cakes and protein crops	9	19	(1987)
Soya beans	–	6	(1987)
Wood and cellulose fibres	48	40	

TABLE II
Production and Self-sufficiency for
EC-grown Cereals (1981-84)

Cereal Grain	Production (1000 tons)	Self-sufficiency (%)
Oats	6.742	98
Rye	2.425	97
Barley	38.980	114
Hard wheat	4.143	104
Soft wheat	53.547	126
Grain maize	19.373	78
Sorghum	360	–
Rice	1.700	–

are large deficits in the production of vegetable oil raw materials, fodder protein and wood and cellulose fibres, all imported on a large scale mainly from the U.S. and Canada.

The total yearly production of grain in the EC 12 is about 165-170 mio tons while EC consumption is about 140 mio tons. It should be remembered that EC imports 10-15 mio tons of cereal substitutes such as maize gluten feed, by-products from the fruit industry, such as citrus pulp, and manioc from the U.S. and some developing countries, which adds to the surplus of cereals. An examination of the individual cereals in the EC (Table II) shows an overproduction of barley and especially soft wheat and a deficit in maize, which has been reduced significantly in recent years. There has been introduced a grain ceiling of 160 mio tons per year in the EC implying that the farmer himself has to help financing the surplus on the world grain market of the cereals produced above this level.

Monetary Considerations

It is obvious that the role of agriculture extends beyond its role to produce food and keep the wheels of industry rolling. One should not forget that agriculture contributes in maintaining a populated countryside and thus makes both a social and an environmental contribution to our society. Additionally, we must remember that fluctuations in climate as well as catastrophes of nature may greatly reduce agricultural yields as seen in 1988 in North America. It is clear that the problem of agricultural surplus can not be compared with the severity of agricultural deficit and that it also must be possible to explain to the community of a rich country the necessity of paying the costs of a reasonable granary to assure food security.

The world's total cereal stocks, reflecting the short-term food security, fluctuate more than we realize. Thus, the drought in the U.S., Canada and China this year means a drop of 114 mio tons from 1987/88 in wheat and coarse grains. Hence, despite substantial production increases in Europe and India, the global net-reduction is estimated to be 86 mio tons below last year's very moderate harvest of 1,340 mio tons. Consequently, world cereal stocks can be expected to be almost halved from 101 days of global consumption to only 54 days. While the net-exporting countries will profit from accompanying price increases, the grain-importing developing countries may fear that the recordlow stocks will trigger a doubling of world market prices as in 1973.

However, we must reconcile the difficulty to balance food security (self-sufficiency) with other considerations, such as 'a fair standard of living' for the farmers and 'reasonable prices' for agricultural products to the townspeople.

In the following we will quote some grotesque examples on the effect of subsidies in the EC. One hectare in the EC producing 7 tons of surplus wheat requires 614 ECU (1 ECU = 1.22 US$) for storing and export. To compete with imported feed protein at world market prices, one hectare of peas (4.5 tons) requires a subsidy of 990 ECU, when substitution for cereals by peas is taken into account. Also rapeseed requires substantial subsidies: one hectare (2.8 tons) requires 805 ECU to be competitive. The EC governments are no longer willing to pay these subsidies and, therefore, interest in alternative uses outside the food sector has been developed (Rexen and Munck, 1984; Munck, 1986a, 1988).

It is thus obvious that in well developed economies such as the EC and the U.S. economies the only alternative is an open economy without direct subsidies to support the price of the products in order to stimulate the farmers to operate efficiently. Supports should be given to diversify the agri-sector into an agri-cultural and agri-productional sector as previously discussed.

Agri-Commodity Production for Industry

Looking at the future needs of the increasing world population, efficient agri-production should reconcile with sound environmental practices. Up till now we have used very crude and clumsy methods to manipulate the crop. A European side-effect of those practices is endangered life in the North Sea, decreasing its economical importance as a source of fish. It is, therefore, mandatory to develop a fertilizer technology to produce more efficient and less polluting slow-release fertilizers, natural N-fixation systems, controlled compost production as well as biological pest control through new biotechnology. Some commercially available slow-release fertilizers are not used because of their present cost. The present agri-supply industry must develop precise and efficient plant control systems through remote-sensing, biodegradable additives, crop composition monitoring and computer control. A stimulation of the use of these opportunities should be contemplated by governments.

During recent years the surplus of cereals in Europe and the U.S. has triggered old and new ideas on how to use agricultural products in the manufacturing industry. Hundred years ago, a wide range of agricultural products from starch and cellulose were used solely for industrial purposes to produce cloth, paper, building materials etc. The exploitation of cheap fossil fuels produced a range of substitutes

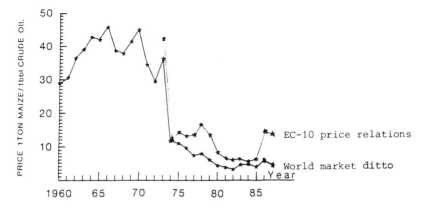

Fig. 2. Maize price versus oil price in the years 1960 to 1987.

for those products. This development is described in Fig. 2 where we look at the ratio between the price of one ton of maize and one barrel of crude oil as an indicator. Competitiveness of agricultural raw materials for industry has increased considerably, especially when we look at world market prices.

Starch, cellulose and oil are the major agricultural components of industrial interest. Table III provides data on starch production in the EC. Traditionally, starch is produced from imported maize

TABLE III
Starch Production in the EC

Raw Material	Number of Plants		Starch Production		Average Size of Plants	
	1984	1988	1984	1988	1984	1988
			mio tons/year		ton starch/year	
Maize	19	19	2.5	3.5	132.000	184.000
Potatoes	17	17	0.78	1.0	46.000	59.000
Wheat	16	30	0.18	1.1	11.500	36.000

TABLE IV
Costs of Raw Materials for Starch after
Valorization of By-Products in 1986
(in ECU/ton of starch)

Cost Sources	Maize	Wheat	Potato
1) Cost of raw material	376	400	298
2) Valorization of by-product	128	270	260
1 - 2) Net price of raw material	248	130	272[a]

[a]This high figure is probably due to the inclusion of extra cost of waste water disposal.

or potato. In the last four years, however, wheat starch factories have expanded both in size and numbers. The reason can be seen in Table IV (Munck and Rexen, 1986). In spite of higher costs of wheat over maize and potato as raw material, the price of wheat within the EC is more favourable when the income from the high-quality wheat gluten by-products is deducted. However, because of overproduction, wheat gluten prices are rapidly decreasing.

In Table V, we review starch production and uses in the EC. The present total is about 1.7 mio tons which is just a small fraction of the 165-170 mio tons cereal produced. The Commission has, however, stimulated the use of starch for non-food purposes by giving a subsidy so that the starch price reaches that of the world market. This will hopefully stimulate industrial uses of starch. Optimistic forecasts indicate a consumption of starch in the year 2000 of 3.7 mio tons. This area should attract investments in research because starch price now is only 25% of synthetic polymers and 1/3 of the price of cellulose (world market prices). There is a wealth of possibilities to utilize the starch polymer after modification by means of organic chemistry or biotechnological transformation (Fig. 3). Other ma-

TABLE V
Forecast for the Consumption of Starch in
the Non-food Industry

Non-food Product	1000 tons/year 1986 Commission Figures	2000 Commission Forecast	2000 1984- Forecast a)
Paper and board	820	1300	900-1550
Textiles	62	62	160-350
Adhesives, chemical, pharmaceutical and polymer industries	451	1025-1785	405-1803
Other	390	457	-
Total	1723	2844-3604	1465-3703

a) Rexen and Munck, 1984

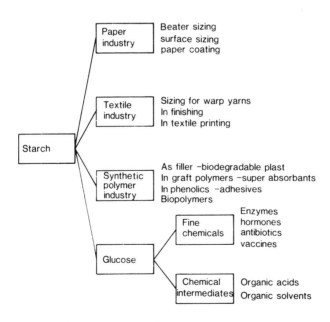

Fig. 3. Non-food applications of starch.

jor sources of biological components to be used in industry are plant oils and cellulose (Rexen and Munck, 1984).

In the future, a total utilization of biomass should be aimed at. Co-products, like straw, whether burnt or degraded by microorganisms in the field will unavoidably in the end contribute to the CO_2 level in the atmosphere. These products should therefore be taken care of as rationally as possible as substitutes for fossil products so that the total emission of CO_2 can be reduced or remain constant. Furthermore, utilization of agricultural raw materials will automatically ensure a better balance between emission and fixation of CO_2 if the plant productivity of the environment is maintained.

Cereal straw is the most neglected agricultural co-product in the world. Hundred years ago, whole straws (leaves and stems) were used for paper-making. Then technology developed in Sweden allowed manufacturing of paper pulp from trees (wood). This attracted large industry investments both in raw material production and in processing. The wood-based pulp factories grew large, and when pollution restrictions appeared they could, thanks to their mere size with large investments, pay the capital needed to make a closed process where the extracted by-products were burnt for energy. Most of the small straw pulp factories could not afford to follow this development and they were, consequently, closed down. Straw is a mediocre raw material and is neither optimal for feed to ruminants nor for paper making. In the wood-based industry the needles and the leaves from the trees are obviously not used for paper, only the stems. We can follow this innovation from the wood-based industry and separate mechanically straw into leaves and stems. We obtain powdered leaves, approaching hay as a nutrient for ruminants, and chips from the stem with nearly the same composition as birch wood. We have designed such a separation process based on the UMS disc mill (Gram et al., 1988) and processed straw from rice, barley, wheat, maize and sorghum (Bjørn Petersen, 1987). We have obtained excellent results in small-scale processing tests for paper in coopera-

TABLE VI
Chemical Constituents in Barley
and Wheat Straw (g/100 g DM)

	Barley		Wheat	
	Leaves + Nodes	Inter-nodes	Leaves + Nodes	Inter-nodes
Weight	49.6	50.4	45.0	55.0
α-Cellulose	28.2	37.7	29.3	42.0
Protein	3.9	2.0	5.2	3.0
Lignin	13.0	18.3	15.4	18.0
Si	1.4	0.5	2.3	1.0

tion with one of the few still operating straw-based paper mills in Denmark.

In internodes in spring barley and winter wheat the α-cellulose content is about 40% (similar to soft wood) while the leaf fraction contains more protein and less cellulose and lignin (Table VI). Digestibility studies in ruminants have shown that the leaf fraction in straw has similar digestibility as hay. The internodes suitable for both paper and particle boards for use in the building and furniture industries are also low in Si which gives less problems in recycling the extraction liquid in the paper pulp factory compared to whole straw.

In Table VII we have calculated the total yield of botanical components of cellulose and starch from cereal crops in the EC. Out of 73 mio tons of internodes we may, theoretically, produce about 36 mio tons of paper pulp. It has been calculated that the total potential demand for internodes for pulp production in the EC of 10 is 10-17 mio tons if all printing and writing paper, corrugated board and solid board produced in the EC by the year 2000 contains 30-50% straw pulp as in Denmark today (Rexen and Munck, 1984).

At present, pulp prices are high (up to 6.0 DKr./kg) reflecting the high costs in collecting the

TABLE VII
Harvest 1985 in EC-10
Total Yield of Botanical Components
and of Cellulose and Starch from
Cereal Crops (mio tons)

Crop	Grain	Starch	Straw	Inter-nodes	Cellulose from In-ternodes	Leaves
Barley	51.3	30.0	37.4	19.0	7.6	15.7
Wheat	71.5	46.0	64.9	37.0	14.1	22.7
Maize	24.4	17.0	18.8	9.0	3.6	7.7
Oats	7.8	3.0	10.5	5.0	2.0	4.4
Rye	3.2	2.0	4.2	3.0	1.2	1.0
Total	157.2	98.0	135.8	73.0	28.5	51.5

Source: Rexen and Munck, 1986.

wood. Starch prices on the world market are, however, low (2.0 DKr./kg) due to the present apparent surplus. In Europe we are piling up the cheap starch-rich seeds while we are burning straw containing the valuable fraction, cellulose. In the Philippines, New Guinea, Brazil etc., precious rain forests are cut every day, milled and sent to paper pulp industries in Europe and Japan. Stem fractions from local straw of rice, maize, sorghum etc. could after preprocessing be exported as substitutes for the valuable rain forests which is badly needed to retain water capacity and climate (Munck and Rexen, 1985). Setting up such projects would also create necessary employment in the rural areas of the developing countries. Thus, a continuation of the benefits of the green revolution in developing countries can only be sustained by developing the agro-industrial option to secure the delivery of foods and other agricultural raw materials from rural to urban areas, substitute for import, and generate purchasing power for the poor population.

It is a paradox that huge areas of the world's natural vegetation are systematically being destroyed by enormous machines using diesel oil and gasoline. We are in fact using non-renewable resources, which future generations will badly miss, to destroy renewable resources that might have helped to reduce the demand for the fossil resources.

A Bridge Between Agriculture and Industry
- The Concept of Agricultural Refineries -

The disadvantages of the present agricultural system in north-western Europe are as follows (Rexen and Munck, 1984; Munck, 1986b, 1987):

- It is unbalanced both from the point of view of products (surpluses and deficits) and from a local production distribution, e.g. imbalance between plant and husbandry agriculture;
- Specialization leads to monoculture and disease problems;
- Mismanagement of plant nutrition is expensive and polluting - excessive and uncontrolled distribution of untreated local animal manure and of excessively soluble chemical fertilizers;
- Disease problems are counteracted at a high price with pesticides;
- Erosion and soil packing problems affect yield;
- Ineffecient use of expensive farming equipment;
- Combine harvesting seeds alone leaves potentially valuable straw in the fields;
- Strong seasonal concentration of plant husbandry work prevents efficient utilization of rural manpower;
- Lack of drying and other conservation facilities makes in many places plant production dependent on weather, resulting in loss of quality. This is a particular problem for industrial utilization of crops where constant, defined quality is important;
- Agricultural land is becoming expensive because of competition with alternative uses;
- Mechanization makes agriculture less labour intensive and the lack of alternative local work leads

to a depopulated countryside with negative side-effects on rural society and the environment;
- Centralized agro-industries such as feed, sugar, and starch factories result in excessive transportation costs;
- Price structures prevent industrial utilization of starch by chemical modification, for example. Such products are imported and make development of new agricultural-based industries more difficult. The 1986 change in the pricing system in the EC for starch is a positive move, but not sufficient for the biotechnological fermentation industry to change from imported molasses to starch;
- The cited inefficiencies result in unnecessarily high energy costs per unit of produce, retard restructuring to deficit crops, increase pollution and bring a higher price of agricultural products than necessary, impeding alternative uses of land and products, which could reduce the surplus.

Today there exist potential, technically and economically viable, partial solutions for most of these problems. Since the solutions are often interdependent, however, their implementation will not move until the whole system moves.

We have previously launched the concept of agricultural refineries in our EC report (Rexen and Munck, 1984) as a way of improving the connection between agriculture and the agriculture-derived industry. The term should not be misunderstood. The refinery as we see it from the northern region of the EC agriculture (Fig. 4) is a relatively small local preprocessing station possibly combined with a field machine station, serving about 2000 ha on contractual basis and deliverring products to the farmers (feed) and to the food and manufacturing industries. Such a unit is designed to increase flexibility in agricultural production just as an oil refinery (though at a much larger scale) can adjust input and output according to raw materials and market prices at a very wide range, each time finding the most profitable combination.

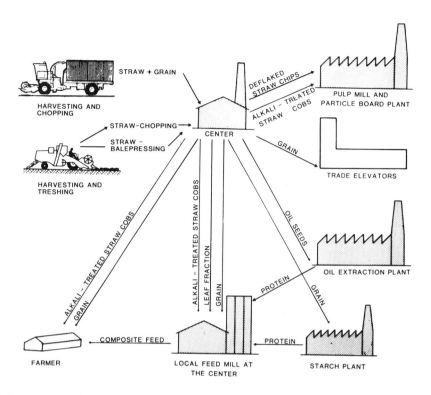

Fig. 4. The agricultural refinery concept.

Whole-Crop Harvesting

An alternative to the combine system for wet northern European harvesting conditions was introduced around 1970 in Denmark and Sweden (Rexen, 1971; Munck, 1972). The idea was to adopt the system used in the green crop drying industry, employing pneumatic drum dryers. The system is based on a self-propelled chopper, which in one step cuts the crop, chops it, and blows it into a container attached to the harvester. After drying, straw and grain are separated in a stationary separator (situated at the drum dryer plant). The whole crop is alternatively dryed and pressed into feed pellets. This system displays some obvious advantages compared with traditional combine harvest: (1) it is independent of weather, (2) the field is cleared in one step, (3) a higher yield of both grain and straw is achieved, (4) the straw quality is much better due to less microbiological degradation, and (5) it can also be used (after some development) to harvest, dry, and process a wide range of non-cereal crops including grass, legumes, and oilseeds.

The Agricultural Refinery

The whole-crop harvesting system is ideal for northern Europe as an overture to industrial processing of cereal crops. The main objectives of a local agricultural refinery are:

- Increased efficiency in agricultural crop harvesting and handling,
- Optimal utilization of the capital invested in machinery and equipment,
- Optimal utilization of all botanical components of the biomass,
- Extended growth season through a diversified plant cropping system enabling a continuous operation - 1-3 shifts for 5-8 months with the broadest possible range of crops.

The following 3-step procedure has been suggested (Rexen, 1977, 1978):

Step 1. The crop is harvested with a whole-crop harvester and transported to the refinery or (alternatively but less optimal) the grain is harvested with a combine, and the straw is collected with a baling machine whereafter both products are transported to the unit.

Step 2. The crop is separated into 3 fractions at the refinery: (1) grain, (2) leaves (light fraction), and (3) stems. Each fraction may be pre-treated according to its end-use. Industrial-scale equipment for this process exists already.

Step 3. The pre-treated products are transported to centrally placed factories, where the main processing takes place. Each factory will be served by several "satellite units".

A local agricultural refinery will typically have a capacity of 10,000 tons of cereal grains (2,000 ha) per season plus approx. 10-13,000 tons of cereal straw. To this can be added the processing of other crops, the amount of which will depend on the local growing conditions. The refinery could e.g. include three harvesters, separator, drum dryer, grain sifter, straw separator and a feed factory. One "parent" factory (a starch factory, a cellulose factory, or an oil mill, etc.) would need approx. 10-100 agricultural refineries to secure an all-around-the-year supply of raw materials.

Cereal grain for food, starch, and malting is dried in a conventional grain dryer (which can utilize excess heat from the drum dryer) to a moisture content of 14-15% and is then purified. Broken and small kernels, weed seeds, etc. are removed and used in fodder mixtures. In the drum dryer, the straw is separated into a "light" fraction, (leaves and ears) and a "heavy" fraction (internodes, nodes and parts of the leaves attached to the internodes). The light fraction is pelletized and used as fuel for the drum dryer (40-70 kg pellets are needed to dry 1000 kg biomass from 30 to 15% moisture content). One ton of straw yields up to 300 kg of light fraction which alternatively can be used as feed after alkali treatment or as raw material for chemicals such as furfural etc. The heavy fraction is either alkali-treated and pelletized for use as fodder or "flaked" and purified for production of straw chips for the board and cellulose industry.

CONCLUSIONS

As scientists from a private laboratory run by an industrial company, we can see a short term conflict between industries and organizations favouring fossil versus renewable resources. In the long run, however, it should be possible to avoid conflicts (Munck, 1988). Why should, for instance, oil companies not stimulate use of agricultural products for industry as a substitute for oil if they could participate in this development themselves? Firstly, it is in their own interest to stimulate the agricultural sector which is one of their basic markets for delivery of chemicals. Secondly, there will always be a need for fossil fuels and it might be economical to leave more of it in the ground to obtain a better price. The present trend that plant breeding companies are sold to the the chemical and oil industries is a sign from the large industrial sector of an interest in this problem. Restructuring from fossil to renewable resources will not be allowed to go too fast by these multinational companies because the present huge investments have to be paid.

It will take time to realize these perspectives in Europe, which depends on the production costs of biomass versus fossil energy. A well organized, consistent research and development programme supporting the plant production chains from breeding and production to agricultural refineries and industry (Munck, 1986b) for both agriculture and agri-commodity production is needed during the coming 20 years. The newly presented ECLAIR development programme for the EC is a good start. This programme should be completed by cooperative technological projects to support the self-sufficiency of developing countries based on development of local agriculture-based industries and endurable technology. Implementation of both tasks on a large scale would be in the interest of all governments and industries, local or multinational. The results from the programmes in the developing countries could later be used in the industrialized countries to support the unavoidable long-term development towards a sustainable agri-industrial production. Wheat will certainly continue as one of the leading agro-commodities in the future agri-industrial development but with extended use of the whole plant for food, feed, starch and cellulose.

LITERATURE CITED

BJØRN PETERSEN, P. 1987. Separation and characterization of botanical components in straw. Agricultural Progress 63 (in press).

GRAM, L.E., ANDERSEN, M., and MUNCK, L. 1988. A short

milling process for wheat (these Proceedings).
MUNCK, L. 1972. Improvement of nutritional value in cereals. Heriditas 72:128.
MUNCK, L. 1986a. The use of surplus 'grain' in the production of fuel and chemical commodities. Proc. Conference 'International Grain Forum', November 10-12, 1986, Amsterdam, The Netherlands.
MUNCK, L. 1986b. Stimulation of the agro-industrial interface - Future perspectives in structure improvements as reflected in a programme suggested for research and development. Proc. 2nd Workshop 'Agricultural Surpluses' EEC - EFB - DECHEMA, June 16-17, 1987, Oberursel/Ts, West Germany.
MUNCK, L. 1987. A new agricultural system for Europe? Trends in Biotechnology 5(1):1-4.
MUNCK, L. 1988. Cereals for industrial use - Towards a renewable source, recycling economy on the basis of plant production. Proc. EUCARPIA Cereal Section Meeting 'Cereal Breeding Related to Integrated Cereal Production', February 24-26, 1988, IAC, Wageningen, The Netherlands.
MUNCK, L., and REXEN, F. 1985. Increasing income and employment in rice farming areas - Role of whole plant utilization and mini rice refineries. In: Impact of Science on Rice; International Rice Research Institute, Manila, The Philippines, pp. 271-280.
MUNCK, L., and REXEN, F. 1986. The needs of a long-range precompetitive development programme within the EC to stimulate a system change in agriculture, opening up economical utilization of agri-products in industry. Proc. Conference 'The Future of Agriculture in Europe', organized by Club de Bruxelles, November 5-6, Brussels, Belgium.
REXEN, F. 1971. Fremtidens tekniske system til høstbjergning og konservering og oplagring af kornafgrøder. Tidsskrift for Landøkonomi 5, pp. 183-233.
REXEN, F. 1977. Neu Ergebnisse bei der Herstellung von Strohzellulose. Allgemeine Papier Rundschau 4: 346-348.
REXEN, F. 1978. Opdeling af halm i fraktioner til fremstilling af spånplader. Proc. Congress 'Nordisk Jordbrugsforskerkongres', Reykjavik, Iceland.

REXEN, F., and MUNCK, L. 1984. Cereal Crops for Industrial Use in Europe, EUR 9617 EN. Report prepared for the Commission of the European Communities, 242 pp.

REXEN, F., and MUNCK, L. 1986. Synthesis of EC work on industrial valorization of cereals. Proc. Symposium 'Non-Food Industrial Use of Wheat and Corn', November 20-21, 1986, CIIA, Paris, France.

5

HOW THE STRUCTURE OF THE WHEAT CARYOPSIS SHOULD BE MODIFIED TO INCREASE ITS END-USE VALUE

Donald B. Bechtel

U.S. Grain Marketing Research Laboratory
United States Department of Agriculture
Agricultural Research Service
1515 College Avenue
Manhattan, Kansas   66502

ABSTRACT

The structure of the wheat caryopsis has important implications in the utilization of wheat. Modification of the wheat grain to increase its end-use value by changing structural components is an area of breeding that has not received a great deal of attention. There are a number of structural changes that could be made to the wheat grain to improve its quality and end-use properties. The elimination of the crease region would increase flour yields because it is this region that prevents easy removal of the bran. Increased flour yields could be obtained and more efficient milling techniques utilized if aleurone-less wheats could be developed. Flour composition could be altered by making changes in endosperm structure. High fiber flours could be produced if thicker cell walls were bred into endosperm cells or flours of differing compositions could be produced if advantage was taken of the nonuniform deposition of storage reserves. The synthesis of different types of starch granules appears to be another area where structural modifications could be made. These are a few of the many possible structural changes that are discussed which could be made to the wheat caryopsis.

INTRODUCTION

The structure of a mature cereal is a product of the genetic and environmental effects acting on the plant during development. Because the structure of wheat, is for all practical purposes, the result of a series of genetic commands that have been carried out during development, we can view the wheat grain as a system waiting to be exploited to meet our needs. Eons of evolutionary changes have resulted in a wheat plant that was best suited to its environment. Man's intervention in this process has given us a plant product that has specific attributes found nowhere else in the plant kingdom. One of the problems we face is how to change the wheat grain to better our use of it. This is not an easy task as little is known about the developmental genes that control the formation of the wheat caryopsis. We are now only beginning to understand the genetics of a few of the wheat storage proteins and how they relate to quality, let alone what genes are responsible for grain shape, size, amount of storage components, etc. Many of the factors that govern the end-use properties of cereals in general are determined by the structure of the caryopsis. Indeed, every structural component of wheat fruit has some effect on grain quality and end-use properties. The development of these components is genetically controlled and, therefore, is available for being changed or altered. Most breeding programs dwell on grain yield, disease resistance, protein content, and quality as the main emphasis while many other features that affect grain quality are considered second or not at all.

This paper offers a structural biologist's views on how the wheat caryopsis might be altered to change a variety of end-use and quality properties.

SELECTED FEATURES FOR POSSIBLE ALTERATION

Crease

The elimination of the crease would have several effects on wheat grain. An increase in total grain volume could result because the crease occupies 0.7 to

1.9 % of the total grain volume (Kingswood, 1975). Second, would be an increase in flour yield. Increased flour yield would result because the difficulty of removing starchy endosperm particles from the crease would be eliminated. Third, would be the elimination of the endosperm cavity (Fig. 1). The endosperm cavity (Kosina, 1979) forms during caryopsis development at the junction of the terminus of crease and starchy endosperm and its presence causes low test weights. Many F1 and F2 crosses exhibit this phenomenon thereby reducing the number of selections from which breeders have to choose.

Some interest has already been shown in breeding for creaseless wheats (Kingswood, 1975). The development of the crease presents an interesting aspect to wheat morphogenesis. The process starts about the time of fertilization of the egg by the sperm nucleus and is manifested by a postion change of the ovule with respect to the maternal tissues (see Evers and Bechtel, 1988). The ovule is connected to the maternal tissues by the funiculus, a short stalk-like structure that houses the vascular tissues. At first, the ovule is situated in the carpel like a lollipop but with differential growth on its dorsal side, the ovule begins to turn down. As the differential growth continues the funiculus comes to rest between the ovary wall and the ovule (Aziz, 1972). This type of development is typical for the cereals in general and certainly aids in the bringing of nutrients to the growing caryopsis. As all cereals possess this anatropous position, the questions are why some cereals possess creases while others do not, and how can the developmental process be altered so the crease could be eliminated?

The crease forms as a result of the lateral growth symmetry of the endosperm. The lateral growth produces a flattened endosperm when viewed in cross section. As development continues the lateral flanks fold down and the middle section containing the vascular tissues becomes trapped between the two flanks (cheeks) forming the crease (Evers, 1970; Figs. 2-4). Some cereals such as rice do not have this lateral growth pattern. Instead, the growth pattern is more dorsal-ventral resulting in the vascular tissues being exterior to the

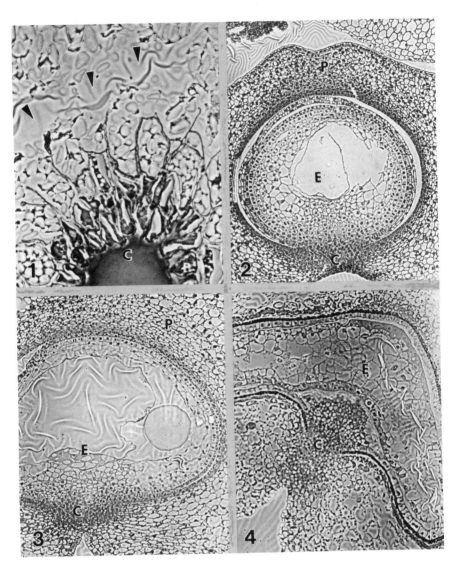

Figs. 1-4. 1. Light micrograph (LM) of cross section through wheat endosperm showing portion of crease (C) with endosperm cavity (arrows) (x 240). 2. Cross section through 4 day old endosperm (E) with incipient crease (C) and pericarp (P) (x 70). 3. Five day old wheat endosperm (E) with incipient crease (C) and pericarp (P) (x 70). 4. Seven day old wheat endosperm showing forming crease (C) (x 70).

endosperm rather than internalized (Hoshikawa, 1973). These types of developmental patterns are certainly under genetic control and thus subject to manipulation.

Aleurone

There are several features of the endosperm that affect end-use properties and quality. The endosperm actually consists of two major cell types; the starchy endosperm and the aleurone layer (Fig. 5). Most wheats possess an aleurone layer that is one cell thick. The aleurone is high in protein and minerals, but is removed from the starchy endosperm during milling. The aleurone is a major component of bran which is botanically a combination of the pericarp, seed coats, nucellus, and aleurone layer. Although the aleurone is endosperm tissue, its structure and composition make it an unwanted component in white flour. Its high oil content, in the form of lipid droplets, high ash content, primarily from the protein bodies called aleurone grains, and its tenacious adherence to the starchy endosperm, make it an undesirable structure as well as a difficult one to remove. Breeding for changing the thickness of aleurone layers is not promising because there is little correlation between it and flour yield (Kosina, 1979). One aspect that has not received any consideration, however, is the total elimination of this layer through breeding. The elimination of aleurone would create several advantages. First, the miller could obtain higher extractions without higher ash content due to the elimination of high ash-containing aleurone. Second, the higher extractions would probably increase the fiber content of the flour due to the inclusion of part of the caryopsis coats and this may be nutritionally important. Third, the smoother surface between starchy endosperm and nucellar tissues would result in greater efficiency of the milling process. It is the irregular inner surface between the aleurone and starchy endosperm that is responsible for the difficulty in removing starchy endosperm from the bran during milling (MacMasters et al., 1971). The milling engineer has trouble in dealing with variations in aleurone thickness and the problem is further complicated by the fact that aleurone and starchy endosperm are botanically the same tissue and intimately in contact with one another (MacMasters, 1962). A smoother surface between endosperm and caryopsis coats or a weakness between layers would assist the milling process. Development of the caryopsis coats has been decribed (Morrison, 1975, 1976) and several areas in the mature coats may provide for their easy separation from the starchy endosperm. In fact, there are several natural weaknesses in the coats that are revealed when wheat sections are prepared for microscopy. Exterior to the aleurone is the nucellus, a tissue that functions during early caryopsis development (see Evers and Bechtel, 1988 for a review). It degenerates rapidly leaving a crushed layer of cells at maturity. This layer represents a weak spot in that the aleurone can be mechanically separated from the nucellus at maturity (Evers and Reed, 1988; Fig. 6). This weakness would allow easy separation of the starchy endosperm from the outer caryopsis coats, if the aleurone layer were absent. Other possible weak points would be where various parts of the caryopsis coats meet cuticles of the seed coats (Fig. 7).

Starchy Endosperm

Storage Proteins

Another aspect of wheat endosperm possibly subject to alteration is the distribution of storage proteins within this tissue. It is a well known observation that the subaleurone region typically contains a higher percentage of protein than does the central region (Gaines et al., 1985; Fig. 8). This high protein subaleurone layer may be difficult to separate from the bran (Pomeranz, 1982) and result in reduced flour yield as well as protein content of the resulting flour. A more even distribution of the storage proteins throughout the endosperm tissue might increase both flour yield and protein content. A view of the development of the endosperm cells reveals that this may not be as simple as it sounds. The subaleurone cells are the youngest (last formed) cells of the

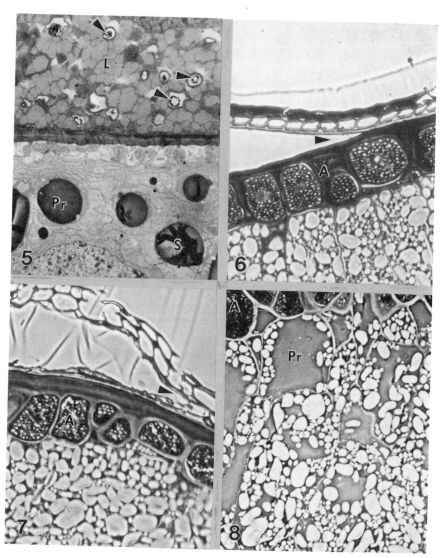

Figs. 5-8. 5. Electron micrograph showing aleurone cell with lipid droplets (L) and aleurone grains (arrows). Starchy endosperm has storage protein (Pr) and starch granules (S) (x 4030). 6. LM of wheat cross section showing separation of caryopsis coats (arrow) from aleurone (A) (x 230). 7. LM of wheat caryopsis with separation of caryopsis coats (arrow) from seed coat and aleurone (A) (x 310). 8. LM of wheat with high protein content (Pr) beneath aleurone (A) (x 270).

starchy endosperm and as a result are the last to differentiate. The spatial location and the timing of formation of these cells certainly plays a role in determining the amount of storage protein to be deposited in the subaleurone region. Coupled with the synthesis of storage proteins is the development of starch granules. One of the reasons the subaleurone region is high in protein is the reduction of the amount of starch in this area. One may not want to increase the uniformity of reserves, because methods are presently available for the production of different protein content flours from normal flours using air classification methods. In fact, wheats with even greater protein content gradients may be better for air classification.

Starch
Starch is the major component of wheat endosperm, comprising approximately 64-74 % of milled endosperm (Pomeranz, 1988). Starch, in the form of distinct granules, is synthesized in an organelle called the plastid. In endosperm tissue, a specialized type of plastid called the amyloplast is responsible for the synthesis and formation of the starch granules. Wheat endosperm plastids typically produce one granule per amyloplast (some evidence indicates that more than one small starch granule can form in some plastids; Parker, 1985) and at least two distinct size classes of granules, large type A and small type B granules (Parker, 1985; May and Buttrose,1959; Fig. 9). A third intermediate class may also exist (Baruch et al., 1983).

Wheat starch is unique as it can not typically be replaced by other noncereal starches to yield a satisfactory baked product. In addition, there are differences in both composition and functional properties of the different types of granules (Kulp, 1973). Plastids are self replicating organelles. They contain DNA that codes for both the RNA used for the production of ribosomes and proteins synthesized within the plastid. The DNA is replicated and the plastid divides by fission as do bacteria. Therefore, the plastids that produced type B granules formed from the plastids that produced type A granules. Differences must exist among the different types of plastids in order for them to produce the different types of starch granules.

The development of new varieties of wheat for commercial use that lack a specific type of starch granule may be exploited by altering the plastid genome and may allow wheat to take on a new prominence in the commercial market.

Fiber

The importance of dietary fiber in our diets is becoming more important. Wheat endosperm cell walls contain small amounts of cellulose and large amounts of arabinoxylans (Lineback and Rasper, 1988). The composition of the cell walls may be open for exploitation as more information becomes available on how the composition and thickness of the endosperm cell walls are controlled during development.

Cell walls are first secreted when the coenocytic endosperm cytoplasm undergoes cytokinesis which proceeds by a two-step process. The first step occurs independent of a phragmoplast (Morrison and O'Brien, 1976; Mares et al., 1977). The second method involves a phragmoplast typical of most other plants undergoing cytokinesis (Fig. 10). The cell wall thickens by the addition of material via cytoplasmic vesicles fusing with the plasma membrane and the deposition of precursors to the cell wall (Figs. 11). The type of cell wall components, amount of components, and the length of time the cell wall is secreted are probably all under close genetic control and therefore subject to manipulation.

Embryo

The recovery of the embryo (called germ by millers) during the milling process is an important step because of its value in the food and pharmaceutical industries (Bass, 1988). Two major factors contribute to how easily the embryo can be removed from the rest of the caryopsis. The first factor is the arrangement of a layer of cells called the scutellar epithelium which separate the scutellum from the endosperm. These cells may provide an irregular surface as well as a cementing

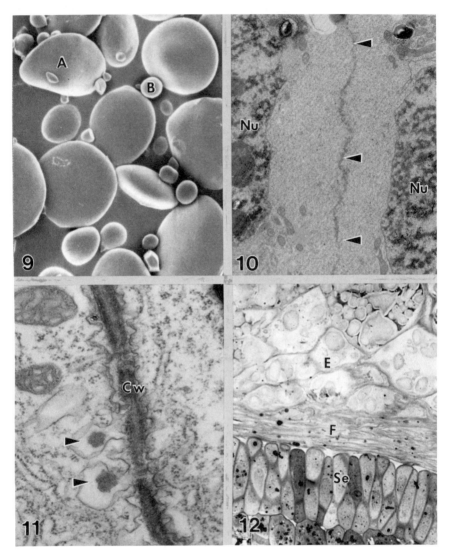

Figs. 9-12. 9. Scanning electron micrograph showing isolated type A and type B starch granules. (x 1200). 10. Electron micrograph showing dividing endosperm cell producing a phragmoplast (arrows) between two telophase nuclei (Nu) (x 4100). 11. Enlarging endosperm cell wall (Cw) via vesicle fusion (arrows) (x 27,900). 12. LM of junction between endosperm (E) and scutellar epithelium (Se) showing fibrous zone (F) (x 428).

material to which the endosperm may bind and prevent embryo removal (Pomeranz, 1982). Second is a layer of crushed cells which forms a fibrous region between the scutellar epithelium and the starchy endosperm (Evers and Bechtel, 1988; Fig. 12). Embryo removal should be facilitated by decreasing the irregular nature of the epithelium and the amount of cementing material and increasing the thickness of the fibrous region.

SUMMARY

The above list discusses only a few of the many possible structural alterations that one day may increase our utilization of the wheat caryopsis. It will be no easy task to develop wheats without creases and aleurone layers, but the advantages of such developments will be readily apparent. Other structural changes will affect milling, and nutritional factors. Much research must yet be conducted at the basic level before any structurally-changed new wheats become commercially available. It is these types of changes, however, that will make wheat even more fundamentally important to man.

LITERATURE CITED

AZIZ, P. 1972. Histogenesis of the carpel in Triticum aestivum L. Bot. Gaz. 133:376-386.

BARUCH, D. W., MEREDITH, P., JENKINS, L. D., and SIMMONS, L. D. 1983. Starch granules of developing wheat kernels. Cereal Chem. 56:554-558.

BASS, E. J. 1988. Wheat flour milling. Pages 1-68 in: Wheat Chemistry and Technology, vol. II., Y. Pomeranz, ed. American Association of Cereal Chemists, St. Paul.

EVERS, A. D. 1970. Development of the endosperm of wheat. Ann. Bot. 34:547-555.

EVERS, A. D., and BECHTEL, D. B. 1988. Microscopic structure of the wheat grain. Pages 47-95 in: Wheat Chemistry and Technology, vol. I., Y. Pomeranz, ed. American Association of Cereal Chemists, St. Paul.

EVERS, A. D., and REED, M. 1988. Some novel observations by scanning electron microscopy on the seed coat and nucellus of the mature wheat grain. Cereal Chem. 65:81-85.

GAINES, R. L., BECHTEL, D. B., and POMERANZ, Y. 1985. Endosperm structural and biochemical differences between a high-protein amphiploid wheat and its progenitors. Cereal Chem. 62:25-31.

HOSHIKAWA, K. 1973. Morphogenesis of endosperm tissue in rice. JARQ 7:153-159.

KINGSWOOD, K. 1975. The structure and biochemistry of the wheat grain. Pages 47-62 in: Bread, Social, Nutritional and Agricultural Aspects of Wheaten Bread, A. Spicer, ed. Applied Science Publishers, Ltd., London.

KOSINA, R. 1979. Association between structure and quality of the wheat grain. Cereal Res. Commun. 7:11-17.

KULP, K. 1973. Characteristics of small-granule starch of flour and wheat. Cereal Chem. 50:666-679.

LINEBACK, D. R., and RASPER, V. F. 1988. Wheat carbohydrates. Pages 277-372 in: Wheat Chemistry and Technology, vol. I., Y. Pomeranz, ed. American Association of Cereal Chemists, St. Paul.

MACMASTERS, M. M. 1962. Important aspects of kernel structure. Trans. ASAE 5:247-248.

MACMASTERS, M. M., HINTON, J. J. C., and BRADBURY, D. 1971. Microscopic structure and composition of the wheat kernel. Pages 51-113 in: Wheat Chemistry and Technology, Y. Pomeranz, ed. American Association of Cereal Chemists, St. Paul.

MARES, D. J., STONE, B. A., JEFFERY, C., and NORSTOG, K. 1977. Early stages in the development of wheat endosperm. II. Ultrastructural observations on cell wall formation. Aust. J. Bot. 25:599-613.

MAY, L. H., and BUTTROSE, M. S. 1959. Physiology of cereal grain. II. Starch granule formation in the developing barley kernel. Aust. J. Biol. Sci. 12:146-159.

MORRISON, I. N. 1975. Ultrastructure of the cuticular membranes of the developing wheat grain. Can. J. Bot. 53:2077-2087.

MORRISON, I. N. 1976. The stucture of the chlorophyll-containing cross cells and tube cells of the inner pericarp of wheat during grain development. Bot. Gaz. 137:85-93.

MORRISON, I. N., and O'BRIEN, T. P. 1976. Cytokinesis in the developing wheat grain division with and without a phragmoplast. Planta 130:57-67.

PARKER, M. L. 1985. The relationship between A-type and B-type starch granules in the developing endosperm of wheat. J. Cereal Sci. 3:271-278.

POMERANZ, Y. 1982. Grain structure and end-use properties. Food Microstruct. 1:107-124.

POMERANZ, Y. 1988. Chemical composition of kernel structures. Pages 97-158 in: Wheat Chemistry and Technology, vol. I., Y. Pomeranz, ed. American Association of Cereal Chemists. St. Paul.

6

NUTRITIONAL POTENTIAL AND LIMITATIONS OF WHEAT

Antoinette A. Betschart

Food Quality Research Unit
Western Regional Research Center
Agricultural Research Service
U.S. Department of Agriculture
Albany, CA 94710

INTRODUCTION

The Surgeon General's Report on Nutrition and Health states "... what we eat may affect our risk for several of the leading causes of death for Americans, notably, coronary heart disease, stroke, atherosclerosis, diabetes and some types of cancer. These disorders now account for more than two-thirds of all deaths in the United States. ... The report's main conclusion is that overconsumption of certain dietary components is now a major concern for Americans. While many food factors are involved, chief among them is the disproportionate consumption of foods high in fat, often at the expense of foods high in complex carbohydrates and fiber that may be more conducive to health. ... As the diseases of nutritional deficiency have diminished, they have been replaced by diseases of dietary excess and imbalance..." (USDHHS, 1988). Thus, one of the recommendations is to "increase consumption of whole grain foods and cereal products, vegetables and fruits".

Emphasis on high-carbohydrate, low-fat diets has brought about a renewed interest in wheat and wheat foods. Early nutritional interest focused on wheat as a good source of protein, calories, selected B-vitamins and minerals. More recent interest has focused on the nutritional and health-related properties of dietary fiber and starch components of wheat. Nutritional quality and the role of wheat in health and disease have been extensively reviewed (Betschart, 1988; Danforth, 1985; Johnson and Mattern, 1987). The purpose of this paper is to discuss the potential of selected nutritional and health-related properties of wheat, and to suggest methods for improving some of the nutritional limitations.

Nutritional priorities continue to evolve as our understanding of the nutrient composition of food, nutrient interactions, bioavailability of nutrients, diet and disease relationships, and human nutritional requirements improves. From 1930 through the 1950's the interest was to increase vitamin content of foods; in the 1950's and 1960's the emphasis was on adequate protein intake; in the 1960's and 1970's there was an increased awareness of the importance of trace elements, quality of fat and dietary fiber. More recently there has been an emphasis on the role of composition of the diet (especially carbohydrate and fat) and quantity of food consumed in the development or deterrence of some major chronic diseases. The latter concerns have resulted in the issuance of dietary guidelines by various nations, agencies and organizations. Agreement on major guidelines and issues is apparent in Table I. The guidelines generally recommend to decrease intake of total calories, calories from fat, saturated fatty acids, cholesterol, simple sugars, sodium, and alcohol. Guidelines recommend an increase in calories derived from carbohydrates, especially complex carbohydrates such as starch and fiber, and the consumption of a wide variety of foods. What are some approaches to optimizing nutritional potential of wheat foods?

OPTIMIZING NUTRITIONAL POTENTIAL

Discussion of some of the traditional nutritional issues will be followed by some health-related

TABLE I

Dietary Guidelines

	US Senate Committee 1977	USDA/ USDHHS 1985	USDHHS 1988	AHA 1988
Fat[a]	30	ATM[b] 30-35	ATM	<30
Saturated	10	10-12	--	<10
Polyunsaturated	10	--		<10
Cholesterol (mg/day)	300	ATM	ATM	<300
Carbohydrates[a]	58			>50
complex	42-45	ADQ[c]	ADQ	EMPH[f]
sugar	15	ATM	ATM	
Protein[a]	12	--	--	15-20
Sodium (g/day)	3	ATM 1.1-3.3	ATM	--
Alcohol (oz/day)	--	MOD[d]	MOD	--
Calories	--	REC WT[e]	REC WT	REC WT
Variety	--	X	X	X

[a] % of Kilocalories. [b] Avoid too much.
[c] Adequate. [d] Moderate. [e] Consume calories to maintain recommended weight. [f] Emphasize

issues. Distribution of nutrients throughout the wheat kernel affords some opportunities for concentrating selected nutrients. Protein quality and the high concentration of minerals and vitamins in the aleurone cells suggest that this cell layer has good potential. The aleurone cell layer could be used to nutritionally enrich other products and fractions such as endosperm flour. The phytate present in aleurone cells, and its effect on bioavailability of other nutrients, must also be considered.

Protein

The issue of protein quality of cereal grains continues to remain prominent. Protein quantity and quality received major emphasis in a recent text on the nutritional quality of cereal grains (Olson and Frey, 1987). The importance of protein quality of wheat is clearer in animal feeds than it is in the mixed diet of humans. Wheat protein is adequate to meet the needs of adults. On the basis of both calculated amino acid score and results from a human metabolic ward study, wheat flour protein was shown to be adequate for human adults (Betschart, 1978; Betschart et al., 1985). White bread was superior to whole wheat in terms of protein digestibility and nitrogen (N) balance when adult men consumed one pound loaves bread/day for 24-day periods. Wheat protein provided 85% of total protein intake. Target groups who need higher quality protein, i.e., infants and children, require that wheat protein be consumed in combination with other protein sources with complementary amino acid profiles. Proteins rich in lysine, threonine, leucine and valine would be necessary for the young child.

Breads and other products could be formulated to contain higher quality protein. Such formulations which could include the higher quality fractions of wheat and other grains do not, however, always result in higher quality. Data from our laboratory with rats have shown that the Protein Efficiency Ratio of wheat flour, multiple grains and whole wheat breads were not significantly different (Table II). These data show that protein quality should be evaluated by some form of biological system since reliance on amino acid data alone is, at times, inadequate.

TABLE II

Protein Content and Quality of Breads
and Breakfast Foods[a]

Wheat Foods	Protein[b]	Protein Efficiency Ratio (adjusted)
Breads		
Enriched White	14.5	0.78[d]
Dark Style	18.2	0.53[d]
Whole Wheat	16.3	0.72[d]
Wheat Berry	15.7	0.59[d]
Seven Grains	14.8	0.98[cd]
Wheat Germ	16.0	0.96[cd]
Breakfast Foods		
Whole Wheat, Processed	12.4	0.18[e]
Natural #1	13.6	1.19[c]
Natural #2	14.3	0.92[cd]
Casein		2.50[a]

[a] Adapted from Betschart (1982). PER values with different superscripts in common are significantly different (P<.05). All PER values adjusted with casein as 2.50.

[b] Percent moisture-free basis.

Minerals and Vitamins

Minerals and vitamins are found in high concentrations in the aleurone cell layer which is removed in the milling process. Thus, whole wheat flour contains from 3 to 10 times the amount of

TABLE III

Selected Mineral Content of Wheat Fractions[a]

Cereal	Fe	Zn	Cu
		(mg/kg)	
Wheat (Range)	18-31	21-63	1.8-6.2
Bran	74-103	56-141	8.4-16.2
Germ	41-58	<100-144	7.2-11.8
Flour	3.5-9.1	3.4-10.5	0.62-0.63

[a] Adapted from Turnlund (1982).

iron (Fe), zinc (Zn) and copper (Cu) present in wheat flour (Table III) (Turnlund 1982). Danish workers reported that 80% extraction wheat flour contained 40-50% of the Fe, Zn and Cu present in the whole kernel, whereas 66% extraction flour contained around 30% (28-33%) of the original level of these minerals (Pedersen and Eggum, 1983). Mineral concentrations are also high in the germ; i.e. two to four times that of whole wheat.

Although percent bioavailability of minerals is less in whole wheat than in wheat flour, the absolute amount available is greater. Van Dokkum et al (1982) showed that, with the exception of calcium (Ca), mineral balance, which reflects availability, is significantly higher for whole wheat than for wheat flour. The significant decrease in Ca absorption reported by Van Dokkum & coworkers was confirmed by Turnlund in a human metabolic ward study using stable isotopes. There was a decrease in Ca retention from +151 to -36 mg/day when subjects changed from white bread to whole wheat bread. These latter data suggest that there may be some cause for concern if potentially vulnerable groups such as postmenopausal women consume large quantities of whole wheat or wheat bran products.

Distribution of B-vitamins in wheat varies but, in general, relatively high concentrations are found in the aleurone cell layer and outer cell layers (Table IV). Although data are mixed, most human studies have shown that whole wheat and bran either have no effect or slightly diminish bioavailability of B-vitamins. Rat studies from our laboratory on vitamin B-6 (Hudson et al, 1988) and folate (Keagy and Oace, 1986) have reported increased availability of the respective vitamins with wheat bran diets.

HEALTH-RELATED POTENTIAL AND LIMITATIONS

The Canadian Expert Advisory Committee Report on dietary fiber concluded "there are three and possibly four specific physiologic effects for different types of dietary fiber: 1) regulating colonic function, 2) normalizing serum lipid levels, 3) attenuating the postprandial glucose response and perhaps 4) suppressing appetite" (Health and Welfare Canada, 1985). Physiological effects most often associated with wheat bran are effects on intestinal function which include increases in fecal bulk and weight and

TABLE IV

Selected Vitamin Content of Wheat Fractions[a]

Cereal	Thiamin	Riboflavin	Niacin	Pyridoxine
		(mg/kg)		
Wheat				
Whole	9.9	3.1	48.3	4.7
Bran	13.2	5.5	171.4	13.0
Flour	0.7	1.5	9.5	0.5

[a] Michela and Lorenz, dwb (1976).

a decrease in intestinal transit time. The effects of wheat bran on colon cancer are not clear and data are inconclusive (Wisker et al, 1985). Wheat bran has, however, continued to be quite effective in the treatment of constipation and diverticular disease. The differences in fecal weight observed with the consumption of white and whole wheat bread and bran of differing particle sizes were reviewed by Wisker et al (1985) (Table V). Whole wheat bread and larger particle sizes bran significantly increase human fecal weight.

Cholesterol

Some of the more water-soluble sources of dietary fiber are effective in decreasing levels of circulating serum cholesterol. Gums and oats (with a

TABLE V

Effect of Wheat Fractions on Fecal Weight[a]

Subject		Fiber Source	Intake Dietary Fiber[b] (g/day)	Fecal Weight (g/day)
Gender	Number			
Female	8	White Bread	--	101
		Whole Wheat Bread	--	176
Male	4	White Bread + Fine Bran	22	102
		Whole Wheat Bread	22	143
		White Bread + Coarse Bran	35	202

[a] Adapted from Wisker et al (1985).
[b] Neutral detergent fiber.

TABLE VI

Effect of Dietary Fiber on Serum Cholesterol[a]

Source (g/day)	Number of Subjects[b]	Length of Study (wk)	Cholesterol (mg/dl)
Pectin			
15	9	3	190 (224)
9	15	5	158 (171)
Wheat Bran			
24	27	5	197 (194)
40	20	4	147 (189)
50	9	4	168 (161)
Whole Grain Bread[c]			
12 (NDF)	8	3	+15
Oat Bran			
100	8	1.5	234 (269)
Oat Flakes			
125	10	3	187 (204)

[a] Adapted from Wisker et al (1985). Control values are in parentheses. [b] All subjects were healthy. [c] Neutral Detergent Fiber.

significant concentration of β-glucans) have been shown to decrease plasma cholesterol in humans and some animals. The influence of more insoluble dietary fiber such as that found in wheat bran holds less promise. Although some human studies have suggested that wheat bran decreases cholesterol levels, others have reported increases or no change (Table VI). Small quantities of pectin compared to larger amounts of oat products produced significant effects. Rat studies conducted by Kahlon et al (1988) showed that wheat bran had no effect on liver cholesterol in cholesterol fed rats. Initial level of plasma cholesterol in subjects, changes in low-density relative to high-density lipoprotein, length of study, and true or long-term effects all need to be considered when interpreting plasma cholesterol data.

Mechanisms responsible for lowering plasma cholesterol levels are not entirely clear. Excretion of bile acids is often proposed as a mechanism. Data for pectin and oat products seem to support this mechanism whereas data for wheat bran do not (Table VII). Wheat bran has been reported to increase as

TABLE VII

Influence of Dietary Fiber on Steroid Excretion and Serum Cholesterol[a]

Dietary Fiber	Changes in Fecal Excretion of Bile Acids (%)	Changes in Total Serum Cholesterol (%)
Pectin	+34	-15
	+33	-13
	+51(m)[b]	
	+53 (f)	-8
Wheat Bran	+90	
	-49	0
	-23 (m)	
	+41 (f)	+8%
Oat		
Bran	+51	-31
Flakes	+35	-8

[a] Adapted from Wisker et al (1985).
[b] m = male; f = female.

well as decrease fecal excretion of bile acids. The correlation of these findings with changes in serum cholesterol is poor.

The implication for wheat and wheat products is that they be formulated to include constituents known to decrease cholesterol. Many products now available include such constituents at levels of 2% or less. Responsible marketing suggests that when claims are made about components which are associated with decreasing cholesterol levels, these components should be included at levels which would be likely to have some impact. The rat study to be reported by Kahlon et al (1988) was designed to determine what effect an oat bran product would have on wheat bran when combined in various ratios.

Carbohydrate Digestion

Research objectives at one time focused mainly on improving carbohydrate digestion. More recently, the health advantages of decreasing the rate of starch digestion have become apparent. Slowing the digestion of starch is associated with a lower glycemic response, reduced insulin secretion and lower serum cholesterol levels. Of 62 commonly consumed foods, the glycemic index (GI) for food types was: vegetables, $70\pm5\%$; breakfast cereals, $65\pm5\%$; fruit, $50\pm5\%$; and dried legumes, $31\pm3\%$. The glycemic index is the cumulative area under the blood glucose curve two hours after ingestion of a 50g carbohydrate sample of food. This is expressed as a percentage of the response to 50g of a reference carbohydrate such as glucose (Jenkins et al, 1981).

Fiber was originally proposed as a factor responsible for lowering the GI. The data in Table VIII show that the quantity of fiber in whole wheat has little, if any, effect and processing appears to be an important factor. Data on breakfast cereals suggest that cereal source, soluble and insoluble fiber and processing affect GI. The GI for corn flakes ($80\pm6\%$) is higher than bread and rice, shredded wheat is about the same, whereas all-bran and porridge oats are similar to pasta. Test meals clearly showed the effects of guar gum,

TABLE VIII

Glycemic Index of Foods[a]

Food	Glycemic Index[b]
Bread	
White	69 ± 5
Whole Wheat	72 ± 6
Rice	
White	72 ± 9
Brown	66 ± 5
Spaghetti	
White	50 ± 8
Whole Wheat	42 ± 4
Pastry	59 ± 6
Sponge Cake	46 ± 6

[a] Adapted from Jenkins et al (1981)
[b] 50 grams carbohydrate.

soybeans and lentils (Table IX). These data suggest that many factors affect the rate of starch digestion in foods including: nature of the starch (amylose:amylopectin ratio, degree of retrogradation and degree of gelatinization), food particle size, the presence of soluble fiber and antinutrients (Jenkins et al, 1988).

There is good correlation between plasma glucose and rate of in vitro hydrolysis with α-amylase (r=0.98). There is also a high correlation between in vitro hydrolysis and degree of starch gelatinization (r=0.96) as reported by Holm et al (1988a) and others. Thus, in vitro hydrolysis is a means of obtaining useful and predictive information on GI and

TABLE IX

Glycemic Index of Meals[a]

Test Meals	Glycemic Index[b]
Bread	
Whole Wheat + Cheese	100
Whole Wheat + Guar Crispbread	51
Whole Wheat + Soya Beans	65
Guar Crispbread + Soya Beans	25
Corn Flakes + Toast	108

[a] Adapted from Jenkins et al (1980).
[b] 42-44 grams carbohydrate/test meal.

degree of gelatinization.

Data for wheat foods suggest that the relatively high GI of breads could be lowered through formulations which include factors associated with a lower GI. Initial data on _in vitro_ hydrolysis of breads suggest that rye flour holds promise. Closer examination of the formulations of these breads, however, reveals that they differ in their content of simple carbohydrates. Whole rye bread contains less sugars and is also digested at a slower rate, especially during the first 30 minutes. Thus, other cereal grains, other sources of soluble fiber and processing conditions which diminish degree of starch gelatinization are all possible means of decreasing the GI of wheat breads and other wheat products.

The common mechanism associated with a lower GI seems to be slower absorption (and uptake) of digested starch by the organism. Thus, any factor which decreases the rate of delivery to the sites of digestion, rate of digestion and/or rate of absorption into the bloodstream would have an effect. The lower GI associated with various foods reflects some of the factors involved: pasta (processing), coarse cracked grain (particle size), oats and barley (soluble dietary fiber, β-glucans), and beans and lentils (starch source, accessibility to enzymatic digestion). Holm and co-workers (1988b) recently reported that when more extreme conditions were used for popped, steam-flaked, and drum-dried wheat products, the availability of starch to α-amylase was greater. When products were produced under milder conditions, a lower degree of gelatinization and digestibility of starch was observed. Extrusion promotes starch gelatinization and these products usually produce higher GI. Due to the multiplicity of factors involved in the production of GI, there is still much to be learned before parameters which have been shown to be correlated with lower GI can be identified as causative factors.

CONCLUSIONS

Wheat continues to be an important food and feed. Often properties which are desirable for foods are not desirable for feeds, e.g., low caloric density, slower rate of starch digestion. Within foods, properties which are desirable for one age group or target population may be less desirable for another, e.g., caloric density. Thus, whether properties are seen as unique and positive or as limitations and negative depends heavily on the end use. Wheat as a high complex carbohydrate, low fat food with low caloric density is desirable for most western diets. In adults, the satiety value of wheat foods, especially whole wheat products, has been associated with weight loss in some human studies (Mickelsen et al, 1979). Wheat may be less desirable for young children in whom appetite may be satisfied before caloric/nutritional needs are met.

The insoluble dietary fiber of wheat has positive

effects on intestinal function and less positive effects on plasma cholesterol. Products could be formulated to include some forms of soluble fiber. Ultimately the balance and ratio of insoluble to soluble fiber in our foods and diets may be more important than the exaggerated emphasis on either total dietary fiber, soluble fiber or one source of either of these.

The uniqueness of wheat may continue to rely more on functional than nutritional properties. However, within the food/feed markets wheat might be further explored as a substrate in the development of other substances which are physiologically functional, i.e., forms of soluble dietary fiber, β-glucans, gums. The fiber, starch and/or protein, and derivatives thereof, of wheat may have some functional potential as components of packaging materials and in various industrial products.

The quantity of wheat which can now be produced may provide the impetus for some of the most creative opportunities for wheat utilization yet envisioned.

LITERATURE CITED

BETSCHART, A. A. 1988. Nutritional quality of wheat and wheat foods. Vol II. pages 91-130. In: Wheat Chemistry and Technology. Y. Pomeranz, ed. American Association of Cereal Chemists, St. Paul, MN (In Press).

BETSCHART, A. A. 1982. Protein content and quality of cereal grains and selected cereal foods. Cereal Foods World 27:395-401.

BETSCHART, A. A. 1978. Improving protein quality of bread- nutritional benefits and realities. Adv. Exp. Med. Biol. 105:702-734.

BETSCHART, A. A., HUDSON, C. A., and TURNLUND, J. R. 1985. Protein digestibility and nitrogen balance of white and whole wheat breads in humans. Cereal Foods World 30:538-539.

DANFORTH, E., Jr. 1985. Wheat Foods: nutritional implications in health and disease: Introduction Amer. J. Clin. Nutr. 41:1069.

HEALTH AND WELFARE CANADA. 1985. Report of the Expert Advisory Committee on Dietary Fiber. Minister of National Health and Welfare, Ottawa.

HOLM, J., LUNDQUIST, I., and BJORCK, I. 1988a. Degree of starch gelatinization and metabolic response in rats. Proc. Intern. Symp. Cereal Sci. Tech. June 13-16. Ystad, Sweden.

HOLM, J., BJORCK, I., and HAGANDER, B. 1988b. Starch availability in vitro and in vivo of wheat products. Proc. Intern. Symp. Cereal Sci. Tech. June 13-16. Ystad, Sweden.

HUDSON, C. A., BETSCHART, A. A., and OACE, S. M. 1988. Bioavailability of vitamin B-6 from rat diets containing wheat bran or cellulose. J. Nutr. 118:65-71.

JENKINS, D. J. A. JENKINS, A. L., WOLEVER, T. M. S., VUKSAN, V., THOMPSON, L. U., and RAO, A. V. 1988. Food processing, starch digestibility and the glycemic response. Proc. Intern. Symp. Cereal Sci. Tech. June 13-16, Ystad, Sweden.

JENKINS, D. J. A., THOMAS, D. M., WOLEVER, M. S., TAYLOR, R. H., BARKER, H., FIELDEN, H., BALDWIN, J. M., BOWLING, A. C., NEWMAN, H. C., JENKINS, A. L., and GOFF, D. V. 1981. Glycemic index of foods: a physiological basis for carbohydrate exchange. Am. J. Clin. Nutr. 34:362-366.

JENKINS, D. J. A., WOLEVER, T. M. S., TAYLOR, R. H., BARKER, H. M., FIELDEN, H., and JENKINS, A. L. 1980. Effect of guar crisp-bread with cereal products and leguminous seeds on blood glucose concentration of diabetics. Br. Med. J. 281:1248-1250.

JOHNSON, V. A., and MATTERN, P. J. 1987. Wheat, rye and triticale. Page 133-182 in: Nutritional Quality of Cereal Grains: Genetic and Agronomic Improvement. R. A. Olson and K. J. Frey, eds. ASA, CSSA and SSS, Pub., Madison, WI.

KAHLON, T. S., CHOW, F. I., CHIU, M. C., and BETSCHART, A. A. 1988. Influence of wheat and oat fractions on rat plasma cholesterol and triglycerides. Cereal Foods World 33:686.

KEAGY, P. M. and OACE, S. M. 1986. Folic acid utilization from high fiber diets in rats. J. Nutr. 114:1252-1259.

MICHELA, P. and LORENZ, K. 1976. The vitamins of triticale, wheat and rye. Cereal Chem. 53:853-861.
MICKELSEN, O., MAKDANI, D. P., COTTON, R. H., TITCOMB, S. T., COLMEY, J. C., and GATTY, R. 1979. Effects of a high fiber bread on weight loss in college-age males. Am. J. Clin. Nutr. 32:1703-1709.
OLSON, R. A., and FREY, eds. 1987. Nutritional Quality of Cereal Grains: Genetic and Agronomic Improvement. American Society of Agronomy, Inc., Crop Science Society of America, Inc., and Soil Science Society of America, Inc. Madison, WI.
PEDERSEN, B., and EGGUM., B. O. 1983. The influence of milling on the nutritive value of flour from cereal grains. 2. Wheat. Qual. Plant. Plant Foods Human Nutr. 33:51-61.
TURNLUND, J. R. 1982. Bioavailability of selected minerals in cereal products. Cereal Foods World 27:152-157.
U.S. DEPARTMENT OF HEALTH AND HUMAN SERVICES. 1988. The surgeon general's report on nutrition and health. U.S. Government Printing Office, Wash., D.C.
VAN DOKKUM, W., WESTRA, A., and SCHIPPERS, F. A. 1982. Physiological effects of fibre-rich bread. I. The effect of dietary fibre from bread on the mineral balance of young men. Br. J. Nutr. 47:451-460.
WISKER, E., FELDHEIM, W., POMERANZ, Y., and MEUSER, F. 1985. Dietary fiber in cereals. Adv. Cereal Science Technol. VII:169-238. Y. Pomeranz, ed. American Association of Cereal chemists, St. Paul, MN.

7

CORRELATION BETWEEN VISUAL ASSESSMENT OF PRE-HARVEST SPROUTING OF WHEAT AND THE α-AMYLASE CONTENT OF GRAIN AND FLOUR

Alicia de Francisco, Lars Munck
Department of Biotechnology
Carlsberg Research Laboratory
Copenhagen, Denmark

Jan Ruud-Hansen
Statistical Services
Carlsberg Research Laboratory
Copenhagen, Denmark

INTRODUCTION

Pre-harvest sprouting of cereal grains is a major problem for farmers, millers, malsters and other members of the food industry (Brookes, 1979; Meredith and Pomeranz, 1985; Munck, 1987). Traditionally, the degree of sprouting in seeds is determined by experienced grain inspectors who visually count the number of sprouted seeds (Jensen et al., 1984). This is a rather tedious and subjective method. To determine the degree of sprouting, several wet chemical and rheological methods are also used and the results expressed as the α-amylase content of flour; either indirectly, as in the amylograph (AACC Methods, 1983), or Falling Number methods (Hagberg, 1960; ICC Standard, 1968) or directly as in the nephelometric (AACC Methods, 1983) or dyed starch procedures (Olered, 1967; AACC Methods, 1983). However, by measuring α-amylase of flours, two different situations may give rise to the same result:

1. a sample with many slightly sprouted grains or
2. a batch with a few highly sprouted ones.

Information about the extent of endosperm modification in the seeds of a population can be of use in improving the quality of flour from sprouted grain by removing most of the sprouted kernels by flotation or gravity separation or by milling techniques (Henry et al., 1986; Liu et al., 1986). This latter approach is based on the fact that during the initial stages of germination, α-amylase is first produced in the scutellum and aleurone adjacent to the embryo and then it moves into the endosperm (Gibbons, 1979a,b 1983; Paleg, 1960; MacGregor, 1983). Thus, removing the germ and bran from slightly sprouted seeds should lower the overall α-amylase in a batch, while this may not necessarily be the case in a sample with a few highly sprouted kernels where the enzyme is distributed throughout a large part of the endosperm.

The rate of transport of α-amylase into the endosperm is another factor that must be considered in the overall analysis of sprouted kernels. It is not only the intrinsic enzyme activity of a seed, which is variety dependent, but how far this enzyme will travel into the endosperm in a given period of time. It is well known that physical factors such as cell walls, endosperm structure and in some cases the presence of active enzyme inhibitors may affect the total amount of α-amylase concentration in the endosperm of a given seed and, therefore, affect its endosperm modification pattern (Palmer and Harvey, 1977; Munck et al., 1981; Munck, 1986).

BASI, the functional α-amylase/subtilisin inhibitor described by Mundy et al. (1984), is specific for both bacterial serine proteases and for the endogenous α-amylase of germinating barley. This latter specificity has suggested the involvement of BASI in the regulation of amylase activity during precocious germination. It is interesting to note, however, that BASI and α-amylase synthesis in the aleurone layer are antagonistically controlled by plant hormones during germination (Mundy, 1984; Leah and Mundy, in preparation).

To get a better understanding of the processes involved in kernel sprouting, the following study correlates the visual evaluation of endosperm modification with the α-amylase content of both grain and milling fractions of several wheat varieties. The possible effect of BASI and its location in the grain is also discussed

Finally, statistical models for the assessment of α-amylase content of grain and flours using the visual method are presented.

MATERIALS AND METHODS

Forty wheat cultivars were used in the development of a statistical model for the correlation of the visual method for modification and α-amylase content. The samples consisted of 20 Australian varieties (# 1-20) supplied by Dr. D.P. Law from Agritech, Toowoomba, and described as hard wheats; and 20 European wheat varieties decribed as soft : 10 Norwegian (# 21-30) supplied by K. Ringlund from the Agricultural University, Oslo, and 10 Polish (# 31-40) ones, supplied by I. Hodowli, Roslin/Laskach.

All samples were analyzed for:

- Degree of modification as determined by the visual Fluorescein Dibutyrate (FDB)-Carlsberg method (Analytica-EBC, 1987); (See below for detailed description of this method).
- α-Amylase content in whole ground grain as well as in milled fractions using an iodine-starch procedure as modified by Olered (1967). The results are expressed in activity units. These analyses were performed by the Svenska Cereal Laboratoriet AB, Svalöv, Sweden;
- α-Amylase inhibitors in whole grain by radial immunodiffusion (Mancini et al., 1965);
- Correlation between these methods, by simple regression and multiple regression analysis.

All 40 samples were milled in a Brabender Quadrumat Jr. II equipped with a 200 μm screen. The three milling fractions collected were: flour, bran and middlings.

In addition, a few samples were analyzed for BASI content in milled fractions by an enzyme-linked immunosorbent assay incorporating biotin and streptavidin (Munck et al., 1984, Vaag, 1985). Both immunochemical methods for BASI were based on antibodies raised to the α-amylase inhibitor from barley. However, these antibodies are known to cross-react strongly with the wheat inhibitor, whose amino acid sequence in turn is similar to that of the barley inhibitor (Mundy et al., 1984; Weslake et al., 1985; Svendsen et al., 1986).

Visual Assessment of Sprouting

The FDB-Carlsberg method was used to estimate the percent germination (modification) in 199-300 longitudinally halved seeds using the procedure and equipment developed by Heltved et al. (1982) and by Jensen and Heltved (1983). The stained samples examined under blue light (495 nm) gave a strong yellow fluorescence (525 nm), indicative of germination (Fig. 1). Subjective measurements of germination were confirmed with a Quantimet 900 image analyzer

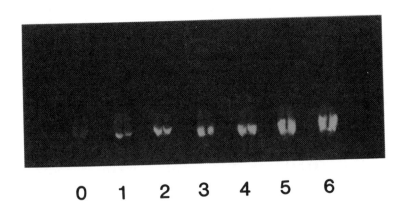

Fig. 1. Wheat modification classes. The modified endosperm becomes fluorescent with the FDB staining under blue light.

TABLE I
Grading Scale for the Evaluation of Pre-harvest Sprouting by the Visual FDB Method

Class	% Modification
0	0
1	1-10
2	11-20
3	21-30
4	31-40
5	41-50
6	> 50

from Cambridge Instruments, England, equipped with a Plumbicon scanner. For single seed measurements as well as for multiple measurements (50 seeds at a time), the procedures outlined by de Francisco and Munck (in press) were used. The results were expressed as:

$$\% \text{ Seed modification} = \frac{\text{Fluorescent area}}{\text{Total endosperm surface area}} \times 100$$

For the evaluation of the half seeds, the grading scale shown in Table I was used (see also Fig. 1). The break points for each class were arbitrarily chosen to facilitate the subjective evaluation as well as the statistical computations.

Seed Flotation

The modified flotation method of Hallgren and Murty (1983) for grading of sorghum was tested to determine whether this method could be used to separate sprouted kernels from sound ones by differences in density. For this purpose, the sodium nitrate solution was adjusted to a density in which no kernels from a known sound batch would float. This density corresponded to 1.264 g/ml as measured with a hydrometer. Then, a given number of the kernels germinated in the laboratory for 2-6 days were marked, added to sound batches, floated and counted.

RESULTS AND DISCUSSION

For each of the 40 wheat samples, the number of sprouted seeds in the six modification classes, total % modification per sample, and grain α-amylase content are shown in Tables II and III.

TABLE II
Hard Wheat Analysis

ID #	\multicolumn{6}{c	}{Sprouted Seeds/ Class}	Total Seeds	α-Amyl.	% Mod.
	C1	C2	C3	C4	C5	C6			
1	12	2	0	0	0	0	199	19.0	0.45
2	24	0	0	0	0	0	199	13.0	0.60
3	22	2	1	0	0	0	200	8.9	0.83
4	28	4	0	0	0	0	200	23.0	1.00
5	37	0	0	0	0	0	199	9.1	0.93
6	13	0	0	0	0	0	200	10.0	0.33
7	26	2	0	0	0	0	200	9.9	0.80
8	15	0	0	0	0	0	199	11.0	0.38
9	11	0	0	0	0	0	200	24.0	0.28
10	18	1	0	0	0	0	199	10.0	0.53
11	4	0	0	0	0	0	300	6.8	0.07
12	6	1	0	0	0	0	200	6.4	0.23
13	18	0	0	0	0	0	200	28.0	0.45
14	12	1	0	0	0	0	200	19.0	0.38
15	20	2	0	0	0	0	200	16.0	0.65
16	2	0	0	0	0	0	200	6.2	0.05
17	28	5	0	0	0	0	300	24.2	0.72
18	46	5	0	0	0	0	200	42.0	1.53
19	8	0	0	0	0	0	200	11.0	0.20
20	9	1	0	0	0	0	200	8.0	0.30

TABLE III
Soft Wheat Analysis

ID #	Sprouted Seeds/ Class						Total Seeds	α-Amyl.	%Mod.
	C1	C2	C3	C4	C5	C6			
21	6	3	3	0	0	0	199	46.0	0.75
22	4	3	3	1	0	0	200	71.0	0.88
23	1	0	0	0	0	0	199	4.4	0.03
24	1	0	0	0	0	0	199	5.8	0.03
25	0	0	0	0	0	0	199	3.4	0.00
26	0	2	0	0	0	0	200	14.0	0.15
27	0	0	1	0	0	0	200	3.5	0.13
28	5	0	1	1	0	0	200	30.0	0.43
29	10	4	0	0	2	1	300	71.0	0.92
30	3	1	0	0	0	0	200	29.0	0.15
31	2	0	0	0	0	0	200	5.8	0.05
32	9	3	0	0	0	0	200	8.6	0.45
33	0	0	0	0	0	0	200	2.2	0.00
34	0	0	0	0	0	0	200	1.5	0.00
35	0	0	0	0	0	0	200	2.2	0.00
36	0	0	0	0	0	0	200	0.3	0.00
37	0	0	0	0	0	0	200	0.6	0.00
38	0	0	0	0	0	0	200	0.6	0.00
39	1	0	0	0	0	0	200	3.8	0.03

The % total modification (%TM) for each of the 40 samples shown was calculated as follows:

$$\%TM = 0.05 \times \%C1 + 0.15 \times \%C2 + .25 \times \%C3 + .35 \times \%C4 + .45 \times \%C5 + .75 \times \%C6$$

Where %C1...%C6 refer to the number of sprouted seeds in each of the categories and the coefficients correspond to the midpoints on each of the classes (Table I). Based on regression analysis, the following equations for α-amylase as a function of %TM could be established:

A. 40 Total samples: α-amylase = 3.6 + 34.4 x %TM
B. 20 Australian wheats: = 7.1 + 15.4 x %TM
C. 20 European wheats: = 1.7 + 68.0 x %TM

The correlations were r = 0.69, 0.60 and 0.96, respectively. These results indicate that to obtain a better correlation between visual assessment of sprouting and grain α-amylase content, the data from the Australian and European wheats should be treated separately. As seen in Fig. 2 the correlations for these population have different slopes which could be related to their difference in endosperm hardness.

It was observed that even though all the Australian wheat kernels examined had some degree of modification (probably due to environmental conditions), most of them were less than 10% modified, that is between 0.05 to 1.53% of the total sample modification, with a contribution of less than 0.93 to the whole population. The range of α-amylase content for these samples was between 6.2 and 42.0 activity units.

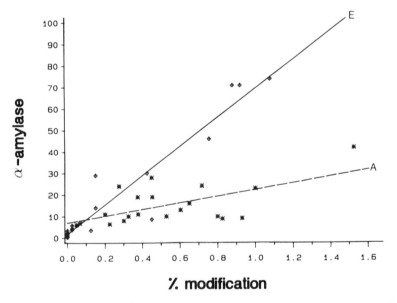

Fig. 2. Correlation between α-amylase content and % modification for European (E) and Australian (A) wheat samples.

In the European varieties a wider range of kernel modification was observed, however, 7 samples did not exhibit any modification at all with the FDB test and had an α-amylase content between 0.6 and 3.4 units again, probably due to climatic conditions (Table III). The sprouted European kernels ranged betweeen 0.03 and 1.08% total modification and between 3.8 to 74.0 α-amylase activity units.

As seen in Fig. 2 where the α-amylase content of the seeds is correlated to modification, the Australian samples exhibit higher modification at a lower α-amylase content in relation to the European varieties.

Since there were no sprouted seeds with a modification higher than 20% for any of the Australian wheats it was not possible to build a model based on classification, due to insufficient data points on the highly modified classes for these 20 samples. For the European wheats the following model was developed:

$$\alpha A = 2.1 + 3.1 \times \%C1 + 5.6 \times \%C2 + 20.3 \times \%C3 + 33.8 \times \%C4 + 51.5 \times \% (C5 + C6)$$

This model explains 93% of the variation of α-amylase.

All 40 wheat samples were examined for the presence of BASI to see whether or not this could explain some of the α-amylase differences among the 40 samples; there was no significant correlation. The inhibitor levels varied between 30 and 60 mg/kg. The lack of correlation between α-amylase content and BASI has been previously reported for barley grain by Munck et al. (1984).

In the 40 samples the correlations between α-amylase of whole grain vs. that of flour, bran and middlings were r = 0.76, 0.72 and 0.70, respectively. To explain these rather low correlations, two soft wheat samples were analyzed for the presence of BASI in the three milling fractions. The results are shown in Table IV. Contrary to our expectations, the highest amount of inhibitor was found in the bran fraction, unlike in barley where most of it is found

TABLE IV
BASI Contents in Milling Fractions

Sample #	Fraction	Mg/kg BASI
A	Whole grain	42
	Flour	20
	Bran	60
	Intermediate	38
B	Whole grain	46
	Flour	20
	Bran	67
	Intermediate	40

in the starchy endosperm (Mundy et al., 1984, Weselake et al., 1985). Additional studies of this aspect will be published elsewhere. We found no significant correlation between BASI and α-amylase content of the milling fractions.

The correlations between flour α-amylase content and % modification were significant but low: $r = 0.63$ for the Australian wheat samples and 0.69 for the European.

To determine whether sprouted kernels could be removed by flotation, a batch of sound Danish grain of the Kraka 84 variety was mixed with 20% of seeds germinated for 2, 4, and 6 days in the laboratory. These seeds were marked on their surface for identification. At a density of 1.264 g/ml no seeds floated in a 100% sound batch. At this density all sprouted seeds could be removed. It is important to notice that different varieties may require different densities.

CONCLUSIONS

Chemical analysis and visual evaluation of preharvest sprouting were successfully correlated and a model to estimate the α-amylase content of sprouted soft wheat by visual assessment was developed.

Comparison between collections of Australian and European wheat varieties of limited numbers indicated a significant difference between their modification pattern and confirmed the idea that a sample with few highly sprouted kernels could contribute more to the total α-amylase in the population than a sample with many slightly sprouted kernels. The signicant but low correlations between the α-amylase content of milling fractions and that of the whole grain could not be explained by the presence of BASI.

Preliminary studies indicated that it is possible to remove sprouted kernels from a mixed batch by flotation in sodium nitrate. Useful information could probably be obtained using the described methods and additional sprouted wheat cultivars.

ACKNOWLEDGMENTS

The authors gratefully acknowledge Pia Vaag for performing the immunosorbent assay for BASI and critical review of the manuscript, Lise Tang Petersen for her excellent technical assistance and sample preparation, and Birthe Pedersen, Regis Cabral and Robert Leah for their helpful suggestions.

LITERATURE CITED

AACC METHODS. 1983. Method 22-10 Diastatic activity of flour with the Amylograph. Method 22-07 α-Amylase activity-nephelometric. Method 22-01 α-Amylase activity of malt. Method 22-06 Cereal α-amylase. Method 22-15 Diastatic activity of flour and semolina. Rev. 9-17-87. Approved Methods of the Am. Assoc. Cereal Chem., 8th ed., The Association, St. Paul, MN., U.S.A.

ANALYTICA-EBC. 1987. 3.7.1 Fluorescein dibutyrate-Carlsberg method. Page E 47, 4th Edition. Analysis Committee of the European Brewery Convention. Brauerei- und Getränke-Rundschau, Zurich, Switzerland.

BROOKES, P. A. 1979. The significance of preharvest sprouting of barley in malting and brewing. Cereal Res. Comm. 8:29.

FRANCISCO, A. de, and MUNCK, L. In press. Practical applications of fluorescence image analysis. In: Fluorescence Analysis in Foods. L. Munck, ed. Longman Scientific and Technical Publishers, London, England.

GIBBONS, G. C. 1979a. Immunohistochemical determination of the transport pathways of α-amylase in germinating barley seeds. Cereal Res. Comm. 8:87.

GIBBONS, G. C. 1979b. On the localization and transport of α-amylase during germination and early seedling growth of Hordeum vulgare. Carlsberg Res. Commun. 44:353.

GIBBONS, G. C. 1983. The action of plant hormones on endosperm breakdown and embryo growth during germination of barley. Page 169 in: Proc. 3rd Intern. Symp. Preharvest Sprouting of Cereals. J. E. Kruger and D.E. LaBerge, eds. Westview Press, Boulder. CO, U.S.A.

HALLGREN, L., and MURTY, D. S. 1983. A screening test for grain hardness in sorghum employing density grading in sodium nitrate solution. J. Cereal Sci. 1:265.

HAGBERG, S. 1960. A rapid method for determining α-amylase activity. Cereal Chem. 37:218.

HELTVED, F., AASTRUP, S., and MUNCK, L. 1982. Preparation of seeds for mass screening. Carlsberg Res. Commun. 47:291.

HENRY, R. H., MARTIN, D. J., and BLAKENEY, A. B. 1986. Reduction of α-amylase content of sprouted wheat by pearling and milling. J. Cereal Sci. 5:155.

ICC STANDARD NO. 107, 1968. Determination of the "Falling Number" according to Hagberg-Perten as a measure of the degree of α-amylase activity in grain and flour. Standard Methods of the International Association for Cereal Science and Technology (ICC), Moritz Schafer, Detmold, FRG.

JENSEN, S. Aa., and HELTVED, F. 1983. An improved method for the determination of pregerminated grains in barley. Carlsberg Res. Commun. 48:1.

JENSEN, S. Aa., MUNCK, L., and KRUGER, J. E. 1984. A rapid fluorescence method for assessment of preharvest sprouting of cereal grains. J. Cereal Sci. 2:187.

LEAH, R. and MUNDY, J. The bifunctional α-amylase/-subtilisin inhibitor (BASI) of barley: nucleotide sequence and patterns of seed-specific expression. Plant Molecular Biol. In preparation.

LIU, R., LIANG, Z., POSNER, E. S., and PONTE Jr., J. G. 1986. A technique to improve functionality of flour from sprouted wheat. Cereal Foods World 31:471.

MACGREGOR, A. W. 1983. Cereal endosperm degradation during initial stages of germination. Page 162 in: Proc. 3rd Intern. Symp. Preharvest Sprouting of Cereals. J. E. Kruger and D.E. LaBerge, eds. Westview Press, Boulder. CO. U.S.A.

MANCINI, G., CARBONARA, A. O., and HEREMANS, J.F. 1965. Immunochemical quantitation of antigens by single radial immunodiffusion. Immunochem. 2:235.

MEREDITH, P., and POMERANZ, Y. 1985. Sprouted grain. Adv. Cereal Science Technol. 7:239.

MUNCK, L. 1986. Breeding for quality in barley. Experiences and perspectives. Page 753 in: Barley Genetics V, S. Yasuda and T. Konishi, eds. Fifth Intern. Barley Genetics Symp., Okyama, Japan.

MUNCK, L. 1987. The control of pre-harvest sprouting in cereals for seeds, malting and milling. Page 176 in: Proc. 4th Int. Symp. Preharvest Sprouting in Cereals, D.L. Mares, ed. Westview Press, Boulder, CO., U.S.A.

MUNCK, L., GIBBONS, G. C., and AASTRUP, S. 1981. Chemical and structural changes during malting. Page 11 in: 18th Eur. Brew. Conv. Proc. Congr., Copenhagen, Denmark.

MUNCK, L., MUNDY, J., and VAAG, P. 1984. Characterization of enzyme inhibitors in barley and their tentative role in malting and brewing. ASBC J. 43:35.

MUNDY, J. 1984. Hormonal regulation of α-amylase inhibitor synthesis in germinating barley. Carlsberg Res. Commun. 49: 439.

MUNDY, J., HEJGAARD, J., and SVENDSEN, I. 1984. Characterization of a bifunctional wheat inhibitor of endogenous α-amylase and subtilisin. FEBS Letters 1235:210.

OLERED, R. 1967. Development of α-amylase and Falling Number in wheat and rye during ripening. Plant Husbandry 33:34.

PALEG, L. G. 1960. Physiological effects of gibberellic acid.II. On starch hydrolyzing enzymes of barley endosperm. Plant Physiol. 35:902.

PALMER, G. H., and HARVEY, A. E. 1977. The influence of endosperm structure on the behaviour of barleys in the sedimentation test. J. Inst. Brew. 58:295.

SVENDSEN, I., HEJGAARD, J. and MUNDY, J. 1986. Complete amino acid sequence of the α-amylase/subtilisin inhibitor from barley. Carlsberg Res. Commun. 51:43.

VAAG, P. 1985. Enzyme linked immunosorbent assay (ELISA) for quality control in beer. Page 547 in: 20th Eur. Brew. Conv. Congress Helsinki, Finland.

WESLAKE, R.J., MACGREGOR, A.W., and HILL, R.D. 1985. Endogenous α-amylase inhibitors in various cereals. Cereal Chem. 62:120.

8

CHALLENGES WITH ALPHA-AMYLASE INHIBITION IN SPROUT-DAMAGED WHEAT

Urszula Zawistowska

ABI Biotechnology Inc.,
A-1150 Waverley Street,
Winnipeg, Canada, R3T 0P4

INTRODUCTION

The germination of wheat grain on the plant at harvest time, is known as field sprouting. Sprouting is accompanied by a number of biochemical changes in the grain, but the most pronounced is the increase in the level of alpha-amylase. Degradation of carbohydrate storage reserves by alpha-amylase results in deterioration of flour quality for production of noodles, macaroni, cakes and cookies as well as bread (Meredith and Pomeranz 1985). Use of flour from sprouted wheat only in bread making is associated with a sticky dough and bread with damp and gummy crumb, and darker in color and inferior in texture compared to bread from sound wheat (Kozmin 1933; Tipples et al 1966; Buchanan and Nicholas 1980; D'Appolonia 1983).

Effects of sprouting are seen by a commercial baker as a problem in bread slicing. For the consumers, the bread crumb is sticky and adhers to the mouth lining and teeth. Thus, sprouted wheat is not acceptable for bread production and is classified as lower grade and only sold for feed. Sprouting, therefore, causes economic losses resulting from the lower value of sprouted grain. Since we can not control weather conditions to avoid field sprouting, we can use a number of approaches in dealing with sprouting problems. In this paper I will focus only on one of them, namely utilization of already sprouted wheat for bread baking.

The utilization of sprout-damaged wheat for bread production has been the topic of interest to scientists for several decades. The effects of sprouting, when not too excessive, can be corrected to some extent by modification of milling procedures (Henry et al 1987), heat treatment of sprouted grain or flour (Ranhotra et al 1977) and blending defective grain with sound grain (Hunecke and Petzold 1979), or by adjusting such processing parameters as water absorption, times of fermentation, mixing and proofing or increasing baking temperature in the first stages of baking.

REDUCING ALPHA-AMYLASE ACTIVITY

Since the increased level of alpha-amylase is a major factor responsible for deterioration of technological quality of sprouted wheat flour, it has been implied by different authors that it should be possible to improve baking properties of sprout-damaged wheat by reducing alpha-amylase activity, especially as it has been shown in many studies (cited in Westermarck-Rosendahl et al 1979) that the starch fraction of sprouted wheat still may be of good quality. A number of physico-chemical factors (such as heat, low pH, high pH, calcium binding substances, surfactants, etc) have been studied as potential alpha-amylase inhibitors and inactivators.

Heat Inactivation

The most common method of enzyme inactivation is by heat treatment. Wheat alpha-amylase is a rather thermostable enzyme and its thermal inactivation can be achieved by raising the temperature to approximately 80°C. According to Hutchinson (1963) steam treatment of sprouted wheat yields flour suitable for use in soups, adhesives and other special uses but unsuitable for breadmaking because of the destruction of the functional properties of heat-sensitive gluten proteins. Heat treatment of flour has a similar effect on functional properties of flour components as heat treatment of grains. Additionally, it results in a loss of moisture from the flour to an extent requiring reconditioning of such flour by steaming. Inactivation of alpha-amylase by microwave energy was proposed by Aref et al (1972) and Edwards (1964); however, heat generated in irradiated material also caused denaturation of gluten proteins and the irradiated flour lost its dough making properties. Effects of microwave and gamma radiations on bread properties were discussed by MacArthur and D'Appolonia (1982). According to the authors, although those two techniques have some advantages, they are not commercially feasible.

Acid Inactivation

The reducing effect of acids on alpha-amylase has long been used in rye bread production where dough making is based on the sour dough fermentation principle. Kozmin (1933) first attempted to improve sprouted wheat bread characteristics by acidification. The crumb of this bread was quite normal, dry and elastic but slightly sour in taste. Also, Fuller et al (1970) and Meredith (1970) used hydrochloric acid treatment to inactivate alpha-amylase in sprouted wheat flour. Their attempts, however, to use an inactivated and neutralized slurry for bread baking were not successful. Acids in addition to reducing alpha-amylase activity, adversely affected other physical dough properties. Alkalization of sprouted wheat flour by the addition of basic amino acids (arginine and lysine), suggested by Dorfer and Kobal (1980), caused some inhibition of alpha-amylase, as detected by changes in falling number and amylograph viscosity, but the inhibitory effect was smaller than that obtained by dough acidification.

Inhibition by Chemical Agents

Many chemical compounds (including ethanol, isopropanol, potassium nitrite, potassium permanganate, 2,4-dinitroflourophenol, phenylisocyanate, and formaldehyde) have been reported to inhibit wheat alpha-amylase (Dorfer and Koball 1980). Silver nitrate was proposed by Fuller et al (1968), Meredith (1970), and Westermarck-Rosendahl et al (1979). Because of their toxicity they can not be considered as food additives. Inhibition of wheat alpha-amylase by ascorbic acid was reported by Palla and Verrier (1974), Osadchaya et al (1979), and Puchkova et al (1980), and isoalloxazine derivatives by Palla and Verrier (1974). Inhibitory effects of phytic acid were studied using the Hagberg penetrometer (Cawley and Mitchell 1968) and by falling number determination (Noll 1985; Westermarck-Rosendahl et al 1979) or amylogram viscosity measurements (Westermarck-Rosendahl et al 1980). Inhibitory properties of phytic acid towards alpha-amylase were attributed to Ca^{+2}-binding properties of this compound by Cawley and Mitchell (1968), and Westermarck-Rosendahl et al (1979), and to lowering the pH of the flour-water slurries by others (Noll 1985).

Extensive studies on improvement of baking properties of sprout-damaged wheat by application of different chemical additives were performed by Westermarck-Rosendahl et al (1979, 1980). The effects of 23 chemical agents on alpha-amylase activity of sprout-damaged wheat were evaluated by falling number and amylograph tests as well as baking tests and determination of dough mixing characteristics. The tested chemicals included inorganic phosphates (sodium tri-, di-, monobasic and polyphosphate), organic acids (citric, lactic, ascorbic), chelators and reagents with Ca^{+2}-binding properties (EDTA-disodium salt, phytic acid, LM-pectin, sorbitol, xylitol), surfactants (sodium dodecyl sulfate and calcium stearoyl lactylate) and other chemicals such as silver nitrate, sodium hexafluorosilicate, sodium ascorbate, maltodextrin, maltose, lactose, sucrose, ethanol and tannic acid. Some of them (monobasic sodium phosphate, sodium ascorbate, LM-pectin, maltodextrin, maltose, lactose, sucrose, sorbitol, xylitol) had no suppressing effect on alpha-amylase as shown by the falling number test. The most effective alpha-amylase inhibitor was silver nitrate. Already a concentration of 2.5 ug/g of meal caused an increase in falling number from 143 to 327. This reagent was used only for purely scientific reasons, since it can not be added to flour because of its toxicity. The other chemicals found to suppress alpha-amylase activity most effectively were tri- and dibasic sodium phosphates, sodium phytate and ethylenediaminetetraacetic acid, disodium salt (Table I). The viscosity increasing effects of tri-

TABLE I
Effect of Some Alpha-Amylase Activity Reducing
Agents on Falling Number and
Amylogram Viscosities[a]

Chemical	Concentration (µeq/g)	Falling Number (s)	Amylogram Peak Viscosity (B.U.)
No addition	-	143	70
Sodium phosphate tribasic	10	238	160
	25	461	890
Sodium phosphate dibasic	25	252	160
	50	302	320
Sodium phytate	5	241	90
	10	416	530
EDTA[b], disodium salt	25	138	40
	50	277	1070

a Modified from Westermarck-Rosendahl et al 1980. Used by permission.

b EDTA - ethylenediaminetetraacetic acid.

basic sodium phosphate, EDTA and sodium phytate, used at levels of 25, 50, and 10 µeq/g of flour, respectively, were quite pronounced. The increase in viscosity was accompanied by an increase in falling number. Slightly smaller effects were observed for dibasic sodium phosphate. All the reagents listed in Table I improved elasticity of the dough. The effects on specific dough mixing characteristics varied. For instance, sodium phytate reduced dough extensibility whereas EDTA and dibasic sodium phosphate affected water absorption and dough development time. Concentrations of EDTA and dibasic sodium phosphate (50 µeq/g) effective in improving mixing and baking characteristics of sprout-damaged wheat flour were applied at levels exceeding those permitted in food products. Tribasic sodium phosphate supressed alpha amylase activity at levels allowed for bread products in Finland. The sodium phosphate salts, however, adversely affected loaf volume.

Thus, use of different chemicals to inhibit alpha-amylase in sprouted wheat flour as food additives is limited in part because of their toxicity and in part because of their adverse effects on functional properties of flour components important in baking performance. Consequently, the chemicals have found no commercial application to date.

Natural Alpha-Amylase Inhibitor

Already Sullivan (1949) and more recently Meredith and Pomeranz (1985) suggested that an ideal alpha-amylase inhibitor would be a simple fool-proof, non-toxic substance that can be safely added to sprouted wheat flour to "alter the properties of harmful proteinaceous alpha-amylase without altering the properties of the beneficial and equally sensitive gluten proteins" (Meredith and Pomeranz 1985). Recent work on natural inhibitors made this possible. In the studies performed at ABI Biotechnology Inc., an attempt has been made to apply a naturally occurring proteinaceous inhibitor (Weselake et al 1983; Mundy et al 1983, 1984) to inhibit excessive activity of alpha-amylase in sprout-damaged wheat.

The natural alpha-amylase inhibitors found in the endosperm of different cereals are low molecular weight, salt soluble proteins with a high content of basic amino acids (14 mole%; Mundy et al 1984) which inhibit the high isoelectric point (pH 6.0-6.5) alpha-amylases. Since, according to Weselake et al (1985), barley contains several fold higher levels of the inhibitor than other cereals, it was chosen as a source for the isolation of alpha-amylase inhibitor.

The inhibitor was isolated from the barley cultivar Bonanza according to the procedure developed by Zawistowska et al (1987) and its effectiveness was evaluated in flour-slurry and dough systems (Zawistowska et al 1988).

The control flour used in our studies was a commercial "all-purpose" flour milled from Canadian hard red spring wheat. In the standard amylograph test it had a peak viscosity of 310 B.U. The high alpha-amylase flour was prepared by adding an appropriate amount of barley malt to the control flour. As a result, the peak amylograph viscosity was reduced to 180 B.U. Baking was done on a fully automated Japanese Bread Bakery - Panasonic, model SD-BT-2P. All the ingredients (Table II) including the inhibitor, where applicable, were added to the mixing-fermentation-baking pan. The entire bread making process is completed within 4 hours.

The effect of the inhibitor on peak amylograph viscosity of the high alpha-amylase flour is shown in Fig. 1. Addition

TABLE II
Basic Bread Formula[a]

Flour	300 g
Sugar	17 g
Milk powder	6 g
Salt	5 g
Butter	11 g
Water	200 ml
Dry yeast (Firmapan)	4.5 g

[a] Reprinted, with permission, from Zawistowska et al 1988.

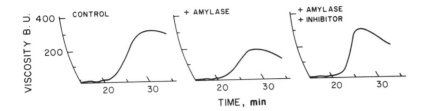

Fig. 1. Effect of inhibitor on peak amylograph viscosity. (Reprinted, with permission, from Zawistowska et al 1988).

of the inhibitor at the same level as that used in the baking experiments recovered most of the loss in peak amylograph viscosity.

Baking results showed that excess alpha-amylase yielded a loaf with caved-in top and side walls and open grain and sticky crumb (Fig. 2). Adding the alpha-amylase inhibitor as one of the bread ingredients completely neutralized the detrimental effect of excess alpha-amylase. The size and appearance of

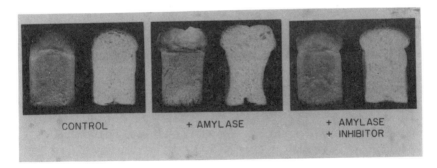

Fig. 2. Effect of inhibitor on external and internal bread appearance. (Reprinted, with permission, from Zawistowska et al 1988).

the loaf were essentially the same as those of the control loaf. Furthermore, the original grain and texture were fully recovered. The results obtained have clearly shown that sprouted wheat flour supplemented by the inhibitor is suitable for breadmaking as it produces loaves which are comparable with those made of sound wheat flour. At the present time, the economic aspects of supplementing sprout-damaged wheat flour with alpha-amylase inhibitor are being evaluated in our laboratory.

ACKNOWLEDGEMENT

The author gratefully acknowledges Dr. A.D. Friesen and Dr. J. Langstaff (ABI Biotechnology Inc.) for the encouragement and constant support in carrying out the inhibitor studies; Dr. W. Bushuk, Mrs. E. Slominska and Mr. R. Zillman (University of Manitoba) for performing baking and amylograph studies; and Dr. J.E. Kruger (Grain Research Laboratory) for the determination of alpha-amylase activity. Miss S. Prystawski is gratefully acknowledged for her patience in typing this manuscript.

LITERATURE CITED

AREF, M.M., NOEL, J-G., and MILLER, H. 1972. Inactivation of alpha-amylase in wheat flour with microwaves. J. Microwave Power, 7:215.

BUCHANAN, A.M., and NICHOLAS, E.M. 1980. Sprouting, alpha-amylase, and breadmaking quality. Cereal Res. Commun. 8:23.

CAWLEY, R.W., and MITCHELL, T.A. 1968. Inhibition of wheat α-amylase by bran phytic acid. J. Sci. Food Agric. 19:106.

D'APPOLONIA, B.L. 1983. "Sprouted" flour coping with damage. Bakers Digest 57(2):6.

DORFER, J., and KOBALL, G. 1980. Untersuchungen zur Inhibierung der Alpha-Amylase durch ausgewahlte anorganische und organische Substanzen. Baeker und Konditor. 2:58.

EDWARDS, G.H. 1964. Effects of microwave radiation on wheat and flour. The viscosity of the flour pastes. J. Sci. Food Agric. 15:108.

FULLER, P., HUTCHINSON, J.B., McDERMOTT, E.E., and STEWART, B.A. 1970. Inactivation of α-amylase in wheat flour with acid. J. Sci. Food Agric. 21:27.

FULLER, P., HUTCHINSON, J.B., and STEWART, B.A. 1968. The effect of silver nitrate upon the Hagberg Falling Time of flour. Flour Milling Baking Res. Assoc. Bull, p. 407.

HENRY, R.J., MARTIN, D.J., and BLAKENEY, A.B. 1987. Reduction of the α-amylase content of sprouted wheat by pearling and milling. J. Cereal Sci. 5:155.

HUNECKE, H., and PETZOLD, H. 1979. Kompensation von Auswuchschaden durch Mischen von Getreide und von Mehl. Getreidewirtschaft 13:204.

HUTCHINSON, J.B. 1963. Steam treatment of wheat: a new type of flour. Chem. Ind., p. 1084.

KOZMIN, N. 1933. Biochemical characteristics of dough and bread from sprouted grain. Cereal Chem. 10:420.

MacARTHUR, L.A., and D'APPOLONIA, B.L. 1982. Microwave and gamma radiation of wheat. Cereal Foods World 27:58.

MEREDITH, P. 1970. Inactivation of cereal α-amylase by brief acidification: the pasting strength of wheat flour. Cereal Chem. 47:492.

MEREDITH, P., and POMERANZ, Y. 1985. Sprouted grain. Adv. Cereal Sci. Technol. 7:239.

MUNDY, J., SVENDSEN, I., and HEJGAARD, J. 1983. Barley α-amylase/subtilisin

inhibitor I. Isolation and characterization. Carlsberg Res. Commun. 48:81.
MUNDY, J., HEJGAARD, J., and SVENDSEN, I. 1984. Characterization of a bifunctional-wheat inhibitor of endogenous α-amylase and subtilisin. FEBS Letters. 167:210.
NOLL, J.S. 1985. Effect of phytate, pH, and acid treatment on the falling number of sound and weathered wheats. Cereal Chem. 62:22.
OSADCHAYA, N.T., PUCHKOVA, L.I., RAKHMANKULOVA, R.G., and SHUB, I.S. 1979. Effect of some additives on alpha-amylase activity. Izv. Vyssh. Ucebn. Zaved. Pisch. Tekhnol. 6:108.
PALLA, J.-C., and VERRIER, J. 1974. Inhibition de l'alpha-amylase de ble par l'acide ascorbique et par des derives de l'isoalloxazine. Ann. Technol. Agric. 23:151.
PUCHKOVA, L.I., SHUB, I.S., and FILINA, M. 1980. Use of ascorbic acid derivatives in the processing of defective flour. Khelbopek. Konditer. Promst. 9:20.
RANHOTRA, G.S., LOEWE, R.J., and LEHMANN, T.A. 1977. Breadmaking quality and nutritive value of sprouted wheat. J. Food Sci. 42:1373.
SULLIVAN, B. 1949. Evaluation of methods and equipment in cereal laboratories. AACC Trans. 7:63.
TIPPLES, K.H., KILBORN, R.H., and BUSHUK, W. 1966. Effect of malt and sprouted wheat. Cereal Sci. Today 11:362.
WESELAKE, R.J., MacGREGOR, A.W., and HILL, R.D. 1983. Endogeneous α-amylase inhibitor in barley kernels. Plant Physiol. 72:809.
WESELAKE, R.J., MacGREGOR, A.W., and HILL, R.D. 1985. Endogeneous alpha-amylase inhibitor in various cereals. Cereal Chemistry. 62:120.
WESTERMARCK-ROSENDAHL, C., JUNNILA, L., and

KOIVISTOINEN, P. 1979. Efforts to improve the baking properties of sprout-damaged wheat by reagents reducing α-amylase activity I. Screening tests by the falling number method. Lebensm.-Wiss. Technol. 12:321.

WESTERMARCK-ROSENDAHL, C., JUNNILA, L., and KOIVISTOINEN, P. 1980. Efforts to improve the baking properties of sprout-damaged wheat by reagents reducing α-amylase activity III. Effects on technological properties of flour. Lebensm.-Wiss. Technol. 13:193.

ZAWISTOWSKA, U., LANGSTAFF, J., McVICAR, L., and FRIESEN, A.D. 1987. Preparation and application of barley protein that inhibits alpha-amylase in sprouted wheat flour. Canadian Patent Pending.

ZAWISTOWSKA, U., LANGSTAFF, J., and BUSHUK, W. 1988. Improving effect of a natural α-amylase inhibitor on the baking quality of wheat flour containing malted barley flour. J. Cereal Sci. 8:207.

9

IMMUNODIAGNOSTIC APPROACHES TO QUALITY AND PROCESS CONTROL IN WHEAT BREEDING AND UTILISATION

John H. Skerritt
CSIRO Wheat Research Unit, Division of Plant Industry
PO Box 7, North Ryde NSW, 2113 Australia

INTRODUCTION

While there has been a general trend by many countries towards export of manufactured goods or at least of "value added" products, opposing trends in the export of cereal crops have taken place, with exports of much more raw grain than grain products such as flour or noodles. Furthermore, declines in commodity prices especially affect returns on these "non-value-added" commodities. These pressures place increasing demands on the cereal producer and marketer, with respect to yield, quality and freedom from natural or man-made toxins in grain and grain products. Nevertheless, moves in several countries toward local deregulation of grain sales make these requirements more difficult to enforce without simple and innovative testing methods such as "cereal diagnostics". The scope of this review is to describe achievements and possible developments in the production of diagnostic tests for use by the cereal growing and processing industries. The term "diagnostics" in this context means simple tests employing probes for specific molecules which have important effects on agronomic factors or the end-use quality of cereal crops.

Diagnostic tests can play many roles. They can serve (a) as breeding tools to confirm the presence or absence of specific gene products, (b) to screen grain crops in the field for mineral deficiencies or pathogen infections, (c) to examine soil for herbicide residues, and (d) to test grain at receival or after storage for variety, quality type, or contamination with mycotoxins or pesticides. During cereal processing, the tests could assess milling quality, monitor the malting and brewing processes or the production of starch and gluten. Thus diagnostic tests can help ensure production and export of high-quality cereal grain and grain products. In addition, development of the diagnostic tests themselves will result in products of high added value.

Three technical factors must firstly be considered in test development:
- Is there a need for specific tests which are improvements over current technology?
- Can the attribute under study be linked to the presence or absence of a specific group of proteins or other molecules?
- Can a sufficiently simple test be developed, feasible for routine use?

The final factor requires the development of novel "delivery systems", that is, immunoassay devices and/or techniques which can simultaneously satisfy the requirements of speed, economy and simplicity.

ANTIBODY VERSUS NUCLEIC ACID PROBES

An immediate decision in designing a diagnostic kit is whether to use an antibody or a nucleotide probe. These two types of probe are competing technologies in medical diagnostics, although monoclonal antibodies are more established in the marketplace of the 1980's. While both often promise to provide similar information, this may not be so. Nucleic acid probes detect the presence of a structural gene, but in many cases the level of expression of the gene (e.g. production of a specific protein), detectable with an antibody probe, is more important. However, nucleic acid probes, unlike antibodies, can detect latent viral infections of plants. Assay formats involving antibodies are usually easier to use (Skerritt 1988a), although recently RNA or DNA probes using colorimetric or chemiluminescent detection have been incorporated into diagnostic kits (Matthews and Kricka 1988). Nucleic acid probes are also easier to modify in specificity than antibodies and are often more specific, since longer base (and resultant amino acid) sequences are involved in hybridisation. The development of nucleic acid-based diagnostic tests for barley yellow dwarf virus infection has especially been successful (Habili et al 1987).

APPLICATIONS OF ANTIBODY-BASED TESTS

Cereal Breeding

The ability to breed the best cultivars for local conditions, with the appropriate combination of yield, disease resistance and end-use quality remains the most fundamental approach to success in cereal production. Molecular biological methods are perceived as being of greatest potential use to breeders by speeding up the incorporation of desirable traits in a particular background, which would otherwise require several years of backcrossing. The introduction of foreign genes into wheat is of most use for single gene traits (such as disease resistance) and for certain quality-related proteins (such as high-molecular weight glutenins, Table I). These factors are also (theoretically) the easiest to quantify using antibody or DNA probe diagnostics. In some cases, it has proven difficult to obtain either DNA probes or antibodies of sufficient specificity, for example, "good" quality and "poor" quality high-molecular weight glutenin subunits differ only over very limited parts of their amino acid sequences. Attempts are being made to develop subunit-specific antibodies by use of synthetic peptides corresponding to the "unique" sequences. Such an approach has been successful in generation of subunit specific antibodies to alpha-zeins (Esen et al 1987) and keratins (Roop et al 1984) Other factors (such as yield) represent a complex interaction of many genes, including those affecting nutrition, photosynthesis, transpiration and translocation, and are likely to be beyond the reach of a single test.

Applied genetic engineering of cereals is being pursued with several aims:
- to enhance the expression of storage protein genes having technologically or nutritionally superior products,
- to modify existing storage protein genes for the above purposes,
- to transfer genes from other species endowing desirable characteristics.

In the case of conventional breeding, these species are related grasses, such as primitive wheats, but much broader opportunities exist with recombinant DNA methods. For example, alien genes from cereal rye, Agropyron or Aegilops have been used as sources of rust resistance, although linked genes in the transferred chromosome segment often bring in yield and quality defects (Sears 1981). An immediate application of molecular biological techniques (once cereal transformation becomes a routine method) will be to enable more specific transfer of desirable genes.

With current successes in transformation of cereal protoplasts by electroporation (Ou-Lee et al 1986) and in pollen transformation and regeneration of plants from protoplast-derived callus in rice (Cocking and Davey 1987, Goodman et al 1987), transfer of specific genes within a much wider range of species into cereals should become possible. Genes producing toxins acting on specific fungi could be introduced. Plants transformed with viral coat protein genes could show resistance to viral infection; tomato and tobacco have in this way been engineered to be resistant to tobacco mosaic virus (Powell-Abel et al 1986). Genetically-engineered herbicide resistance in cereals is another area of tremendous potential. For example, genes for mutant enzymes, only weakly affected by herbicides, could be introduced in the same manner as the engineering of glyphosate resistance in dicotyledons has been performed (Della-Cioppa et al 1987).

Diagnostics can play an important role in monitoring foreign-gene incorporation by conventional or molecular plant breeding. DNA probes can test for

TABLE I
Quality-related Proteins in Wheat Endosperm

Class	Examples	Role or Function
Storage proteins	High- and low-molecular-weight glutenins, Gliadins	Dough strength Extensibility
Starch proteins	Granule-associated proteins	Milling quality Starch synthesis, deposition and functional properties
Grain enzymes	Alpha-amylase Proteases	Weather damage Dough fermentation

actual transfer of the genes under study, while the expression of these genes (which is not automatic!) can be measured by antibodies. These methods are being used in the research applications of transformation described above, but will assume greater importance as attempts are made to transform large populations of germplasm. An antibody test for incorporation (and expression) of rye chromosomal material in 1B/1R translocation wheats is under development for this purpose (Dhaliwal et al 1988).

Variety and Species Identification Tests

Differences between the various monoclonal antibodies in their specificities for storage proteins from the different cereals allow their potential use in a range of tests. Antibodies with specificity for bread- and durum-wheats, rye and barley may be useful in tests for gluten (toxic to gluten-intolerant or celiac individuals) in foods; such antibodies form the basis of test kits commercially developed from work in our laboratory (Skerritt 1988b). The use of antibodies which did not bind barley proteins but do bind proteins from wheat, rye, maize and/or rice grains would enable detection of the contamination of barley malt-based brewers' grists with other cereal adjuncts. Such adjuncts are illegal in certain countries such as Germany; for this purpose, a mixture of some of our monoclonal antibodies such as clones 227/22 and 236/9 (Skerritt and Underwood 1986) would be useful. Polyclonal antibody-based immunological methods for the identification of these adjuncts in finished beer have been described by German workers (Ehrenstorfer and Gunther 1978, Wagner et al 1986). On the other hand, antibodies with strong selectivity towards bread wheat would be suitable for detection of bread wheat in durum pasta; illegal use of bread wheat for this purpose remains widespread in Italy, although such use is legal in other countries such as Australia.

Some experiments have also been performed to investigate the potential of monoclonal antibody methods for varietal identification of wheat and barley. Polymorphism is greatest among the prolamin storage protein fraction, and it is this fraction which has proven most useful for variety identification. Barley has only one genome, consisting of seven chromosome pairs, and all of the structural genes for storage proteins are found on one chromosome pair. In contrast, bread wheat is an allohexaploid and has storage protein gene loci on six chromosomes. With fewer genes coding for fewer proteins, barley gives simpler patterns on protein electrophoresis gels. The use of antibodies for varietal identification in barley has been described elsewhere by ourselves (Burbidge et al 1986, Skerritt et al 1988) and others (Deichl and Donhauser 1985). The presence in wheat of homologous proteins coded for by genes on homoeologous chromosomes probably enables a smaller proportion of antibodies to demonstrate varietal variation in binding compared with barley.

A number of approaches for antibody-based varietal identification of wheat have been investigated. Immunoblotting of gliadins, following acidic-buffer polyacrylamide gel electrophoresis of grain extracts made with 1M urea (du Cros et al 1984), can detect varietal differences in gliadin composition that, due to clustering of protein bands, are hard to detect from simple Coomassie Blue staining of polyacrylamide gels (Skerritt, Temperley and Metakowsky, unpublished). However, while this method is useful to identify allelic "blocks" of components (coded for by genes on a particular chromosome arm) in different varieties, these findings are of greater use in genetic studies than in routine varietal analysis, since they are slower and involve even more steps than polyacrylamide gel electrophoresis. Of simple types of immunological assay investigated, indirect, antigen-competition and sandwich ELISA assays using microwell plates and dot assays using nitrocellulose have been investigated as well as assays involving sectioned grains.

Both spot-tests and quantitative enzyme-immunoassays require the preparation of liquid extracts from meals of individual grains or half-grain samples. However, with the use of modern sample preparation equipment such as multichannel automated pipettes and high-speed microcentrifuges, simultaneous screening of large numbers of samples with several antibodies is possible. With the availability of automated microplate loaders, washers and photometers and the increasing sophistication of computer software for on-line analysis of ELISA data, this approach is attractive for the identification of large numbers of samples by seed testing laboratories. A relative binding profile for several antibodies with extracts to each grain sample could readily be obtained and then be matched to those of authentic samples. Indeed, this strategy for cultivar identification would be very similar to HPLC methods for varietal identification, using automated sample injectors and software for comparison of elution profiles by reference to "libraries" on computer discs.

The simplest type of test is the indirect ELISA in which grain extracts are applied to uncoated microwells, and after washing steps, an enzyme-labelled monoclonal antibody is incubated with bound protein. In a quantitative assay such as an ELISA, varietal differences in storage protein content or composition structure or sequence can only be manifest as differences in apparent antigen concentration.

However, as the antigen concentration-versus-absorbance response curves are relatively flat in these indirect ELISA assays, few consistent varietal differences were seen, in a study of 8 Australian cultivars grown at five sites. However steeper antigen concentration-response curves were found in "sandwich" assays (Skerritt 1988a) and using certain monoclonal antibody pairs, reproducible varietal differences were detected. In addition, some antibodies demonstrated varietal differences in antibody binding in both antigen-competition immunoassays (using different varieties of bread and durum wheats) and in rapid spot-tests (bread wheats).

Quality Type Identification

In many instances for domestic or international grain grade, wheat, barley and other grains are classified and marketed on the basis of "quality type" rather than actual variety. Wheat, for example, may be classified on the basis of glume colour, hardness and as winter or spring habit and/or as being in bread-making, multi-purpose, cookie or durum grades.

Much of the challenge of the 1980's and 1990's remains in obtaining a much firmer picture of the relationship between the presence or absence of specific proteins in wheat or barley grain and their effects on end-use quality (Payne 1987). Success will come from application of several techniques, including protein electrophoresis, dough reconstitution and antibody-binding studies using genotypes having wide ranges of backgrounds and with "research" cereals bred especially to lack specific polypeptides. Over the last decade, electrophoresis methods have made the most substantial contribution in this area (Payne and Lawrence 1983, Campbell et al 1987). Links between other protein components and quality are also emerging. For example, the presence of three M_r 15,000 proteins on the starch granule surface seems closely related to endosperm softness, but a cause-and-effect relationship will be hard to establish (Greenwell and Schofield 1986). For diagnostics to screen for soft milling endosperm and possibly to predict milling starch damage in wheats, establishing such a relationship may not be important, provided the correlations between the presence of these components and the quality attribute are very tight.

Our initial attempts to develop antibody-based quality tests for wheat centered around the measurement of quality loss resulting from grain sulfur-deficiency (Skerritt et al 1987). In sulfur-deficient grain, there are decreases in dough extensibility accompanied by changes in the relative proportions of prolamins; in particular, there are decreases in the amounts of certain higher-mobility gliadins and increases in the proportions of omega-gliadins and of high-molecular-weight glutenin polypeptides. The best antibody indicator of quality loss due to sulfur-deficiency was clone 227/22, specific for beta-gliadins which are relatively rich in sulfur. In a competition enzyme-immunoassay, the binding of this antibody was correlated significantly and positively with both flour sulfur (eg. for Egret cookie wheat $r = 0.86$, $n = 18$, $P<0.001$) and the extensibilities of doughs prepared from these flour samples ($r = 0.94$, $P<0.001$). These correlations were seen for samples of each of three wheat cultivars that were examined, namely one cookie-, one breadmaking- and one multipurpose wheat.

Pedigree-related differences in protein composition present a greater challenge in quantification with specific antibodies. Seven antibodies with gliadin-binding specificities and four antibodies binding to high-molecular-weight (HMW) glutenin polypeptides were used to screen a set of 71 wheat varieties from 26 countries; dough- and baking-quality data and protein electrophoretic profiles have been determined for this set of samples (Campbell et al 1987). Three types of antibody-binding pattern were seen:
1) no varietal differences in binding other than that due to differences in total protein content,
2) varietal differences in binding noted, but no correlations found between antibody binding and the quality parameters,
3) varietal differences in binding, correlating with quality parameters.

This final group of antibodies may be of use in

quality testing. Of seven gliadin-specific antibodies tested by competition enzyme-immunoassay using flour extracts made with 1M urea, two bound in a manner which correlated significantly (P<0.02) with dough resistance and dough-mixing work input. These were antibodies 230/9 and 404/6, which are specific for alpha/beta- and beta/gamma-gliadins, respectively.

Quality Control Tests in Cereal Processing

Test methods for quality control in production of starch and gluten from wheat are needed to monitor the efficiency of their separation and to reduce gluten losses in plant effluent. Without specific probes, it is hard to distinguish traces of gluten protein from salt-soluble proteins and other nitrogenous material. Gluten analysis is important to ensure that starch is low in gluten, so that it can be incorporated into special dietary foods and so that its pasting properties and the sugars produced upon industrial hydrolysis are not detrimentally affected by high protein levels (Fritschy et al 1985, Skerritt 1985, 1988b,c). These antibodies also form the basis of a simple test for gluten not only in flour, starches and baking mixes but also in foods that had been cooked or extensively processed. These antibodies bound to low-sulfur heat-insensitive storage proteins from the coeliac-toxic cereal species, but not to proteins from non-toxic species such as rice or maize (Skerritt 1985, Skerritt and Underwood 1986). Such a test will have application for food analysts and others in determining "safe" and "toxic" foods for cereal-sensitive individuals. A quantitative test for gluten would be used by food manufacturers to monitor or quality-control cereal protein incorporation into foods. A test kit for these purposes which incorporates monoclonal antibodies, is now on the market. Other kits such as tests for gluten quality or heat damage during drying, grain enzyme levels, flour starch damage or germ content are technically possible but require both the development of suitable antibodies and the demand for an alternative test method in each case, arising from inadequacies of the existing technology.

There are many applications of antibody diagnostics in barley malting and brewing being investigated in a number of laboratories (Vaag and Munck 1987). Antibody-based methods have been used for many years in research on malting (Grabar and Nummi 1967, Hejgaard and Carlsen 1977, Daussant and Skakoun 1983) but attempts to apply these methods to routine analysis have not been made. With the knowledge that qualitative and quantitative differences in protein composition between malt samples can markedly influence endosperm hydrolysis (modification) and subsequent processing of malts during brewing (Smith and Lister 1983, Moonen et al 1987), antibody-based tests have been developed to quantitate such modification. Using monoclonal antibodies with specificities for certain storage protein components, it has been possible to rapidly assess the degree of protein modification in test malt samples using an enzyme-immunoassay (Skerritt and Henry 1988). Immunoassays for non-permitted enzymes and cereal adjuncts in beers have been developed by European workers (Wagner et al 1986), and antibody tests to monitor proteolysis during mashing are also under development. Immunoassays can also quantitate residual protein components causing haze- and head-stability problems in finished beers and predict such problems in production, allowing their rectification before bottling (Asano and Hashimoto 1980, Asano et al 1983). These methods have also been used extensively to measure levels of carbohydrate-degrading enzymes in cereals, especially barley (Hejgaard and Carlsen 1977, Daussant and Skakoun 1983). It would thus be possible to construct a kit containing immunoassays for amylases, glucanases, proteases and protease substrate (hordein) modification, for use by barley breeders and maltsters (McLeod et al 1987).

Diagnostics for Crop Disease or Stress

The last twenty years have seen considerable use of conventional antisera in taxonomic analyses of viral, bacterial and fungal pathogens of cereals and other crops. Several antibody-based diagnostic tests for plant viruses have been on the market since the mid-1980's (Gugerli 1983, Halk and de Boer 1985, Sheppard et al 1986). These tests can be used to prevent the spread of viral infection by ensuring that seed and nursery stock are disease-free. The diagnostics could either use an antibody specific for an epitope common to all known strains of the virus or else a mixture of antibodies such that the known serological variants of the virus are detected.

Immunoassays have also been developed for cereal viruses, such as maize dwarf mosaic virus and barley yellow dwarf luteovirus (BYDV) (Diaco et al 1983, Hsu et al 1984). The need for such a diagnostic is evidenced in the devastating effects of BYDV on yield of a wide range of cereal crops, and the difficulty in distinguishing BYDV-infected cereals from those that are simply nitrogen-deficient. Earlier "research" serological tests such as radioimmunoassay and serological electron microscopy have been replaced by rapid field ELISA tests or DNA probe-based diagnostics. Diagnostics may also be of use in identifying BYDV-resistant lines. For example, BYDV-resistant barleys somehow slow replication of the virus in their leaves. Recently, polypeptide markers of virus resistance and of susceptibility have been identified in barley coleoptiles and roots (Holloway 1987), but development of a diagnostic test will be difficult as the resistance and susceptibility polypeptides are very similar in properties. A search for similar markers of cereal cyst nematode resistance is underway.

While a series of elegant diagnostics for turfgrass fungal diseases have now reached the market (Miller et al 1986), substantial progress in the development of diagnostics for cereal fungal infections should be made in the 1990's. First, however, fundamental work is needed to determine which fungal antigens are 1) species-specific, 2) specific for pathogenic strains, and 3) exposed and stable to extraction. A major methodological problem is the need to protect pathogen-specific proteins from destruction by endogenous leaf proteolytic enzymes. Monoclonal antibodies to the Tilletia wheat bunt fungus teliospores and to Fusarium cereal rusts have been prepared (Banowetz et al 1984, Hardham et al

1986) but their specificities are not suitable for diagnostic use. The application of diagnostics to bacterial infections of plants is so far extremely limited, compared with their application in human medicine. In both cases, work must be done first to ensure that "seropositive" bacteria are actually pathogenic.

A more speculative application of diagnostics is the detection of stress proteins in cereal crop leaves, which either would allow remedial treatment of the crop during growth or segregation of grain at harvest. The results might also help in formulating strategies to prevent stress in future seasons. Some possible "stress" markers are proteins that are newly synthesized following salt stress of barley and maize (Ramagopal 1986, 1987) and heat shock proteins in wheat and other cereals (Sachs and Ho 1986, Mansfield and Key 1987, Necchi et al 1987). Furthermore, a particular epicotyl protein has also been identified in cold-tolerant wheat varieties (Sarhan and Perras 1987). Immunologically-related "pathogenesis-related" proteins have been detected in leaves of mildew-infected maize (White et al 1987). The function of some of these proteins (Schlesinger 1986) may be to provide the plant with stress tolerance; in this case diagnostic tests for these proteins could be used as a breeding tool to screen for stress-tolerant progeny. It must be noted, however, that in some cases the appearance of these proteins is tissue-specific; most appear in roots and coleoptiles.

SMALL MOLECULES:
MYCOTOXINS, PESTICIDES AND PLANT HORMONES

Since 1980, there has been an increasing realisation that the speed, sensitivity and specificity of immunological methods can be readily applied to the detection of a range of small molecules. Coupling of the molecule (termed a hapten) to a carrier protein before immunisation can yield high-titer antisera or high-affinity monoclonal antibodies.

Mycotoxins

Mycotoxins are toxic substances produced by fungi and may be present in cereal grains and foods even if there are no obvious signs of mold. These toxins can have a variety of effects on the growth, health and reproduction of farm animals, causing a range of symptoms in humans including immune suppression, leukopenia, bone marrow degeneration and hepatic cancers. Mycotoxin contamination of cereals is not uncommon where grain development, harvest or storage has been under relatively damp conditions. The type of toxin formed depends on the species of fungus infecting the grain. Aflatoxins are most likely to occur in warm humid climates, while the other mycotoxins (vomitoxin, T-2 toxin, zearalenone and ochratoxins) are more characteristic of temperate climates following wet harvests. ELISA assays for use with cereals have been developed for each class of mycotoxin (Pestka et al 1981, Morgan et al 1983, Chu 1984, Gendloff et al 1984, Liu et al 1985, Kemp et al 1986, Ram et al 1986, Warner et al 1986), although many of these mycotoxin assays have either only been developed for within-laboratory use or are still in the stages of conversion into commercial test kits. However, recently both quantitative and rapid semiquantitative enzyme-immunoassays for aflatoxins became commercially available and at least one product has received Association of Official Analytical Chemists approval after interlaboratory collaborative trials. Some of these have been found very accurate for use with cereal grains, yet are cheaper, faster and simpler to use than chromatographic techniques.

Pesticides and herbicides

Increasing consumer awareness and demand for "additive-free" foods has coincided with a world commodity glut in the mid to late 1980's, enabling potential foreign purchasers of grain and grain products to insist on stringent tolerances on the levels of grain protectant residues. While greater use of fast-metabolising pesticides on stored grain has reduced residue problems, increasing consumption of wholemeal or wholegrain breads and other fiber-rich products has exposed the consumer to greater potential levels of pesticide residues, since many of the pesticides partition into bran and germ. In addition, deregulation of grain trading in many countries, with grain producers selling produce directly to customers, has reduced the role of centralised residue testing and made the requirements for rapid field testing of grain more critical. Meanwhile, the prescribed residue tolerances in these products tend to become more stringent with time. The inappropriate use of pesticides (either in dosage or timing of application, or the nature of substance used) or storage of grain in contaminated silos also increases the need for accurate and routine monitoring of residues. There is also a need for antibody tests to quantitate herbicides. In this case, soil rather than grain analysis is more relevant to the cereal-production industries. The use of broadleafed-weed herbicides such as chlorsulfuron is often very effective in improving barley and wheat yields and in reducing the contamination of grain and hay with undesirable weed species, but persistence of these agrochemicals in soil can be hard to predict, and their toxicities to rotation crops such as legumes can be severe.

Other sensitive and specific test methods for pesticides based on gas-, high performance liquid- or thin-layer- chromatography suffer from several major disadvantages for use in rapid testing. The high capital cost of equipment restricts analyses to central laboratories. While such equipment may be automated, the chromatography time for individual samples and the need for considerable pre-analysis "clean-up" of samples greatly restricts sample throughput. The need for skilled analysts also leads to high costs for each analysis. These factors, together with the slow times for analyses (usually over one hour), make these methods inappropriate for the analysis, at silo receival, of grain from individual farms.

A reasonable number of publications describing immunoassays for pesticides and herbicides has appeared in the scientific literature (see Hammock and Mumma 1981, Van Emon et al 1985, Skerritt 1988c for reviews). However, many of the assays described in the scientific literature by the late 1980's have been developed by university scientists for research applications, and only some of the assays have been adapted to cereals. Most work has been done on herbicides, such as paraquat, atrazine and 2,4-dichlorophenoxyacetic acid (2,4-D). Study of the pesticides of greatest relevance to stored grain (organophosphates and synthetic pyrethroids) has been far less thorough. Test kits for some of these herbicides and pesticides are already available commercially. Some are adapted for field testing, providing a "yes/no" answer in under 5 minutes. In the case of paraquat and paraoxon, monoclonal antibody tests have also been developed although many groups have obtained acceptable specificities using polyclonal antisera. It should be noted that often in screening applications it is more desirable to detect the presence of any of a large number of compounds, so that grain providing positive results could be segregated rapidly and studied in more detail.

Plant hormone diagnostics

Reasons for a poor understanding of the role of many plant hormones relate largely to the very low levels of hormones that are found naturally. Hormones such as abscisic acid may also be important in screening of wheat samples for susceptibility to preharvest sprouting (Walker-Simmons and Sesing 1986, Walker-Simmons 1987.) Better measurement of these hormones in cereal crops could provide valuable information on several processes which affect cereal productivity and quality; some, such as the role of gibberelins in the germination of barley during commercial malting, are well understood. Several groups have developed sensitive immunoassays for plant hormones in cereals and other plants (Weiler 1982, Wang et al 1986, Knox et al 1987, Trione et al 1987). One American company (Idetek) specialises in their manufacture.

ASSAY FORMATS

The nature of the sample in cereal diagnostics presents in many cases greater challenges than in medical diagnostics. Many of the latter assays are routinely performed on body fluids such as blood, serum or urine, that is, fluids that are relatively consistent in physicochemical properties such as viscosity, pH and ionic composition. In contrast, the types of substrates to be examined in the broad field of cereal diagnostics include soil, whole grain and grain sections, aqueous and organic-solvent grain extracts, leaf sap from growing plants, green malt, dry flours, wholemeal and wide ranges of food types.

Despite advances in sample handling, in recent years, most cereal chemists and analytical chemists determining pesticide residues or mycotoxins in food products would agree that sample extraction and "work-up" is the most labor-intensive and often the most time-consuming part of the analysis. Immunoassays by virtue of their biological specificity are often free of major interference problems, reducing the work-up that is necessary but for quantitative analysis, sample extraction is still required.

The various users of "cereal quality" diagnostics require different types of assay formats ("delivery systems") refecting the differing demands of their work (Table II). Plant breeders screen large numbers of small samples and thus would prefer not to have to extract each sample. In many cases, "yes/no" results for the presence or absence of specific components in progeny from a cross (or otherwise transformed and regenerated plants) is sufficient.

At receival, sample size is clearly not limiting, but speed is. Most results need only be qualitative (such as variety or quality type) or at most semi-quantitative (eg. enzyme levels, reflecting possible weather damage). Quality control laboratories in flour mills, starch/gluten plants, maltings or breweries also have larger samples to analyze and analyses taking one or more hours are often quite satisfactory. However, while some of the qualitative receival tests may also be used in this setting, other quantitative tests (eg. gluten content, gluten quality, malt protein modification, malt enzymes) may also be required.

TABLE II
Sample and Test Requirements in Cereal Quality Diagnostics

	Plant Breeding	Grain Receival	Product Quality Control
Site	field or laboratory	field	laboratory
Sample size	small	large	large
Sample type	individual grains	representative	representative
Sample numbers	large	moderate	small
Time per set of tests	½ day	minutes "on-line"	minutes to 1 hour
Type of result required	usually qualitative	qualitative or quantitative	quantitative
Destructive testing appropriate?	No: require sowing of germ half	Yes	Yes

Assay formats in which the sample requires extraction (Table III) are usually based on either enzyme-immunoassay or particle agglutination (arising from the antibody-antigen interaction). While agglutination reactions are commonly recognised to be rapid, commercial enzyme-immunoassays can be completed in under 10 minutes. A rapid "home" test for gluten in foods based on a test-tube sandwich ELISA (developed in our laboratory) and spot-test competition ELISA tests for pesticides and aflatoxins are two examples of relevance in commercial production. Biosensors (North 1985) using bound

TABLE III
Modern Immunoassay Formats

Assay Type	Format	Solid Phases	Results
ELISA (soluble product)	indirect competition sandwich	microwells beads tubes dipsticks	qualitative or quantitative
ELISA (insoluble product)	indirect competition sandwich	teststrip or card or immuno-filtraton "dry chemistry"	qualitative or semi-quanti-tative
Immunochromatog-raphy	indirect ELISA	paper strips	quantitative
Particle Agglutination	antibody-coated latex beads or red blood cells	synthetic beads or cells	fast, simple but artefact-prone
Biosensors	analyte binding causes capacitance, refractive index, conductivity change	electrodes crystals testcards	on-line monitoring capability disposable sensors now developed

antibodies or antigen are under active development but are of most obvious application to quality control in processes where the compound to be analysed is usually already present in solution, for example wort analysis in brewing, effluent analysis in starch/gluten production or pesticide and herbicide detection in groundwater.

Also simpler, is the use of in situ tests, in which large numbers of grain are mounted on a Cernit block, lightly sanded, then dipped into the antibody reagent (antibody coupled to a fluorescent tag) before visualisation in a small portable viewer. The grains remain viable for planting-out of lines predicted to have desirable quality characteristics. This method is possibly the least labor-intensive method of immunological analysis of large numbers of grains, and is already in use in the antibody screening for high-lysine barleys by the Carlsberg laboratories in Denmark (Rasmussen 1985). It also has considerable application for the screening of cereal grain samples for assessment of homogeneity of variety or uniformity of modification in malting.

CONCLUSIONS

This review has described some of the ways that antibodies can be used in the management of cereal-crop breeding, production and processing. Crop management can be assisted (Lankow et al 1987) by new tests which 1) rapidly diagnose and detect crop disease before symptoms appear, 2) detect pathogens in soil, stubble or grain before sowing, or 3) influence choice of chemical treatments (pre- or post-harvest) influencing crop rotation. Crop breeding will also be aided by these diagnostic tests, allowing development of wheats for specific markets, such as 1) soft-milling yet high protein, strong-doughed wheats, 2) hard-grained, high-yielding hexaploid wheats with yellow endosperm colour for pasta manufacture, 3) specialised confectionary flours containing sweet peptides such as thaumatin or monellin or salty peptides (Tada et al 1984) for cracker biscuits. The antibody probes themselves can be modified. Single-chained antibodies produced by expression of recombinant DNAs may be cheaper to produce, of higher specific activity and are more stable than conventional monoclonal antibodies (Klausner 1986). Monoclonal antibodies which bind enzyme-substrate transition-state analogues can catalyze specific biochemical reactions (Pollack et al 1986, Tramontano et al 1986). In the cereal processing industry, such substances could be used to specifically inactivate poor-quality components, or once the celiac-toxic sequences in gluten are identified, to specifically detoxify gluten while retaining its functional properties in special dietary foods.

However, despite the excitement of these results and of their application to cereal breeding and technology, realistic commercial considerations must always apply. Development and successful application of cereal diagnostics firstly requires satisfaction of several technical criteria (Skerritt 1988c) then the application of several commercial criteria. There must be sufficient product demand in the target market; the nature of the diagnostic must be such that the test it performs is considered important for kits to be routinely used. The cost per assay that the market will bear depends on the importance of the test result and the test's application. Breeders' screens or analysis of foundation seed or of grain at receival needs large numbers of tests, necessarily demanding a low cost per test. However, the actual cost of materials in many monoclonal antibody tests is often only 1/5 to 1/20 of the retail price of the kit. Many cereal diagnostic markets are either highly fragmented or too specialised for the development of a particular diagnostic to be economically viable. For example, there are many pesticides used on stored cereals, but only some are used sufficiently widely to justify development of a diagnostic. Also, particular varietal or quality-type identification problems or pathogen infection problems may be limited to one small geographical region, not economically justifying development of a specific test.

Another consideration is the nature of the interaction between the companies and government bodies developing, manufacturing and marketing the tests and those who will become users of the diagnostics. In the US especially, but also in the UK and Europe, three healthy means of commercial involvement in agricultural or food diagnostics have occurred. Firstly, small, specialised agri-diagnostic companies have formed. Secondly, many large multinational agrichemical companies have formed diagnostic divisions (for example, companies can combine development and marketing of a herbicide or fungicide with a herbicide or fungal infection diagnostic). Finally, several biotechnology and food companies have formed alliances in the area of diagnostics. Hopefully these trends will soon lead to the development of an increased range of useful diagnostic tests for use by the cereal industry.

LITERATURE CITED

ASANO, K., and HASHIMOTO, N. 1980. Isolation and characterization of foaming proteins in beer. J. Amer. Soc. Brew. Chem. 38:129.

ASANO, K., SHINAGAWA, K., and HASHIMOTO, N. 1983. Characterization of haze- forming proteins of beer and their roles in chill haze formation. Rep. Res. Lab. Kirin Brewing Co. 26:45.

BANOWETZ, G.M., TRIONE, E.J., and KRYGER, B.B. 1984. Immunological comparisons of teliospores of two wheat bunt and fungi, Telletia species using monoclonal antibodies and antisera. Mycologica 76:51.

BURBIDGE, M., BATEY, I.L., CAMPBELL, W.P., SKERRITT, J.H., and WRIGLEY, C.W. 1986. Distinction between barley varieties by grain characteristics, electrophoresis, chromatography and antibody reaction. Seed Sci. Technol. 14:619.

CAMPBELL, W.P., WRIGLEY, C.W., CRESSEY, P.J., and SLACK, C.R. 1987. Statistical correlations between quality attributes and grain-protein composition for 71 hexaploid wheats used as breeding parents. Cereal Chem. 64:293.

CHU, F.S. 1984. Immunoassays for analysis of mycotoxins. J. Food Protect. 47:562.

COCKING, E.C., and DAVEY, M.R. 1987. Gene transfer in cereals. Science 236:1259.

DAUSSANT, J., and SKAKOUN, A. 1983. Immunochemistry of seed proteins. Page 103 in: Seed Proteins. J. Daussant, J. Mosse and J. Vaughan, eds. Academic Press., London.

DEICHL, A., and DONHAUSER, S. 1985. Die immunochemische Bestimmung der Sortenreinheit und des Vermischungsgrades von Braugerste und

Braumalz. Proc. Eur. Brewery Cong. Conv. 20:611.
DELLA-CIOPPA, G., BAUER, S.C., TAYLOR, M.L., ROCHESTER, D.E., KLEIN, B.K., SHAH, D.M., FRALEY, R.T., and KISHORE, G.M. 1987. Targeting a herbicide- resistant enzyme from Escherichia coli to chloroplasts of higher plants. Bio/Technology 5:579.
DHALIWAL, A.S., MARES, D.J., MARSHALL, D.R., and SKERRITT, J.H. 1988. Protein composition and pentosan content in relation to dough stickiness of 1B/1R translocation wheats. Cereal Chem. 65:143.
DIACO, R., LISTER, R.M., DURAND, D.P., and HILL, J.H. 1983. Production of monoclonal antibodies against three isolates of barley yellow dwarf virus. Phytopathology 73:788 (Abstract).
du CROS, D.L., CAMPBELL, W.P., SKERRITT, J.H., WRIGLEY, C.W., and CRESSEY, P.J. 1984. Characterization of quality-related gluten proteins using chemical methods and monoclonal antibodies. Page 155 in: Gluten Proteins. A. Graveland and J.H. Moonen, eds. TNO, Wageningen, The Netherlands.
EHRENSTORFER, I., and GUNTHER, H.O. 1978. Nachweis von Enzymen und Rohfrucht in Bier. Brauwelt 118:52.
ESEN, A., BIETZ, J.A., PAULIS, J.W., and WALL, J.S. 1987. A 23.8kD alpha-zein with N-terminal sequence and immunological properties similar to 26.7kD alpha-zeins. Plant Mol. Biol. 9:421.
FRITSCHY, F., WINDEMANN, H., and BAUMGARTNER, E. 1985. Quantitative determination of wheat gliadins in foods by enzyme-linked immunosorbent assay. Z. Lebensm.-Unters. Forsch. 181:379.
GENDLOFF, E.H., PESTKA, J.J., SWANSON, S.P., and HART, L.P. 1984. Detection of T-2 toxin in Fusarium sporotrichioides - infected corn by enzyme-linked immunoassay. Appl. Environ. Microbiol. 47:1161.
GOODMAN, R.M., HAUPTLI, H., CROSSWAY, A., and KNAUF, V.C. 1987. Gene transfer in crop improvement. Science 236:48.
GRABAR, P., and NUMMI, M. 1967. Recent immunoelectrophoretic studies on soluble proteins in their transformation from barley to beer.

Brewers Digest 42(2):68.
GREENWELL, P., and SCHOFIELD, J.D. 1986. A starch granule protein associated with endosperm softness in wheat. Cereal Chem. 63:379.
GUGERLI, P. 1983. Use of enzyme-immunoassay in phytopathology. Page 369 in: Immunoenzymatic Techniques. S. Avrameas, P. Druet, R. Masseyeff and G. Feldmann, eds. Elsevier, Amsterdam.
HABILI, N., McINNES, J.L., and SYMONS, R.H. 1987. Nonradioactive photobiotin-labelled DNA probes for the routine diagnosis of barley yellow dwarf virus. J. Virol. Methods 16:225.
HALK, E.L., and DE BOER, S.H. 1985. Monoclonal antibodies in plant disease research. Ann. Rev. Phytopathol. 23:321.
HAMMOCK, B.D., and MUMMA, R.O. 1981. Potential of Immunochemical Technology for Pesticide Analysis. Page 321 in: Pesticide Analytical Methodology. J. Harvey, Jr and G. Zweig, eds. Amer. Chem. Soc. Symp. Ser. 136., Washington, D.C.
HARDHAM, A.R., SUZAKI, E., and PERKIN, J.L. 1986. Monoclonal antibodies to isolate-, species-, and genus-specific components on the surface of zoospores and cysts of the fungus Phytophthora cinnamomi. Can. J. Bot. 64:311.
HEJGAARD, J., and CARLSEN S. 1977. Immuno-electrophoretic identification of a heterodimer beta-amylase in extracts of barley grain. J. Sci. Food Agric. 28:900.
HOLLOWAY, P. 1987. Identification of polypeptide markers of barley yellow dwarf virus-resistance in barley. Proc. Aust. Barley Tech. Symp. (Wagga), 251.
HSU, H.T., AEBIG, J., and ROCHOW, W.F. 1984. Differences among monoclonal antibodies to barley yellow dwarf viruses. Phytopathology 74:60.
KEMP, H.A., MILLS, E.N.C., and MORGAN, M.R.A. 1986. Enzyme-linked immunosorbent assay of 3-deoxynivalenol applied to rice. J. Sci. Food Agric. 37:888.
KLAUSNER, A. 1986. Single-chain antibodies become a reality. Bio/Technology 4:1041.
KNOX, J.P., BEALE, M.H., BUTCHER, G.W., and MacMILLAN, J. 1987. Preparation and characterization of

monoclonal antibodies which recognise different gibberellin epitopes. Planta, 170:86.

LANKOW, R.K., GROTHAUS, G.D., and MILLER, S.A. 1987. Immunoassays for crop management systems and agricultural chemistry. Page 228 in: Biotechnology In Agricultural Chemistry. H. M. LeBaron, R.O. Mumma, R.C. Honeycutt and J.H. Duesing, eds. Amer. Chem. Soc. Symp. Ser. 334., New York.

LIU, M-T., RAM, B.P., HART, L.P., and PESTKA, J.J. 1985. Indirect enzyme-linked immunosorbent assay for the mycotoxin, zearalenone. Appl. Environ. Microbiol. 50:332.

McLEOD, L.L., LANCE, R.C.M., SKERRITT, J.H., and HILL, A.S. 1987. An immunological technique for the rapid assessment of enzymes associated with malting quality in barley. Chem. Aust. 54:336. (Abstract).

MANSFIELD, M.A., and KEY, J.L. 1987. Synthesis of the low molecular weight heat shock proteins in plants. Plant Physiol. 84:1007.

MATTHEWS, J.A., and KRICKA, L.J. 1988. Analytical strategies for the use of DNA probes. Anal. Biochem. 169:1.

MILLER, S.A., GROTHAUS, G.P., PETERSEN, F.P., and PAPA, S.L. 1986. Detection of pythium blight in turfgrass using a monoclonal antibody-based diagnostic test. Phytopathology 76:1057 (Abstract).

MOONEN, J.H.E., GRAVELAND, A., and MUTS, G.C.J. 1987. The molecular structure of gel proteins from barley. Its behaviour in wort filtration and analysis. J. Inst. Brew. 93:125.

MORGAN, M.R.A., McNERNEY, R., and CHAN, H.W.S. 1983. Enzyme-linked immunosorbent assay of ochratoxin A in barley. J. Assoc. Off. Anal. Chem. 66:1481.

NECCHI, A., POGNA, N.E., and MAPELLI, S. 1987. Early and late heat shock proteins in wheat and other cereals. Plant Physiol. 84:1378.

NORTH, J.R. 1985. Immunosensors: antibody-based biosensors. Trends Biotechnol. 3:180.

OU-LEE, T-M., TURGEON, R., and WU, R. 1986. Expression of a foreign gene linked to either a plant virus or a Drosophila promoter, after

electroporation of protoplasts of rice, wheat and sorghum. Proc. Nat. Acad. Sci., USA 83:6815.
PAYNE, P.I. 1987. Genetics of wheat storage proteins and the effect of allelic variation on breadmaking quality. Ann. Rev. Plant Physiol. 38:141.
PAYNE, P.I., and LAWRENCE, G.J. 1983. Catalogue of alleles for the complex loci, Glu A1, Glu B1, Glu D1 which code for high-molecular weight subunits of glutenin in hexaploid wheat. Cereal Res. Commun. 11:29.
PESTKA, J.J., LEE, S.C., LAU, H.P., and CHU, F.S. 1981. Enzyme-linked immunosorbent assay for T-2 toxin. J. Amer. Oil. Chem. Soc. 58:940A.
POLLACK, S.J., JACOBS, J.W., and SCHULTZ, P.G. 1986. Selective chemical catalysis by an antibody. Science 234:1570.
POWELL-ABEL, P., NELSON, R.S., DE, B., HOFFMANN, N., ROGERS, S.G., FRALEY, R.T., and BEACHY, R.N. 1986. Delay of disease development in transgenic plants that express the TMV coat protein gene. Science 232:738.
RAM, B.P., HART, L.P., SHOTWELL, O.L., and PESTKA, J.J. 1986. Enzyme-linked immunosorbent assay of aflatoxin B1, in naturally contaminated corn and cottonseed. J. Assoc. Off. Anal. Chem. 69:904.
RAMAGOPAL, S. 1986. Protein synthesis in a maize callus exposed to NaCl and mannitol. Plant Cell Rep. 5:430.
RAMAGOPAL, S. 1987. Differential mRNA transcription during salinity stress in barley. Proc. Natl. Acad. Sci. USA. 84:94.
RASMUSSEN, U. 1985. Immunological screening for specific protein content in barley seeds. Carlsberg Res. Commun. 50:83.
ROOP, D.R., CHENG, C.K., TITTERINGTON, L., MEYERS, C.A., STANLEY J.R., STEINHERT, D.M., and YUSPA, S.H. 1984. Synthetic peptides corresponding to keratin subunits eicit highly specific antibodies. J. Biol. Chem. 259:8037.
SACHS, M.M., and HO. T-H.D. 1986. Alteration of gene expression during environmental stress in plants. Ann. Rev. Plant Physiol. 37:363.
SARHAN, F., and PERRAS, M. 1987. Accumulation of a high molecular weight protein during cold

hardening of wheat (Triticum aestivum L.). Plant Cell. Physiol. 28:1173.

SCHLESINGER, M.J. 1986. Heat shock proteins: the search for functions. J. Cell Biol. 103:321.

SEARS, E.R. 1981. Transfer of alien genetic material to wheat. Page 75 in: Wheat Science - Today and Tomorrow. L.T. Evans and W.J. Peacock, eds. University Press, Cambridge.

SHEPPARD, J.W., WRIGHT, P.F., and DESAVIGNY, D.H. 1986. Methods for the evaluation of EIA tests for use in the detection of seed-borne diseases. Seed Sci. Technol. 14:49.

SKERRITT, J.H. 1985. Detection and quantitation of cereal protein in foods using specific enzyme-linked monoclonal antibodies. Food Tech. Aust. 37:570.

SKERRITT, J.H. 1988a. Immunochemistry of Cereal Grain Storage proteins. Page 263 in: Advances in Cereal Science and Technology, Volume IX. Y. Pomeranz, ed. American Association of Cereal Chemists, St Paul, Minnesota.

SKERRITT, J.H. 1988b. Immunoassays of non-meat protein additives in foods. in: Development and Application of Immunoassay for Food Analysis. J. Rittenburg, ed. Elsevier Applied Science Publishers, London. in press.

SKERRITT. J.H. 1988c. Enzyme immunoassay technology-application in the production and processing of cereals. in: Biotechnology-Principles and Application to the Food Industry. G.H. Fleet and P. Rogers, ed. Gordon and Breach, New York. in press.

SKERRITT, J.H., and HENRY, R.J. 1988. Hydrolysis of barley endosperm storage proteins during malting. 2) Quantification by enzyme- and radio-immunoassay. J. Cereal Sci. 7:265.

SKERRITT, J.H., and UNDERWOOD, P.A. 1986. Specificity characteristics of monoclonal antibodies to wheat grain storage proteins. Biochim. Biophys. Acta. 874:245.

SKERRITT, J.H., MARTINUZZI, O., and WRIGLEY, C.W. 1987. Monoclonal antibodies in agricultural testing: quantitation of specific wheat gliadins affected by sulfur deficiency. Can. J. Plant Sci. 67:121.

SKERRITT, J.H., WRIGLEY, C.W., and HILL, A.S. 1988. Prospects for the use of monoclonal antibodies in the identification of cereal species, varieties and quality types. Seed Sci. Technol., in press.

SMITH, D.B., and LISTER, P.R. 1983. Gel-forming proteins in barley grain and their relationships with malting quality. J. Cereal Sci. 1:229.

TADA, M., SHINODA, I., and OKAI, H. 1984. L-ornithyltaurine, a new salty peptide. J. Agric. Food Chem. 32:994.

TRAMONTANO, A., JANDA, K.D., and LERNER, R.A. 1986. Catalytic antibodies. Science 234:1566.

TRIONE, E., BANOWETZ, G.M., KRYGIER, B.B., KATHREIN, J.M., and SAYAVEDRA-SOTO, C. 1987. A quantitative fluorescence enzyme-immunoassay for plant cytokinins. Anal. Biochem. 162:301.

VAAG, P., and MUNCK, L. 1987. Immunochemical methods in cereal research and technology. Cereal Chem. 64:59.

VAN EMON, J.M., SCIBER, J.N., and HAMMOCK, B.D. 1985. Applications of immunoassay to paraquat and other pesticides. ACS Symp. Ser. 276: 307.

WAGNER, N., KRUGER, E., and RUBACH, K. 1986. Nachweis der Mitverwendung von Reis und Mais bei der Bierherstellung mittels enzyme-linked immunosorbent assay (ELISA). Brauwissenschaft 39:104.

WALKER-SIMMONS, M. 1987. ABA levels and sensitivity in developing wheat embryos of sprouting - resistant and susceptible cultivars. Plant Physiol. 84:61.

WALKER-SIMMONS, M., and SESING, J. 1986. Development of a sensitive immunoassay for abscisic acid in wheat grain using a monoclonal antibody. Page 591 in: 4th International Symposium of Pre-Harvest sprouting in cereals. D. Mares, ed. Westview, Boulder, CO.

WANG, T.J., GRIGGS, P., and COOK, S. 1986. Immunoassays for plant growth regulators - a help or a hinderance? Page 26 in: Plant Growth

Substances 1985. M. Bopp, ed. Springer-Verlag, Berlin.
WARNER, R., RAM, B.P., HART, L.P., and PESTKA, J.J. 1986. Screening for zearalenone in corn by competitive direct enzyme-linked immunosorbent assay. J. Agric. Food Chem. 34:714.
WEILER, E.W. 1982. Plant hormone immunoassay. Physiol. Plant. 54:230.
WHITE, R.F., RYBICKI, E.P., VON WECHMAR, M.BV., DEKKER, J.L., and ANTOVIW, J.F. 1987. Detection of PR 1-type proteins in Amaranthaceae, Chenopodiaceae, Graminae and Solanaceae by immunoelectroblotting. J. Gen. Virol. 68:2043.

10

NOVEL APPROACHES TO GRAIN QUALITY AND PROCESS CONTROL

Ian L. Batey, Colin W. Wrigley and Peter W. Gras
CSIRO Division of Plant Industry,
Wheat Research Unit, P.O. Box 7,
North Ryde NSW 2113, Australia.

Gertrude Stein wrote, ".. a rose is a rose..." and to most people, wheat is wheat. The advertising industry has made some people aware that there is something called durum wheat from which better pasta may be made than from common wheat. Few of these people would know what the difference is! To most consumers, the concept of wheat quality is too remote for concern, and they proceed through life oblivious of the multitude of tests that ensure that the bread, the cakes, the noodles and other cereal food products meet their expectations of taste and texture.

The quality attributes of wheat are many. Traditional methods of measuring flour and wheat quality have used instruments such as the Farinograph, the Mixograph, the Extensigraph, the Amylograph and the Alveograph. These procedures require reasonably large quantities of material and are thus mainly applicable to samples taken from harvest deliveries or to cultivars in advanced stages of breeding programmes. Advances in electronics and instrumentation, such as microprocessor control of equipment, have enabled the scale of dough-quality testing to be reduced. New approaches to the prediction of grain quality have been developed permitting tests to be made on much smaller samples, even as little as half a kernel. Depending on the final product to which the wheat will be converted during processing, different parameters are determined in the prediction and measurement of quality. Behind these different aspects of product quality, there are four main thrusts which can be considered - the measurement of actual quality in the final product ("the proof of the pudding"), process control, the prediction of raw material quality by small scale testing, and maintaining grain quality during storage. The CSIRO Wheat Research Unit has been mainly involved in the latter two aspects through the development of instrumentation and techniques, and in studying the effects of controlled atmospheres on the quality of stored grains (Banks et al., 1988). This paper describes aspects of this work in the area of small scale testing and prediction of grain quality.

RAPID VISCO ANALYSER (RVA)

This instrument was originally developed to measure sprout damage in wheat at the receival point after harvest, but it has also proved wider in its versatility. The 1983-84 Australian wheat crop was severely affected by rain to the extent that almost a quarter of the record crop of 20.8 million tonnes was downgraded from milling grade to the general purpose category. In view of the extent of this damage, we were asked to develop a rapid test for weather damage, requiring a minimum of operator training. The test had to be suitable for receival conditions, using only the limited facilities available at the silo, and because of the large numbers of samples that would be tested, it had to be reasonably inexpensive.

After examining a number of options, we chose to measure the resistance to stirring of a gelatinized mixture of wheatmeal and water. In the version finally adopted, changes in viscosity are determined from the power consumption of an electric motor driving a paddle stirrer. Heating of the sample is achieved with a split metal block which was clamped tightly around an aluminum cup when the machine is in operation. When the RVA is used as a weather-damage tester, at the end of the three-minute cycle, the power demand is converted to a numerical value (between 0 and 255), and this "Stirring Number" is displayed. The instrument then resets itself ready to start the next cycle. After allowing time for one sample cup to be removed and the next to be inserted, at least 15 samples can be tested in a hour, permitting load-by-load testing under Australian receival conditions.

During the recent harvests, sets of prototype and production RVA instruments have been tested in field and laboratory trials. They were shown to be robust, easy to operate, and, most importantly, reliable. In comparisons with the Falling Number procedure over a range of samples and operators, the RVA gave favourable results with comparable coefficients of variation. In many field and laboratory trials on wholemeal samples, Stirring Number values gave correlations to Falling Number results of 0.95 or better (e.g., Ross et al., 1987).

In refining the accuracy of the RVA for weather damage testing, its potential for measuring the pasting viscosity of starch and or starch-based products was developed in parallel. Traditionally, these properties have been measured by the Brabender Visco/Amylograph, an instrument which requires a relatively large sample, and takes from 45 minutes to 2 hours for each analysis, depending on whether the peak viscosity only is required, or whether the set-back curve (obtained on cooling) also needs to be measured.

The basic RVA can be used without modification to provide rapid pasting curves that can distinguish between many starches (Walker et al. 1988). To achieve results equivalent to the Visco/Amylograph, the RVA requires a slightly different configuration to that used for weather damage. Modifications are made by the addition of cooling coils to the heater block, and by changing the heating procedure to permit variations in block temperatures. The program cycle is also changed so that, instead of terminating the cycle at the end of three minutes, the heating is continued for a preset time, after which the heat is turned off and the block is cooled to $50^{\circ}C$. Whereas the RVA (as a grain soundness analyser) displays a single result at the end of its cycle, in the modified form power consumption is continuously monitored and the output is displayed on a chart recorder or video display.

The heating and cooling rates, and the maximum and minimum temperature may be matched to those used by the Amylograph. When this is done, the profiles are almost identical to those obtained with the Amylograph (Figure 1). Sample size, however, is much smaller (3g) with the RVA. In addition, it is more portable, less expensive and generally easier to use.

A major advantage of the RVA when used to measure paste viscosities is the speed with which the

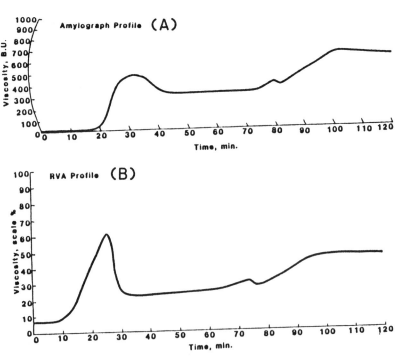

Figure 1. Pasting curves for a commercial hard red winter wheat all purpose flour. (a) Conventional Visco/Amylograph profile, heating from 50°C to 95°C at 1.5°C/min., 30 min. holding at 95°C, cool from 95°C to 50°C at 1.5°C/min., 30 min. hold at 50°C. (b) RVA profile following identical temperature-time sequence as in (a). (From Walker et al., 1988. Reproduced with permission of the authors and the publisher.)

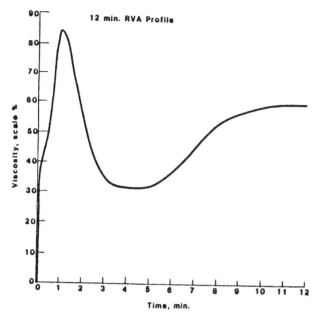

Figure 2. Pasting curve for a commercial hard red winter wheat all purpose flour. Rapid RVA profile, 5 min. at 95°C followed by 7 min. at 50°C. (From Walker et al., 1988. Reproduced with permission of the authors and the publisher.)

procedure may be carried out. Figure 2 shows a twelve minute pasting curve, including viscosity changes on cooling. A rapid cycle of heating and cooling, such as this, is probably more appropriate in mimicking processing methods than the longer heat-cool cycle used routinely to assess pasting properties. Currently we are collaborating with other laboratories to establish optimized procedures to assess a combination of sprout damage and starch properties.

MIXOGRAPH

The Mixograph has served a useful purpose in measuring dough properties in relatively small flour samples (10 or 35g) for several decades. Measurements of the time to peak dough-development enable prediction of optimum dough-mixing conditions. Baking performance may be predicted by such parameters as peak height, bandwidth at peak development and breakdown.

These values must be measured manually from the Mixograph curve, but the complex shape of the curve makes this labour-intensive and subject to operator "bias". While an experienced operator is usually quite consistent in technique from one curve to another, there can be substantial variations in estimates between different operators when measuring the same mixing curve. This means that Mixograph results and measurements are not readily transferable between laboratories, restricting its use to essentially "in-house" application. This is particularly true for problem flours which do not have a simple mixing curve. Automation of the measurement of mixing curves would enable the derived results to be obtained more easily and with greater consistency.

To this end, a standard 35g Mixograph has been modified by the addition of a load cell (or strain gauge), interfaced to an IBM AT compatible computer through a signal conditioner and an analog to digital converter. Data is presently recorded using Labtech Notebook software, although there are a number of possible software packages with which to perform this task.

The results may be displayed in a number of forms. A mixing curve, analogous to the standard mixogram curve, may be drawn on the screen or printed to paper. This trace usually has more noise than a conventional curve, as the damping effect of the mechanical arm is absent. Alternatively, the data may be smoothed to varying degrees by the use of walking averages or similar procedures, before display or printing (Figure 3). Algorithms have been devised to determine automatically the time to peak development based on the average resistance, the time to and magnitude of peak dough resistance, as well as estimates of the decrease in dough resistance after the peak. The measurements determined by the computer have a good correlation with those of individual operators measuring by hand. Similar procedures using

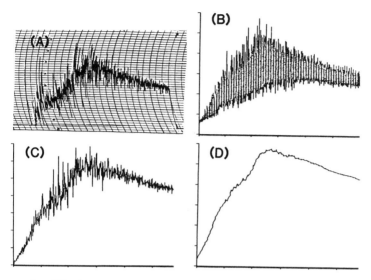

Figure 3. Mixing curves of a prime hard flour recorded on (a) a standard 35g Mixograph, (b) 35g Mixograph with electronic recording, (c) 1 second averages of data from (b), and (d) 25 second walking averages of the data from (b).

the upper and lower edges of the mixing trace have also been developed. Further research will be required to determine whether these measurements provide improved estimates of baking performance. Estimation of the change in width of the trace are also made much simpler by this technique.

Another approach to the automated measurement of Mixograph curves has been to measure the power consumption of the motor driving the mixer. The motor fitted to the 35g Mixograph is overpowered and, consequently, there is little change in the power consumption as the dough resistance changes. A much smaller motor has been fitted to a bowl and mixing head designed for use with 2g of flour. With this arrangement, the trace obtained resembles the standard trace obtained with a 35g instrument (Figure 4). Further development of this miniaturized instrument is under way. At this level of flour requirement, there is great potential for testing early-generation

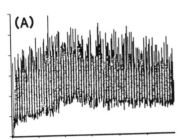 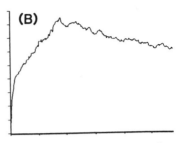

Figure 4. Mixing curve of a standard bakers' flour recorded on a 2g mixograph by measurement of the power demand; (a) complete mixogram and (b) 25 second walking averages of the same data.

breeders' samples where the amount of flour is limited. It will also be valuable in testing functional properties for small amounts of highly purified flour fractions (Gras et al., 1988).

At the other extreme, the same technique and algorithms may be applied to measuring the power consumption on a large scale mixer. Using a laboratory-scale Eberhardt mixer (2 Kg), traces were obtained and peak development time and peak height were determined. Experienced bakers independently assessed the time the dough "cleared". Time to peak power demand and time to dough clearing were closely correlated. Thus, the technique of electronic recording has the facility to predict optimum mixing times for larger scale mixers. Application to full-scale commercial mixers is an area with a potential to improve the quality of production in large bakeries and to permit adaptation to variations in flour specifications.

CULTIVAR IDENTIFICATION

Knowledge of variety is one of the first steps towards measuring the potential quality of a wheat sample. While poor agronomic practice, environmentally-caused damage by rain or drought, or bad handling after harvest may result in a particular sample of a variety not achieving the quality expected of it, under normal conditions knowledge of varietal composition and protein content are the most satisfactory indications of quality type.

Grain morphology is the easiest to apply but it is not able to discriminate between many varieties, especially if there has been no requirement for a registered variety to be visually distinguishable from cultivars of other quality classes. Image analysis, in which the external characteristics of individual grains may be measured and compared by computer, offers a possible solution to the problem. This technique is currently being investigated by several research institutions around the world. We have found the technique to be reliable in distinguishing wheat from foreign seeds (Ronalds et al., 1987), but distinguishing between wheat varieties is not so simple. A set of discriminators has been determined and this shows reasonable success in identifying individual grains from samples of five Australian wheat cultivars (Myers and Edsall, 1988; Wrigley et al., 1989). Further work is necessary before the usefulness of this technique in field situations can be assessed, but promising reports of its use have been given (Keefe and Draper, 1986; Neuman et al., 1987).

Antibody-based testing of grain proteins is another area which we are investigating (Skerritt et al., 1988). Our present studies are directed towards developing a laboratory test but the potential of antibody-based tests for speed and convenience may mean that such tests could eventually be applied at the point of receival. Antibodies which show selectivity to a range of barley varieties of differing quality have been produced (Wrigley et al., 1987). The use of a set of antibodies showing differing reactivities with grain proteins from different varieties could permit identification in a short time. Equipment already available for medical diagnostics could be used to automate the test with interpretation of the results being processed by computer.

Gel Electrophoresis

Both of the new techniques described above are in the early stages of development and are not yet suitable for routine use. The main approach available to grain handling authorities for practical variety identification in laboratories is still gel electrophoresis of the gliadin proteins (Wrigley et al., 1982). Pre-formed gradient polyacrylamide gels have been used for many years in Australia as the basis of routine identification, and the results have been consistent and reliable (du Cros and Wrigley, 1979; RACI, 1988). Some recently released wheat varieties, however, have electrophoretic patterns for their gliadin proteins similar enough to those of other varieties to make distinction between them very difficult. In these cases, it is necessary to use other techniques to permit the distinction to be made. Gel electrophoresis in the presence of sodium dodecylsulfate (SDS) and a reducing agent (2-mercaptoethanol) has been adapted to provide distinction among nearly all Australian wheat varieties. With this method, the high molecular weight glutenins are also visualized in the gels and these provide the basis for the distinction.

High-Performance Liquid Chromatography (HPLC)

HPLC has also been investigated as a means of routine identification of wheat cultivars. It has the advantage of being quick (less than 1 hour including sample preparation, or 30 minutes for chromatography) and the procedure may be carried out continuously using automatic sample injectors. Rapid ion-exchange chromatography of gliadin proteins on a Mono-Q column (Pharmacia, Uppsala, Sweden) was found to give distinctions between many Australian wheat varieties (Figure 5), including Egret and Condor (Batey, 1984). These two varieties are difficult to distinguish by gradient gel electrophoresis of their gliadin proteins, although SDS gel electrophoresis does permit the distinction to be made. The separation of these varieties is important because they differ in processing quality. Ion-exchange HPLC resolves a limited number of peaks, but there are some grain cultivars which are not distinguished by other forms of HPLC or by electrophoresis (Burbidge et al., 1986).

Reversed-phase HPLC (cf. Bietz, 1986) gives many more peaks (approximately 50, compared to 10-15 for anion-exchange chromatography), and thus better distinction between varieties is possible with this method. The speed of separation is similar for both these types of HPLC and most equipment may be used for

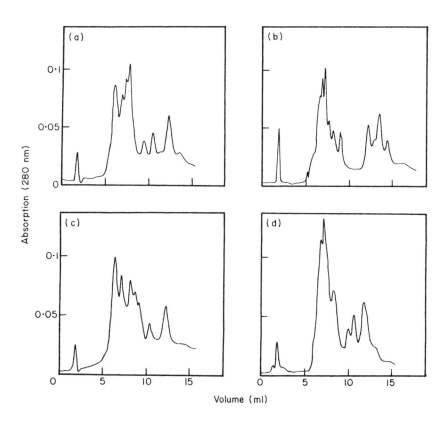

Figure 5. Mono-Q ion-exchange chromatography of the wheat varieties (a) Condor, (b) Durati, (c) Egret and (d) Eagle. The elution buffer contained in 1M urea, 0.01M 3-(cyclohexylamino)-propane-sulphonic acid (pH 10.4) with a gradient of 0-0.5M sodium acetate. (From Batey, 1984. Reproduced with permission.)

either type of column. The Pharmacia Mono-Q column requires a much lower pressure and may be utilized with some pumps which do not operate at the pressures required for reversed-phase HPLC columns.

Variety identification by HPLC has mainly been carried out on reversed-phase C_{18} columns (Figure 6) with a large pore size (30 nm), although some workers have used C_8 columns (Bietz et al., 1984; Bietz and Cobb, 1985; Marchylo et al., 1988). (C_8 and C_{18} refer to the length of the alkylated silanyl group which is bound to the silica support packed into the column.) The pore size is important as larger proteins may suffer a size exclusion effect on HPLC column packings of standard pore size (usually 5-6 nm). Our experience with columns containing packings with this

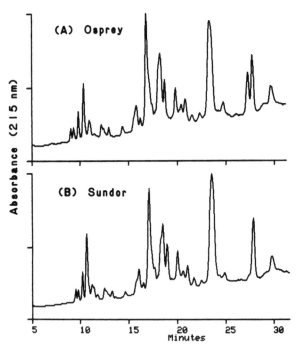

Figure 6. Reversed-phase HPLC traces of 70% ethanol extracts of wheat flours from the cultivars (a) Osprey, and (b) Sundor. Chromatography was on a Vydac C_{18} column at 50°C in 0.1% tri-fluoroacetic acid with a gradient of 30% to 55% acetonitrile over 25 minutes.

smaller pore size has been that the number of peaks in a gliadin extraction is reduced and the peaks are much broader. Distinction between varieties is thus made more difficult with no distinction between some varieties which show significant differences with other techniques.

HPLC also offers the advantage that the system may be interfaced to a computer for control of the pumps for solvent delivery and acquisition of data from the detector. This data can be subsequently reprocessed by the computer; this step can include the automatic identification of a sample by comparison with a library of known varieties. A number of algorithms has been devised and tested for this purpose (Batey, 1986). The one which has been found to be fastest involves the placing of a sample into one of two groups of possibilities on the basis of the presence or absence of a particular feature. This feature need not be a single peak; in fact, it is more commonly two or more peaks with a particular difference in retention times. When a sample is placed into one group, it is scanned for the presence or absence of another feature which enables it to be placed into an even smaller group, and so on until there is only one possibility left. At this point, the chromatogram of the unknown sample is compared to the stored chromatogram of the suspected variety and identification is made if these match. Occasionally, it is not possible for the computer to distinguish reliably between two varieties without "operator assistance". For example, a broad peak is not recognized by the program as being different from a sharp, narrow one of the same height. In this case, the operator can make the distinction if this difference is always observed. It may be possible for this sort of distinction to be made by the computer by measuring such parameters as width at half height or area, but this extension of the work does not seem warranted routinely.

There are some pairs of varieties, and in the case of oats, a group of five varieties, which are closely related and are not distinguishable by any method of HPLC. If they are of the same quality class, this failure to make a positive identification is of no consequence to the end user of the wheat. For certification of pure seed, or if the varieties are in different quality grades, some other method of identification must be employed. If standard laboratory procedures are not successful, there are other possibilities such as the resistance of the variety to a particular strain of rust or electrophoresis of restriction-fragment length polymorphisms from the DNA of leaf (May and Appels, 1987). These methods are time-consuming and would only be used as a last resort when positive identification is needed.

CONCLUSION

The cereal industry has been using tried and tested methods for many years. These methods have been successful in improving quality in grain and in processed products, and the consumer has benefited as a result. The growth of technology that has influenced all areas of scientific endeavour has not passed by the cereal industry. Away from the publicity surrounding many of the glamour areas of scientific research, new techniques are being investigated in order to provide more accurate, more rapid or more convenient methods of measuring and ensuring grain quality. Advances in biotechnology, electronics and computer software and hardware are being utilized in various aspects of cereal quality. As new techniques are developed and proven, their application is one way that the profitability of all sections of the cereal industry, from grower to retailer, may be maintained.

LITERATURE CITED

BANKS, H.J., GRAS, P.W., BASON, M.L., and ARRIOLA, L.P. 1988. A quantitative study of the influence of temperature, water activity and storage atmosphere on the yellowing of milled rice. J. Cereal Sci. 6: in press.

BATEY, I.L. 1984. Wheat varietal identification by

rapid ion-exchange chromatography of gliadins. J. Cereal Sci. 2: 241.

BATEY, I.L. 1986. Identification of cereal varieties by computer analysis: a strategy. Proc. 36th Aust. Cereal Chem. Conf., 187.

BIETZ, J.A. 1986. High performance liquid chromatography of cereal proteins. Adv. Cereal Sci. Technol. 8: 105.

BIETZ, J.A., BURNOUF, T., COBB, L.A., and WALL, J.S. 1984. Wheat varietal identification and genetic analysis by reversed-phase high-performance liquid chromatography. Cereal Chem. 61: 129.

BIETZ, J.A., and COBB, L.A. 1985. Improved procedures for rapid wheat varietal identification by reversed-phase high performance liquid chromatography of gliadin. Cereal Chem. 62: 332.

BURBIDGE, M.J., BATEY, I.L., CAMPBELL, W.P., SKERRITT, J.H., and WRIGLEY, C.W. 1986. Distinction between barley varieties by grain characteristics, electrophoresis, chromatography and antibody reactions. Seed Sci. Technol. 14: 619.

DU CROS, D.L., and WRIGLEY, C.W. 1979. Improved electrophoretic methods for identifying cereal varieties. J. Sci. Food Agric. 30:785.

GRAS, P.W., HIBBERD, G.E., and RASPER, V.F. 1988. Modernising the Mixograph. Proc. 38th Aust. Cereal Chem. Conf. (in press)

KEEFE, P.D., and DRAPER, S.R. 1986. The measurement of new characters for cultivar identification in wheat using machine vision. Seed Sci. Technol. 14: 715.

MARCHYLO, B.A., HATCHER, D.W., and KRUGER, J.E. 1988. Identification of wheat cultivars by reversed-phase high-performance liquid chromatography. Cereal Chem. 65: 28.

MAY, C.E., and APPELS, R. 1987. Variability and genetics of spacer DNA sequences between the ribosomal-RNA genes of hexaploid wheat (Triticum aestivum). Theor. Appl. Genet. 74: 617.

MYERS, D.G., and EDSALL, K.J. 1988. The application of image processing techniques to the identification of Australian wheat varieties. Report publ. by Curtin University, Western Australia.

NEUMAN, M., SAPIRSTEIN, H.D., SHWEDYK, E., and BUSHUK, W. 1987. Discrimination of wheat class and variety by digital analysis of whole grain samples. J. Cereal Sci. 6: 125.

RACI. 1988. Official Testing Method E2. Electrophoretic identification of cereal varieties. Official Testing Methods, Cereal Chemistry Division, Royal Australian Chemical Institute, Melbourne.

RONALDS, J.A., MADDOCKS, I.G., and JACKSON, N. 1987. Application of image analysis of quality control in the cereal industry. Proc. 36th Aust. Cereal Chem. Conf. RACI, Melbourne, Pp 157-162.

ROSS, A.S., WALKER, C.E., BOOTH, R.I., ORTH, R.A., and WRIGLEY, C.W. 1987. The Rapid Visco-Analyzer: A new technique for the estimation of sprout damage. Cereal Foods World 32: 827.

SKERRITT, J.H., WRIGLEY, C.W., and HILL, A.S. 1988. Prospects for the use of monoclonal antibodies in the identification of cereal species, varieties and quality types. Seed Sci. Technol. (in press).

WALKER, C.E., ROSS, A.S., WRIGLEY, C.W., and MCMASTER, G.J. 1988. Accelerated starch-paste characterization with the Rapid Visco-Analyzer. Cereal Foods World 33:491.

WRIGLEY, C.W., AUTRAN, J.C., and BUSHUK, W. 1982. Identification of cereal varieties by gel electrophoresis of the grain proteins. Adv. Cereal Sci. Technol. 5: 211.

WRIGLEY, C.W., BATEY, I.L., CAMPBELL, W.P., and SKERRITT, J.H. 1987. Complementing traditional methods of identifying cereal varieties with novel procedures. Seed Sci. Technol. 15: 679.

WRIGLEY, C.W., TOMLINSON, J.D., SKERRITT, J.H., BATEY, I.L., and SING, W. 1989. Efficient identification of wheat varieties by established and novel procedures. Cereal Foods World (in press)

11

CHARACTERIZATION OF WHEAT STARCH AND GLUTEN
AS RELATED TO END-USE PROPERTIES

Ann-Charlotte Eliasson

Department of Food Technology, University of Lund
Box 124, 2-221 00 Lund, Sweden

INTRODUCTION

Wheat starch and gluten constitute a considerable part of our daily food intake; wheat flour is used as an ingredient in so many food products. In many of these products it is difficult, or even impossible, to use anything else than wheat flour, a truth which persons suffering from coeliac disease have experienced. Wheat starch and gluten are also used as additives, both in food and in a wide range of non-food products. Wheat starch, as well as gluten, are used because they give desirable properties to the product, properties that might be described in terms of texture, viscosity, water holding capacity or foam stability. We thus make use of the physico-chemical properties or in other words, the functional properties of wheat starch and gluten. Which functional properties are then unique for wheat starch and gluten? To describe all properties that might be of relevance is not possible in a short paper, and the discussion will therefore be limited to the following topics:
- granule size distribution of wheat starch,
- interactions between wheat starch and other components (polar lipids and gluten),
- surface properties of gluten proteins.

WHEAT STARCH

When starch is present in a product it is usually because of its gel-forming ability and its water-holding properties. We make use of the changes occurring during gelatinization of starch. The quality of the product will thus depend on the gelatinization process, and on the rheological properties which are obtained during this process (Fig. 1). A starch gel is not a system in equilibrium, and changes occurring in the gel with time, retrogradation, might also be of importance for product quality.

The above discussion is valid not only for wheat starch, but for all starches. As seen in Fig. 1 the gelatinization process is affected by starch properties such as starch granule composition, crystallinity and size distribution. These properties depend on the botanical origin of the starch (e.g. cereal starches differ from tuber starches) but they might depend also on the wheat variety. It is also possible to change the performance of wheat starch (and of all other starches) by adding certain substances to the starch during gelatinization.

Wheat Starch Granule Size Distribution

The granule size distribution of wheat, rye and barley starches is unique in that it is bimodal with large, lenticular A-granules, and small, spherical B-granules (Fig. 2). The variation in starch granule size distribution among wheat varieties has been investigated (Soulaka and Morrison 1985a). It is possible to fractionate wheat starch into different size classes as shown in Fig. 2.

The size class "small granules" (B-granules) is comprised mainly from granules with a diameter below 10 µm, whereas the size class "large granules" (A-granules) is comprised mainly from granules with a diameter above 10 µm. An obvious question is then if the functional properties differ between these size classes. Investigations using differential scanning calorimetry (DSC) show that the gelatinization properties as well as the retrogradation behavior are

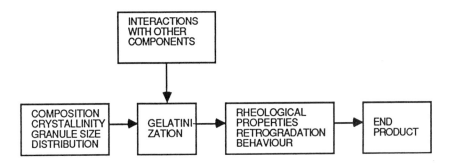

Fig. 1. Relation between some starch properties and quality of the end product.

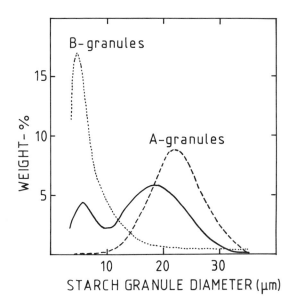

Fig. 2. Size distribution of starch granule fractions. The unbroken line shows the size distribution of unfractionated starch. (from Eliasson and Karlsson 1983, with permission)

related to the granule size (Table 1). Gelatinization starts at a somewhat lower temperature in case of B-granules and at a somewhat higher temperature in case of A-granules compared to the control. The gelatinization enthalpy (ΔH_G), however, is similar for

TABLE I
Gelatinization and Retrogradation Parameters of Large and Small Wheat Starch Granules Measured by Differential Scanning Calorimetry[a,b]

Starch	Gelatinization[c]				Amylose-lipid complex[d]		Retrogradation[e] after storage (23°C)					
							1 day		3 days		7 days	
	T_0 (°C)	T_G (°C)	ΔT (°C)	ΔH_G (J/g)	T_{CX} (°C)	ΔH_{CX} (J/g)	T_C (°C)	ΔH_C (J/g)	T_C (°C)	ΔH_C (J/g)	T_C (°C)	ΔH_C (J/g)
Unfractionated starch	53.7	59.7	37	12.3	117.4	2.5	59.2	4.1	59.5	6.2	62.7	7.0
Large granules	55.7	60.2	36	10.7	116.7	1.1	59.0	3.8	60.4	6.4	64.9	7.5
Small granules	51.8	61.6	34	11.4	115.8	3.2	59.2	2.5	58.6	4.6	60.8	6.3

[a] Starch-to-water ratio 1:1, heating rate 10°C/min.
[b] Eliasson and Karlsson 1983 with permission, Eliasson 1987.
[c] T_0 = start of gelatinization, T_G = temperature at peak maximum, ΔT = width of endotherm, ΔH_G = enthalpy of gelatinization.
[d] T_{CX} = temperature at peak maximum, ΔH_{CX} = enthalpy of transition
[e] T_C = temperature at peak maximum, ΔH_C = enthalpy of melting retrograded starch,

all the three classes in Table I. The most evident difference between A-granules and B-granules is the much higher enthalpy due to the amylose-lipid complex (ΔH_{CX} in Table I) of B-granules. This higher ΔH_{CX}-value seems to be related to the higher lipid content of the B-starch fraction (Soulaka and Morrison 1985a). There is also a tendency to a slower retrogradation of B-granules (Table I). The rheological properties of a starch gel are very sensitive to the starch granule size distribution; an increased part of small granules causes the gel to become stiffer (Eliasson and Karlsson 1983). The loaf volume in bread baking is related to the proportion of B-granules in the flour (Soulaka and Morrison 1985b). The highest specific loaf volume was obtained with 25-35% B-granules.

Interactions Between Wheat Starch and Other Components

Wheat starch contains polar lipids. This might be observed as the presence of a DSC-endotherm due to a transition of the amylose-lipid complex (Table I, Kugimiya et al. 1980). However, it is difficult to elucidate if and how wheat starch properties are affected by the presence of these lipids as it has not yet been possible to obtain a lipid-free wheat starch. Complete extraction of wheat starch lipids has to be performed under conditions which cause at least partial gelatinization of the starch (Morrison and Coventry 1985). Instead polar lipids are added to wheat starch, and the effect of the added lipids on functional properties are studied.

When polar lipids are added to wheat starch the gelatinization parameters determined by DSC change as shown in Table II. The gelatinization temperature (T_G) is decreased or increased depending on the lipid added, sodium dodecylsulphate (SDS) decreases T_G, whereas sodium stearoyllactylate (SSL) increases T_G. SDS greatly affects also the onset of gelatinization (T_0 in Table II). The gelatinization enthalpy (ΔH_G) decreased, especially at the highest level of addition (5% (w/w) in Table II). Such a decrease might be explained by an exothermic complex formation occurring during gelatinization. The extent of complex

TABLE II
Gelatinization Parameters of Wheat Starch in the Presence of Lipid Additives as Measured by Differential Scanning Calorimetry[a]

Lipid Additive	Gelatinization				Amylose-Lipid-Complex	
	T_0 (°C)	T_G (°C)	ΔT (°C)	ΔH_G (J/g)	T_{CX} (°C)	ΔH_{CX} (J/g)
-	59.5	60.2	36	12.7	114.9	1.50
1% SDS[b]	56.6	61.2	36	11.6	113.1	2.26
5% SDS[b]	52.5	59.2	36	7.8	111.9	4.44
1% CTAB[c]	57.2	62.1	35	11.8	113.5	2.93
5% CTAB[c]	56.8	61.8	34	8.2	112.6	4.69
5% SSL[d]	59.0	63.5	34	8.7	115.0	5.20
5% lecithin	56.0	59.9	32	11.5	104.1	1.05

[a]Starch-to-water ratio was 1:1.
SDS-conditions as described in Eliasson 1986.
[b]SDS = sodium dodecylsulphate
[c]CTAB = cetyltrimethylammonium bromide
[d]SSL = sodium stearoyllactylate

formation is indicated by the increased ΔH_{CX}-values in Table II. Direct comparisons between the amount of complex formed by different additives is not possible as the ΔH_{CX}-value depends not only on the amount of complex formed but also on the type of lipid in the complex.

The swelling pattern of the starch granule and the release of amylose during gelatinization are both affected by the presence of polar lipids (Eliasson 1985). As a consequence, the rheological properties of the wheat starch gel are changed (Osman and Dix 1960, Krog 1973). The presence of polar lipids affects the retrogradation of starch, a well-known phenomenon which takes place in baking (Krog and Nybo Jensen 1970).

Not only polar lipids but also gluten affects the gelatinization of starch, and if gluten is added to wheat starch T_G increases (Table III). From this follows that the gelatinization behavior of starch is somewhat different in a wheat-flour-water mixture compared to a wheat starch-water mixture as illustrated by the two DSC-thermograms in Fig. 3.

TABLE III
Gelatinization Temperature (T_G) of Wheat Starch in the Presence of Gluten[a]

Gluten:Starch Ratio	T_G (°C)
0.00	60.4
0.10	61.4
0.20	61.9
0.40	63.5

[a] 0.90 g water/g starch, heating rate = 10°C/min (Eliasson 1983 with permission).

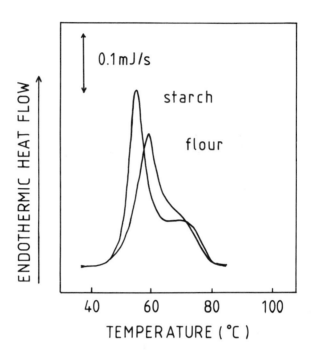

Fig. 3. DSC-thermograms of a wheat flour-water mixture and the corresponding wheat starch-water mixture. Starch:water ratio was 1:1, heating rate was 10°C/min (Eliasson unpublished results).

Although the properties of a wheat starch gel might depend on the starch/gluten ratio the most important properties of gluten are other than its effects on starch gelatinization, as will be discussed in the following paragraphs.

GLUTENS

Some functional properties of gluten that are of importance for its end use are illustrated in Fig. 4. Solubility, rheological properties and surface properties are related to gluten composition and to the molecular weight of gluten proteins. "Composition" in Fig. 4 means both the proximate composition of gluten in terms of protein, fat and

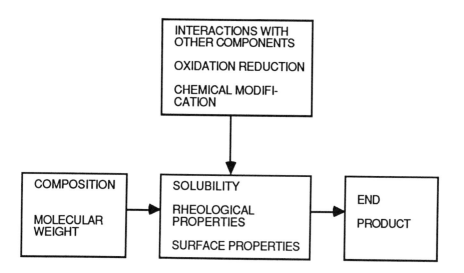

Fig. 4. Relations between some gluten properties and quality of the end product.

carbohydrate, and the composition in terms of different proteins. "Composition" might also mean the amino acid composition of an individual protein molecule.

The high molecular weight of gluten proteins (Fig. 5) complicates the investigation of these proteins in the unreduced state. As a consequence most of our knowledge concerning molecular weight and protein composition is restricted to the subunits. The molecular weight distribution of gluten shown in Fig. 5 is for the unreduced gluten. However, gluten is not completely soluble in the buffer used, so the chromatogram still represents only a part of total gluten. The high molecular weight is one reason for the low solubility, another reason is the characteristic amino acid composition (low level of charged amino acids).

The viscoelastic properties of gluten are well recognized as being unique; they contribute to the unique baking properties of a wheat flour. Gluten might be characterized as a viscoelastic liquid (Mita and Matsumoto 1984), and fundamental rheological

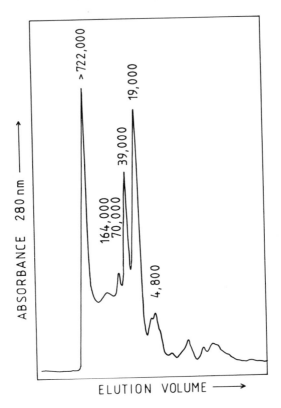

Fig. 5. High performance liquid chromatography of wheat gluten in phosphate buffer, pH 7, containing 1% SDS (Eliasson and Lundh unpublished results).

measurements indicate that the rheological behavior is correlated to baking performance of the wheat flour (Abdelrahman and Spies 1986, Lindahl and Eliasson 1987).

Surface Properties of Gluten Proteins

Gluten proteins show surface activity. This means that gluten proteins are able to stabilize air/water interfaces in foams, for example in a wheat flour dough. The surface properties of wheat flours, as well as of wheat protein fractions, have been studied by the surface balance technique (Lundh et al.

1988). In this method the protein powder (or flour) is spread at a water surface and the surface pressure (Π) is calculated from forces acting on a plate immersed in the water (the Wilhelmy plate method, Bull 1947). The surface pressure is defined as

$$\Pi = \gamma_0 - \gamma$$

where γ_0 is the surface tension of pure water, and γ is the surface tension in presence of a surface active substance. The changes in surface pressure during compression of the protein film are measured and represented as Π-A isotherms, where A is the area per molecule or per mg of protein (Fig. 6).

The surface properties of a wheat protein fraction might be represented in several ways according to the following description.
- The amount of sample that will give a preselected starting pressure (for example 8 mN/m) expressed as a surface concentration (mg/m²).
- The time from adding the sample onto the surface balance until the preselected starting pressure is reached.
- The appearance of the Π-A isotherm during compression of the protein film (Fig. 6).

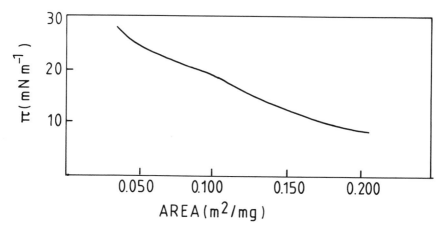

Fig. 6. The Π-A isotherm obtained when a wheat gluten protein film is compressed at the air/water interface (Eliasson and Lundh, unpublished results).

- The surface-concentration at the final surface pressure.
- The protein film is compressed and expanded several times in order to study if cross-linking occurs.

Increases in Π-values at a fixed area might indicate cross-linking, whereas a decrease indicates that losses of material into the bulk have occurred. After compression the protein film might be picked up from the surface balance and studied by SDS polyacrylamide gel electrophoresis (SDS-PAGE) or high-performance liquid chromatography (HPLC) in order to detect changes in molecular weight. The surface concentration and spreading tie of some protein fractions are shown in Table IV. Low molecular weight (LMW) proteins spread quickly and a small amount was required to give the pre-selected Π-value (Eliasson and Lundh 1988). However, when a wheat protein fraction composed of HMW-material was compressed several times the Π-value increased for each subsequent compression, and SDS-PAGE showed that cross-linking had occurred in the film. When a protein film composed of LMW proteins (gliadin

TABLE IV
Surface Behaviour of some Wheat Protein Fractions differing in Molecular Weight Distribution[a]

Wheat protein fraction[b]	Amount of protein at $\Pi = 8$ mN/m (mg/m^2)	Equilibrium time (min)	Amount of protein at $\Pi = 28$ mN/m (mg/m^2)
LMW	4.6	30	26.7
MMW	33.3	110	113.6
HMW	37.8	180	211.9

[a]Eliasson and Lundh 1988
[b]LMW = low molecular weight; MMW = medium molecular weight; HMW = high molecular weight

type) was treated in the same way no such cross-linking was observed, instead the Π-A isotherm indicated a loss of material into the bulk (Lundh et al. 1988).

The surface properties of the gluten proteins are unique in several aspects. Firstly, the Π-values obtained for the gluten films are much higher than is usually observed for protein films. Gluten proteins decrease the surface tension of water to almost the same extent as polar lipids do (Larsson 1986). Secondly, the plateau region observed in the Π-A isotherm of gluten proteins and wheat flours is not observed for other proteins. Usually a smooth, sigmoid curve is observed (Gaines 1966). Such a behavior has also been reported for A-gliadin (Carlson 1981). Plateau regions observed when monolayers of polar lipids are compressed correspond to phase transitions in three dimensions (Gaines 1966, Larsson 1986). It is not yet known what the plateau region in the Π-A isotherm of wheat proteins represents. Thirdly, cross-linking of protein films has not previously been shown, although the possibility has been discussed (MacRitchie 1978).

The wheat flour dough was described as a foam above. Surface active substances are required both for foam formation and for foam stabilization. The best foam formation is achieved with proteins that quickly lower interfacial tension, one example of such proteins is LMW gluten proteins. The best foam stabilization, on the other hand, is obtained from proteins of high molecular weight which give thick and elastic films at the interface. One example of such proteins are HMW gluten proteins. These two requirements are usually not met by the same protein. Gluten protein fractions might thus be very suitable for foam formation and stabilization, as in one fraction both types of proteins might be present. The surface properties of gluten protein fractions depend on the composition (Table IV), and it should thus be possible to obtain protein fractions that are more suitable for a certain application. As indicated in Fig. 4 there are also possibilities to change the properties of gluten by chemical modification.

CONCLUSIONS

In summary, both wheat starch and wheat proteins have properties that differentiate them from other starches and proteins. There also exist many possibilities to modify wheat starch as well as gluten in order to obtain more suitable properties in relation to a certain end use. With increasing knowledge about the functional properties of these components our possibilities to use them and to improve their performance will increase.

LITERATURE CITED

ABDELRAHMAN, A., and SPIES, R. 1986. Dynamic rheological studies of dough systems. Page 87 in: Fundamentals of Dough Rheology. H. Faridi and J.M. Faubion, eds. AACC, St. Paul, MN.
BULL, H. B. 1947. Spread monolayers of protein. Adv. Protein Chem. 3:95.
CARLSON, T. L.-G. 1981. Law and order in wheat flour dough. Thesis, Lund University, Sweden.
ELIASSON, A. -C. 1983. Differential scanning calorimetry studies on wheat starch-gluten mixtures. I. Effect of gluten on the gelatinization of wheat starch. J. Cereal Sci. 1:199.
ELIASSON, A.-C. 1985. Starch gelatinization in the presence of emulsifiers. A morphological study of wheat starch. Starch/Stärke 37:411.
ELIASSON, A. -C. 1986. On the effects of surface active agents on the gelatinization of starch - a calorimetric investigation. Carbohydr. Polym. 6:463.
ELIASSON, A.-C. 1987. Starch retrogradation in bread staling. 23. Nordiske Cerealkongres, Carlsberg Forskningscenter, Copenhagen, Denmark, 17-20 August.
ELIASSON, A. -C. and KARLSSON, R. 1983. Gelatinization properties of different size classes of wheat starch granules measured with differential scanning calorimetry. Starch/Stärke 35:130.

ELIASSON, A.-C. and LUNDH, G. 1988. Rheological and interfacial behavior of some wheat protein fractions. Paper presented at Cereals' 88, Lausanne, Switzerland, May 30-June 3.

GAINES, G. L. 1966. Insoluble Monolayers at Liquid-gas Interfaces. Interscience, New York.

KROG, N. 1973. Influence of food emulsifiers on pasting temperature and viscosity of various starches. Starch/Stärke 25:22.

KROG, N., and NYBO JENSEN, B. 1970. Interaction of monoglycerides in different physical states with amylose and their anti-firming effects in bread. J. Food Technol. 5:77.

KUGIMIYA, M., DONOVAN, J. W., and WONG, R. Y. 1980. Phase transitions of amylose-lipid complexes in starches: a calorimetric study. Starch/Stärke 32:265.

LARSSON, K. 1986. Physical properties - structural and physical characteristics. Page 321 in: Lipid Handbook. F.D. Gunstone, J.L. Harwood, and F.B. Padley, eds. Chapman & Hall, London.

LINDAHL, L., and ELIASSON, A.-C. 1987. Viscoelastic properties of starch gels with added gluten and of dough systems. Page 507 in: Gluten Proteins. R. Lasztity, and F. Bekes, eds. World Scientific, Singapore.

LUNDH, G., ELIASSON, A.-C., and LARSSON, K. 1988. Cross-linking of wheat storage protein monolayers by compression/expansion cycles at the air/water interface. J. Cereal Sci. 7:1.

MACRITCHIE, F. 1978. Proteins at interfaces. Adv. Protein Chem. 32:283.

MITA, T., and MATSUMOTO, H. 1984. Dynamic viscoelastic properties of concentrated dispersions of gluten and gluten methyl ester: contributions of glutamine side chain. Cereal Chem. 61:169.

MORRISON, W. R., and COVENTRY, A. M. 1985. Extraction of lipids from cereal starches with hot aqueous alcohols. Starch/Stärke 37:83.

OSMAN, E. M., and DIX, M. R. 1960. Effects of fats and nonionic surface-active agents on starch pastes. Cereal Chem. 37:464.

SOULAKA, A. B., and MORRISON, W. R. 1985a. The amylose and lipid contents, dimensions, and gelatinization characteristics of some wheat starches and their A- and B-granule fractions. J. Sci. Food Agric. 36:709.

SOULAKA, A. B., and MORRISON, W. R. 1985b. The bread baking quality of six wheat starches differing in composition and physical properties. J. Sci. Food Agric. 36:719.

12

UNIQUENESS OF WHEAT STARCH

William R. Morrison
Food Science Division,
Department of Bioscience and Biotechnology,
University of Strathclyde,
Glasgow G1 1SD, Scotland, UK

There are several aspects of wheat starch that are distinctive, but it would be misleading to claim that it is an entirely unique starch because in many ways it closely resembles the normal starches of the other members of the Triticeae - barley, rye and triticale. However, the Triticeae starches as a group are quite different from other cereal starches of similar amylose content, and they are not at all like the major food starches obtained from legumes, roots and tubers.

STARCH GRANULE DEVELOPMENT

The endosperm starches of the mature Triticeae have characteristic bimodal granule size populations. The larger-sized granules are called A-granules and the smaller ones B-granules. The morphological evolution of the developing wheat endosperm and its starch granules has been reviewed in detail elsewhere (Evers, 1971, 1974; Simmonds and O'Brien, 1981) and only the salient features are discussed here.

From stereological analysis of the developing wheat endosperm it has been shown that, during the early stages when there is active cell division, amyloplast numbers per cell (and per endosperm) increase, but from the stage when the aleurone cells have differentiated amyloplast numbers remain essentially constant (Briarty et al., 1979). Each amyloplast produces one A-granule which increases in size throughout the grain-filling period. In addition, from approximately 14-20 days after anthesis (DAA) each amyloplast begins to produce small B-granules in evaginations of the amyloplast envelope (Buttrose, 1963; Parker, 1985). The B-granules increase in number until grain maturity, but they are always much smaller in size than the A-granule. Small A-granules from the subaleurone cells (which are formed last and consequently are least developed) may be similar in size to the oldest B-granules in the central endosperm cells, but in practice this overlap is not evident in size analyses of A- and B-granule populations. The data in Table I, from a study of changes in the composition of starch in developing wheats (Morrison and Gadan, 1987), are entirely consistent with these observations.

In wheat (Evers, 1971, 1974) and in the other *Triticeae* (Czaja, 1982) the A-granules develop asymmetrically, and this is relevant to some of their properties discussed below.

TABLE I

Starch Content and Granule Dimensions in Developing Wheat Endosperm (var. Flanders).

Property	10	20	DAA 30	40	50
Starch, mg/endo.	0.2	8.1	21.5	29.9	30.4
A-granules, 10^6/endo.	11	14	14	15	16
B-granules, 10^6/endo.	0	45	161	220	224
Mean vol. A, μm^3	20	692	1604	1928	1920
Mean vol. B, μm^3	0	22	36	43	45
B-granules, % (no.)	0	69	91	93	93
B-granules, % (wt.)	0	6	19	22	24

From Morrison and Gadan (1987)

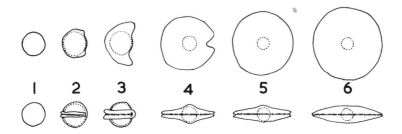

Figure 1. Six stages in the development of an A-type wheat starch granule (dotted nucleus is constant size, hence scale reduces from 1 to 6) Source: Evers (1971).

The granule is built from an initial spherical nucleus by forming a disc in the equatorial plane (Figure 1), and then enlarging the granule mostly above and below the plane until it is filled out into the familiar lenticular or flattened spheroidal shapes seen in scanning electron micrographs (Evers, 1971; Morrison and Gadan, 1987). The edge of the equatorial plane is marked by a deep groove in immature granules that is still visible at maturity. By contrast, the B-granules appear to develop symmetrically, but, like the A-granule, they can acquire flattened surfaces from pressing against adjacent granules during the later stages of grain-filling.

COMPOSITION OF STARCH GRANULES

Wheat starch granules are comprised of two major polysaccharides, amylose (AM) and amylopectin (AP), and small amounts of lipids and proteins. AM is often described as a linear α-(1,4)-glucan, but a recent report (Takeda et al., 1987) shows that wheat AM has a number average degree of polymerization (DP_n) of 1290 and a mean chain length of 270, giving an average of 4.8 α-(1,6)-branch points per molecule. When the starch granules are heated in water, lipid-free AM is leached progressively as the temperature is raised (R.F. Tester and W.R. Morrison, unpublished), and, by analogy with other starches

(Banks and Greenwood, 1975), this AM probably has a lower DP and less branching than the residual AM in the swollen granules.

Wheat AP closely resembles AP in waxy and normal types of barley and maize starches, and its structure may be described by one of the models of French (1973), Robin et al. (1974), Nikuni (1978), Kainuma (1980), or Manners and Matheson (1981). Gel permeation chromatography of α-(1,4)-glucans produced by enzymic debranching indicates two components, DP_n = c.15-20 and c. 45-60 (Lii and Lineback, 1977; Atwell et al., 1980; Sargeant, 1982; Morrison et al., 1984), but recent analyses show that there is a third component in barley (MacGregor and Morgan, 1984), and a fourth higher-DP component has been claimed in wheat (Kobayashi et al., 1986).

Carefully prepared wheat starch has small quantities of proteins that are quite different from the major prolamins (storage proteins) and the other albumins and globulins normally described in wheat and flour. The granule surface proteins are readily desorbed with salt solutions, sodium dodecylsulfate (SDS), and sodium laurate at ambient temperature, whereas the integral proteins, which occur near the surface and throughout the granule, require extraction with SDS at 50°C or at higher temperatures (Lowy et al., 1981; Gough et al., 1985; Greenwell and Schofield, 1986; Schofield and Greenwell, 1987; N. Matsoukas and W.R. Morrison, in preparation; J.B. South and W.R. Morrison, in preparation).

On SDS-polyacrylamide gradient gel electrophoresis (SDS-PAGE, 12-20% PAGE) the surface proteins show two major bands with apparent molecular weights of c. 15kD and c. 30kD. The 15kD band, which is weak in starches from hard bread and durum wheats and strong in starches from soft wheats, has been named friabilin because its presence is associated with soft friable endosperm texture. The integral proteins all have higher apparent molecular weights with a single major band of c. 59kD, and minor bands up to c. 150kD (Greenwell and Schofield, 1986; P. Greenwell, personal communication; N. Matsoukas and W.R. Morrison, in preparation). The 59kD protein has been identified as granule-bound starch synthase

(Schofield and Greenwell, 1987), the enzyme that synthesizes the long α-(1,4)-glucan chains of AM.

Wheat starch has 700-1400 mg% lipids inside the granules. These lipids are almost exclusively lysophospholipids (LPL) comprised of c. 70%, 20% and 10% respectively lysophosphatidylcholine, -ethanolamine and -glycerol (Morrison, 1978a, 1988a, 1988b). LPL account for nearly all the phosphorus in carefully isolated starch (LPL = P x 16.3), but without adequate precautions during starch isolation (by washing) it may contain appreciable amounts of inorganic phosphate also. The starch lipids contain c. 30-40% palmitic, 1-2% stearic, 6-11% oleic, 44-56% linoleic and 1-4% linolenic acids, and the cis-unsaturated acids are exceptionally well protected against oxidation by the dense surrounding structure of the granule (Morrison, 1978b, 1988a, 1988b).

There are also variable quantities of free fatty acids (FFA) and other monoacyl lipids which are artifacts of starch isolation procedures, and these have been termed starch surface lipids (Morrison, 1981). FFA content is correlated with surface area in granule size fractions (Morrison and Gadan, 1987) and may occur as amylose-inclusion complexes (Morrison, 1988a). There is no reason to suppose that they are part of a natural lipoprotein complex with the surface proteins. Clean starches have < 40 mg% FFA, while impure starches commonly have 50-200 mg% FFA (Morrison, 1988a; Morrison and Gadan, 1987; Soulaka and Morrison, 1985a).

Treatment of starch with pronase or toluene to remove surface proteins also removes some lipid phosphorus, as shown by electron spectroscopy (Russell et al., 1987). Chemical analysis of starches washed with SDS or sodium laurate shows that these treatments remove surface proteins (but no integral proteins), together with the surface FFA and a little LPL (J.B. South and W.R. Morrison, unpublished). This suggests that a small proportion of the LPL may be surface lipids, or that the comparatively small LPL molecules near the surface of the starch granules can be solubilized in this way.

DISTRIBUTION OF COMPONENTS IN STARCH GRANULES

Some information on the quantitative distribution of components within starch granules can be deduced from changes in the composition of developing A-granules (Morrison and Gadan, 1987) by relating these to morphological development, discussed above. In the starches of the major diploid cereals exhibiting mutations affecting starch, there is a linear correlation between AM content and internal lipid content (LPL in barley, FFA and LPL in rice and maize), and it has been suggested that lipids regulate some aspect of AM biosynthesis (Morrison and Milligan, 1982; Morrison et al., 1984; Morrison, 1985, 1988a, 1988b; Morrison and Gadan, 1987). This assumes that the lipids remain more or less quantitatively in the isolated starch granules together with the AM and integral protein, whereas the other enzymes for starch synthesis, which are soluble or are in the amyloplast envelope, disappear.

The AM and LPL contents of the A- and B-granules in wheat and barley increase during grain development (Table II). The AM-LPL relationship is different between the A- and B-granules in each of the three wheat varieties and in the four barleys.

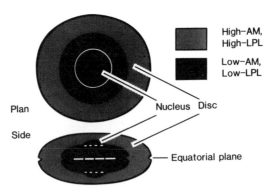

Figure 2. Model of a fully-developed A-type granule from mature wheat showing low-amylose low-lipid central region deposited initially (cf. Fig. 1) and high-amylose high-lipid region deposited later (calculated from data of Morrison and Gadan, 1987)

TABLE II

Amylose and Lipid Contents of Immature and Mature A and B Starch Granules in Wheat and Barley Varieties

	Immature 20-21 DAA		Mature 50-60 DAA	
	AM %	LPL mg%	AM %	LPL mg%
Wheat, A and B				
M. Huntsman, A	23.4	514	28.4	699
M. Huntsman, B	22.0	434	26.7	818
Brigand, A	20.5	445	27.2	793
Brigand, B	20.1	374	25.9	982
Flanders, A	21.6	553	30.4	794
Flanders, B	21.7	462	29.1	1017
Barley, A only				
Waxy O'brucker	2.0	83	5.8	386
Oderbrucker	18.0	357	29.2	1096
Glacier	19.0	219	28.2	797
Glacier High-AM	29.2	329	38.8	1037

Data from McDonald et al. (in press), Morrison (1987), Morrison and Gadan (1987)

These results can be explained if the starch synthesized initially has a low AM content and little or no LPL, but from c. 20 days after anthesis there is a switch to synthesis of high-AM, high-LPL starch. A model of the structure of an A-granule, developed from Figure 1, is shown in Figure 2.

Kassenbeck (1975, 1978) has also used selective chemical treatment of thin sections of granules for transmission electron microscopy to demonstrate gradients, from center to periphery, of radially oriented AM, amorphous AM, and crystalline AP. Thus, it is evident that the A-granules are asymmetrical in the quantitative distributions of AP, AM and LPL, and in the crystallinity of the AP. These all contribute to the swelling behaviour of the granules discussed below.

ENZYME DIGESTIBILITY OF GRANULES

Undamaged starch granules are very resistant to attack by β-amylase and this enzyme is therefore used in sensitive assays for damage. Undamaged granules are, however, digested by endogenous α-amylases in vivo during germination and by other amylases (e.g. salivary α-amylase, amyloglucosidase) in vitro. The initial rate of digestion is affected by granule surface properties, and is greatly accelerated by mechanical damage. Digestion of native A-granules eventually leads to pitting so that the enzyme penetrates the granule where it digests the more amorphous regions preferentially. Scanning electron micrographs show pits all over the A-granule surface, often concentrated in the equatorial groove, and when the granules are fractured concentric shells of undigested crystalline material are revealed (Evers and McDermott, 1970; Meredith and Pomeranz, 1985; Lorenz and Meredith, 1988). The B-granules appear to be digested by surface erosion, and there is little evidence of pitting and internal digestion.

Size-fractionated starches from developing wheat (Morrison and Gadan, 1987) provided material for studying the effects of some properties on the initial rate of digestion by amyloglucosidase. The samples constituted a matrix within which AM, LPL and

TABLE III

Correlations Between Rates of Starch Granule Digestion by Amyloglucosidase and Granule Specific Surface Area, and Ranges in Some Other Properties of These Granules Not Correlated with Rate of Digestion

	M. Huntsman[b]	Brigand[c]	Flanders[b]
Linear corr., r	+0.993	+0.992-0.999	+0.951
No. samples	21	6-7	18
Surface, m^2/g	0.18-0.90	0.17-0.83	0.18-0.70
AM, %	21.5-28.4	20.3-27.2	22.1-30.5
LPL, mg%	440-910	390-1070	460-1106
GT, °C	57.6-59.0	56.7-60.8	not detd.

[a] W.R. Morrison and J. Karkalas, unpublished
[b] Samples at various DAA combined as one set
[c] Samples at various DAA as separate sets (see text)

specific surface area varied independently. In Maris Huntsman and Flanders wheats the rate of digestion was very highly correlated with specific surface area, and was totally independent of AM and LPL contents and stage of grain development (DAA). A single linear regression line fitted all Maris Huntsman samples, while a separate regression line fitted the Flanders samples (Table III). In these samples gelatinization temperature (GT) showed no anomalies. Brigand wheat required separate regression lines for each set of starches (by DAA). Samples at 30 and 40 DAA were less susceptible than those at 20 and 50 DAA, while those at 60 DAA were most susceptible to digestion. These differences were evidently caused by some artifactual changes in the crystallinity of the granules, because GT was consistent in all size-fractions at each stage (DAA) but not between stages (discussed below).

The digestibility of a selection of starches from Brigand, Bounty, Daws, Scout 66 and CBWC wheats (also discussed below) covering the range 0.35-0.48 m^2/g gave a lower correlation ($r = +0.470$, $n = 11$) which may be attributed to varietal differences (cf. Maris Huntsman and Flanders, above) and there were no correlations with AM or LPL contents. Similar results and conclusions have been reported for barley starches (Morrison et al., 1986). In all these studies the starches were isolated using a protease to facilitate removal of proteins, including the granule surface proteins which, if present, might affect the digestibility of the soft wheat starches, compared with starches from the hard wheats which would have much less of these particular proteins.

SWELLING AND GELATINIZATION OF GRANULES

Gelatinization, as discussed here, is restricted to starch granules heated in excess water and is now commonly measured by differential scanning calorimetry (DSC), using a heating rate of 10°C/minute.

Gelatinization is an interactive process involving uptake of water, granule swelling, disordering of AP crystallites with loss of birefringence, leaching of part of the AM into solution (Blanshard, 1987; R.F. Tester and W.R. Morrison, in preparation), and possibly formation of some insoluble AM-LPL inclusion complexes within the swollen granules (Biliaderis et al., 1986a, 1986b; Kugimiya et al., 1980). Gelatinization and swelling properties are determined primarily by AP, and comparing waxy and normal lines of barley, maize and rice it seems that AM acts partially as a diluent of AP and partially as an inhibitor of swelling, especially when it is complexed with lipids (R.F.Tester and W.R. Morrison, in preparation). Thus, in normal starches there will be an effect from the usual complement of starch LPL plus the variable amount of surface FFA which may be acquired from the wheat non-starch lipids during processing, or by adding an AM-complexing surfactant such as monoglyceride to bread doughs. The swelling curve of

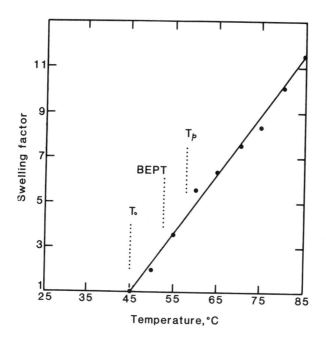

Figure 3. Swelling curve of wheat starch heated for 30 minutes at various temperatures in excess water. T^O and T^P = onset and peak gelatinization temperatures (DSC), BET = birefringence end-point temperature. From R.F. Tester and W.R. Morrison (unpublished).

wheat starch (Figure 3) shows very limited swelling below GT, then rapid swelling (with temperature-dependent leaching of AM) above GT and a final plateau at 90-95°C (not shown in figure). The swelling curves of starches from various types of wheat are very similar, and are conveniently compared by measurements of swelling factors (SF) at 70°C or 80°C.

The GT of Maris Huntsman starch used by Morrison and Gadan (1987) decreased by 1°C over the period of grain development (21-63 DAA), and also decreased by 1.2-0.2°C from the smallest to the largest granules at each stage of development (W.R. Morrison and A.M. Coventry, unpublished). Thus, at 21 DAA the GT of the largest A granules was 59.0°C, smallest A and all B was 60.2°C, while at 63 DAA the GT of largest A was

58.1°C, smallest B was 58.3°C. In general the GT and enthalpy (ΔH) of developing starch change very little, so that there is probably uniform ordering in the AP crystallites throughout the granules, and the degree of swelling is therefore dictated primarily by AP distribution which is the inverse of AM distribution indicated in Figure 2. Thus, A-granules swell preferentially in the equatorial plane so that they contort and buckle in a characteristic fashion. This behaviour is common to all the <u>Triticeae</u> A-granules, but is not shown by their B-granules, other cereal starches, or by legume, root and tuber starches (Bowler et al., 1980; Williams and Bowler, 1982). This unique swelling may be part of the reason why wheat starch can be substituted by normal barley, rye and triticale starches for baking, but not by other starches (D'Appolonia and Gilles, 1971; Hoseney et al., 1971; Sollars and Rubenthaler, 1971; Ciacco and D'Appolonia, 1977).

Gelatinization properties can be altered by holding starch for several hours just below GT (Gough and Pybus, 1971; Lorenz and Kulp, 1978, 1980, 1981, 1982; Al-Jawad and Lorenz, 1982). This allows "annealing" or rearrangement of AP crystallites so the peak GT is raised, GT range (onset to conclusion temperature) is narrowed, and enthalpy is slightly increased. Corn starches extracted after steeping maize kernels in dilute sulfite at 50-53 °C are partially annealed (Kreuger et al., 1987), but commercial processes for wheat starch do not use warm steeping and the starches are much more like native granules.

VARIABILITY OF STARCH PROPERTIES

Most properties of pure wheat starches are remarkably uniform among varieties, although atypical values can be obtained from wheats grown under stress, and from impure and enzymically damaged samples. Mean values from a survey of starch in five wheat varieties, each grown at several locations in their main areas of cultivation, are given in Tables IV and V with coefficients of variation (CV) indicated by symbols. For each type of wheat differences were due to both site and year-to-year effects, but nearly as much variation has been observed in starch from Brigand wheat grown at one site over four years (Morrison and Scott, 1986).

As expected, winter wheats had larger kernels and more starch/endosperm than spring wheats. The two British varieties had more A- and B-granules, probably related to the longer grain-filling period, but the granules were of similar sizes in all types of wheat except CBWC which would have the shortest period of grain-filling. B-granule numbers and dimensions were most variable. It is thought that errors due to selective loss of small B-granules during starch isolation were minimal, but, nevertheless the percentage by weight and by number of the B-granules should perhaps be a little greater, especially for Scout 66 samples.

The three winter wheats had less AM and phosphorus (i.e. LPL), and slightly greater swelling factors than the spring wheats, while soluble glucan figures were consistently low only in CBWC starches. However, differences between samples of each type were nearly as great as between types.

Similar data, almost entirely within the limits given in Tables IV and V have been obtained using the same analytical methods on starches from 13 Syrian, 8 French, and 6 Canadian durum wheats, and 8 Syrian and 9 Canadian breadwheats (A.B. Soulaka and W.R. Morrison, unpublished).

Statistical analysis of the raw data for Tables IV and V showed significant correlations ($p < 0.001$) between AM and P (or LPL, +ve), AM and SF (-ve), AP and SF (+ve), P and SF (-ve), and SF and soluble glucan (or sol. AM, +ve). Very similar results were obtained with starch from Triumph barley grown at 16 UK sites (Morrison et al., 1986). Some of these correlations should be interpreted with caution.

Weather data were included in the data matrix for the Brigand and Bounty wheat starches and this revealed a significant positive correlation between phosphorus and accumulated temperature, solar radiation (MJ/m^2) and sunshine hours over the grain-filling period.

In a recent experiment with barley grown in

TABLE IV

Starch Content and Dimensions of Starch Granules from Five Types of Wheats Grown at Several Locations in the Same Year[a]

Property[b]	Wheat type or variety[c]				
	Bounty	Brigand	Daws	Scout 66	CBWC
Years	1981-2-4	1981-2-4	1982	1982	1981
Sample/year	9+6+1	9+7+3	6	9	9
Kernel wt., mg	54.2*	56.4**	41.8*	37.5*	35.3*
Starch, wt. % in grain	66.1	67.4	67.3	62.9	59.9
Starch, mg/endosperm	30.9**	32.5**	24.2*	20.3*	18.2*
A-granules, 10⁶/endosperm	12.4*	13.0**	8.7*	9.1*	9.1*
B-granules, 10⁶/endosperm	170°	177°°	123**	86**	143°°
Mean vol. A, μm³	2263*	2378**	2594*	2299*	1696*
Mean vol. B, μm³	70.3**	70.1°	72.2*	64.6*	58.0*
B-granules, % (no.)	93.1	92.7	93.3	90.3	93.8
B-granules, % (wt.)	29.4*	27.9**	28.3**	20.9*	34.7*
A surface, m²/g	0.25	0.24	0.24	0.25	0.28
B surface, m²/g	0.74	0.74	0.70	0.73	0.76
Total surface, m²/g	0.39	0.38	0.37	0.35	0.44

[a] W.R. Morrison and D.C. Scott (unpublished), with permission of Min. Agric. Fisheries and Food.
[b] Mean values, coeff. var. < 5% except *(5-10%), **(10-15%), °(15-20%), or °°(> 20%).
[c] Bounty = UK hard winter, Brigand = UK soft winter, Daws = US NW Pacific soft white winter, Scout 66 = US hard red winter CBWC = Canadian hard red spring composite.

TABLE V

Composition and Properties of Starches[a] from Five Types of Wheat Described in Table IV.

	Wheat type or variety				
	Bounty	Brigand	Daws	Scout 66	CBWC
Total amylose, %	28.3	27.6	28.4	31.5	30.1
Apparent amylose, %	22.9	22.3	22.5	25.8	23.7
Phosphorus, mg %	50.2*	49.5*	54.2*	56.1*	60.9
Swelling factor at 70°C	9.2	9.5	9.2	9.1	8.8
Sol. glucan at 70°C	4.9*	5.7*	5.6**	4.9**	4.0*

[a] See notes at foot of Table IV.

TABLE VI

Some Properties of Starch from Barley Grown at Three Constant Temperatures[a]

Property	var. Triumph			var. Golden Promise		
	10°C	15°C	20°C	10°C	15°C	20°C
AM,%	28	28	27	28	27	28
LPL, mg%	767	816	1210	1000	1084	1687
GT,°C	48.8	53.6	64.5	49.6	51.8	62.8
ΔH, J/g	9.1	9.3	10.3	6.4	8.3	10.1
SF_{70}	7.4	6.7	4.8	6.6	6.2	4.1
SF_{80}	10.0	9.6	7.3	9.0	7.9	4.2

[a] R.F. Tester, W. R. Morrison and R.P. Ellis, unpublished results

three constant temperature chambers at 10, 15 and 20°C (R.F. Tester, W.R. Morrison and R.P. Ellis, unpublished), AM contents were nearly constant, but phosphorus contents and GT were much higher in starches from grain grown at 20°C and conversely at 10°C (Table VI). There is every reason to suppose that wheat would behave similarly. These observations show that phosphorus or LPL need not be tightly linked to total AM content. They also show a relationship between GT and environmental temperature, already observed in rice starch (Asaoka et al., 1984, 1985a, 1985b; Morrison and Nasir Azudin, 1987), that merits further study in wheat. The ranges of GT found in starches from a selection of 23 bread wheats and 26 durum wheats were nearly identical (57-65°C, 58-64°C, resp.) and ΔH values were also similar (Soulaka and Morrison, 1985a).

However, more careful study may reveal minor effects of variety and environmental temperature interactions which could be important since bread loaf volume, initial crumb firmness and rate of crumb firming are all inversely related to the GT of wheat starch (Soulaka and Morrison, 1985b). These properties are also affected by the proportions of A- and B-granules (Soulaka and Morrison, 1985b) which, as discussed above, are very much affected by environmental conditions during grain filling.

CONCLUSIONS

The physical and chemical properties of wheat starch are similar to those of normal barley, rye and triticale starches, but are quite different from other cereal and non-cereal starches. Wheat starch properties are comparatively uniform and generally show small variations attributable to type or variety and to site and year-to-year effects. However, in sensitive processes such as breadmaking small differences in properties can cause perceptible differences in performance and product quality. Present indications are that the minor lipid and protein components inside the granules and on the surface of the granules relate to fundamental aspects of starch, but their full implications for technological properties have yet to be established, although the importance of the lipids is not in doubt (Galliard and Bowler, 1987).

LITERATURE CITED

AL-JAWAD, N., and LORENZ, K. 1982. Steeping of poor quality wheat. Effects on physicochemical properties and functional characteristics of starch. Starch 34:198.

ASAOKA, M., OKUNO, K., SUGIMOTO, Y., KAWAKAMI, J., and FUWA, H. 1984. Effect of environmental temperature during development of rice plants on some properties of endosperm starch. Starch 36:189.

ASAOKA, M., OKUNO, K., and FUWA, H. 1985a. Effect of environmental temperature at the milky stage on amylose content and fine structure of amylopectin of waxy and nonwaxy endosperm starches of rice. Agric. Biol. Chem. (Tokyo) 49:373.

ASAOKA, M., OKUNO, K., and FUWA, H. 1985b. Genetic and environmental control of starch properties in rice seeds. Pages 29-38 in: New Approaches to Research on Cereal Carbohydrates. R.D. Hill and L. Munck, eds. Elsevier Sci. Publ., Amsterdam.

ATWELL, W.A., HOSENEY, R.C., and LINEBACK, D.R. 1980. Debranching of wheat amylopectin. Cereal Chem. 57:12.

BANKS, W., and GREENWOOD, C.T. 1975. Starch and its Components, Edinburgh Univ. Press, Edinburgh.

BILIADERIS, C.G., PAGE, C.M., MAURICE, T.J., and JULIANO, B.O. 1986a. Thermal characteristics of rice starches: a polymeric approach to phase transitions of granular starch. J. Agric. Food Chem. 34:6.

BILIADERIS, C.G., PAGE, C.M., and MAURICE, T.J. 1986b. On the multiple melting transitions of starch/monoglyceride systems. Food Chem. 22:279.

BLANSHARD, J.M.V. 1987. Starch granule structure and function: a physicochemical approach. Pages 16-54 in: Starch: Properties and Potential. T. Galliard, ed. John Wiley and Sons, Chichester.

BOWLER, P., WILLIAMS, M.R., and ANGOLD, R.E. 1980. A hypothesis for the morphological changes which occur on heating lenticular wheat starch in water. Starch 32:186.

BRIARTY, L.G., HUGHES, C.E., and EVERS, A.D. 1979. The developing endosperm of wheat - a stereological study. Ann. Bot. 44:641.

BUTTROSE, M.S. 1963. Ultrastructure of the developing wheat endosperm. Aust. J. Biol. Sci. 16:305.

CIACCO, C.F., and D'APPOLONIA, B.L. 1977. Characterization of starches from various tubers and their use in bread. Cereal Chem. 54:1096.

CZAJA, A.Th. 1982. Critical examination of the development of Triticeae starch grains. Starch 34:109.

D'APPOLONIA, B.L., and GILLES, K.A. 1971. Effect of various starches in baking. Cereal Chem. 48:625.

EVERS, A.D. 1971. Scanning electron microscopy of wheat starch. III. Granule development in the endosperm. Starch 23:157.

EVERS, A.D. 1974. The development of the grain of wheat. Proc. 4th Int. Congr. Food Sci. Technol., Madrid 1:422.

EVERS, A.D., and McDERMOTT, E.E. 1970. Scanning electron microscopy of wheat starch. II. Structure of granules modified by alpha-amylolysis - a preliminary report. Starch 22:23.

FRENCH, D. 1973. Chemistry and biochemistry of starch. Page 267 in: Biochemistry, Series One, Vol.5. Biochemistry of Carbohydrates. W.J. Whelan, ed. Butterworths: London. University Park Press: Baltimore.

FRENCH, D. 1984. Organization of starch granules. Pages 183-247 in: Starch, Chemistry and Technology. R.L. Whistler, J.N. BeMiller and E.F. Paschall, ed. Academic Press, Orlando.

GALLIARD, T., and BOWLER, P. 1987. Morphology and composition of starch. Pages 55-78 in: Starch: Proprties and Potential. T. Galliard, ed. John Wiley and Sons, Chichester.

GOUGH, B.M., and PYBUS, J.N. 1971. Effect on the gelatinization temperature of wheat starch granules of prolonged treatment with water at 50°C. Starch 23:210.

GOUGH, B.M., GREENWELL, P., and RUSSELL, P.L. 1985. On the interaction of sodium dodecylsulphate with starch granules. Pages 99-108 in: New Approaches to Research on Cereal Carbohydrates. R.D. Hill and L. Munck, eds. Elsevier Science Publishers B.V.: Amsterdam.

GREENWELL, P., and SCHOFIELD, J.D. 1986. A starch granule protein associated with endosperm softness in wheat. Cereal Chem. 63:379.

HOSENEY, R.C., FINNEY, K.F., POMERANZ, Y., and SHOGREN, M.D. 1971. Functional (breadmaking) and biochemical properties of wheat flour components. VII. Starch. Cereal Chem. 48:191.

KAINUMA, K. 1980. Chori Kagaku 13:83 (cited by French, 1984).

KASSENBECK, P. 1975. Electron microscopic contribution to the study of the fine structure of wheat starch. Starch 27:217.

KASSENBECK, P. 1978. The distribution of amylose and amylopectin in starch grains. Starch 30:40.

KOBAYASHI, S., SCHWARTZ, S.J., and LINEBACK, D.R. 1986. Comparison of the structures of amylopectins from different wheat varieties. Cereal Chem. 63:71.

KRUEGER, B.R., KNUTSON, C.A., INGLETT, G.E., and WALKER, C.E. 1987. A differential scanning calorimetry study on the effect of annealing on gelatinization behavior of starch. J. Food Sci. 52:715.

KUGIMIYA, M., DONOVAN, J.W., and WONG, R.Y. 1980. Phase transitions in amylose-lipid complexes in starches: a calorimetric study. Starch 32:265.

LII, C.-Y., and LINEBACK, D.R. 1977. Characterization and comparison of cereal starches. Cereal Chem. 54:138.

LORENZ, K., and KULP, K. 1978. Steeping of wheat at various temperatures - effects on physicochemical characteristics of starches. Starch 30:333.

LORENZ, K., and KULP, K. 1980. Steeping of starch at various temperatures - effects on functional properties. Starch 32:181.

LORENZ, K., and KULP, K. 1981. Heat-moisture treatment of starches. II. Functional properties and baking potential. Cereal Chem. 58:49.

LORENZ, K., and KULP, K. 1982. Cereal and root starch modification by heat-moisture treatment. Starch 34:76.

LORENZ, K., and MEREDITH, P. 1988. Insect damaged wheat. Effects on starch characteristics. Starch 40:136.

LOWY, G.D.A., SARGEANT, J.G., and SCHOFIELD, J.D. 1981. Wheat starch granule protein: the isolation and characterisation of a salt-extractable protein from starch granules. J. Sci. Food Agric. 32:371.

MacGREGOR, A.W., and MORGAN, J.E. 1984. Structure of amylopectins isolated from large and small starch granules of normal and waxy barley. Cereal Chem. 61:222.

MANNERS, D.J., and MATHESON, N.K. 1981. The fine structure of amylopectin. Carbohydr. Res. 90:99.

McDONALD, A.M.L., STARK, J.R., MORRISON, W.R., and ELLIS, R.P. 1989. Composition of starch granules from developing barley genotypes. J. Cereal Sci., in press.

MEREDITH, P., and POMERANZ, Y. 1985. Sprouted grain.

Adv. Cereal Sci. Technol. 7:239.

MORRISON, W.R. 1978a. Cereal lipids. Adv. Cereal Sci. Technol. 2:221.

MORRISON, W.R. 1978b. Stability of wheat starch lipids in untreated and chlorine-treated cake flours. J. Sci. Food Agric. 29:365.

MORRISON, W.R. 1981. Starch lipids: a reappraisal. Starch 33:408.

MORRISON, W.R. 1985. Lipids in cereal starches. Pages 61-70 in: New Approaches to Research in Cereal Carbohydrates. R.D. Hill and L. Munck, eds. Elsevier Sci. Publ., Amsterdam.

MORRISON, W.R. 1987. Lipids in wheat and barley starch granules. Pages 438-445 in: Cereals in a European Context. I.D. Morton, ed. Ellis-Horwood: Chichester.

MORRISON, W.R. 1988a. Lipids in cereal starches: a review. J. Cereal Sci. 8:1.

MORRISON, W.R. 1988b. Lipids. In: Wheat, Chemistry and Technology, 3rd edn., Vol.1. Y. Pomeranz, ed. Am. Assoc. Cereal Chem., St. Paul. MN.

MORRISON, W.R., and GADAN, H. 1987. The amylose and lipid contents of starch granules in developing wheat endosperm. J. Cereal Sci. 5:263.

MORRISON, W.R., and MILLIGAN, T.P. 1982. Lipids in maize starches. Pages 1-18 in: Maize: Recent Progress in Chemistry and Technology. G.E. Inglett, ed. Academic Press, New York.

MORRISON, W.R., and NASIR AZUDIN, M. 1987. Variation in the amylose and lipid contents and some physical properties of rice starch. J. Cereal Sci. 5:35.

MORRISON, W.R., and SCOTT, D.C. 1986. Measurement of the dimensions of wheat starch granule populations using a Coulter Counter and 100-channel analyzer. J. Cereal Sci. 4:13.

MORRISON, W.R., MILLIGAN, T.P., and AZUDIN, M.N. 1984. A relationship between the amylose and lipid contents of starches from diploid cereals. J. Cereal Sci. 2:257.

MORRISON, W.R., SCOTT, D.C., and KARKALAS, J. 1986. Variation in the composition and physical properties of barley starches. Starch 38:374.

NIKUNI, Z. 1978. Studies on starch granules. Starch 30:105.

PARKER, M.L. 1985. The relationship between A-type and B-type starch granules in the developing wheat endosperm. J. Cereal Sci. 3:271.

ROBIN, J.P., MERCIER, C., CHARBONNIERE, R., and GUILBOT, A. 1974. Lintnerized starches. Gel filtration and enzymic studies of insoluble residues from prolonged acid treatment of potato starch. Cereal Chem. 51:389.

RUSSELL, P.L., GOUGH, B.M., GREENWELL, P., FOWLER, A., and MUNRO, H.S. 1987. A study by ESCA of the surface of native and chlorine-treated wheat starch granules. The effect of various surface treatments. J. Cereal Sci. 5:83.

SARGEANT, J.G. 1982. Determination of amylose:amylopectin ratios of starches. Starch 34:89.

SCHOFIELD, J.D., and GREENWELL, P. 1987. Wheat starch granule proteins and their technological significance. Pages 407-420 in: Cereals in a European Context. I.D. Morton, ed. Ellis-Horwood, Chichester.

SIMMONDS, D.H., and O'BRIEN. 1981. Morphological and biochemical development of the wheat endosperm. Adv. Cereal Sci. Technol. 4:5.

SOLLARS, W.R., and RUBENTHALER, G.L. 1971. Performance of wheat and other starches in reconstituted flours. Cereal Chem. 48:397.

SOULAKA, A.B., and MORRISON, W.R. 1985a. The amylose and lipid contents, dimensions and gelatinisation characteristics of some wheat starches and their A- and B-granule fractions. J. Sci. Food Agric. 36:709.

SOULAKA, A.B., and MORRISON, W.R. 1985b. The breadbaking quality of six wheat starches. J. Sci. Food Agric. 36:719.

TAKEDA, Y., HIZUKURI, S., TAKEDA, C., and SUZUKI, A. 1987. Structure of branched molecules of amyloses of various origins, and molar fractions of branched and unbranched molecules. Carbohydr. Res. 165:139.

WILLIAMS, M.R., and BOWLER, P. 1982. Starch gelatinization: a morphological study of Triticeae and other studies. Starch 34:221.

13

PROPERTIES OF WHEAT STARCH COMPARED TO NORMAL MAIZE STARCH

Yong-Cheng Shi and Paul A. Seib

Department of Grain Science and Industry
Kansas State University
Manhattan, KS 66506

INTRODUCTION

Corn, potato, wheat, tapioca and waxy maize dominate the tonnage of starch produced around the world (Swinkels, 1985). Among those five, normal corn and wheat resemble each other most in properties, yet they differ in granule size, whiteness, internal lipids, gelatinization temperature, paste consistency, gel strength, and surface contaminants. In this paper, we describe the uniqueness of wheat starch compared to normal corn starch, and sometimes compared to potato and tapioca starches.

COLOR, GRANULE SIZE AND DISTRIBUTION

Wheat starch is white in color, while yellow-dent corn starch is slightly tinted and must be bleached with oxidizing agents. Whiteness of starch is important in bakery applications. Examples include coating and dusting compounds for donuts and bread doughs and starch added to white cakes.

Wheat starch has two populations of granules, whereas corn starch has one (Fig 1). The large A-granules are disc-shaped and have an average diameter of 25 μm as seen in the microscope. The small B-granules appear spherical and are less than 15 μm in diameter (Meredith, 1981). The B-granules comprise approximately 30% by weight of wheat starch (Meredith, 1981; Dengate, 1984). A-granules

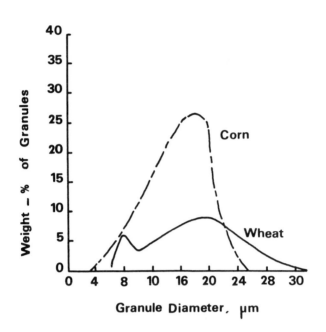

Fig. 1. Size distribution by weight of corn (Wolf et al, 1964) and wheat starches (Eliasson and Karlsson, 1983).

develop first in endosperm cells followed by late bursts of B-granules (Dengate, 1984). The small B-granules have lower amylose content (18%) than A-granules (28%), and contain more lipid and protein (Morrison and Laignelet, 1983; Dengate, 1984; Eliasson, 1988; Armstrong and Fulcher, 1988).

MOLECULAR COMPOSITION AND STRUCTURE

Wheat and corn starches contain roughly one-third amylose (AM), while potato and tapioca are one-fifth AM (Table I).

The fine structure of amylose (AM) and amylopectin (AP) from many starches has been examined in recent years by Professor Susumu Hizukuri (1988). Wheat and corn AM are much

TABLE I
Properties of Wheat Amylose
Compared to Others[a]

Source	Amylose Content, %	Number-Average Degree of Polymerization DP_n	Intrinsic Viscosity, $[\eta]$, mL/g	Iodine Binding Capacity, g I_2/100g
Wheat	28[b]	1300	80[c]	20.5
Corn	28[b]	930	179	20.1
Tapioca	17[d]	2600	384	20.0
Potato	20[d]	4900	384	20.5

[a] Data from Hizukuri (1988) unless otherwise referenced.
[b] Morrison and Laignelet (1983).
[c] In 1M sodium hydroxide solution at 22.5°C.
[d] Young (1984)

smaller molecules than tapioca and potato AM (Table I). In spite of the one-third larger molecular weight of wheat vs corn AM, corn AM showed more than double the intrinsic viscosity of wheat AM. The β-limit dextrin from wheat AM was much more highly branched (15 vs 5 chains/molecule) than from corn, and its unit chains were one-half the average length of those in corn β-limit AM (Takeda et al, 1987). The low viscosity of wheat AM may be explained by the presence of a large number (73 mole %) of short linear molecules mixed with a small number but high weight-percent of moderately branched molecules (Takeda et al, 1987).

Wheat amylopectin (AP) has one-half the number average molecular weight of corn AP (Hizukuri 1988). Moreover, wheat AP is more highly branched than corn AP as shown by its shorter average chain length, and its external and internal chain lengths (Table II). The more highly ramified structure of AP in wheat and rice compared to corn AP may explain why rice and wheat starch gels are more tolerant to cold-temperature storage. A word of caution may be in order concerning the molecular structures of AM and AP reviewed here, since the structures of wheat and corn AM and AP were from data on a few samples of grain.

Wheat AP (Kobayashi et al, 1986) and corn AP (Praznik et al, 1987) both showed a tetramodal distribution of unit-chains upon debranching. Hizukuri (1986) found a tetramodal distribution in many other amylopectins, which led to his revised cluster model for AP.

TABLE II
Properties of Wheat Amylopectin Compared to Others[a]

Source	Number-Average Degree of Polymerization, $\overline{DP}_n$	Average Chain Length		
		Overall	External	Internal
Wheat	4800	19	13	5
Corn	8200	22	15	6
Potato	9800	24	15	8

[a] Data from Hizukuri (1988)

GELATINIZATION AND GLASS-TRANSITION TEMPERATURES

The crystalline phase in granules of wheat and corn starch has been estimated to be 25-30% and 38-39% by x-ray diffraction, respectively. At the same time, ^{13}C CP/MAS NMR indicated 48 and 42% double helix in wheat and corn starch, respectively (Gidley and Bociek, 1985 and Gidley, 1987). As the moisture is increased in starch, the glass-transition temperature (Tg) of the amorphous phase, and the melting temperature (T_M) of the crystalline phase (gelatinization) decline to a limiting value when moisture reaches approximately 30% and 70% by weight, respectively.

In the presence of excess moisture (> 70%, wet basis) wheat starch gelatinizes 6-8°C below corn starch (Table III). It is tempting to speculate that the higher degree of branching and the shorter external chain length of wheat AP (Table II) result in less perfect crystals compared to those in corn starch AP. It is well known (Derby et al 1975; Ghiasi et al, 1982b) that wheat starch with <30% moisture does not gelatinize below

TABLE III
Gelatinization Temperatures of Various Starches in Excess Water[a]

Starch	Birefringence End-Point Temperature, °C
Wheat	58-61-64
Potato	50-53-68
Tapioca	59-65-70
Corn	62-67-72
Waxy Corn	63-68-72

[a]Snyder (1984)

100°C, and furthermore that sodium chloride and sucrose increase gelatinization temperature (Ghiasi et al, 1983). The results of those findings explain the different degrees of gelatinization of wheat starch in various bakery foods (Hoseney et al, 1977, Varriano-Marston, et al, 1980).

The glass-transition temperature (Tg) of the amorphous phase of starch is of fundamental importance in understanding gelatinization and retrogradation of starch, and the texture of dry snack foods and instant breakfast cereals (Slade and Levine, 1987). At 11% MC the molecular chains in wheat starch are immobile at room temperature, but they become mobile at MC $\geq$ 20% (Table IV). As the starch chains become more and more hydrated, Tg decreases until it reaches a limiting value of -5°C at MC $\geq$ 28%. After full plasticization of the polymer chains at ~27%

TABLE IV
Glass Transition Temperature (Tg) of Wheat Starch at Different Moisture Content[a]

Sample	Moisture Content %, Wet Basis	Tg, °C
Native	0	(125)[b]
Native	11	-
Native	15	60
Native	20	28
Pregelatinized	11	80
Pregelatinized	15	42
Pregelatinized	20	25
Pregelatinized	55[c]	-5[c]

[a] All data from Zeleznak and Hoseney (1987) except where noted.
[b] Extrapolated value (Slade and Levine, 1987).
[c] Experimental value (Slade and Levine, 1988).

MC, any additional water phase separates as ice crystals upon cooling while Tg of the hydrated starch remains at -5°C (Slade and Levine, 1987). Those concepts have been applied universally to all starches.

PASTE CONSISTENCY

It is well known (Takahashi and Seib, 1988) that the paste consistency of wheat starch is below that of corn starch (Fig 2). The difference in consistencies of those starches may be explained by examining the properties of the two phases that comprise the paste (Miles et al, 1985). The swollen gel phase, which constitutes the discontinuous phase, has lower swelling power in wheat than in corn starch (Table V). The solubles (Table

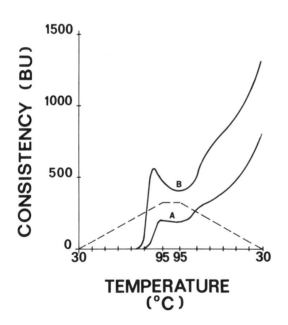

Fig. 2. Amylograms of (A) wheat and (B) corn starch at 7.5% solids in water. The dashed line is the temperature profile (Takahashi and Seib, 1988).

TABLE V
Swelling Power and Solubility of
Wheat and Corn Starches[a]

Starch	Swelling Power, g/g			Solubility, %		
	75°C	85°C	95°C	75°C	85°C	95°C
Wheat	8.8	11.4	23.0	5.8	9.6	20.6
Corn	10.0	14.5	23.3	6.7	9.7	14.5
Low-Lipid Wheat	11.2	15.5	19.7	13.6	21.9	23.3
Low-Lipid Corn	13.5	17.9	22.4	15.5	17.2	18.4

[a]Determined using slow stirring rate, 3.0% starch solids in water, and heating at 10°C/min.

V), which are dissolved in the continuous phase, are practically pure amylose for both corn and wheat, since the iodine complex of the solubles in wheat and corn shows absorbance maxima at 645-650 nm and 625-640 nm, respectively.

The low swelling power of the gel phase in wheat starch provides extra water in its continuous phase, which results in lower consistency for wheat starch pastes compared to corn starch. In addition, wheat amylose that has dissolved in the continuous phase has a lower intrinsic viscosity compared to corn (Table I). The low swelling power of wheat starch and its amylose's low intrinsic viscosity explain the low paste consistency of wheat starch.

PURE AMYLOSE LEACHED FROM WHEAT AND CORN STARCH

Doublier et al (1987) reported that high yields of practically pure amylose could be leached from wheat and corn starch. Ghiasi et al (1982a) found high iodine-binding capacity (IBC) for the solubles leached from wheat starch at 85-95°C. In recent work, we obtained pure amylose in 28% yield from low-lipid wheat starch that was gently agitated at 2% solids and heated to 95°C at 10°/min (Fig 3). Furthermore, by annealing native wheat starch before leaching, we doubled the yield of amylose from 8 to 16% in a 4.5% slurry at 95°C. Annealing reduced starch swelling and facilitated separation of the soluble phase from the swollen gel. In contrast, we obtained 15-20% amylose by crystallization of its n-butanol complex.

High-performance size-exclusion (HPSE) chromatograms of leached wheat amylose (Fig 3) showed a bimodal distribution of chains and no high molecular weight amylopectin. A markedly

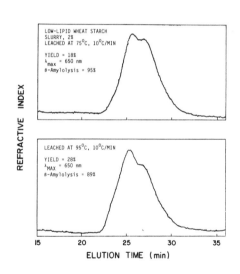

Fig. 3. High-performance size exclusion (HPSE) chromatograms of AM leached from wheat starch. Stationary phase, PL gel, 10 μm, 75 x 300 mm; mobile phase, dimethyl sulfoxide containing 0.05M sodium nitrate; column temperature 80°C (Chuang and Sydor, 1987).

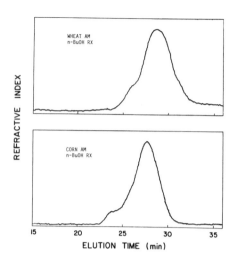

Fig. 4. HPSE chromatograms of wheat and corn AM isolated as n-butanol complex. Separation conditions the same as in Fig. 3.

different bimodal distribution (Fig 4) was shown by wheat AM isolated from solubilized starch and crystallized as its butanol complex. The excluded volume on the HPSE column in Figs 3 and 4 was 22.0 mL. Additional data will be needed to differentiate the molecular structures of leached vs butanol-complexed AM from wheat.

STARCH LIPIDS

Wheat starch is thought to be the purest source of lysolecithin in nature (Wren and Merryfield, 1970). Morrison and Milligan (1982) reported that corn starch contains 0.59-0.76% lipids, and wheat starch 0.77-1.17%. Over 90% of the lipid in wheat starch is lysolecithin (Meredith and Dengate, 1978), whereas the lipid in corn starch is approximately 60% free fatty acids and 25% lysolecithin (Tan and Morrison, 1979). The predominant fatty acids in wheat and corn starches are palmitic and linoleic, plus some oleic acid. Acker and Schmitz (1967) found that wheat lysolecithin is 75% α- and 25% β-acyl ester.

The location of lysolecithin in native wheat starch granules is not known with certainty. Biliaderis (1986) cited X-ray and DSC evidence indicating the lysolecithin in native wheat starch is not complexed as the guest molecule in the cylindrical cavity of the α-helix of amylose. Morrison and South (1988) proposed the lipids are associated with protein in the granule, and that they complex with AM after gelatinization. Gidley and Bociek (1988) presented ^{13}C solid-state nmr data that is consistent with a α-(1→4)-glucan lysolecithin complex. These data suggest that lysolecithin may be complexed with AM in the native granule, and the complexed molecules are widely dispersed and do not crystallize until mobilized by heating the starch in water.

Lysolecithin from wheat starch forms liposomes in aqueous dispersion (Larsson 1980) and also contains a high equilibrium concentration of monomeric molecules (Eliasson, 1986). Lysolecithin is an amphophilic molecule and is an excellent emulsifying agent (Larsson 1980). Thus, pregelatinized wheat starch might be expected to exhibit emulsifying properties due to release of lysolecithin, even though much of the lysolecithin is complexed with amylose near room temperature. Enthalpy measurements (Kugimiya and Donovan, 1981) show amylose can bind up to 14% by weight of lysolecithin. Wheat starch with 30% amylose could bind up to 4.2% of its weight as lysolecithin, which agrees with the estimate of Wren and Merryfield (1970).

Morrison and Coventry (1985) reported the removal of internal lipids from starches using hot water/alcohol mixtures. Takahashi and Seib (1988) confirmed that three successive extractions with a boiling (81°C) mixture (3/1, v/v) of ethanol/water removed 90% of the lipids from wheat and corn starches. Those facts are attractive; a pure, powerful surfactant can be obtained from wheat starch using an edible alcohol at atmospheric pressure. Moreover, low-lipid wheat starch gives especially strong gels (Takahashi and Seib, 1988), and should be readily converted to sweetener (Bowler et al, 1985).

Lysolecithin is responsible for the two-stage "swelling" observed in the amylograms of wheat and corn starches. The first stage increase in paste consistency between 55° and 65°C for wheat starch in Fig 5 is due to the rapid release of amylose and the accompanying partial swelling of the granules. The second stage rise in consistency between 85° and 95°C

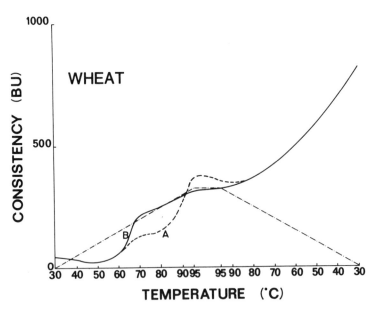

Fig. 5. Amylograms of (A) native and (B) low-lipid wheat starch at 5% starch in water containing 1% carboxymethycellulose (Takahashi and Seib, 1988).

appears to be due to dissociation of the amylose-lysolecithin complex, which triggers more release of amylose into the continuous phase and more swelling of the gel phase. Most of the two-stage swelling disappears (Fig 5) when 90% of the lipids in wheat starch are removed (Takahashi and Seib, 1988).

Increasing the level of wheat starch lipids from 1 to 3% increases paste consistency, but decreases gel strength (Takahashi and Seib, 1988). If wheat starch lipids are added to wheat starch after cooking to $95^{\circ}C$ in the amylograph, the paste consistency decreases substantially during the cooling cycle.

GEL STRENGTH

Force-deformation experiments were used to examine the strength of corn and wheat starch gels (Takahashi and Seib, 1988). Below 6% solids, corn starch gave stronger gels than wheat starch, but above 7%, and up to 30%, wheat starch gels were considerably more elastic (Fig 6). This difference at > 7% solids may be due to the relatively low levels of amylose leached from corn starch compared to wheat starch. It will be recalled that the strength of normal starch gels aged at $25^{\circ}C$ for 24-48 h is due mainly to retrogradation of amylose in the continuous phase of the cooked paste (Miles et al, 1985). With increasing concentration of starch in the cooking water, the amount of amylose leached from corn starch declined much faster than for wheat starch (Table VI). Krusi and Neukom (1984) reported that freshly prepared wheat starch gels at 40 and 50% dry solids contained 5.4% and 2.6% cold-water solubles.

SURFACE OF WHEAT STARCH

The surface of a wheat starch granule is not pure carbohydrate; instead, the surface

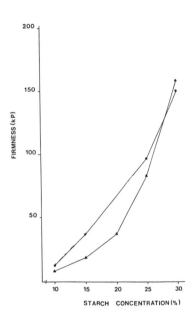

Fig. 6. Firmness of wheat (•) and corn starch (▲) gels after aging 24 h at 25°C (Takahashi and Seib, 1988).

contains protein (Simmonds et al. 1973) and lipids as well (Morrison, 1981 and 1987). Those contaminants may play a pivotal role in the hardness of wheat kernels and the behaviour of wheat doughs. Soft-textured wheat kernels have been associated with the

TABLE VI
Solubles (%) Leached from Starches Heated to 95°C in Water

Starch	Starch Concentration, %		
	1.5	3.0	4.5
Wheat	29	21	9
Corn	28	14	3

presence of a small protein (friabilin) on the surface of its isolated starch (Schofield and Greenwell 1987).

Hoseney et al (1971) showed in fractionation/reconstitution studies that only wheat, rye, and barley starches performed in breadmaking. Furthermore, the rheology of bread dough changed when potato (more slack) or corn starch (more stiff) was substituted for wheat starch (Hoseney, 1988). Sollars and Rubenthaler (1971) found in Cl_2-treated starches, that wheat, rye and especially barley gave good cakes with a lean formula, while rice, potato and corn starches were inferior. In cookies, wheat, rye and barley performed much better than corn and rice starches, while potato was of intermediate value. Because wheat, rye and barley starches share the same gelatinization and swelling properties (Williams and Bowler, 1982), those starches may perform best in baking formulas developed by bakers through trial and error.

CONCLUSIONS

Wheat starch appears bright and white, and contains two populations of granules. The large granules are more readily purified from contaminating protein and lipid than the small granules. Wheat starch contains 1% lipid, most of which is lysolecithin that can be extracted with boiling ethanol/water (75/25, v/v). Wheat starch gelatinizes 6-8°C below corn starch. Pastes of wheat starch are lower in consistency than corn starch, and its gels are weaker below 6% solids but stronger between 8-30% solids. Practically quantitative yields (28%) of amylose are leached from a 2% suspension of wheat starch under mild agitation in water at 95°C. Wheat amylopectin contains smaller molecules than corn AP, and both wheat AM and AP are more highly branched.

LITERATURE CITED

ACKER, L., and SCHMITZ, H. J. 1967. On the lipids of wheat starch. Starch 19:233.

ARMSTRONG, E. and FULCHER, R. G. 1988. Electrophoretic characterization of proteins from large and small wheat starch granules, Abstract No 30, American Assoc. Cereal Chem. 73rd Annual Meeting, Oct. 9-13, 1988, San Diego, CA.

BILIADERIS, C. G. 1986. On the multiple melting transition of starch/monoglyceride systems. Food Chem. 22:279.

BOWLER, P., TOWERSEY, P. J., WAIGHT, S. G., and GALLIARD, T. 1985. Minor components of wheat starch and their technological significance. In: New Approaches to Research on Cereal Carbohydrates. R. D. Hill and L. Munck, eds. Elsevier, New York, p. 71.

CHUANG, J.-Y., and SYDOR, R. J. 1987. High performance size exclusion chromatography of starch with dimethyl sulfoxide as the mobile phase. J. Appl. Polymer Sci. 34:1739.

DENGATE, H. N. 1984. Swelling, pasting, and gelling of wheat starch. Adv. Cereal Sci. Technol. 6:49.

DERBY, R. I., MILLER, B. S., MILLER, B. F., and TIMBO, H. B. 1975. Visual observation of wheat starch gelatinization in limited water system. Cereal Chem. 52:702.

DOUBLIER, J.-L., PATON, D., and LAMAS, G. L. 1987. A rheological investigation of oat starch pastes. Cereal Chem. 64:21.

ELIASSON, A.-C. 1986. On the effects of surface active agents on the gelatinization of starch - a calorimetric investigation. Carbohydr. Polymers 6:463.

ELIASSON, A.-C. 1988. Characterization of wheat starch and gluten as related to end-use properties. Wheat Industry Utilization Conference, San Diego, CA.

ELIASSON, A.-C., and KARLSSON, R. 1983. Gelatinization properties of different size classes of wheat starch granules measured with differential scanning calorimetry. Starch 35:130.

GHIASI, K., HOSENEY, R. C., and VARRIANO-MARSTON, E. 1982a. Gelatinization of wheat starch. I. Excess-water system. Cereal Chem. 59:81.

GHIASI, K., HOSENEY, R. C., and VARRIANO-MARSTON, E. 1982b. Gelatinization of wheat starch. III. Comparison by differential scanning calorimetry and light microscopy. Cereal Chem. 59:258.

GHIASI, K., HOSENEY, R. C., and VARRIANO-MARSTON, E. 1983. Effects of flour components and dough ingredients on starch gelatinization. Cereal Chem. 60:58.

GIDLEY, M. J. 1987. Personal communication.

GIDLEY, M. J., and BOCIEK, S. M. 1985. Molecular organization in starches: A ^{13}C CP/MAS NMR study. J. Am. Chem. Soc. 107:7040.

GIDLEY, M. J., and BOCIEK, S.M. 1988. ^{13}C CP/MAS NMR studies of amylose inclusion complex, cyclodextrins, and the amorphous phase of starch granules: relationship between glycosidic linkage conformation and solid-state ^{13}C chemical shifts. J. Am. Chem. Soc. 110:3820

HIZUKURI, S. 1986. Polymodal distribution of the chain lengths of amylopectins, and its significance. Carbohydr. Res. 147:342.

HIZUKURI, S. 1988. Recent advances in molecular structures of starch. J. Jpn. Soc. Starch Sci. 35:185.

HOSENEY, R. C. 1988. Personal communication.

HOSENEY, R. C., ATWELL, W. A., and LINEBACK, D. R. 1977. Scanning electron microscopy of starch isolated from baked products. Cereal Foods World 22:56.

HOSENEY, R. C., FINNEY, K. F., POMERANZ, Y., and SHOGREN, M. D. 1971. Functional (breadmaking) and biochemical properties of wheat flour components. VIII. Starch. Cereal Chem. 48:191.

KOBAYASHI, S., SCHWARTZ, S. J., and LINEBACK, D. R. 1986. Comparison of the structures of amylopectins from different wheat varieties. Cereal Chem. 63:71.

KRUSI, H., and NEUKOM, H. 1984. Investigation on the retrogradation in concentrated wheat starch gels. Part I. Preparation of concentrated gels, influence of starch concentration and conditions of preparation on starch retrogradation. Starch 36:40.

KUGIMIYA, M., and DONOVAN, J. W. 1981. Calorimetric determination of the amylose content of starches based on formation and melting of the amylose-lysolecithin complex. J. Food Sci. 46:765.

LARSSON, K. 1980. Inhibition of starch gelatinization by amylose-lipid complex formation. Starch 32:125.

MEREDITH, P., and DENGATE, H. N. 1978. The lipids of various sizes of wheat starch granules. Starch 30:119.

MEREDITH, P. 1981. Large and small starch granules in wheat - Are they really different? Starch 37:1.

MILES, M. J., MORRIS, V.-J., ORFORD, P. D., and RING, S. G. 1985. The roles of amyloses and amylopectin in the gelation and retrogradation of starch. Carbohydr. Res. 135:271.

MORRISON, W. R. 1981. Starch lipids: a reappraisal. Starch 33:408.

MORRISON, W. R. 1987. Lipids in wheat and barley starch granules. In Cereals in a European Context, Morton, I. D., ed., VCH Publishers, New York, p. 438.

MORRISON, W. R., and COVENTY, A. M. 1985. Extraction of lipids from cereal starches with hot aqueous alcohols. Starch 37:83.

MORRISON, W. R., and LAIGNELET, B. 1983. An improved colorimetric procedure for determining apparent and total amylose in cereal and other starches. J. Cereal Sci. 1:9.

MORRISON, W. R., and MILLIGAN, T. P. 1982. Lipids in maize starches. In Maize: Recent Progress in Chemistry and Technology. Inglett, G. ed. Academic Press, pp. 1-18.

MORRISON, W. R., and SOUTH, J. B. 1988. Comparative aspects of starch granule composition and structure, In: International Symposium on Cereal Carbohydrates. Edinburgh, Scotland.

PRAZNIK, W., SCHILLINGER, H., and BECK, R. H. F. 1987. Changes in the molecular composition of maize starch during kernel development. Starch 39:183.

SCHOFIELD, J. D., and GREENWELL, P. 1987. Wheat starch granule proteins and their technological significance. In Cereals in a European Context, Morton, I. D. ed., VCH Publishers, New York, p. 407.

SIMMONDS, D. H., BARLOW, K. K., and WRIGLEY, C. W. 1973. The biochemical basis of grain hardness in wheat. Cereal Chem. 50:553.

SLADE, L., and LEVINE, H. 1987. Recent advances in starch retrogradation. In: Recent Developments in Industrial Polysaccharides, S. S. Stivala, V. Crescenzi and I. C. M. Dea, eds. Gordon and Breach Science, New York, p. 387.

SLADE, L., and LEVINE, H. 1988. Non-equilibrium melting of native granular starch: Part I. Temperature location of glass transition associated with gelatinization of A-type cereal starches. Carbohydr. Polymers 8:183.

SNYDER, E. M. 1984. Industrial microscopy of starches. In: Starch: Chemistry and Technology, R. L. Whistler, J. N. Bemiller, and E. F. Pashall, eds. Academic Press, NY. p. 661.

SOLLARS, W. F., and RUBENTHALER, G. L. 1971. Performance of wheat and other starches in

reconstituted flours. Cereal Chem. 48:397
SWINKELS, J. J. M. 1985. Composition and properties of commercial native starches. Starch 37:1.
TAKAHASHI, S., and SEIB, P. A. 1988. Paste and gel properties of prime corn and wheat starches with and without native lipids. Cereal Chem., in press.
TAKEDA, Y., HIZUKURI, S., TAKEDA, C., and SUZUKI, A. 1987. Structures of branched molecules of amylose of various origins, and molar fractions of branched and unbranched molecules. Carbohydr. Res. 165:139.
TAN, S. L., and MORRISON, W. R. 1979. The distribution of lipids in the germ, endosperm, pericarp and tip cap of amylomaize, LG-11 hybrid maize and waxy maize. J. Am. Oil. Chem. Soc. 56:531.
VARRIANO-MARSTON, E., KE, V., HUANG, G., and PONTE, J. JR. 1980. Comparison of methods to determine starch gelatinization in bakery foods. Cereal Chem. 57:242.
WILLIAMS, M. R., and BOWLER, P. 1982. Starch gelatinization of Triticeae and other starches. Starch 34:221.
WOLF, M. J., SECKINGER, H. L., and DIMLER, R. J. 1964. Microscopic characteristics of high-amylose corn starches. Starch 16:376.
WREN, J. J., and MERRYFIELD, D. S. 1970. Firmly bound lysolecithin of wheat starch. J. Sci. Food. Agric. 21:254.
YOUNG, A. H. 1984. Fractionation of starch. In: Starch: Chemistry and Technology. R. L. Whistler, J. N. Bemiller, and E. F. Paschall, eds. Academic Press, NY, p. 249.
ZELEZNAK, K. J., and HOSENEY, R. C. 1987. The glass transition in starch. Cereal Chem. 64:121.

Contribution No. 89-171-A from the Kansas Agricultural Experiment Station, Manhattan, KS 66506.

14

MODIFICATION OF WHEAT STARCH

Stuart A. S. Craig
Fundamental Science
Nabisco Brands Inc.
RMS Technology Center
E. Hanover, NJ 07936

INTRODUCTION

Corn starch is the major starch produced in the US. There is, however, increasing interest in the use of wheat starch, either native or as a raw material for modification. Economics plays a significant role in the choice of starches as raw material for modification, as evidenced by the almost exclusive use of wheat as a source for starch in Australia and New Zealand, and increased use in the EEC. Inherent differences in the properties of native starches are reduced by modification. This paper will address two means of modifying starch - chemical and physical.

CHEMICAL MODIFICATION

Chemical modification of starch includes a subdivision commonly termed derivatization (Rutenberg and Solarek, 1984). The chemical structure of some D-glucopyranosyl units within the molecules of amylose or amylopectin will have been changed by derivatization. These modifications include esterification, etherification and oxidation. Other methods of modification include hydrolysis with acid and/or enzymes to reduce the molecular weight of starch chains. Only esterification and etherification will be discussed here.

Monosubstitution (ester or ether) may be carried out with reagents such as acetic anhydride, propylene oxide, succinic anhydride and tertiary and quaternary ammonium compounds. The reagent molecules bind covalently with one or more hydroxyl groups per D-glucopyranosyl unit (Moore et al, 1984). This changes the rheological properties of the starch which are

seen as -
1) a decrease in pasting temperature,
2) reduced gel formation,
3) improved freeze-thaw and water-holding properties,
4) increased paste clarity.

Cross-linking can be achieved with di- or polyfunctional reagents such as phosphorus oxychloride or epichlorohydrin. These reagents bind covalently with the hydroxyl groups of starch to form a bridge between the polymer chains. Very low levels (0.0005%) of cross-linking can alter rheological properties. The functional effects are seen as -
1) an increased pasting temperature,
2) increased stability to temperature, shear and pH,
3) desirable texture and mouth-feel (by stabilizing swollen granules during cooking),
4) improved freeze-thaw and water-holding properties,
5) reduced paste clarity.

In addition, double-derivatization (a combination of monosubstitution and cross-linking) can produce starches with desirable properties that are "fine-tuned".

Beyond the effects of these substituents on observed functionality of starches, there has been some detailed work on changes at a molecular level. Differential Scanning Calorimetry (DSC) has been used (Craig et al, 1987) to measure the effect of substitution on the gelatinization of wheat starch (Fig. 1, Table I). An increasing degree of monosubstitution, using a quaternary ammonium compound, caused a decrease and broadening of the temperature (To, Tp, Tc) and endotherm (ΔH) of gelatinization. It was shown that the changes were due to the presence of these cationic groups and not the other reaction conditions. In addition, large granular wheat starch was used to produce monosubstituted (cationic, adipate-acetate) and cross-linked (phosphorus oxychloride) derivatives. The monosubstituted starches showed (Table I) a decrease in gelatinization temperature and endotherm, but the cross-linked starch did not.

These results suggested that the monosubstituting groups were able to facilitate melting of amylopectin crystallites. Hood and Mercier (1978) and Biliaderis

TABLE I
DSC Properties of Native and Modified Wheat Starches

Starch	M.S.	ΔH (cal/g)	To (°C)	Tp (°C)	Tc (°C)
Native	0	2.7	58	64	71
Control	0	2.7	58	63	69
Cationic	0.015	2.6	58	63	69
Cationic	0.026	2.4	54	60	65
Cationic	0.055	2.4	49	56	73
Large granule	0	2.9	58	62	67
Cationic	0.048	2.4	47	53	60
Cross-linked (0.03%)	-	2.9	59	62	65
Cross-linked (0.10%)	-	2.8	59	63	66
Adipate/acetate	-	2.4	51	57	63

(Source: Craig et al, 1987).

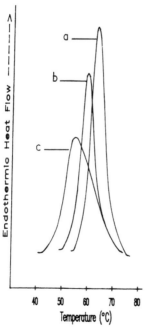

Fig. 1. DSC thermograms of native and cationic wheat starch. a = native; b = M.S. 0.026; c = M.S. 0.055. (Source: Craig et al, 1987).

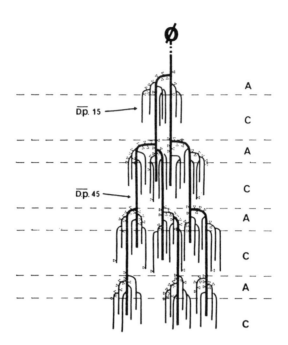

Fig. 2. Proposed structure of modified amylopectin. Δ indicates location of groups. (A) Amorphous region, (C) crystalline region. (Source: Hood and Mercier, 1978).

(1982) proposed that substituting groups were located in the amorphous regions of amylopectin within the granule, i.e. predominantly around the branch points, with some groups near the non-reducing chain ends of the clusters (Fig. 2). The substituents may have increased the hydration of starch molecules in the amorphous regions and provided extra water to the crystallites during melting. The extra water, acting as a plasticizer (Maurice et al, 1985), lowered both the gelatinization temperature and endotherm.

Fig. 3 shows a more complete DSC curve of native and cationic wheat starch. This includes the endotherm of gelatinization (crystallite melting) and the endotherm of amylose-lipid complex melting. Note that this second endotherm occurred at around 100°C for native starch and at 90°C for cationic starch. This trend was consistent for all cationic and adipate-acetate starches in this study. Cross-linked

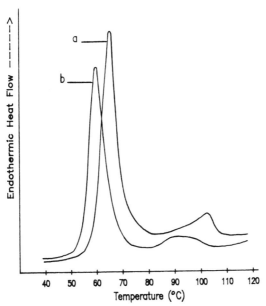

Fig. 3. DSC thermograms of native and cationic wheat starch. a = Native; b = M.S. 0.026 (Source: Craig et al, 1987).

starches did not show this feature. This suggests that amylose contained substituent groups which inhibited the formation of lipid single helices (V-complex). These imperfect crystals melted at a lower temperature.

TABLE II
Iodine Complexing Properties of
Native and Modified Starches

Starch	M.S.	Blue value	λ_{max} (nm)
Native wheat	0	0.47	633
Cationic wheat	0.048	0.16	600
Native corn	0	0.43	605
Cationic corn	0.071	0.27	585

(Source: Craig et al, 1987).

Another V-complex which was inhibited by the presence of substituting groups was iodine-complexing (Table II). The presence of these groups lowered the blue value and λ_{max} of starches. Also in that study, fractionation of cationic wheat starch into amylose and amylopectin was attempted by solvent precipitation using thymol/1-butanol (Cowie and Greenwood, 1957) or n-amyl alcohol (Schoch, 1945). The amylose complex could not be precipitated by centrifugation, suggesting that the V-complex formation of amylose and alcohol was inhibited by the small number of substituents on the amylose chain.

Table III shows the effect of various derivatization treatments on the clarity of starch pastes (1%) as measured by % light transmittance (%T) at 650 nm (Craig et al, 1988). Note that

TABLE III
Light Transmittance of Modified Wheat Starch Pastes

Starch (1%)	Transmittance (% at 650 nm)
Native	62
Cationic (M.S. 0.000)[b]	62
Cationic (M.S. 0.015)	89
Cationic (M.S. 0.026)	91
Cationic (M.S. 0.048)[a]	93
Cationic (M.S. 0.055)	93
Phosphorylated (D.S. 0.0039)[a]	76
Phosphorylated (D.S. 0.0056)[a]	85
Succinylated[a]	78
Cross-linked (0.01%)[a]	56
Cross-linked (0.02%)[a]	52
Cross-linked (0.03%)[a]	52

[a]Large granules
[b]Control

(Source: Craig et al, 1987).

monosubstitution (phosphate, succinate, cationic) improved %T but cross-linking decreased it. Also, the whiteness of the starch pastes was reduced by monosubstitution. One of the reasons that potato starch pastes exhibit good clarity is the presence of native phosphate groups covalently attached to amylopectin in the granule. This clarity can be obtained from wheat starch by monosubstitution of the granules with phosphate (or other groups), as shown in Table III. Repulsion and/or steric hinderance of ionic or non-ionic substituents prevents starch molecules from hydrogen-bonding to each other, collapsing and ultimately retrograding. This helps keep the molecules fully hydrated, thus promoting light transmittance and decreasing whiteness.

PHYSICAL MODIFICATION

Starches can be modified in a number of "physical" ways. The most obvious is by milling to produce damaged starch. The extent to which starch granules are damaged is often a consequence of the relative "hardness" of the grain from which flour must be produced. Thus, the harder grains require more physical force to produce flour of the required particle size and thereby are inflicted with the most starch damage. The extent of starch damage is determined rapidly either by measuring the increased susceptibility to amylolytic enzymes (AACC, 1972) or the increased cold-water solubility (McDermott, 1980). For bakery products there is an optimum level of starch damage (in terms of absorption and bread quality) depending on the flour protein content, level of α-amylase, the type of baking process used and the final product desired. Table IV shows the desired state of starch granule disorganization in some baked goods (Greenwood, 1979).

Other changes that occur to starch granules as a result of physical damage include -
1) loss of birefringence
2) loss of x-ray crystallinity (Lelievre, 1974)
3) loss of endothermic heat of gelatinization as measured by DSC (Stevens and Elton, 1971)
4) decrease in molecular weight (Meuser et al, 1978).

TABLE IV
Extent of granular disorganization in baked goods.

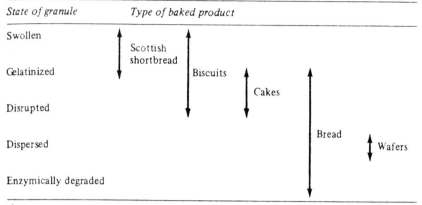

(Source: Greenwood, 1979; reprinted with permission of the publishers, Butterworth & Co. Ltd. ©)

Changes to the molecular structure of wheat starch granules after physical damage have been studied (Craig and Stark, 1984). Table V shows the result of treatment of wheat starch granules in a McCrone Micronizing Mill for 35 minutes, followed by extraction with cold water. This severe milling treatment rendered previously insoluble starch granules to be 53% cold water-soluble. The blue value, λ_{max} and viscosity ($[\eta]$) values indicated that the cold water-soluble extract was either mostly amylopectin - type material or severely degraded

TABLE V
Properties of Wheat Starch Fractions

Fraction	Blue value	λ_{max} (nm)	Cold H_2O sol. (%)	Visc. $[\eta]$
Wheat Starch (A)	0.47	633	0	188
Damaged Wheat Starch (B)	0.47	630	53	132
Cold-water extract of B	0.18	603	sol.	34
Insoluble residue of B	0.58	634	insol.	156
Wheat-flour α-D-glucan	0.04	537	sol.	64

(Source: Craig and Stark, 1984)

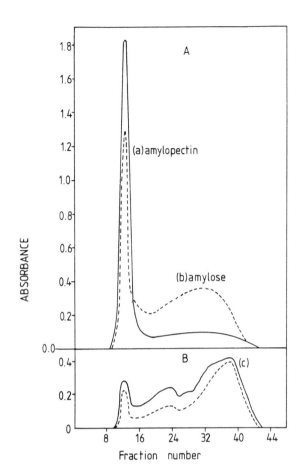

Fig. 4. Chromatography on Sepharose 2B of wheat starch (A) and the cold-water extract of damaged wheat starch (B): total carbohydrate analyzed by the phenol-sulfuric acid method at 490 nm (———), and iodine-staining absorbance (– – –) at 600 nm. (Source: Craig and Stark, 1984).

amylose. Fig. 4A shows a gel permeation profile of wheat starch with amylopectin (peak a, weak iodine-stain) separated from amylose (peak b, strong iodine-stain). Comparison of this profile with that for the cold-water extract of damaged wheat starch (Fig. 4B) showed that the latter contained mostly amylopectin-type material. Further enzymic characterization showed that the cold-water soluble material was 87% low molecular weight amylopectin.

The above findings raised questions regarding the long held belief that the cold-aqueous extract from wheat starch present in flour is amylose (Lampitt et al, 1941, Williams and Fegol, 1969 and McDermott, 1980). Craig and Stark (1984) turned their attention to commercially milled wheat flour which undergoes milder stresses than those imposed in the laboratory micronizing mill. The properties of a purified α-D-glucan extracted from commercial wheat flour are shown in Table V. Note that the yield from flour was only <1%. The iodine-staining and viscosity values

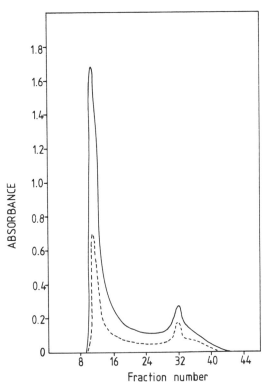

Fig. 5. Chromatography on Sepharose 2B of the α-D-glucan component of wheat-flour extract. Total carbohydrate (———) and iodine-staining (— — —) as in Fig. 4. (Source: Craig and Stark, 1984).

suggested a low molecular weight amylopectin-like polymer. This was confirmed by further enzymic characterization and a gel permeation profile (Fig. 5). This profile showed that the extract had a molecular weight greater than that of the extract from laboratory-milled wheat starch, but that the iodine-staining profile remained low. The extract was found to be 99% amylopectin-type. The final conclusion of that paper was that cold-aqueous extracts of damaged starch granules consisted mainly of branched material with a molecular weight lower than that of normal amylopectin.

Starch manufacturers can derivatize starches by physical means. Pre-gelatinized starches have long been available for food applications where an instant paste without heat is required. These starches are powdered, but have lost their native granular structure. A recent development by Staley Mfg. Co. has been the use of a high-temperature, aqueous alcohol system to produce starches under the brand name Miragel (Eastman and Moore, 1984). These starches are unique because they are cold-water soluble and granular integrity is retained.

A study by Jane et al (1986) characterized the properties of these starches (including a wheat starch version). They concluded that the cold-water solubility of granular cold-water soluble (GCWS) starch results from the disappearance of the native, A-starch crystal structure and the formation of a structure with the same x-ray diffraction pattern as V-hydrate amylose (a water-soluble structure). GCWS starch gave a pattern equivalent to a 1:1 mixture of crystalline V-amylose and amorphous waxy maize (mainly amylopectin). Since the GCWS starch contained only about 27% amylose, the amylopectin component must participate in the V-crystal structure. Under the light microscope, GCWS starch was birefringent but no Maltese cross was observed, indicating that the native organization of the crystallites has been disrupted. Fig. 6 shows their proposed mechanism for conversion of native to GCWS starch. Heating with aqueous alcohol to a high enough temperature converts the native double helical structure into single helices. This transformation occurs in both amylose and

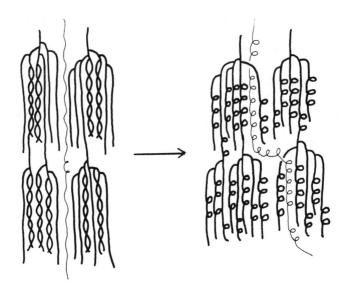

Fig. 6. Proposed conversion of an A-type starch granule to a V-type granule. Amylose (———) and (———) amylopectin. (Source: Jane et al, 1986).

amylopectin molecules. The granules remain intact either due to the association (entanglement) of amylose with amylopectin molecules, or participation of amylose molecules in several crystallites. Removal of the solvents leaves the starch in a semi-stable state that is soluble in cold water.

There is great current interest in "Resistant Starch" (RS) and its presence in food processing (Berry, 1986; Englyst et al, 1987). RS is starch which is resistant to human digestion through the small intestine. RS is produced by certain physical processing techniques such as baking and extrusion. Different structures may form, dependant on heat, time, moisture, pH, heating/cooling cycles, drying, milling, presence of complexing agents, etc. Extraction of native lipids from starch with aqueous alcohol resulted in an increase in RS content (Morrison, 1981). RS may be solubilized in DMSO or 2M

KOH and subsequently hydrolyzed by amylolytic enzymes. Thus it is believed that the resistance is due to physical rearrangement, rather than chemical modification, of starch molecules (Berry et al, 1988). Englyst (1988) has proposed the following classification of starch for nutritional purposes -

RAPIDLY DIGESTIBLE STARCH, e.g. as found in freshly cooked food.

SLOWLY DIGESTIBLE STARCH, e.g. as found in raw cereals.

RESISTANT STARCH, subdivided into 4 types:
1) Physically inaccessible starch, e.g. as found in partly milled grains and seeds.
2) Resistant starch granules, e.g. as found in raw potato and banana.
3) Retrograded amylopectin, e.g. as found in cooled cooked potato.
4) Retrograded amylose, e.g. as found in white bread and cornflakes.

It is clear that starch can be modified physically to reduce its digestibility. As extrusion becomes an even more popular tool for the preparation of wheat based foods, more work on RS will be forthcoming to better define the physical structures produced and their physiological significance.

CONCLUSION

Wheat starch is used in a wide variety of ways and may be modified both chemically and physically. There is still much work that needs to be done to increase our understanding of what structural changes we are imparting to granules.

LITERATURE CITED

AMERICAN ASSOCIATION OF CEREAL CHEMISTS. 1972. Damaged starch. Cereal Laboratory Methods. 76-30A.
BERRY, C.S. 1986. Resistant Starch: Formation and Measurement of Starch that Survives Exhaustive

Digestion with Amylolytic Enzymes during the Determination of Dietary Fibre. J. Cereal Sci. 4:301.
BERRY, C.S., I'ANSON, K., MILES, M.J., MORRIS, V.J. and RUSSELL, P.L. 1988. Physical Chemical Characterisation of Resistant Starch from Wheat. J. Cereal Sci. 8:203.
BILIADERIS, C.G. 1982. Physical characteristics, enzymic digestibility, and structure of chemically modified smooth pea and waxy maize starches. J. Agric. Food Chem. 30:925.
COWIE, J.M.G., and GREENWOOD, C.T. 1957. Physicochemical studies on starches. V. The effect of acid on potato starch granules. J. Chem. Soc. 2658.
CRAIG, S.A.S., and STARK, J.R. 1984. The effect of physical damage on the molecular structure of wheat starch. Carbohydr. Res. 125:117.
CRAIG, S.A.S., SEIB, P.A., and JANE, J.-L. 1987. Differential scanning calorimetry properties and paper-strength improvement of cationic wheat starch. Starch 39:167.
CRAIG, S.A.S., MANINGAT, C.C., SEIB, P.A., and HOSENEY, R.C. 1988. Starch paste clarity. Cereal Chem. In press.
EASTMAN, J.E., and MOORE, C.O. 1984. Cold-Water-Soluble Starch for Gelled Food Compositions. U.S. Patent 4,465,702.
ENGLYST, H. 1988. New concepts in starch digestion in man. International Symposium on Cereal Carbohydrates, Edinburgh, Scotland.
ENGLYST, H.N., TROWELL, H.W., SOUTHGATE, D.A.T. and CUMMINGS, J.H. 1987. Dietary Fiber and Resistant Starch. Am. J. Clin. Nutr. 46:873.
GREENWOOD, C.T. 1979. Observations on the structure of the starch granule. Page 129 in: Polysaccharides in Food. J.M.V. Blanshard and J.R. Mitchell, eds. Butterworths, London.
HOOD, L.F., and MERCIER, C. 1978. Molecular structure of unmodified and chemically modified manioc starches. Carbohydr. Res. 61:53.
JANE, J.-L., CRAIG, S.A.S., SEIB, P.A., and HOSENEY, R.C. 1986. Characterization of granular cold water-soluble starch. Starch 38:258.

LAMPITT, L.H., FULLER, C.H.F., and GOLDENBERG, N. 1941. The fractionation of wheat starch. Part 1. The process of grinding. J. Soc. Chem. Ind. 60:1.

LELIEVRE, J. 1974. Starch damage. Starch 26:85.

MAURICE, T.J., SLADE, L., SIRETT, R.R., and PAGE, C.M. 1985. Polysaccharide-water interactions - thermal behavior of rice starch. Page 211 in: Properties of Water in Foods. D. Simatos and J.L. Multon, eds. Martinus Nijhoff, Dordrecht, The Netherlands.

McDERMOTT, E.E. 1980. The rapid, non-enzymic determination of damaged starch in flour. J. Sci. Food Agric. 31:405.

MEUSER, F., KLINGLER, R.W., and NIEDIEK, E.A. 1978. Characterization of mechanically modified starch. Starch 30:376.

MOORE, C.O., TUSCHHOFF, J.V., HASTINGS, C.W., and SCHANEFELT, R.V. 1984. Application of starches in foods. Page 575 in: Starch: Chemistry and Technology. 2nd ed. R.L. Whistler, J.N. BeMiller and E.F. Paschall, eds. Academic Press: New York.

MORRISON, W.R. 1981. Starch Lipids - A Reappraisal. Starch 33:408.

RUTENBERG, M.W., and SOLAREK, D. 1984. Starch Derivatives: Production and Uses. Page 312 in: Starch: Chemistry and Technology. 2nd ed. R.L. Whistler, J.N. BeMiller and E.F. Paschall, eds. Academic Press: New York.

SCHOCH, T.J. 1945. The fractionation of starch. Page 247 in: Advances in Carbohydrate Chemistry. Vol. I. W.W. Pigman and M.L. Wolfrom, eds. Academic Press: New York.

STEVENS, D.J., and ELTON, G.A.H. 1971. Thermal properties of the starch/water system. Part 1. Measurement of heat of gelatinization by differential scanning calorimetry. Starch 23:8.

WILLIAMS, P.C., and FEGOL, K.S.W. 1969. Colorimetric determination of damaged starch in flour. Cereal Chem. 46:56.

15

MINOR COMPONENTS OF WHEAT STARCH AND THEIR TECHNOLOGICAL SIGNIFICANCE

T. Galliard, P. Bowler and P.J. Towersey

RHM Research and Engineering Ltd,
Lord Rank Research Centre, Lincoln Road,
High Wycombe, Bucks., England, U.K. HP12 3QR

INTRODUCTION

The 1980's have seen a dramatic increase in the utilization of wheat for starch production in the European Economic Community (EEC). In 1980, only 3% of starch and starch-derived products were from wheat. By 1986, 0.6 million tonnes of products (starch equivalent basis) were from wheat, accounting for 13.5% of the total; 2.8 and 0.6 million tonnes were from maize and potatoes, respectively (Galliard 1987; Jones 1987). Currently (1988) it is estimated that wheat starch accounts for 18% of the products of the starch processing industry in the EEC (R.G. Jones, Tenstar Products Ltd, personal communication).

The reasons for this dramatic increase in wheat utilization are two-fold:-
1) the vastly increased demand for vital wheat gluten in the baking industry where it is more economical to use gluten-supplemented flour from lower-protein European wheat than to use higher-protein, non-EEC breadmaking wheat. Gluten consumption in the EEC has increased from 30,000 tonnes in 1982 to a current (September 1988) estimated level of 140,000 tonnes on an annualised basis (R.G. Jones, personal communication).
2) during the early 1980's the economics of starch (and its gluten co-product) production from wheat were

favourable when compared with starch production from imported, non-EEC maize (Wookey and Melvin 1981, Galliard 1987). However, the increasing use of European maize and the recent fall in the price of wheat gluten have largely eroded the economic advantage of wheat starch production.

Analysis of the major uses of starch in the U.K. (Table I) shows that, of the approx. 0.9 million tonnes of starch and its derivatives used, around 75% goes into the food and beverage industry and, of this, 84% is in the form of starch hydrolysates - mainly as glucose or maltose syrups; unlike in N. America, iso-glucose production is small, due to restrictions imposed by the EEC and designed to protect the beet sugar industry.

The dramatic increase in the availability of wheat starch has heightened interest in its potential and in its characteristics *vis-a-vis* those of maize and potato starches.

TABLE I
Major Uses of Starch in the U.K. (thousand tonnes)[a]

	Food and Beverages	Non-Food	Total
Starch, unmodified	70	130	200
Modified starch	33	67	100
Glucose, syrups	400	40	440
Glucose, solid	50	0	50
iso-Glucose	50	0	50

[a] tonnes expressed on a commercial basis; (Source: R.G. Jones, Tenstar Products Ltd, personal communicatio

Much of the recent work in our laboratories has been based on the premise that the differences in properties between wheat and maize starches have more to do with the minor components of these materials, rather than to basic differences in polysaccharide chemistry and architecture. Because starch hydrolysates represent the major current and potential uses of starch in the U.K., we have paid particular attention to the role of minor components of starch in glucose syrup production. However, the minor components also affect the behaviour of starch *per se* and some of these will be reviewed briefly.

MINOR COMPONENTS OF UNMODIFIED STARCHES

Commercial samples of wheat and maize starches contain similar amounts and classes of minor components (Bowler et al. 1985a, Galliard and Bowler 1987; see also Table III) none of which exceed 1% of starch by weight. However, because substantial proportions of these materials are associated with the surface of starch granules, where they act at the interface between starch granules and their environment, their effects on starch properties can be proportionately greater.

The minor components of technological significance are:-

non-starch polysaccharides, mainly insoluble pentosans derived from endosperm cell wall materials and carried through the starch separation process.

proteins (including enzymes) from three sources:- a) contaminating endosperm materials, b) proteins associated with the surface of starch granules and c) proteins buried within the granule matrix.

lipids from the same three sources as the proteins mentioned above.

The effects of these minor components on starch properties and the results of experiments in which these were removed selectively from starch have been reviewed elsewhere (Galliard 1985, Bowler et al. 1985a). The following paragraphs describe some recent observations on the nature and significance of components at the surface of starch granules.

'Friabilin'

At least 5 different polypeptides, with M_r values between 5,000 and 30,000 have been separated from wheat starch surface proteins and one of these, with a M_r of 15,000 and named 'friabilin' has been studied in depth at the Flour Milling and Baking Research Association in the U.K. (Schofield and Greenwell 1987). In a survey of over 300 wheat cultivars with widely different genetic backgrounds, it has been shown that starch from all soft wheats has a strong M_r 15,000 band on electrophoresis, whereas this band is absent or very weak in hard wheat varieties. Different cytogenetic approaches have established that the gene controlling friabilin synthesis is on the short arm of the 5D chromosome and is very closely associated with genes controlling the phenotypic characteristics of endosperm texture (hard/soft) and, as recently demonstrated, of polar lipid extractability (Morrison et al. 1988).

Alpha-Amylase

Studies in our laboratory, using different extractants to remove selectively the various components of the material on the surface of starch granules, have shown large differences in the pasting properties of water-washed starches from different varieties of wheat. As expected, unwashed starch that had been prepared from flour of high alpha-amylase activity gave low viscosities, compared with starch from low alpha-amylase wheat. However, whereas water-washing was sufficient to remove residual enzyme from some varieties, it was necessary to increase the ionic strength of the wash water to remove alpha-amylase activity from others (see Table II).

These observations have not been explored in further detail, but may indicate varietal differences in affinity for alpha-amylase binding to starch granules.

TABLE II

Effect of Starch Treatment on Peak Viscosities of Pastes of Starch from Different Varieties of Wheat

Wheat variety	Flour alpha-amylase Activity (Farrand units)	Peak Viscosity of 10% Starch Pastes (Brabender units)			
		Unwashed	Water-washed	Salt-washed[a]	Acid-treated[b]
Maris Huntsman	38	40	60	420	500
Bounty	7	100	280	600	700
Copain	1	180	520	640	780
Aquila	1	560	860	880	980

[a] 0.25M NaCl. [b] pH 2.2 (HCl). Data adapted from Bowler et al. (1985a)

Surface Lipids and Proteins

Lipids and proteins at the surface of starch granules have been implicated in the behaviour of starch with respect to pasting properties, wetting and dispersion, stability of starch suspensions and in the improving action of chlorine gas treatment of cake flours (for references, see Russell et al. 1987). X-ray photoelectron spectroscopy has been used recently to probe the nature of the starch granule surface and recent results (Russell et al. 1987) have shown:- a) concentration of both lipids and proteins in a thin surface layer, b) close association between these components and c) involvement of both in the reactions with chlorine.

MINOR COMPONENTS OF STARCH IN GLUCOSE SYRUP PRODUCTION

The economics of the wheat starch/gluten separation industry demand maximum yield of starch, commensurate with quality for end-use. The bimodal distribution of starch granules in wheat offers the commercial opportunity to produce two fractions of starch:- prime 'A' starch, comprising mainly intact large starch granules, and 'B' starch containing a mixture of smaller granules, damaged starch granules and, a higher proportion of non-starch impurities.

As with unmodified starch discussed above, the minor components that affect the processing and product quality of starch syrups derive from contaminating endosperm components, such as pentosan-rich cell wall materials and particulate proteins, and from the non-polysaccharide compounds of starch granules, i.e. the surface and internal lipids and proteins.

The contaminating pentosans and proteins are concentrated in the 'B' starch fraction and, hence, prime 'A' starch has much lower levels of these materials than does the unfractionated starch (Table III).

In prime 'A' starch, as with maize starch, the main non alpha-glucan materials are the lipids and proteins of starch granules. Although these are

TABLE III

Minor Components in Wheat and Maize Starches and in Residues Produced in Glucose Syrup Production[a]

Raw material	Hydrolysis conditions	Content (g/kg syrup, dry basis)						
		Residue after hydrolysis	Components of residue					
			Protein	Pentosan	Lipid	Starch	Acid-resistant starch	
Unfractionated wheat starch	None	-	7.7	14.6	8.6	-	-	
	Low shear	18.5	2.3	1.2	0.8	5.3	6.5	
	Jet-cooker	6.6	2.2	1.4	0.5	0.6	1.0	
Wheat 'A' starch fraction	None	-	2.8	1.4	6.6	-	-	
	Low shear	24.0	0.1	0.03	0.1	10.9	11.4	
	Jet-cooker	0.8	0.4	0.01	0.1	0.07	0.02	
Maize starch	None	-	4.1	1.4	9.9	-	-	
	Low shear	15.0	1.8	0.1	3.9	4.7	5.2	
	Jet-cooker	7.9	2.1	0.1	6.0	0.3	0.2	

[a] Data adapted from Bowler et al. (1985b)

present in similar amounts in wheat and maize starches, they respond very differently in syrup production.

In a series of experiments designed to investigate the roles of the individual minor components of starch on process characteristics and syrup quality, we used two experimental hydrolysis systems (Bowler et al. 1985b). Both involved 'all-enzyme' conversion, using thermostable alpha-amylase for liquefaction and glucoamylase for saccharification. One, a laboratory-scale process was a low-shear (stirred) system, whereas the other was a pilot-scale process using jet-cooking and high shear rates. The two systems were different with respect to the amount and nature of the insoluble residues (Table III) which affected both process characteristics (e.g. filterability) and product quality (clarity, colour, flavour, etc.).

Pentosans

As shown in Table III, the residues from the hydrolysates of unfractionated wheat starch contained substantial amounts of pentosan from contaminating cell wall material. Pentosan levels in the residues of maize starch were 10-fold lower and were very low in both experimental systems when prime 'A' wheat starch was used as the raw material.

Acid-resistant Starch

In the low-shear, laboratory-scale system all three types of raw material gave residues that contained 1 to 2% (syrup solids basis) of alpha-glucans (Table III). Approximately one half of the alpha-glucan was incompletely hydrolysed starch but the remainder was resistant to the usual acid conditions (3% H_2SO_4, 108°C) used in the analysis of starch. Much stronger acid conditions (82% w/v H_2SO_4, 30°C, 2h) as used for cellulose hydrolysis, were necessary to hydrolyse the material. However, further investigation (Bowler et al. 1985b) indicated that it was, in fact, a complex of incompletely hydrolysed starch and monoacyl lipid (mainly unesterified fatty acid). We presume that the enzyme- and acid-resistant material is similar to the amylose-lipid complexes that have been described by many research groups. This acid-resistant starch-lipid complex probably forms during the liquefaction process by complexing of glucan chains, formed during hydrolysis, with starch lipids that are released from the starch granules during processing. However, it should be noted that the amounts of acid resistant starch in residues of jet-cooked starches were much lower than those in the laboratory-scale preparations. (Table III).

Lipids and Proteins

Commercial wheat and maize starches contain up to 1% of lipid, of which at least half is located within the granule matrix (Morrison 1988). In wheat starch, the internal lipid is predominantly *lyso*-phosphatidylcholine (a monoacyl phospholipid), whereas in maize starch the major lipid is unesterified fatty acid. Analysis of residues formed during syrup production showed important differences between the behaviour of wheat and maize starch lipids.

Only minor amounts of the original starch lipids separated with the syrup residue when wheat starch was hydrolysed (6-9% for unfractionated starch and only 1-2% for prime 'A' starch) whereas 40% and 60% of maize starch lipids were recovered in the low- and high-shear processes, respectively (Table III). Lipid represented 76% of the residue obtained in the jet-cooked maize syrup. Commercially, lipids are removed from maize syrups at an early stage during processing, whereas with wheat starch the lipid remains within the crude syrup.

Similarly, a higher proportion of starch protein is removed in maize syrup residues than in wheat starch hydrolysates (Table III).

Syrups from unrefined wheat starch generally have more colour and flavour than corresponding syrups from wheat 'A' starch or maize starch. No doubt, the higher proportions of wheat lipids and proteins that remain in the syrups during hydrolysis contribute to these characteristics. For applications in which colour and flavour are undesirable, use of prime 'A' starch or a greater degree of post-hydrolysis refining (decolorization, deionization, etc.) is required to produce colourless syrups similar to those from maize (Bowler et al. 1985b). In some of the major applications, e.g. fermentation of glucose syrups for use in brewing or chemical/pharmaceuticals production, the use of unrefined wheat starches or lower levels of post-hydrolysis refining may be acceptable.

CONCLUSIONS

Wheat has recently become an important source of starch products in Europe, and now accounts for 18% of the EEC starch industry's products.

The minor components are responsible for several important technological features of wheat starch:

a) as components of wheat grain (e.g. friabilin on starch granules may affect endosperm texture).
b) as contaminants of unrefined starch (e.g. pentosans that affect subsequent processing, such as filterability of glucose syrups).
c) as materials on the surface of starch granules (e.g. bound alpha-amylase that affects both the yield of starch and its pasting properties).
d) as integral components of starch granules (lipids and proteins) that, together with surface components and contaminants, affect both the processing and product characteristics during starch hydrolysis (e.g. filterability, colour and flavour of glucose syrups).

The minor components account for most of the technologically important differences between unrefined starch from wheat and maize starch. For starch hydrolysates (the main products of the starch industry), the use of refined wheat starch or appropriate post-hydrolysis refining of syrups from less pure starch leads to syrups that are very similar to those from maize.

For applications in which syrup purity is less important, e.g. as a fermentation substrate, unrefined wheat starch may be an appropriate raw material.

ACKNOWLEDGEMENTS

The authors are grateful to N. Wookey and R. G. Jones of Tenstar Products Ltd for information and advise in the preparation of this paper.

The research work reported above from our laboratories was supported by the U.K. Ministry for Agriculture, Fisheries and Food and the results are Crown Copyright (c) 1989.

LITERATURE CITED

BOWLER, P., TOWERSEY, P. J., WAIGHT, S. G. and GALLIARD, T. 1985a. Minor components of wheat starch and their technological significance. In: New Approaches to Research on Cereal Carbohydrates. R.D. Hill and L. Munck, eds. Elsevier, Amsterdam, pp. 71-79.

BOWLER, P., TOWERSEY, P.J. and GALLIARD, T. 1985b. Some effects of the minor components of starch on glucose syrup production. Starch 37: 351-356.

GALLIARD, T. 1985. Bulk chemicals from plants: starch and starch-derived products. Ann. Proc. Phytochem. Soc. Eur. 26: 103-115.

GALLIARD, T. 1987. Starch availability and utilization. In: Starch: Properties and Potential. T. Galliard, ed. John Wiley and Sons, Chichester, U.K. pp. 1-15.

GALLIARD, T. and BOWLER, P. 1987. Morphology and composition of starch. In: Starch: Properties and Potential. T. Galliard, ed. John Wiley and Sons, Chichester, U.K. pp. 55-78.

JONES, R.G. 1987. Quality requirements for wheat starch and gluten extraction. Aspects of Applied Biology 15: 38-48.

MORRISON, W.R. 1988. Lipids in cereal starches: a review. J. Cereal Sci. 8: 1-15.

MORRISON, W.R., LAW, C.N., WYLIE, L.J., COVENTRY, A.M. and SEEKINGS, J. 1988. J. Cereal Sci. 8 (in press).

RUSSELL, P.L., GOUGH, B.M., GREENWELL, P., FOWLER, A. and MUNRO, H.S. 1987. A study by ESCA of the surface of native and chlorine-treated wheat starch granules. J. Cereal Sci. 5: 83-100.

SCHOFIELD, J.D. and GREENWELL, P. 1987. Wheat starch granule proteins and their technological significance. In: Cereals in a European Context. I.D. Morton, ed. Ellis Horwood, Chichester, U.K. pp.407-420.

WOOKEY, N. and MELVIN, M.A. 1981. The relative economies of wheat and maize as raw materials for starch manufacture. In: Cereals: A Renewable Resource. Y. Pomeranz, ed. Am. Assoc. Cereal Chem., St. Paul MN. pp.55-68.

16

WHEAT GLUTEN IS GOOD NOT ONLY FOR BREADMAKING

W. Bushuk and C. Wadhawan

Food Science Department
University of Manitoba
Winnipeg, MB R3T 2N2

INTRODUCTION

Wheat gluten is the first protein of plant origin to be described in the technical literature. The Italian biochemist Becarri first prepared it in 1748 by washing out the starch and solubles from dough (Bailey, 1941). Before that time, it was believed that proteins were present only in materials of animal origin. In 1810 Taddei introduced the term "gliadin" to refer to the portion of gluten that was soluble in alcohol solution and the term "glutenin" to the protein that remained after extraction of the gliadin. However, it was not until the beginning of the present century that the foundation for much of the future research on gluten was laid. In 1907, Osborne published the first comprehensive monograph on the nature of proteins in the wheat kernel (Osborne, 1907). The original fractionation of Osborne, based on solubility, is still extensively used today with just minor modifications. Since Osborne's publication, gluten has been the subject of many excellent reviews; the most recent is that of

---

Publication No. 142 of the Food Science Department, University of Manitoba with financial assistance of the Natural Science and Engineering Research Council of Canada

Lasztity (1984). While considerable progress has been made on the industrial exploitation of the unique properties of gluten, much remains to be discovered.

COMMERCIAL PRODUCTION OF GLUTEN

Gluten can be simply prepared in the laboratory by mixing wheat flour with an appropriate amount of water into a dough and washing out the starch and the solubles in a stream of water. This process yields "wet gluten" which can then be dried and ground into a product known in commerce as "vital wheat gluten". Large scale commercial production of gluten is similar, in principle, to the simple laboratory procedure.

Industrial processes for preparing gluten can be divided into two general classes, the dough process and the batter or slurry process. In the first type of process, flour is mixed into dough with full development of gluten and then the starch and solubles are washed out by kneading the dough in a stream of water. The Martin process (Fig. 1), first reported in France in 1835, is the best known example of the dough process (Mittleider et al., 1978).

In the batter or slurry processes, flour is mixed with a greater amount of water so that the gluten does not have an opportunity to form. The separation of starch and gluten particles in aqueous suspension is achieved by centrifuges or by hydrocyclones. The so-called Raisio process (Fig. 2; Maijala, 1976) is a commercial example of the slurry technology. A major advantage of this process over the dough processes and the earlier batter processes is the much lower water usage per unit of flour processed.

The final stage of gluten manufacturing is the drying of the wet gum gluten. The most effective commercial gluten driers operate on the flash-drying principle and are called ring dryers because the product moves in a stream of warm air around the

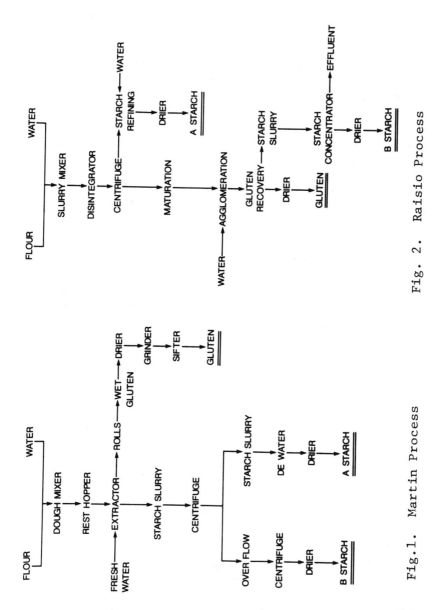

Fig. 1. Martin Process

Fig. 2. Raisio Process

periphery of the drying chamber (Barr and Barr, 1976). Freeze-drying, which produces gluten of the highest vitality, is not commercially economical. While the initial separation of the starch and gluten is important in relation to product yield and composition, it is the final drying stage that determines the quality of the gluten for most current uses. Wet gluten is extremely sensitive to denaturation, whereby it loses its natural vitality, by elevated temperatures. For further details on commercial and laboratory processes for separating wheat starch and gluten, the reader is referred to a recent doctoral thesis presented at the University of Manitoba (Wadhawan, 1988).

Total world wheat gluten production for 1987/88 was 240,000 metric tons (mt; IWGA, private communication). Production has shown a steady growth especially in the last decade - production for 1980/81 was 88,000 mt. Major producers are EC-12 (the 12 members of the European Economic Community), Australia, USA and Canada (Table I). Major consumers are the EC-12, USA and Australia. Australia and Canada are the major exporters.

CHEMICAL AND PHYSICAL PROPERTIES

A recent analytical survey of 27 commercial gluten samples (provided by Mr. Matt Hesser, IWGA) from international sources and one sample prepared in the laboratory from Canadian hard red spring wheat produced the results given in Table II (Wadhawan, 1988).

The range of variation in composition of commercial glutens is extremely wide. Of the various constituents, the inter-sample variation in pentosan, lipid and sodium contents may be functionally significant. The almost 400 fold variation in sodium content suggests a wide variation in the usage of salt in the wash water in commercial gluten manufacturing. Normally salt is added to tighten up the gluten and decrease the binding of lipid

TABLE I
World Gluten Supply/Demand for 1985/86[a]

	Production	Consumption	Import (Net)	Export (Net)
EC-12	98	91	-	7
Australia	42	19	-	23
USA	31	56	25	-
Canada	15	4	-	11
Argentina	7	6	-	1
Japan	6	9	3	-
Finland	5	1	-	4
Others	19	37	18	-
Total	223	223	-	-

[a]Data provided by International Wheat Gluten Association; figures in 1000 mt.

TABLE II
Composition of Commercial and Laboratory-Prepared Glutens[a]

	Commercial		Laboratory
	Range	Average	
Moisture,%	5.5-9.4	7.45[b]	4.7
Ash,%	0.59-1.2	0.83[b]	0.68
Protein,%	73.3-81.8	77.6[b]	76.0
Starch,%	2.7-15.0	10.7[b]	11.5
Pentosan,%	0.60-1.18	0.90[c]	0.86
Lipid, free (FL),%	0.51-1.80	1.04[c]	0.52
Lipid, bound (BL),%	4.80-6.90	6.07[c]	7.50
BL:FL	3.61-9.41	6.63[c]	14.4
Sodium, p.p.m.	60-2150	687[c]	110

[a]Data on dry basis
[b]N=27
[c]N=8

during the separation process. For the commercial samples, the glutens with relatively higher sodium contents generally had lower free lipid content and hence higher BL:FL ratios.

The unique functional properties of gluten derive from its major protein components, gliadin, glutenin and the insoluble residue protein, as determined by the modified Osborne solubility fractionation (Chen and Bushuk, 1970). In the commercial glutens that we analyzed, these fractions were present in the ratio 3:3:4. Devitalization by thermal denaturation generally led to an increase of the insoluble residue and a decrease of the gliadin fractions. Gliadins are single polypeptide chain proteins with molecular weights in the range 30-75 kDa. Glutenins comprise many polypeptide chains (the actual number as yet unknown) linked in various arrays by disulfide bonds. Accurate molecular weights of the polymeric glutenin have not been determined. The polypeptide subunits obtained after reduction of the disulfide groups of native glutenin fall into two distinct groups of molecular weight, 90 to 140 kDa for the high molecular weight (HMW) subunits and 30-75 kDa for the low molecular weight (LMW) subunits (Ng and Bushuk, 1988).

The amino acid composition of gluten is typical of cereal storage proteins, characterized by relatively high contents of glutamine, proline and hydrophobic amino acids, and low contents of amino acids with ionizable side groups. The low content of polar groups gives gluten proteins a resultant positive electrostatic charge compared with most other proteins which have a net negative charge. Wheat gluten is a most unique protein in terms of amino acid composition.

Data for physical properties considered important to functionality for the samples referred to above are given in Table III (Wadhawan, 1988).

Commercial glutens showed an unexpectedly wide range of particle size. Eleven of the 27 samples contained more than 20% coarse material

TABLE III
Selected Physical Properties of Commercial and
Laboratory-Prepared Glutens

	Commercial		Laboratory
	Range	Average[a]	
Particle size			
>149 μm	2.4-28.7	15.8	0
149-75 μm	38.8-83.1	58.1	58.5
>75 μm	3.5-45.0	24.5	41.5
Water abs., g/100	200-287	228	263
Hydration time, min	0.2-5.0	2.1	0.3
SDS Sed. volume, mL	65-157	96	170
Loaf volume, mL[b]	515-615	566	650

[a] N = 8
[b] AACC straight dough procedure for a 5:95 gluten-flour blend.

(>149 μm). Some of the coarse particles appeared semi-crystalline in structure; these would probably require considerable time to hydrate properly in a dough system.

All but one of the commercial gluten samples were inferior to the laboratory-prepared gluten in water absorption, hydration time, and SDS sedimentation volume. All were inferior in volume of the test loaves obtained for soft wheat flour (8.5% protein) supplemented with 5% gluten (Wadhawan, 1988).

In relation to functionality of gluten in breadmaking, highly significant (99% level) correlations were obtained between several physical properties and vitality as measured by the baking test (Table IV; Wadhawan, 1988).

Of the five parameters in Table IV, three (wet gluten stretching force, amount of acetic acid soluble protein and intrinsic fluorescence

TABLE IV
Highly Significant (99% Level) Correlations Between
Physical Properties and Vitality (Loaf Volume)

Parameter	Correlation Coefficient
Wet gluten stretching force[a]	0.73 (n=27)
Osborne residue protein	-0.57 (n=26)
Acetic acid soluble protein	0.88 (n=8)
Fluorescence of dry gluten	-0.74 (n=15)
Fluorescene of acetic acid extract	-0.98 (n=8)

[a] According to Matsuo (1978)

of dry gluten) would be suitable for measuring gluten vitality during the manufacturing process.

The unique functional properties of wheat gluten are only now being appreciated and exploited. In baking applications, the key properties are the high water absorption (two to three times its own weight) and the ability to form viscoelastic films which contribute to dough handling properties and improved loaf characteristics. The film forming properties are useful for many other food applications such as sausage casings and meat patty glazings. Other interesting and potentially useful functional properties are its thermosetting behavior and its bland or light "wheat" flavor. When heated to about 85°C, hydrated gluten coagulates into a chewy product which is stable under a wide variety of food preparation conditions. Wheat gluten alone ranks low on the scale of nutritional quality because of its low lysine content. However combinations with some other proteins (e.g. soy bean) have nutritional quality that is higher than that of either protein alone.

UTILIZATION

The major use of vital gluten has been traditionally and continues to be in the baking industry. However, as a result of increasing awareness of the unique functional properties of gluten, its use is rapidly expanding. The U.S. figures (Table V) are typical of the growth in applications other than baked goods.

The usage figures (Table V) indicate that there has been a significant shift in gluten utilization to applications other than baking. Much of the credit for this expansion in gluten utilization must go to the International Wheat

TABLE V
Utilization of Wheat Gluten in the U.S.A.[a]

Application	1980/81	1987/88
	%	%
Baking	77	64
Milling	4	10
Pet foods	10	8
Meat	0	4
Breakfast cereals	7	4
Seafood analogs	0	1
Pasta	0	1
Cheese analogs	0	1
Aquaculture feed	0	2
Snacks	0	2
Breadings and batters	0	1
Devitalized	1	1
Others	1	1

[a]Data provided by International Wheat Gluten Association.

Gluten Association and especially its energetic Executive Director Matt Hesser.

In relation to baking, there has been a considerable increase in the supplementation of flours at the mill level, presumably to meet the protein content specification of the bakers. A possible explanation of this trend is that the new wheat varieties which have been bred to give higher yields, especially under intensive agronomic management, generally produce grain of lower protein content. This is a general situation in Australia where the production of the higher protein grades Prime Hard and Hard is quite small and much of the wheat goes into export markets. An interesting situation has evolved in the United Kingdom. Here plant breeders have succeeded in improving the protein quality of their cultivars by selecting for improved glutenin subunit composition (Bingham, private communication). At the same time dramatic yield increases have been achieved but it has not been possible to increase the protein content. The deficiency in protein content has been largely corrected by supplementing the flour with gluten separated from the same wheat instead of using imported high protein grain in the milling grists. Imports of high protein wheat into the United Kingdom have declined accordingly.

Wheat gluten offers many benefits to the food industry. In the baking industry, the major user of gluten, the following benefits have been clearly demonstrated (IWGA, 1981).

1. Increased water absorption and thereby improved dough handling properties, bread yield, bread quality and extended shelf life.

2. Improved rheological properties of dough and thereby easier processing and improved bread quality.

3. Improved loaf characteristics and stability due to the film-forming and thermosetting properties of gluten.

4. Improved bread flavor.

5. Improved nutritional quality due to in-

creased protein content.

6. Improved quality of specialty products such as high-fiber bread where gluten improves the carrying capacity of flour for "dead weight" ingredients.

7. Decreased flour inventory in bakery; one flour with varying amounts of gluten can meet the requirements of many different baked products.

Significant quantities of vital gluten are used in the pet food industry. In canned and intermediate moisture content products, the high water absorption and excellent fat binding properties of gluten improve product yield and quality (Magnuson, 1985). Incorporation of gluten into dog biscuits decreases breakage during packaging and shipment, and brings the protein content to the specified level.

Gluten is also used in a variety of breakfast cereals because of its binding or adhesive properties and ability to improve nutritional quality, especially since most cereals are consumed with milk. The most familiar product of this group is Kellogs Special K® where gluten helps to bind the vitamin/mineral enrichment to the rice and contributes to the strength and crispness of the flaked product (Magnuson, 1985).

In extruded and deep fried snacks, gluten improves nutritional quality, crispness, texture and resistance to breakage during handling and shipping. Many gluten-based snacks are becoming popular among vegetarian consumers (IWGA, 1981). In pasta products, added gluten improves strength of the dry product, cooking quality, and sensory qualities of the cooked product. Stability during canning and retorting is also improved. Gluten is added to pizza crust flour to strengthen the crust, provide body, improve chewiness, and reduce moisture transfer from the topping to the crust (Magnuson, 1985).

The meat and fish industry is beginning to exploit the unique adhesive and thermosetting properties of wheat gluten in products such as extended ground meats, textured meats, meat analogs, canned "integral" hams, poultry rolls, fish sausage products, seafood analogs, etc. (for review see Wadhawan, 1988). Addition of gluten to breadings and batter mixes reduces cooking losses and improves yield and hot and cold adhesion (IWGA, 1981).

Profitable utilization of wheat gluten has extended to a number of other interesting products. The unique viscoelastic properties of hydrated gluten can be exploited in the manufacture of synthetic cheese products with sensory properties similar to those of natural cheese (Magnuson, 1985). Wheat gluten, alone or in combination with soybean protein, can replace up to 30% of the more expensive sodium caseinate in imitation American and Mozzarella cheese products (IWGA, 1983). Additionally, the wheat gluten industry has witnessed a rapid growth in the use of gluten in aquaculture feed. In addition to providing an essential nutrient, the gluten improves the stability of the pellets and can be used to control the density as required for surface and bottom-feeding varieties of fish (IWGA, 1981). Further growth in this usage is anticipated.

Wheat gluten, in natural and modified form, has many potential applications in the non-food industry such as adhesives, coatings, detergents, etc. (see review by Wadhawan, 1988). Usage in this sector has been precluded by the availability of cheaper other natural and synthetic materials. However, from the foregoing it should be quite obvious that wheat gluten is unsurpassed for many uses, not only for breadmaking.

LITERATURE CITED

BAILEY, C.H. 1941. A translation of Beccari's lecture "concerning grain" (1729). Cereal Chem. 18:555.
BARR, B.J., and BARR, D.J. 1976. Drying. Page 71 in: Starch Production and Technology. J.A. Radley, ed. Applied Science Publishers Ltd., London.

CHEN, C.H., and BUSHUK, W. 1970. Nature of proteins in <u>Triticale</u> and its parental species. I. Solubility characteristics and amino acid composition of endosperm proteins. Can. J. Plant Sci. 50:9.

IWGA (International Wheat Gluten Association). 1981. Wheat gluten: A natural protein for the future-today. IWGA, Overland Park, KS.

IWGA. 1983. Use of vital gluten in imitation cheese. Product Application Bulletin. IWGA, Overland Park. KS.

LASZTITY, R. 1984. Wheat proteins. Page 13 in: The Chemistry of Cereal Proteins. CRC Press, Inc., Boca Raton, FL.

MAGNUSON, K. 1985. Uses and functionality of vital wheat gluten. Cereal Foods World 30:179.

MAIJALA, M. 1976. Environment and products gain in new wet wheat process. Food Eng. 48:73.

MATSUO, R.R. 1978. Note on a method for testing gluten strength. Cereal Chem. 55:259.

MITTLEIDER, J.F., Anderson, D.E., McDonald, C.E., and Fisher, N. 1978. An Analysis of Economic Feasibility of Establishing Wheat Gluten Processing Plants in North Dakota. ND Agric. Exp. Station, ND State University and U.S. Dept. of Comm., E.D.A. Bulletin No. 508.

NG, P.K.W., and BUSHUK, W. 1988. Relationship between high molecular weight subunits of glutenin and breadmaking quality of Canadian grown wheats. Cereal Chem. 65:000.

OSBORNE, T.B. 1907. The proteins of the wheat kernel. Carnegie Institution of Washington. Publication No. 84.

WADHAWAN, C.K. 1988. Fundamental Studies on Vitality of Gluten for Breadmaking. Ph.D. Thesis, University of Manitoba, Winnipeg, MB.

17

GLUTENIN STRUCTURE IN RELATION TO WHEAT QUALITY

Donald D. Kasarda

USDA-ARS
Western Regional Research Center
800 Buchanan Street
Albany, California 94710, USA

INTRODUCTION

Bread dough properties result from a balance of components--starch, water, gluten proteins, lipids, and so forth--and their interactions. Without all these components in suitable balance, the viscoelasticity and surface activity of the system would be unsatisfactory for breadbaking. When a shipment of flour fails to perform according to expectations in a bakery, however, it is most likely the protein fraction that is responsible. The flour may be low in protein, or the protein quality may not be adequate. The former is less of a problem because protein content can be specified and easily checked. The latter is a more difficult problem. There are intrinsic quality differences among wheat varieties that are independent of the amount of protein (Finney, 1954), but poor quality is more difficult to recognize before test mixing and baking is carried out.

Most of the flour protein consists of the complex mixture of gluten proteins--high-glutamine, high-proline storage proteins [*prolamins* from a compositional standpoint (Shewry et al., 1986)]. Gluten proteins, in a dough, are cohesive, extensible, and yet exhibit a moderate amount of elastic recovery. Reconstitution studies (MacRitchie, 1984) have indicated that the gluten proteins are primarily responsible for quality differences among different varieties. Despite many years of study, however, we do not yet have a detailed understanding at the molecular level of the basis for the viscoelastic properties of doughs, regardless of what variety the flour came from. Neither can we explain why intrinsic protein quality differs among varieties.

Things are looking up, however. A major increase in our knowledge of gluten protein structures has come about in recent years through application of DNA sequencing techniques that permit deduction of protein sequences from DNA sequences. Such sequence information was combined with results of physical-chemical studies by Tatham, Miflin, and Shewry (1984) to hypothesize that an only recently-defined structural element, the β-turn spiral (β-spiral), may be present in gluten proteins and contribute importantly to the properties of doughs (also see: Shewry and Tatham, 1987).

These recent developments have stimulated cooperations between molecular biologists and physical chemists that are certain to result in more efficient breeding of new wheat varieties with good quality. Traditional approaches to breeding will benefit from provision of detailed information about protein components important for quality, along with relatively simple qualitative and quantitative tests for these components that can be applied early in a cross. It is also likely that, within a few years, genetic engineering approaches will be used to improve wheat quality through direct modification of the wheat genome--perhaps by insertion of desirable genes, increase in the amounts of proteins produced by such genes, and by removal of undesirable genes.

Here, I plan to review our current knowledge of gluten protein structure with emphasis on the glutenin fraction, attempt to integrate information from various studies, and indicate areas where I think research is needed to achieve a detailed molecular understanding of protein quality in breadmaking. I will focus on major protein components that have a prolamin composition. I acknowledge that much of what I have to say is highly speculative.

GLIADINS AND GLUTENINS

The gluten proteins traditionally have been divided into roughly equal solubility fractions, the *gliadins* (soluble in alcohol-water solutions) and the *glutenins* (insoluble in alcohol-water solutions, but partially soluble in other solvent systems). Solubility fractions are relatively crude and our understanding of the major components of these fractions has been advanced by a wide range of physical and chemical studies. Some generally accepted concepts of the nature of gliadins and glutenins follow.

Gliadins: Intramolecular Disulfide Bonds

The gliadins are monomeric proteins in which disulfide bonds, when present (as in α-, β-, and γ-gliadins), link one part of a polypeptide chain to another part of the same chain, that is, they have *intramolecular* disulfide bonds (these are absent, however, from ω-gliadins, which have no cysteine in their primary structure). Gliadins contribute mainly to the extensibility of the dough system. Purified, hydrated gliadins have little elasticity and are less cohesive than glutenins (Wall, 1979).

Glutenins

Inter- and Intramolecular Disulfide Bonds

The glutenins, in contrast to the gliadins, have disulfide bonds connecting polypeptide chains to one another, that is, they have *intermolecular* disulfide bonds. Glutenin polypeptide chains may also have intramolecular disulfide bonds. In native glutenin, intermolecular disulfide bonds link polypeptide chains together to form polymers of polymers (each polypeptide chain is itself a polymer of amino acids), which I shall refer to as *glutenin polymers*.

These polymers may be largely linear with each polypeptide chain joined to two other polypeptide chains, but branching is not ruled out (Ewart, 1968; 1987; 1988). I shall refer to the monomeric polypeptide chains of gliadins as *gliadin proteins*, and arbitrarily distinguish the polypeptide chains that are linked in glutenin polymers from gliadins by calling them *glutenin subunits* (although they are also proteins). Glutenin subunits can be separated from linkage to other glutenin subunits only by breaking inter-molecular disulfide bonds by reduction or strong oxidation.

The major distinction then between gliadin proteins

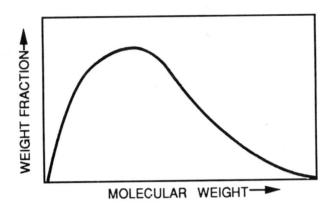

Fig. 1 Hypothetical MW distribution of glutenin polymers.

and glutenin subunits is that the latter are crosslinked into higher order structures. Native glutenin contains various glutenin polymer molecules that vary in the number of glutenin subunits joined together; this latter number ranges from two (dimers) to a great many subunits (there are apparently no glutenin monomers--D. D. Kasarda and H. W. Jones, unpublished results; Graveland et al., 1985; Bietz and Wall, 1980). The molecular weight (MW) distribution of the glutenin polymers in native glutenin isn't known, but may look something like that of Fig. 1. The high-end tail of the distribution (Fig. 1) must correspond to glutenin polymers with MW's in the millions (Schofield et al., 1983).

Although unreduced glutenin for the most part is too large to enter the gel when SDS-polyacrylamide gel electrophoresis (SDS-PAGE) is carried out, at the low end of the MW distribution of Fig. 1 dimers, trimers, and slightly higher-order oligomers of glutenin subunits are small enough to enter the gel. These usually show up as streaking near the origin, but individual bands can sometimes be seen (D. D. Kasarda and H. W. Jones, unpublished results; Khan and Bushuk, 1979; Bietz and Wall, 1980). These appear to make up only a very small part of the total glutenin.

Glutenins contribute elasticity to the dough system and it seems reasonable to attribute this property largely to the crosslinked nature of the glutenin subunits. Hydrated glutenin polymers, free of gliadins, are highly cohesive and elastic (Bietz and Huebner, 1980; Wall, 1979).

Glutenin Subunits

Glutenin subunits consist mainly of two types, the *high-molecular-weight-glutenin subunits (HMW-GS)* and the *low-molecular-weight-glutenin subunits (LMW-GS)*. The HMW-GS are so designated because they form a slower-moving group of components when reduced glutenin subunits are separated by SDS-PAGE with apparent MW's ranging from about 80,000-120,000 (Fig. 2). Most varieties have 4 or 5 HMW-GS, which frequently differ in electrophoretic mobility among varieties and which can be divided into *x types* and *y types* (Payne et al., 1984). In varieties with 5 subunits, the 3 slower-moving (higher-MW) subunits are x types and the two faster-moving are y types. In varieties with 4 subunits, the slowest two are x and the faster two are y.

Sequences of HMW-GS

HMW-GS sequences have been reported by a number of laboratories (most of the work is cited in the recent paper

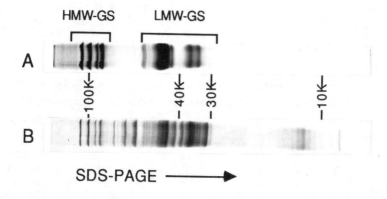

Fig. 2. SDS-PAGE of purified glutenin (A) compared with a total protein extract (B). Both from wheat flour of the variety Chinese Spring. Points on the gels corresponding to MW's of 10,000 (10K), 30K, 40K, and 100K are indicated.

of Halford et al., 1987). Both types of HMW-GS are characterized by relatively small N-terminal and C-terminal domains (about 100 amino acids in length at the N-terminus and about 50 at the C-terminus) separated by a long series of repeating amino acid sequences. These repeats are not perfect and also differ somewhat between the two types of subunits, but both types have many repeats with the following sequence of amino acids:

[-pro-gly-gln-gly-gln-gln-]

where pro = proline, gly = glycine, and gln = glutamine.

Cysteine Residues of HMW-GS

An apparently important aspect of HMW-GS structure is the occurrence of cysteine residues only near the ends of the polypeptide chain--within about 100 residues of each end. These cysteines gives rise to the disulfide bonds (cystine) of native gliadins and glutenins. The regions containing cysteine residues are separated by roughly 400-600 residues of repeating sequences. X-type subunits usually have 4 cysteine residues per polypeptide chain, three near the N-terminal end and 1 near the C-terminal end, whereas y-type subunits have 7 cysteine residues, 5 near the N-terminal end and 2 near the C-terminal end. Which cysteine residues participate in which types of disulfide bonds isn't known, but it seems likely that at least the y-type subunits have some cysteine in the form of intramolecular disulfide bonds (Graveland et al., 1985).

β-Turn Spiral Structures in HMW-GS

Tatham et al. (1984) suggested that the repeating amino acid sequences that make up most of HMW-GS form many β-turns and that these, in turn, fold into a spiral or helical structure that is somewhat rod-like in shape. A similar structure was suggested also for ω-gliadins, which are likely to be made up almost entirely of similar repeating sequences; physical chemical studies provided support for the hypothesis (see Shewry and Tatham, 1987). Thus, for HMW-GS, the structures might be represented as in Fig. 3. [I emphasize that the structures of Fig. 3 are not based on much other than the dimensions that repeating sequences would be likely to have if they formed a perfect β-spiral

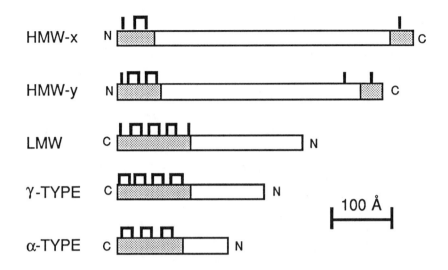

Fig. 3. Schematic representations of x-type and y-type HMW-GS, LMW-GS, γ-type gliadins, and α-type gliadins: patterned areas, unique sequence; plain areas, repeating sequences; black bars, intermolecular disulfide linkages; linked black bars, intramolecular disulfide linkages; N, N-terminal end; C, C-terminal end. Representations are very approximately to scale.

(Field et al., 1987); we have little information about the arrangement of the rest of the polypeptide chain of each molecule. The assignment of disulfide linkages to intermolecular and intramolecular is largely a guess--except for the α-type and γ-type gliadins, which are known to have only intramolecular disulfide bonds.]

Tatham et al. (1984) suggested also that, under the stress of dough mixing, the β-spirals of the HMW-GS might become extended and, when extended, have a thermodynamically-determined tendency to return to the coiled state. This could contribute elasticity to the dough system. Although this is an intuitively satisfying notion, it must be emphasized that β-spirals along with their uncoilings and recoilings remain a hypothesis, one difficult to test.

Structure of LMW-GS

LMW-GS make up a faster moving group of proteins in SDS-PAGE patterns of glutenin subunits (Fig. 2). These often show up as heavily staining bands with apparent MW's of 40,000-55,000 (40-55K) and some less intense bands in the range 30-40K.

There is only one complete sequence published that appears to be for a LMW-GS. This is a cDNA-based sequence (Okita et al., 1985) and it does not seem completely characteristic of the major LMW-GS in that it has a slightly different N-terminal amino acid sequence (Kasarda et al., 1988). Also, the coding sequence corresponds to a MW of only about 33K. The apparent MW's determined by SDS-PAGE for HMW-GS are too high, however, by about 20% and this discrepancy may arise because the rod-like shape found for HMW-GS (Field et al., 1987) is maintained in SDS solution.

It is conceivable that the apparent MW's of the LMW-GS are also too high as estimated by SDS-PAGE if LMW-GS have a substantial region of repeats similar to those of HMW-GS that can assume a β-spiral structure. However, A-gliadin has an apparent MW of 31,000 by SDS-PAGE, which is close to the true MW based on sequence data (Kasarda et al., 1984), yet it has a region of repeats amounting to almost one third of the polypeptide chain. The A-gliadin repeat region may not form a rigid β-spiral, or perhaps the spiral length is too short to affect the mobility of the protein in SDS-PAGE. At present, the true MW's of the LMW-GS with apparent MW's in the range 40-55K remain to be established. I suspect that the clone of Okita et al. (1985) does code for a LMW-glutenin subunit, albeit a minor one, and I shall assume that the structure is essentially descriptive of LMW-GS.

The Okita et al. (1985) sequence indicates that LMW-GS are quite similar to γ-type gliadins. LMW-GS have a large part of the N-terminal half of the molecule in the form of repeating sequences. The rest of the polypeptide chain has a unique sequence that includes three or four disulfide bonds (γ-gliadins usually have four--Bartels et al., 1986; Rafalski, 1986). This C-terminal part of the LMW-GS molecule is highly similar to the C-terminal part of the γ-type gliadins. The repeating sequences differ from those of γ-gliadins, however, and are generally similar to the following sequences:

(γ-gliadin) [-pro-gln-gln-pro-phe-pro-gln-]

(LMW-GS) [-pro-gln-gln-pro-pro-phe-ser-]

where phe = phenylalanine and ser = serine.

In general, the LMW-GS correspond to a distinctive grouping of more basic proteins relative to γ-gliadins and other gliadins in two-dimensional electrophoresis (Payne et al., 1984), which establishes them as a unique group of proteins despite their considerable similarity in structure to γ-gliadins.

Cysteine Residues of LMW-GS

Because LMW-glutenin subunits are found only in glutenin polymers they must have the ability to form at least one intermolecular disulfide crosslink. It is highly probable that they usually participate in two linkages to other glutenin subunits, either to other LMW-GS or to HMW-GS-- this being the minimum requirement for the polymer chain to grow. The other cysteine residues of each subunit are likely to participate in intramolecular disulfide bonds (Graveland et al., 1985).

The sequence of Okita et al. (1985) has a cysteine residue quite close to the N-terminal end of the polypeptide chain, but we have not found indication that this is a major sequence by direct sequencing of mixtures of LMW-GS (Shewry et al., 1983; H. P. Tao and D. D. Kasarda, unpublished results). Although further investigation of this question is required, I shall assume for the purposes of this discussion that most LMW-GS have all cysteine residues in the unique sequence (C-terminal) part of the molecule, all in the form of inter- or intramolecular disulfide bonds, and that the N-terminal regions do not participate in disulfide bonding [these regions may be rod-like (as depicted in Fig. 3) if the repeating sequences form β-spiral structure].

If we assume that LMW-GS, like most γ-gliadins, have 8 cysteine residues per subunit, then 6 of them are likely to form 3 intramolecular disulfide bonds that stabilize the C-terminal part of the molecule while the other two link the subunit to 2 other subunits.

Formation of Glutenin Polymers: Arrangement of Subunits

Most researchers agree that glutenin polymers consist of HMW- and LMW-GS joined by intermolecular disulfide bonds and that there is a range of sizes of the resulting glutenin polymer molecules. There is no general agreement on the arrangement of the subunits in glutenin polymers.

Graveland et al. (1985) proposed a model in which HMW-GS were linked end to end to form the backbone of glutenin polymers; strings of LMW-GS branched off from y-type subunits, which were connected to x-type subunits. Graveland et al. (1985) were apparently influenced by the findings of Lawrence and Payne (1983), who carried out partial reduction experiments in which dimers of HMW-GS were observed; Lawrence and Payne (1983) showed furthermore that x-type subunits seem to be joined mainly to y-type subunits. Also, Graveland et al. (1985) indicated that they had evidence for long strings of HMW-GS, but this evidence was referred to as unpublished and I am not aware of other evidence for strings of HMW-GS.

Ewart (1968, 1987, 1988), on the other hand, suggested a linear arrangement of subunits in glutenin polymers and has provided evidence for such an arrangement. He assumed a random linkage of subunits in the glutenin polymers. Singh and Shepherd (1985) concluded from their electrophoretic analyses of glutenin that all glutenin subunits are associated randomly and that the occurrence of polymers including only LMW-GS results from chance association. They also postulated that there are no native glutenin polymers that include HMW-GS that do not also have LMW-GS, but they did not offer an explanation for this.

Synthesis of Gliadin Proteins and Glutenin Subunits

It seems likely that wheat storage proteins are synthesized randomly by ribosomes bound to the rough endoplasmic reticulum of wheat endosperm cells. Leader sequences guide the polypeptide chains through the membrane and are cleaved off on the other side of the membrane (Greene et al., 1985; Kreis et al., 1985). The newly-synthesized polypeptides probably begin to fold as they are being threaded through the membrane.

Gliadins

Gliadin proteins probably fold rapidly into conformations that permit pairs of cysteine residues to come into close association, whereupon they are oxidized to disulfide bonds--perhaps with help from enzymes like disulfide isomerase (Roden et al., 1982; Bulleid and Freedman, 1988). Gliadins usually have an even number of cysteine residues, all of which become incorporated into intramolecular disulfide bonds. There evidently are some "gliadins," however, that do not have an even number of cysteine residues on the basis of published DNA sequences (Kasarda et al., 1988). I shall return to the possible role of such proteins.

Glutenins

It appears that glutenin subunits as synthesized cannot combine all of their cysteine residues to form intramolecular disulfide bonds; some must form intermolecular disulfide bonds. This must happen either because they have an uneven number of cysteine residues or because folding of the polypeptide chain fails to bring all pairs of cysteine residues sufficiently close to one another for disulfide formation to occur.

If glutenin polymers have a linear arrangement in which each subunit forms two intermolecular disulfide linkages (this is likely, at least to some degree), it is most likely that this comes about because only two cysteine residues from an even numbered set fail to link up internally and thus become involved in intermolecular crosslinking. If more than two cysteine residues are available for formation of intermolecular crosslinking, branching may occur from that subunit.

As synthesis of subunit polypeptide chains is completed and the chains are released from the membrane, the subunits will diffuse and collide with gliadin proteins and with one another. Some collisions between glutenin subunits (or between subunits and a facilitating enzyme, which could still be membrane associated) will result in formation of intermolecular disulfide bonds. This process is likely to continue until all cysteine residues that have not already formed intramolecular linkages have reacted with similar cysteine residues of other polypeptide chains or with low-molecular-weight thiol compounds. The latter process has been suggested (Ewart, 1985) as a mechanism that might affect the MW distribution of glutenin polymers by blocking (terminating) glutenin subunit polymerization.

Termination of Growing Glutenin Polymer Chains by Proteins Having a Single Free Cysteine

Another mechanism that may act to limit the MW distribution of glutenin polymers in a manner similar to low-molecular-weight thiol compounds is termination of growing polymer chains through reaction with a protein that would normally be a gliadin, but which has lost or gained a cysteine residue through mutation in such a way that all cysteine residues but one form normal intramolecular disulfide bonds (Kasarda et al., 1988). The odd cysteine residue must by satisfied by reaction with another subunit (or a thiol compound). Likely candidates for such proteins have been reported by Okita et al. (1985), who described an α-type gliadin, and by Sugiyama et al. (1986) and Scheets and Hedgcoth (1988), who described γ-type gliadins. These proteins have low MW's in the range 30-40K.

When glutenin polymers were fractionated according to size by gel-permeation chromatography and the fractions were examined by SDS-PAGE, changes in the proportions of various subunits were noted (D. D. Kasarda and A. E. Adalsteins, unpublished results). As the size of the glutenin polymers decreased, proteins with MW's in the 30-40K range increased in proportion relative to those glutenin subunits with MW's in the range 40-55K. The 30-40K group of LMW-GS subunits may correspond in part to the chain-terminating type of protein just described. Implicit in this explanation is the assumption that 40-50K subunits form at least two intermolecular disulfide linkages and thus are chain-extending subunits. Chain-terminating 30-40K subunits would constitute a greater part of the lower-MW fractions of glutenin polymers than of the higher-MW, because they only occur at the ends of chains. This would be observable in SDS-PAGE patterns only if the 30-40K group of proteins is largely composed of chain terminators and the 40-55K subunits are primarily chain extenders.

This discussion should be qualified by stating that no special mechanisms for chain terminations are inherently required. Polymerization could just continue until all monomer has reacted.

LMW-GS Are in Excess over HMW-GS

Although good quantitative information that covers all glutenin proteins in flour is not available, most studies of extracted proteins indicate that the molar ratio of LMW-GS to HMW-GS is 2:1 (Huebner and Bietz, 1985) or greater. From examination of various gel patterns, I think that the estimate of Huebner and Bietz is probably on the low side. [Our glutenin pattern of Fig. 2 is not typical because this particular glutenin preparation was a very HMW fraction that we obtained by gel-permeation chromatography, a void volume fraction in which the proportion of the HMW-GS was enhanced (see Payne and Corfield, 1979).] On this basis, and if synthesis and polymerization of glutenin subunits is random, it would not be expected that HMW-GS would be as likely to become linked to one another as to LMW-GS. In order to create linear backbone polymers of the sort suggested by Graveland et al. (1985), it would have been necessary for the wheat plant to have developed special mechanisms for the segregation of HMW-GS from LMW-GS during synthesis, and for bringing the separately polymerized backbone polymers into contact with the LMW-GS at the appropriate stage of glutenin formation.

Because all glutenins and gliadins appear to serve only as storage proteins, synthesized to be deposited in the endosperm cells until needed as a source of nitrogen and amino acids by the newly developing plant upon germination of the embryo, it is difficult for me to imagine why such mechanisms would have evolved. I do not rule out, however, the possibility of segregation of some proteins synthesized in the endosperm, and there is some puzzling evidence that seems to suggest this. For example, the endosperm albumins described by Gupta and Shepherd (1988) that migrate near the ω-gliadins in SDS-PAGE may be aggregated through disulfide bonding, but independently from the prolamin-type glutenin polymers (D. D. Kasarda and A. E. Adalsteins, unpublished results)

Hypothesis for Formation of Glutenin Polymers

I suggest that gliadins and glutenins are synthesized randomly, and that glutenin polymers undergo extension by incorporating HMW- and LMW-GS in a random way until the growing polymer chains are terminated by proteins with single unreacted cysteine residues or low-molecular-weight thiols. The resulting polymers are deposited along with gliadins in a random fashion in protein bodies. Furthermore, I suggest that strings of LMW-GS are more common than strings of HMW-GS in glutenin polymers.

Although higher-MW fractions (as separated by gel-permeation chromatography) of glutenin polymers also have greater proportions of HMW-GS, as noted first by Payne and Corfield (1979), this might result simply from the fact that HMW-GS do indeed have higher MW's than do LMW-GS. A glutenin polymer molecule with a greater number of HMW-GS incorporated into the chain will necessarily be higher in MW than one with an equivalent number of subunits, but with fewer HMW-GS. Additionally, when strings of HMW-GS occur through statistical chance, they may act like much larger molecules because of their having rod-like structures resulting from β-spiral conformations of their repeating sequence domains.

Repeating Sequences and Gluten Protein Structure

Because all gliadin proteins and glutenin subunits have substantial parts of their primary structure in the form of repeating sequences, it seems possible that all may have regions of β-spiral structure. The ω-gliadins are almost all repeats, as are HMW-GS, and these types of proteins almost certainly are rod-like in shape (Field et al., 1987). The N-terminal regions of gliadins and LMW-GS are largely repeats, but studies must be carried out to determine if these regions are fairly rigid. If we assume that they are at least partially rod-like as a consequence of having some β-spiral structure, the various gliadin proteins and glutenin subunits might have shapes something like those of Fig. 3. For the sake of simplicity, I have depicted the unique sequence domains as also rod-like, but we don't have any evidence for this.

We might ask the question: if all these prolamin-type storage proteins should have β-spiral structure to some degree, might not the β-spiral regions tend to interact with one another, perhaps through side-by-side alignment of the spirals? This may not be too far-fetched a suggestion. I have noted that many glutenin preparations seem to be contaminated by ω-gliadins, perhaps as a consequence of such interactions.

If I next assume that most glutenin subunits participate in two intermolecular disulfide linkages, that the average MW of glutenin polymers is about 1 million, and that the ratio of LMW-GS to HMW-GS is about 6:1 on a molar basis, then I suggest that a typical glutenin molecule might look something like the simplified drawing of Fig. 4A. These glutenin polymers might tend, on the average, to reside in a plane. This might facilitate sheet formation through interaction of the glutenin polymers with gliadins, proteins and lipids. Salts, such as NaCl, which strengthen doughs (Danno and Hoseney, 1982), might be important, as well, in promoting ordered interactions of such structures with gliadins and lipids (and lipid-like molecules such as SDS). Finally, J. E. Bernardin (personal communication) has suggested that doughs might have liquid crystalline aspects. It is conceivable to me that planar aggregates of the sort I suggest might interact to create domains with

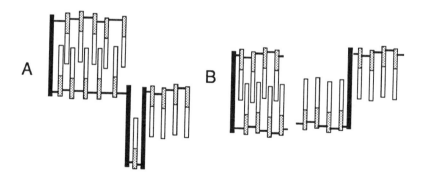

Fig. 4. Hypothetical structures. A. Single glutenin polymer molecule; B. Possible geometries for disulfide bonding of a HMW-GS to LMW-GS.

some characteristics of cholesteric liquid crystals. Could the observations of Danno and Hoseney (1982) on the ability of salt or SDS to bring about the recovery of overmixed doughs result from transition of a nematic (fibrous) liquid crystalline structure to a cholesteric (sheetlike) liquid crystalline structure?

VARIETAL DIFFERENCES IN MIXING AND BAKING QUALITY

In this section, I wish to discuss some possible bases for varietal differences in mixing and baking quality of bread wheats. All are speculative and require investigation. Quality differences may, of course, result from a combination of factors.

Glutenin/Gliadin Ratio

Does the ratio of glutenin to gliadin in a flour determine its mixing and baking quality? This has long been considered likely, and has been demonstrated for reconstituted mixtures of gliadins and glutenins (Belitz et al., 1988; MacRitchie, 1984), but I feel that solid evidence is lacking in regard to unmodified flours of bread wheats. The failure of all nonreducing (or strongly oxidizing) extractants to extract all proteins from flour is a complicating factor in many studies. The work of Autran et al. (1988) has provided valuable evidence, however, that good quality durum wheats do in fact have more glutenin than poor quality durum wheats. Further studies with bread wheats seem to be in order.

The Ratio of HMW-GS to LMW-GS in Glutenin

The HMW-GS seem to play a special role in determining the mixing and baking properties of a wheat variety. When Lawrence et al. (1988) studied a series of wheat lines in which the number of HMW-GS subunits ranged in number from a full complement of 5 to none, the mixing strength and loaf volume of these lines declined in proportion to the quantity of HMW-GS present as indicated by densitometric scanning of SDS-PAGE patterns. The null line with no HMW-GS was extremely lacking in elasticity. Lawrence et al. (1988) put forward the possibility that quantitative effects might be more responsible for apparent differences in quality associated with different forms (alleles) of HMW-GS. That is, one allele or pair of alleles might appear more effective in strengthening a dough or improving loaf volume because of greater protein production by their genes rather than because of important intrinsic differences between or among the allelic genes and their protein products.

Possible changes in the LMW-GS of the lines studied by Lawrence et al. (1988) that might have accompanied changes in the HMW-GS composition were not investigated and might complicate the interpretation of results. Payne and coworkers (1984) have demonstrated that certain HMW-GS in a variety correlate with good quality and others with poor quality. The question of the relative importance of quantity of protein produced by various alleles as compared to intrinsic quality differences stemming from the structures of individual allelic proteins needs to be addressed in future work.

MW Distribution of Glutenin Polymers

If the glutenin polymers of a given variety differ in their MW distribution (see Fig. 1), such differences could give rise to quality differences. A shift in the distribution curve towards higher MW's producing a larger value for the average MW of the mixture of glutenin polymers should result in stronger dough with greater resistance to mixing, a longer time to peak, and increased stability. It should also produce an increase in loaf volume up to a point. A dough that is too elastic might decrease loaf volume because it would resist expansion of gas bubbles formed during mixing and fermentation.

Huebner and Wall (1976) concluded that there are more higher-MW glutenin polymers in doughs from good quality varieties than from poor quality varieties. This interpretation of results was based on peaks appearing at the void volume of gel permeation columns and the assumption that optical absorption at a wavelength of 280 nm provided a measure of the amount of protein in the peaks. These peaks can be deceptive because, at the void volume, turbidity of the eluting solutions appears as light absorption--giving the appearance of increased amounts of protein. In addition, UV-absorbing, but nonproteinaceous material might be eluted at the column void volume of crude flour extracts.

Although it may well be that better quality wheats have more higher MW glutenin polymers, studies supporting this would be valuable. When void volume peaks are analyzed it would be desirable to analyze solids and their nitrogen content to avoid possible errors from light scattering or from UV-absorbing non-proteinaceous materials.

Branching of Glutenin Polymers

Ewart (1988) has provided evidence that glutenin polymers appear to be largely linear. His work does not rule out the possibility that some branching occurs in glutenin polymers. Different degrees of branching among wheat varieties might produce differences in viscoelasticity. In this regard, it may be noted that Greene et al. (1988) have found an extra cysteine residue in the primary structure of a particular HMW-GS (Payne's number 5; see Payne et al., 1984, for numbering system) that is one of an allelic pair (Payne numbers 5,10) associated with good quality. The equivalent protein (Payne number 2) of an allelic pair associated with poor quality (Payne numbers 2,12) was very similar in structure, but had one less cysteine residue. The extra cysteine residue might permit branching of the glutenin polymer chain.

A simple interpretation is precluded at this time, however, because the proteins compared by Greene et al. (1988) are members of allelic pairs. HMW-GS 5 and 10 are x types and their equivalent pairs 2 and 12 are y types. The genes coding for these x,y pairs are genetically linked and usually appear together in a wheat variety. It is always possible that it is the y type of each pair that is responsible for the quality differences associated with these pairs of proteins rather than the x types. Indeed, Pogna et al. (1987) have presented evidence for the y types being responsible for the quality differences. Their conclusions were based on SDS-sedimentation volumes, however, and it will be necessary to carry out actual mixing and baking tests to provide more confidence in the general validity of their results. Elucidation of the inter- and intramolecular disulfide bond structures of glutenin subunits along with definition of the actual regions of the polypeptide chains connected by these disulfide bonds should have high priority in future research. The task is made difficult, however, by the number of different protein subunits (at least 15) incorporated into the glutenin polymers.

Integrity and Elasticity of β-Spirals in HMW-GS

Tatham et al. (1984) proposed that β-spiral structures of HMW-GS might contribute elasticity to dough systems. They might also contribute to varietal differences in viscoelasticity. Goldsbrough et al. (1988) compared the DNA-derived primary structures (amino acid sequences) of the y-type subunits 10 and 12 of the two allelic pairs referred to in the previous section. In contrast to the pairs studied by Greene et al. (1988), both subunits had the same number of cysteine residues per molecule. Goldsbrough et al. (1988) found that the y-type subunit 10 from the pair associated with high quality had a few more perfect repeating sequences than the y-type subunit 12 from the pair associated with poor quality. [Subunit 10 is the subunit that appeared responsible for good quality in the study by Pogna et al. (1987).] Goldsbrough et al. (1988) hypothesized that, because of the greater regularity of repeating sequences in subunit 10, it might have a more perfect β-spiral structure, and, as a consequence, greater elastic recovery when extended during dough mixing.

It is difficult to test their hypothesis, but again, intuitively, I can see that irregularities or breaks in β-spiral structures might have an effect on the properties of HMW-GS. These rod-like subunits might become less rigid and, as a consequence, key geometric arrangements might be lost. The regions of the chain corresponding to the irregularities might also affect the thermodynamic free energy in such a way as to diminish elastic recovery. Finally, if β-spirals should have a tendency to align and interact, as I suggested earlier, then imperfections in the β-spiral region might diminish such interactions for the HMW-GS and their interactions with gliadins and LMW-GS, which also have repeating sequences (and perhaps β-spiral structure).

Geometric Factors

If HMW-GS have rod-like structures (Field et al., 1987) with a considerable amount of structural rigidity, then the arrangement of the intermolecular disulfide bonds extending out from the ends of the HMW-GS might affect the properties of the resulting glutenin polymers.

For example, consider a HMW-GS participating in two intermolecular disulfide bonds. If these bonds, one near each end, projected out in planes approximately perpendicular to the major axis of the subunit and if these bonds projected at the same angle in the plane (say at 0^0 on a 360^0 scale) perpendicular to the major axis, this might promote interactions between the two projecting chains (Fig. 4B).

On the other hand, if one bond projected at 180^0 to the other, then the chains would stick out in opposite directions (Fig. 4B) and would be less likely to interact with one another. They might have more of a tendency to interact with other glutenin polymers, with gliadins, or with lipid molecules. This is but one of many possibilities and it is difficult to predict exactly how such geometric differences might affect glutenin polymer interactions. The resulting arrangements might be important to formation of sheet-like structures of gluten proteins (Grosskreutz, 1960).

Of course, the idea of geometric factors being important is dependent on structural rigidity extending throughout the repeating-sequence (β-spiral?) regions of the HMW-GS to the region of the polypeptide chains where cysteine residues are located. Some cysteine residues are found in the repeating-sequence domain near the unique-sequence domains at the ends. These cysteine residues might be more likely to have rigid geometries as a consequence of any repeat-region structural rigidity than those located in the end domains. Reasonably stable geometries are, however, a possibility for any of the cysteines of HMW-GS. Geometric factors arising from HMW-GS structure could contribute to the different correlations with quality that have been observed for different HMW-GS and thus to varietal differences in quality. On a cautionary note, it might be well to refer once again to the interesting possibility discussed in the paper of Lawrence et al. (1988) that correlations with quality might arise more from the quantity of each subunit synthesized in the grain rather than from intrinsic differences in the subunits themselves.

Intrinsic Differences in Glutenin Subunit Interactions

The glutenin subunits of different varieties might differ in their intrinsic aggregation potential. Certain subunits might have greater potential to interact with one another or with other gluten protein subunits through secondary forces, such as hydrogen bonding, hydrophobic bonding, or ionic bonding. Such intrinsic differences might be responsible in some degree for varietal differences in quality.

CONCLUSIONS

There is clearly much to be done before we achieve an understanding at the molecular level of breadmaking quality in wheat flour and of the differences in this quality among varieties. In the next few years, there is likely to be a resurgence in physical chemical studies (X-ray diffraction, transmission electron microscopy, nuclear magnetic resonance, and many other techniques) that focus not only on the natural components of wheat endosperm but also on specifically designed proteins and peptides produced by genetic engineering techniques. As a consequence, new developments will appear at an accelerating pace, and it should not be long before these new developments are used to improve wheat quality.

ACKNOWLEDGEMENT

I wish to acknowledge helpful discussions with P. R. Shewry, A. S. Tatham, O. D. Anderson, F. C. Greene, , R. B. Flavell, P. I. Payne, V. Colot, J. E. Bernardin, and J. A. Bietz. Those acknowledged may not, of course, agree with all my speculations in this paper.

LITERATURE CITED

AUTRAN, J-C., LAIGNELET, B., and MOREL, M. H. 1987. Characterization and quantification of low-molecular-weight glutenins in durum wheat. In: Gluten Proteins: Proc. 3rd Int. Workshop. R. Lasztity and F. Bekes, eds. World Scientific, Singapore, pp. 266-283.

BARTELS, D., ALTOSAAR, I., HARBERD, N. P., BARKER, R. F., and THOMPSON, R. D. 1986. Molecular analysis of γ-gliadin gene families at the complex *Gli-1* locus of bread wheat (*T. aestivum* L.). Theor. Appl. Genet. 72:845-853.

BELITZ, H-D., KIM, J-J., KIEFFER, R., SEILMEIER, W., WERBECK, U., and WEISER, H. 1987. Separation and characterization of reduced glutelins from different wheat varieties and importance of the gliadin/glutelin ratio for the strength of gluten. In: Gluten Proteins: Proc. 3rd Int. Workshop. R. Lasztity and F. Bekes, eds. World Scientific, Singapore, pp. 189-205.

BIETZ, J. A., and HUEBNER, F. R. 1980. Structure of glutenin: achievements at the Northern Regional Research Center. Ann. Technol. Agric. 29:249-277.

BIETZ, J. A., and WALL, J. S. 1980. Identity of high molecular weight gliadin and ethanol-soluble glutenin subunits of wheat: relation to gluten structure. Cereal Chem. 57:415-421.

BULLEID, N. J., and FREEDMAN, R. B. 1988. Defective co-translational formation of disulphide bonds in protein-disulphide-isomerase-deficient microsomes. Nature 335:649-651.

DANNO, G., and HOSENEY, R. C. 1982. Effect of sodium chloride and sodium dodecylsulfate on mixograph properties. Cereal Chem. 59:202-204.

EWART, J. A. D. 1968. A hypothesis for the structure and rheology of glutenin. J. Sci. Food Agric. 19:617-623.

EWART, J. A. D. 1985. Blocked thiols in glutenin and protein quality. J. Sci. Food Agric. 36:101-112.

EWART, J. A. D. 1987. Calculated molecular weight distribution for glutenin. J. Sci. Food Agric. 38:277-289.

EWART, J. A. D. 1988. Studies on disulfide bonds in glutenin. Cereal Chem. 65:95-100.

FIELD, J. M., TATHAM, A. S., and SHEWRY, P. R. 1987. The structure of a high-M_r subunit of durum wheat

(*Triticum durum*) gluten. Biochem. J. 247:215-221.
FINNEY, K. F. 1954. Contributions of the hard winter wheat quality laboratory to wheat quality research. Trans.American Association of Cereal Chemists 12:127-142.
GOLDSBROUGH, A., ROBERT, L., SCHNICK, D., and FLAVELL, R. B. 1988. Molecular comparisons between the bread making quality determining HMW glutenin subunits of wheat gluten - evidence for the nature of good and poor quality at the protein level. Proc. 7th Int. Wheat Genetics Symp. In press.
GRAVELAND, A., BOSVELD, P., LICHTENDONK, W. J., MARSEILLE, J. P., MOONEN, J. H. E., and SCHEEPSTRA, A. 1985. A model for the molecular structure of the glutenins from wheat flour. J. Cereal Sci. 3:1-16.
GREENE, F. C., ANDERSON, O. D., LITTS, J. C., and GAUTIER, M-F. 1985. Control of wheat protein biosynthesis. Cereal Chem. 62:398-405.
GREENE, F. C., ANDERSON, O. D., YIP, R. E., HALFORD, N. G., MALPICA-ROMERO, J-M., and SHEWRY, P. R. 1988. Analysis of possible quality-related sequence variations in the 1D glutenin high molecular weight subunit genes of wheat. Proc. 7th Int. Wheat Genetics Symposium. In press.
GROSSKREUTZ, J. C. 1960. The physical structure of wheat protein. Biochim. Biophys. Acta 38:400-409.
GUPTA, R. B., and SHEPHERD, K. W. 1987. Genetic control of LMW-glutenin subunits in bread wheat and association with physical dough properties. In: Gluten Proteins: Proc. 3rd Int. Workshop. R. Lasztity and F. Bekes, eds. World Scientific, Singapore, pp. 13-19.
HALFORD, N. G., FORDE, J., ANDERSON, O. D.,GREENE, F. C., and SHEWRY, P. R. 1987. The nucleotide and deduced amino acid sequences of an HMW glutenin subunit gene from chromosome 1B of bread wheat (*Triticum aestivum* L.) and comparison with those of genes from chromosomes 1A and 1D. Theor. Appl. Genet. 75:117-126.
HUEBNER, F. R., and BIETZ, J. A. 1985. Detection of quality differences among wheats by high-performance liquid chromatography. J. Chromatography 327:333-342.
HUEBNER, F. R., and WALL, J. S. 1976. Fractionation and quantitative differences of glutenin from wheat varieties varying in baking quality. Cereal Chem.

53:258-269.
KASARDA, D. D., ADALSTEINS, A. E., and LAIRD, N. F. 1987. In. Gluten Proteins: Proc. 3rd Int. Workshop. R. Lasztity and F. Bekes, eds. World Scientific, Singapore, pp. 20-29.
KASARDA, D. D., OKITA, T. W., BERNARDIN, J. E., BAECKER, P. A., NIMMO, C. C., LEW, E. J-L., DIETLER, M. D., and GREENE, F. C. 1984. Nucleic acid (cDNA) and amino acid sequences of α-type gliadins from wheat (*Triticum aestivum*). Proc. Natl. Acad. Sci. USA 81:4712-4716.
KASARDA, D. D., TAO, H. P., EVANS, P. K., ADALSTEINS, A. E., and YUEN, S. W. 1988. Sequencing of a protein from a single spot of a 2-D gel pattern: N-terminal sequence of a major wheat LMW-glutenin subunit. J. Exp. Botany 39:899-906.
KHAN, K., and BUSHUK, W. 1979. Studies of glutenin. XII. Comparison by sodium dodecyl sulfate-polyacrylamide gel electrophoresis of unreduced and reduced glutenin from various isolation and purification procedures. Cereal Chem. 56:63-68.
KREIS, M., SHEWRY, P. R., FORDE, B. G., FORDE, J, and MIFLIN, B. J. 1985. Structure and evolution of seed storage proteins and their genes with particular reference to those of wheat, barley, and rye. Oxford Surv. Plant Mol. Cell Biol. 2:253-317.
LAWRENCE, G. J., MACRITCHIE, F., and WRIGLEY, C. W. 1988. Dough and baking quality of wheat lines deficient in glutenin subunits controlled by the *Glu-A1*, *Glu-B1*, and *Glu-D1* Loci. J. Cereal Sci. 7:109-112.
LAWRENCE, G. J., and PAYNE, P. I. 1983. Detection by gel electrophoresis of oligomers formed by the association of high-molecular-weight glutenin protein subunits of wheat endosperm. J. Exp. Botany 34:254-267.
MACRITCHIE, F. 1984. Baking quality of wheat flours. Adv. Food Res. 29:201-277.
OKITA, T. W., CHEESBROUGH, V., and REEVES, C. D. 1985. Evolution and heterogeneity of the α/β-type and γ-type gliadin DNA sequences. J. Biol. Chem. 260:8203-8213.
PAYNE, P. I., and CORFIELD, K. G. 1979. Subunit composition of wheat glutenin proteins isolated by gel filtration in a dissociating medium. Planta 145:83-88.
PAYNE, P. I., HOLT, L. M., JACKSON, E. A., and LAW, C. N. 1984. Wheat storage proteins: their genetics and their potential for manipulation by plant breeding. Phil.

Trans. Royal Soc., London, B304:359-371.
POGNA, N. E., MELLIM, F., and DAL BELIN PERUFFO, A. 1987. Glutenin subunits of Italian common wheats of good bread-making quality and comparative effects of high molecular weight glutenin subunits 2 and 5, 10 and 12 on flour quality. In: Agriculture: Hard Wheat: Agronomic, Technological, Biochemical, and Genetic aspects: Report EUR 11172. B. Borghi, ed. Commission of the European Communities, Milan, pp. 53-69.
RAFALSKI, J. A. 1986. Structure of wheat gamma-gliadin genes. Gene 43:221-229.
RODEN, L. T., MIFLIN, B. J., and FREEDMAN, R. B. 1982. Protein disulphide-isomerase is located in the endoplasmic reticulum of developing wheat endosperm. FEBS Lett. 138:121-124.
SCHEETS, K., and HEDGCOTH, C. 1988. Nucleotide sequence of a γ gliadin gene: comparisons with other γ gliadin sequences show the structure of γ gliadin genes and the general primary structure of γ gliadins. Plant Sci. 57:141-150.
SCHOFIELD, J. D., BOTTOMLEY, R. C., TIMMS, M. F., and BOOTH, M. R. 1983. The effect of heat on wheat gluten and the involvement of sulphydryl-disulphide interchange reactions. J. Cereal Sci. 1:241-253.
SHEWRY, P. R., MIFLIN, B. J., LEW, E. J-L., and KASARDA, D. D. 1983. The preparation and characterization of an aggregated gliadin fraction from wheat. J. Exp. Botany 34:1403-1410.
SHEWRY, P. R., and TATHAM, A. S. 1987. Recent advances in our understanding of cereal seed protein structure and functionality. Comments Agric. & Food Chem. 1:71-94.
SHEWRY, P. R., TATHAM, A. S., FORDE, J., KREIS, M., and MIFLIN, B. J. 1986. The classification and nomenclature of wheat gluten proteins: a reassessment. J. Cereal Sci.4:97-106.
SINGH, N. K., and SHEPHERD, K. W. 1985. The structure and genetic control of a new class of disulphide-linked proteins in wheat endosperm. Theor. Appl. Genet. 71:79-92.
SUGIYAMA, T., RAFALSKI, A., and SÖLL, D. 1986. The nucleotide sequence of a wheat γ-gliadin clone. Plant Sci. 44:205-209.

TATHAM, A. S., SHEWRY, P. R., and MIFLIN, B. J. 1984. Wheat gluten elasticity: a similar molecular basis to elastin? FEBS Lett. 177:205-208.
WALL, J. S. 1979. The role of wheat proteins in determining baking quality. In: Recent Advances in the Biochemistry of Cereals. D. L. Laidman and R. G. Wyn Jones, eds. Academic Press, NY, pp. 275-311.

18

TELLING DIFFERENCES AMONG WHEAT PROTEINS MAY MAKE A DIFFERENCE IN MARKETING

Jerold A. Bietz
Northern Regional Research Center
U.S. Department of Agriculture
Agricultural Research Service
1815 N. University St.
Peoria, IL 61604

ABSTRACT

Gluten proteins strongly influence wheat's functional and nutritional properties. There are many gluten proteins; they differ among wheats, but are nearly constant in any variety. Electrophoresis or high-performance liquid chromatography can quickly and accurately analyze these proteins to identify varieties and predict some quality characteristics. This approach is becoming useful in breeding, classification, marketing and utilization. Breeders, under pressure to improve yield, protein content, and disease resistance, may now use wide crosses, novel germplasm, and earlier selection to produce new varieties. Such strategies complicate classification, however, since seed types no longer always specify class. Early screening for desirable proteins during breeding can indicate ideal parents or offspring, and improve final quality. These methods should also, by more objectively defining

---

This chapter is in the public domain and not copyrightable. It may be freely reprinted with customary crediting of the source, American Association of Cereal Chemists, 1988.

wheat classes, strengthen markets. By identifying varieties, millers, bakers and other processors might predict grain quality, and optimize ingredients, processes and products. Such analyses should enhance sales by defining quality specifications for each class, and certifying characteristics for specific uses. Variety identification and quality guaranteed through protein analysis will clearly predict end-use, assuring buyer satisfaction. This should maximize profit and sales for producers, while enhancing wheat quality, utilization, and standing in domestic and export markets.

INTRODUCTION

Wheat has been our most important cereal grain for a very long time. Breeders, producers, millers and bakers everywhere are experts in identifying wheat and judging its quality. Or are they?

The United States has long been the bread basket of the world. Its wheat production and marketing systems are unique: America is large, and many types of wheat are grown. Marketing was once much simpler than it is now, however. All new varieties were bred to be at least as good as existing ones, and kernel and agronomic characteristics, which indicated end use and quality, were uniform. Wheat could therefore be marketed by class alone.

Changes in breeding and marketing strategies are making this system obsolete, however. In the U.S., varieties are developed by federal, state, and private breeders, who now use novel methods to improve varieties and speed their release. There is no control over what is released or produced, or where varieties are grown. More than 100 major varieties are produced at any time. Some varieties may differ in appearance or performance from that which is typical of their class, and some have more than one genetically distinct biotype.

These breeding approaches are, of course, valid, but they have led to some problems in marketing and classification. New methods can help solve these problems.

These methods analyze wheat's gluten proteins. This idea is not new - wheat proteins have unique properties (Osborne 1907), as has also been shown by electrophoresis (e.g., Elton and Ewart 1964, Bietz 1975). Such analyses of proteins can significantly aid breeding.

Gluten's two major constituents, gliadin and glutenin, each have many components (see Wrigley and Bietz 1988). These proteins give wheat its nutritional and functional properties, but they are also important in another way. There are many wheat proteins, which differ between varieties, thus providing specific "fingerprints" of most lines.

Differentiating wheats by protein analysis can significantly influence marketing, breeding, classification, and utilization. This article reviews the available methods, shows how they can be applied, and considers their future.

FINGERPRINTING OF CULTIVARS

Gluten proteins are qualitatively the same within any genotype, but differ between lines. They can thus accurately identify varieties (Wrigley et al 1982). While many cultivars can be identified by agronomic properties, methods based on protein differences are often faster and easier. Analysis of gliadin is especially convenient since it is easily extracted. Electrophoresis and chromatography are often used for such analyses.

Electrophoresis

In gel electrophoresis, proteins migrate through a porous matrix in an electric field. Mobility depends on size and charge. Woychik et al (1961) showed by starch gel electrophoresis (SGE) that gliadins differ among varieties; similarly, glutenin subunits differ among wheats (Huebner and Wall 1976). Polyacrylamide gel electrophoresis (PAGE) has largely replaced SGE for varietal identification (Wrigley et al 1982); the method has shown 4% of wheats to be incorrectly identified (Jones et al 1982).

Isoelectric focusing (IEF), based on protein charge, can also identify varieties (Wrigley 1968).

PAGE in the presence of the detergent sodium dodecyl sulfate (SDS) is especially useful for protein analysis. Bietz and Wall (1972) showed glutenin to contain high molecular weight (MW) subunits which influence quality. These subunits differ among varieties, and biotypes are sometimes present (Fig. 1). SDS-PAGE analysis of glutenin is widely used to predict quality (Payne et al 1981). Electrophoresis methods can also be combined into "two-dimensional" procedures; resulting high resolution separations can differentiate most varieties (Anderson and Anderson 1987, Tkachuk and Mellish 1987, Wrigley and Shepherd 1973).

Chromatography

Chromatography is also widely used to study wheat proteins. Proteins typically bind to a solid

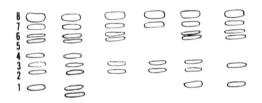

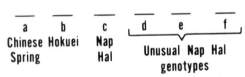

Fig. 1. SDS-PAGE analysis of high MW glutenin subunits in the varieties Chinese Spring (a), Hokuei (b), Nap Hal (c), and unusual Nap Hal genotypes (d-f). From Bietz (1975).

support, and are displaced by varying the solvent composition. Several types of chromatography are possible. Using ion-exchange chromatography, Huebner and Wall (1966) showed that gliadins differ among varieties. Gel filtration chromatography of wheat proteins also differentiated varieties (Huebner and Wall 1976), and suggested that optimal ratios of protein classes determine breadmaking quality.

Another major chromatography method is reversed-phase (RP-) HPLC, first applied to wheat by Bietz (1983). RP-HPLC separates proteins by surface hydrophobicity, complementing separations based on charge or size. Fig. 2 shows a high-resolution separation of gliadins from one kernel. RP-HPLC is

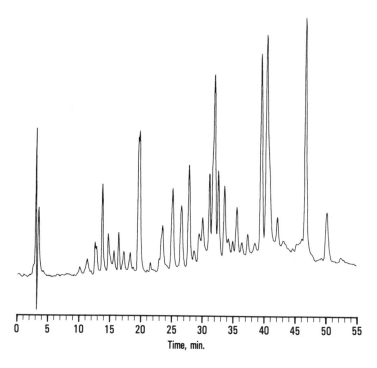

Fig. 2. Separation of 70% ethanol-soluble gliadins from the hard red winter wheat Siouxland by RP-HPLC on a Vydac C_{18} column at 60 C. Gradient, 23-45% CH_3CN (+ 0.05% trifluoroacetic acid) during 50 min.

fast, sensitive, and reproducible. It can be automated, used in a preparative mode, and data can be quantified.

RP-HPLC of gliadins permits identification of nearly all varieties (Fig. 3) (Bietz et al 1984). Maximum resolution requires about one hour, but more rapid separations may also differentiate varieties. RP-HPLC can also reveal heterogeneity within land races (Bietz 1983), and biotypes within varieties (Lookhart et al 1986).

RP-HPLC also gives excellent separations of other protein classes. Glutenin, wheat's "strength protein", can be analyzed after its disulfide bonds are cleaved. RP-HPLC separates these subunits into

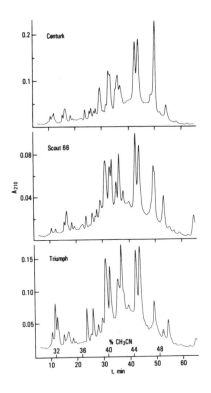

Fig. 3. RP-HPLC of hard red winter wheat gliadins. Different peak heights and elution positions show unique protein compositions of each variety.

two distinct regions (Fig. 4) (Burnouf and Bietz, 1984b). Early peaks contain high MW subunits, which differ between varieties and may directly influence wheat's breadmaking quality (Payne et al 1981).

Both electrophoresis and RP-HPLC can identify varieties by gliadin or glutenin analysis (Lookhart et al 1986). The methods complement each other, separating proteins on the basis of different chemical properties. Both methods have advantages. HPLC is faster for single samples (15-60 minutes) and can be automated; electrophoresis, however, can analyze 40 samples in four hours. Resolution of RP-HPLC is usually better than that of one-dimensional electrophoresis. Both methods have significant roles and applications.

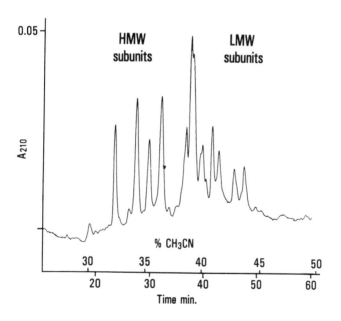

Fig. 4. RP-HPLC of Chinese Spring glutenin subunits after reduction and pyridylethylation of disulfide bonds. High MW subunits elute first, followed by low-MW subunits.

BREEDING

U.S. wheats have traditionally all had very good quality. New varieties were not released unless they were at least as good as the best existing variety for an area. Traditional breeding was highly successful; there was little reason, especially since wheat was marketed by class rather than variety, to guarantee varietal purity, or to use biochemical analyses. Classes were uniform, and users could purchase wheat for particular uses (such as breadmaking) by class alone. Class inferred quality, protein content, and end-use; it did not guarantee these factors, however.

This situation has now begun to change. Breeders use exotic germplasm and cross wheats of different classes to improve disease resistance, protein content and yield. Different wheat types, such as feed wheats, are also bred: these may have excellent agronomic properties, but inferior breadmaking characteristics. Increased costs and pressure for release may also lead to earlier selection of varieties during breeding.

These changes have sometimes led to varieties having more than one biotype. "Multilines" may be agronomically desirable. If kernels vary in appearance, however, morphology may no longer suffice for classification; class may no longer guarantee quality.

Breeders recognize the challenge of providing marketable high-quality varieties. Electrophoresis and chromatography can help them achieve this goal. Protein analysis provides specific markers which can indicate desirable genotypes for crosses or be used for early generation screening or to ensure varietal purity.

Many examples of this approach can be cited. Differences in protein composition between genotypes, as of high MW glutenin subunits in bread wheats (Bietz and Wall 1972, Payne et al 1981) or gamma-gliadins in durums (Burnouf and Bietz 1984a), either directly influence quality or are excellent markers of quality characteristics. Such attributes might not be apparent from kernel phenotype or plant agronomic properties, however. The ability to select promising genotypes quickly through germplasm screening or in early (F1 or F2) crosses can be very helpful. Such selection is possible because protein analyses are very sensitive: half kernels can be analyzed, so germplasm is not destroyed. An example of how HPLC can screen durum progeny was presented by Burnouf and Bietz (1984a).

Biochemical analyses should also help the breeder select useful genotypes. For example, Nap Hal is a land race containing genes for both high protein and high lysine. These characteristics have been bred into hard red winter wheats. Both electrophoresis (Fig. 1) and HPLC (Bietz et al 1984), however, showed that Nap Hal contains many genotypes, differing in quality-associated proteins. Which genotype was used? Breeders can now answer that question.

Genetic studies are also intimately involved with breeding. Wheat aneuploids can be analyzed by electrophoresis or HPLC (Bietz and Burnouf 1985, Burnouf and Bietz 1985), revealing locations of genes which code gliadin and glutenin polypeptides. These proteins can then serve as markers of genes or chromosome segments desirable in breeding. This capability may be especially important in selecting desirable progeny from wide crosses.

Varietal identification should also be useful for plant protection. Protein fingerprints are accurate indicators of pedigree. When considered with physical and agronomic traits, varieties can be uniquely described. This should likewise be applicable in seed certification, and for identification or registration of hybrid wheats.

CLASSIFICATION

In the United States, inspectors from the Federal Grain Inspection Service classify grain for foreign markets, guaranteeing its identity and quality. The classification criteria used are the official U.S. grain standards. According to these rules, only visual kernel features, such as germ, brush, bran texture, crease, and shape of cheeks, are examined. Using these characteristics, trained inspectors can identify most varieties, thereby classifying wheat in marketing channels. Inspectors also grade wheat based on kernel soundness and sample purity. Amount of protein may be reported on certificates, but type or quality of protein is not determined.

Classification by varietal identification is more difficult in the U.S. than it would be in other countries. More than 120 varieties are grown on more than 100,000 acres each, and there is no regulation on what is grown. Both public and private breeders release new lines. Differences in size, shape, color, and other kernel characteristics are not always clear-cut. Nevertheless, accuracy in classification is essential in marketing: for example, "hard" and "spring" imply quality as well as class to some buyers.

Classification today is becoming even more difficult. Inspectors usually can still accurately determine class. But because of changes in breeding (e.g., wide crosses and earlier selection), kernel morphology alone may no longer specifically designate variety or class, especially if varieties have more than one biotype. An example is the variety Arkan, an excellent hard red winter wheat. Its pedigree includes both soft (Arthur) and hard (Sage) wheat parents. Arkan may be difficult to identify visually, however, since some kernels may resemble its soft wheat parent. HPLC of individual kernels of this variety also revealed more than one distinct genotype (Bietz and Cobb 1985). These biotypes do not relate directly to kernel appearance, but do reveal genetic heterogeneity, which is also expressed as differences in kernel appearance.

Because of such considerations, there has been increasing pressure to modify wheat classification. U.S. wheat is as good as or better than it has ever been; but class sometimes no longer predicts end-use. We could lose markets because of perceived quality problems. In response to these challenges, the USDA's Agricultural Research Service is conducting a collaborative study of wheat classes.

Many buyers prefer traditional labels, which have implied end-use. Our aim is to identify objective criteria which define each class, and which can be quickly and accurately measured. It is probable that measuring grain protein content and hardness, and sometimes protein quality, can specify class and imply end-use. Near infrared analysis can be part of such a classification system.

HPLC and electrophoresis serve important roles in defining those criteria which differentiate classes. Analysis of different wheats is revealing proteins which are closely related to hardness, or to environment. Such proteins may directly influence these traits, or, because their coding genes are near those which influence quality, serve as excellent marker proteins. Once such proteins are identified, it may be possible to devise rapid tests useful in classification.

HPLC and electrophoresis may also be used for classification. As the ability to rapidly and automatically identify varieties through protein analysis improves, such a classification method may become practical and cost-effective.

MARKETING

The ability to identify wheat varieties should be especially important in production and marketing. Producers need to know which varieties are best for their areas, which are most marketable, and which will maximize profit. In the grain trade, sellers need to know the quality of the wheat they sell and its potential for various products to identify and maintain markets, meet the competition, and maximize profits. And buyers or processors must know which wheats are suitable for their applications.

Occasionally, there are problems in marketing. Buyers do not always realize that class does not guarantee end-use quality. The ability to accurately identify wheats should strengthen markets and enhance sales. Defined quality characteristics should clearly show end-use potential, assuring buyer satisfaction.

The ability to classify wheat objectively and to identify varieties has great potential. For example, contracting is becoming common to ensure product quality. Rapid specific varietal identification during marketing could also fill this role.

For these reasons, "fingerprinting" of endosperm storage proteins has great promise in marketing. In many countries, such analyses are used routinely to identify varieties <u>not</u> suitable for specific applications, such as breadmaking. The <u>existence</u> of such methods is itself an important deterrent to incorrect labeling of marketed wheat.

The U.S. situation is somewhat different. More varieties are marketed, and most have good quality. Also, since we usually market by class, shipments may contain many varieties. Nevertheless, varietal differentiation can guarantee identity of contracted shipments, and could be used to certify seeds of varieties or hybrids. Since HPLC gives quantitative data, it may also become possible, through computers, to identify and measure amounts of varieties in mixtures (Bietz and Cobb 1985). Such methods also permit analysis of hybrid wheats (Bietz and Huebner 1987).

UTILIZATION AND QUALITY

Because of changes in breeding and the complexity of our classification and marketing system, there is a critical need to identify those varieties at the mill or bakery which are suitable for their intended use, and to optimize processes and product quality.

Several studies have shown relationships of wheat proteins to quality or end-use potential. For example, high MW glutenin subunits, as analyzed by electrophoresis (Fig. 1) or HPLC (Fig. 4), reflect breadmaking quality. Similarly, durum pasta strength can be predicted by gliadins "45" and "42" (Burnouf and Bietz 1984a).

Since proteins significantly influence wheat's functional properties, further relationships of this type probably exist, and will be discovered. For example, HPLC has revealed specific late-eluting components in gliadins which relate to breadmaking potential (Huebner and Bietz 1986). Molecular weight ranges of gluten proteins also relate to breadmaking quality (Huebner and Bietz 1985). Additional useful relationships will undoubtedly be discovered through objective data analysis.

Another important application of these methods is only beginning to be explored. Protein interactions and structures change during processing, and HPLC or electrophoresis can monitor such changes. Resulting information should be useful in optimizing process or product quality. In initial studies, changes have been detected in products or grains which relate to how they have been processed. Learning how to best interpret the information provided by HPLC or electrophoresis could significantly enhance quality and profits, and suggest new uses.

CONCLUSIONS

How important, then, is the ability to tell the difference between varieties? In the U.S. today, since most wheat is marketed by class, the methods I have described are not yet used routinely. As varietal identification becomes easier and faster, however, and methods of data interpretation improve, HPLC and electrophoresis will be increasingly used in marketing.

These methods are already providing fundamental information for further developments. They may be especially valuable in breeding to choose the best parents for new crosses, to select lines for increase and release, and to ensure varietal purity. Fingerprinting of varieties may also be useful for registration. Breeders are receptive to such methods, especially when they are simple, rapid, and useful, and when they can enhance marketability.

Protein analysis will become valuable for optimizing grain, product and process quality, and should contribute to continued excellence in U.S. wheat production, marketing, processing, and utilization.

LITERATURE CITED

Anderson, N. G., and Anderson, N. L. 1987. Application of high-resolution, two-dimensional electrophoresis to the analysis of wheat, milk, and other agricultural products. ACS Symp. Series 335:132-142.

Bietz, J. A. 1975. Protein electrophoresis aids in wheat breeding. pp. 61-75 in: Proceedings of the 9th National Conference on Wheat Utilization Research, Seattle, October 8-10, 1975. Agricultural Research Service ARS-NC-40.

Bietz, J. A. 1983. Separation of cereal proteins by reversed-phase high-performance liquid chromatography. J. Chrom. 255:219-238.

Bietz, J. A., and Burnouf, T. 1985. Chromosomal control of wheat gliadin: analysis by reversed-phase high-performance liquid chromatography. Theor. Appl. Genet. 70:599-609.

Bietz, J. A., Burnouf, T., Cobb, L. A., and Wall, J. S. 1984. Wheat varietal identification and genetic analysis by reversed-phase high-performance liquid chromatography. Cereal Chem. 61:129-135.

Bietz, J. A., and Cobb, L. A. 1985. Improved procedures for rapid wheat varietal identification by reversed-phase high-performance liquid chromatography of gliadin. Cereal Chem. 62:332-339.

Bietz, J. A., and Huebner, F. R. 1987. Prediction of wheat quality by computer evaluation of reversed-phase high-performance liquid chromatograms of gluten proteins. pp. 173-188 in: Proc. 3rd Int. Workshop on Gluten Proteins, Budapest, Hungary, May 9-12, 1987. R. Lasztity and F. Bekes, eds. World Scientific, Singapore.

Bietz, J. A., and Wall, J. S. 1972. Wheat gluten subunits: molecular weights determined by sodium dodecyl sulfate-polyacrylamide gel electrophoresis. Cereal Chem. 49:416-430.

Burnouf, T., and Bietz, J. A. 1984a. Reversed-phase high-performance liquid chromatography of durum wheat gliadins: relationships to durum wheat

quality. J. Cereal Sci. 2:3-14.
Burnouf, T., and Bietz, J. A. 1984b. Reversed-phase high-performance liquid chromatography of reduced glutenin, a disulfide-bonded protein of wheat endosperm. J. Chrom. 299:185-199.
Burnouf, T., and Bietz, J. A. 1985. Chromosomal control of glutenin subunits in aneuploid lines of wheat: analysis by reversed-phase high-performance liquid chromatography. Theor. Appl. Genet. 70:610-619.
Elton, G. A. H. and Ewart, J. A. D. 1964. Electrophoretic comparison of cereal proteins. J. Sci. Food Agric. 15:119-126.
Huebner, F. R., and Bietz, J. A. 1985. Detection of quality differences among wheats by high-performance liquid chromatography. J. Chrom. 327:333-342.
Huebner, F. R., and Bietz, J. A. 1986. Assessment of the potential breadmaking quality of hard wheats by reversed-phase high-performance liquid chromatography of gliadins. J. Cereal Sci. 4:379-388.
Huebner, F. R., and Wall, J. S. 1966. Improved chromatographic separation of gliadin proteins on sulfoethyl cellulose. Cereal Chem. 43:325-335.
Huebner, F. R., and Wall, J. S. 1976. Fractionation and quantitative differences of glutenin from wheat varieties varying in baking quality. Cereal Chem. 53:258-269.
Jones, B. L., Lookhart, G. L., Hall, S. B., and Finney, K. F. 1982. Identification of wheat cultivars by gliadin electrophoresis: electrophoregrams of the 88 wheat varieties most commonly grown in the United States in 1979. Cereal Chem. 59:181-188.
Lookhart, G. L., Albers, L. D., and Bietz, J. A. 1986. Comparison of polyacrylamide gel electrophoresis and high-performance liquid chromatography analyses of gliadin polymorphism in the wheat cultivar Newton. Cereal Chem. 63:497-500.
Osborne, T. B. 1907. The proteins of the wheat kernel. Carnegie Inst., Wash. Publ. No. 84.

Payne, P. I., Corfield, K. G., Holt, L. M., and Blackman, J. A. 1981. Correlations between the inheritance of certain high-molecular weight subunits of glutenin and bread-making quality in progenies of six crosses of bread wheat. J. Sci Food Agric. 32:51-60.

Tkachuk, R., and Mellish, V. J. 1987. Use of two-dimensional electrophoresis procedures to characterize wheat proteins. pp. 111-124 in: Proc. 3rd Int. Workshop on Gluten Proteins Budapest, Hungary, May 9-12, 1987. R. Lasztity and F. Bekes, eds. World Scientific, Singapore.

Woychik, J. H., Boundy, J. A., and Dimler, R. J. 1961. Starch gel electrophoresis of wheat gluten proteins with concentrated urea. Arch. Biochem Biophys. 94:477-482.

Wrigley, C. W. 1968. Gel electrofocusing - a technique for analyzing multiple protein samples by isoelectric focusing. Science Tools 15:17-23.

Wrigley, C. W., Autran, J. C., and Bushuk, W. 1982. Identification of cereal varieties by gel electrophoresis of the grain proteins. Adv. Cereal Sci. Technol. 5:211-259.

Wrigley, C. W., and Bietz, J. A. 1988. Proteins and amino acids. Pages 159-275 in: Wheat: Chemistry and Technology. Vol. 1. 3rd ed. Y. Pomeranz, ed. Am. Assoc. Cereal Chem., St. Paul, MN.

Wrigley, C. W., and Shepherd, K. W. 1973. Electrofocusing of grain proteins from wheat genotypes. Ann. N. Y. Acad. Sci. 209:154-162.

19

WHEAT LIPIDS ARE UNIQUE

William R. Morrison
Food Science Division,
Department of Bioscience and Biotechnology,
University of Strathclyde
Glasgow G1 1SD, Scotland, UK.

Lipids in wheat are qualitatively the same as in other cereals, but they are unique in the ways they affect technological properties, most notably flour quality for breadmaking, but also for many other end-uses (Morrison, 1978a, 1979). Functionality has been demonstrated by adding lipids to bread doughs, and by extracting and fractionating flour lipids so that bread can be made using flours reconstituted with various quantities of particular lipid fractions (Hoseney and Finney, 1971; Chung and Pomeranz, 1977; MacRitchie, 1981, 1983; Chung, 1986; Frazier and Daniels, 1986). These experiments have shown that lipids probably rank next in importance after the gluten-forming prolamins (gliadins and glutenins) as factors affecting breadmaking quality, and, indeed, lipids seem to exert most of their effects through interactions with dough proteins.

COMPOSITION OF WHEAT LIPIDS

The structures of the various types of lipid and their fatty acid compositions have been described in detail elsewhere (Morrison, 1978a, 1978c, 1983, 1988a). For most purposes, including this discussion, lipids can be grouped together either according to structure and polarity, or by solvent extractability. All major and most minor lipids are

TABLE I

Fatty Acid Composition of Lipid Groups in Wheat Flour

	Fatty acids (wt. %)					Principal lipid classes in group[b]
	16:0	18:0	18:1	18:2	18:3	
Non-polar lipids, NL (non-starch)	17-19	1-2	14-15	60-62	4-5	TG, FFA, DG, MG
Glycolipids, GL (non-starch)	12-14	1-2	8-9	71-74	4-5	DGDG, MGDG, DGMG, MGMG
Phospholipids, PL (non-starch)	23-27	<1	9-10	59-63	2-4	NAPE, NALPE, diacyl PL
Starch lipids	36-37	1-2	8-9	49-51	2-4	Lyso PL

[a]Adapted from Morrison (1978a)   [b]Abbreviations given in text

acyl lipids, i.e. they contain the ubiquitous fatty acids found in most plant lipids (Table I). The other minor lipids include sterols, steryl glycosides, and glycosphingolipids which are of no practical importance, tocopherols which are useful antioxidants and a rich source of Vitamin E, and carotenoids which impart a yellow color tint to flour.

The non-polar or neutral lipids (NL) consist of triglycerides (TG), with lesser amounts of steryl esters, diglycerides (DG), monoglycerides (MG), and minor acylated glycolipids. Free fatty acids (FFA), which are moderately polar in some chromatographic solvent systems, are included in the NL. The polar lipids (PoL) are comprised of several glycolipids (GL) and phospholipids (PL). The principal GL are the di- and mono-galactosyldiglycerides (DGDG, MGDG) with lesser amounts of the corresponding monoacyl lipids (DGMG, MGMG). The principal PL include phosphatidyl-choline, -ethanolamine and -glycerol (PC, PE, PG) and two unusual N-acyl derivatives of PE (NAPE, NALPE). There are also substantial quantities of the corresponding monoacyl or lysoPL (LPC + LPE + LPG).

Solvents for the extraction of lipids must be chosen to suit the material (flour, freeze-dried dough or gluten, bread, starch) being extracted (Morrison, 1978a, 1988a, 1988b; Morrison and Coventry, 1985; Morrison et al., 1975, 1980). When dealing with flour, lipids extracted with a very low polarity solvent such as hexane or diethyl ether are termed "free" (prefix F) and the remaining non-starch lipids, which can be extracted with water-saturated butanol and similar solvents at ambient temperature, are termed "bound". Starch lipids, located inside the starch granules, require extraction with hot alcohol-water mixtures and normally they are not considered part of the bound lipids.

DISTRIBUTION OF LIPIDS IN THE GRAIN

The distribution of lipids in the principal parts of the mature wheat grain is given in Table II. TG and most of the other NL are storage lipids found

TABLE II

Distribution of Acyl Lipids in Wheat Kernel Fractions

Fraction	Weight mg	Lipid in fraction µg/kernel	wt.%	% of total lipid in fraction NL	Gl	PL
Germ	0.9-1.1	270-320	7-30.5	79-85	0-3.5	14-17
Aleurone	1.5-3.6	220-387	7-10.6	72-83	2-10	14-18
Pericarp	2.5-3.2	c.40	1.3	86	7	7
Starch	17.9-28.6	220-258	8-1.2	4-6	1-7	89-94
Non-starch endosperm	24.3-35.3	139-387	9-1.1	33-47	20-38	22-35
Whole kernel	30.3-42.4	916-1244	8-3.2	44-57	8-14	35-42

Adapted from Hargin and Morrison (1980)

in oil droplets or spherosomes (Hargin et al., 1980), while the GL and PL are structural lipids in various membranes (Morrison, 1978a, 1984, 1988a).

During the later stages of grain development there is extensive senescence of pericarp tissues, with loss of all starch and most protein and lipid reserves, and at maturity the lipids consist of remnants of TG, with some FFA and partial glycerides.

Germ and aleurone lipids are very similar, being mostly TG with some PL and negligible GL. With incipient sprouting levels of GL appear to increase (Hargin and Morrison, 1980), and there is active synthesis of PL during the first days of germination (Mirbahar and Laidman, 1982). Germ lipids include nearly all the α- and β-tocols in the grain, while the aleurone lipids have most of the α-tocotrienol and some β-tocotrienol but no α- or β-tocols, and the starchy endosperm has a little α-tocotrienol and some β-tocotrienol. Hence the tocopherols can be used as lipid-soluble markers for these tissues (Morrison and Barnes 1983; Morrison et al., 1982).

The non-starch lipids in the endosperm consist of TG and other NL in oil droplets located mostly in the sub-aleurone cells, some PC, PE and PG, and all of the NAPE, NALPE, DGDG, MGDG, DGMG and MGMG of the grain (Hargin and Morrison, 1980; Hargin et al., 1980). The starch lipids are exclusively LPC, LPE and LPG, but some FFA may become bound to the granules as surface lipids artifacts when starch is recovered by warm steeping and washing processes (Morrison, 1981; 1988b; Morrison et al., 1984a; Soulaka and Morrison, 1985a; Morrison and Gadan, 1987). Lipids inside the starch granules comprise about half of the phospholipids in the whole grain, and they are exceptionally stable (Morrison, 1978b, 1988b).

There is no evidence that GL or PL content is greatly affected by wheat variety or environmental factors, unless the grain is seriously stressed by adverse growing conditions. However, NL content is more susceptible to variation, and conditions favoring accumulation of reserves should give grain with more TG and starch.

REDISTRIBUTION OF LIPIDS ON MILLING

Flour millstreams are essentially starchy endosperm contaminated with varying amounts of aleurone, germ and pericarp material, which always includes some lipids. The endosperm itself is not uniform - the older cells towards the center have more starch and less protein and lipids than the subaleurone cells which have much more TG, GL and PL (Hargin et al., 1980). Thus, the lowest-grade millstreams which include subaleurone material have more GL, but their higher TG and PL contents are mostly attributable to very variable quantities of oil transferred from germ and aleurone (Morrison and Hargin, 1981; Morrison et al., 1982). It has been estimated that less than one quarter of the transferred oil comes from the aleurone (Morrison et al., 1982), and it has been shown that the moisture content of the conditioned wheat has a considerable influence on the amounts of germ oil and α- and β-tocols appearing in the millstreams (Morrison and Barnes, 1983). Millstreams can be described quite simply as having nearly constant quantities of endosperm lipids and very variable quantities of transferred material of nearly constant composition (Table III). This transferred material is essentially wheat germ oil, probably accompanied by fragments of tissue that contain protein bodies which would account for the transferred ash and protein (Morrison et al., 1982).

In all millstreams nearly all of the TG and other NL are free, while approximately half of the GL, APE, ALPE, PC, PE and PG and only one quarter of the LPC, LPE and LPG are free (Morrison et al., 1982). Thus, the free lipids in the lower grade millstreams will have proportionately less GL, and the FNL/FPoL ratio will be higher (Table IV).

LIPIDS IN STORED WHEAT AND FLOUR

Changes in the lipids of whole wheat stored under sound conditions are very slow, and consist of gradual hydrolysis to FFA with little loss of baking quality (Morrison, 1978a, 1988a). In damp stored

TABLE III

Composition of Basic Endosperm Lipids and Lipid-rich Material Transferred to it from Aleurone and Germ During Milling

Component	Basic endosperm, mg/100g solids	Transferred material, g/100g lipid
TG	300	83.8
FFA	105	5.0
Other NL	107	–
MGDG, MGMG	91	–
DGDG, DGMG	272	–
NAPE, NALPE	145	–
Diacyl PL	65	4.8
Lyso PL	54	4.4
α-T	–	0.1257
β-T	–	0.0603
α-T-3	0.09	0.096
β-T-3	1.94	0.0128
Ash	380	58.2
Protein	11800	129.9

Adapted from Morrison et al. (1982)

TABLE IV

Free (hexane-extractable) and Total Non-starch Lipids in Four Millstreams[a]

Lipid	Flour C_1		Flour F		Flour G		Flour J_1	
	Free	Total	Free	Total	Free	Total	Free	Total
TG	363	365	1132	1223	1583	1587	2711	2654
FFA	92	94	129	140	157	155	234	223
Other NL	116	110	114	124	139	127	181	168
MGDG, MGMG	63	94	56	89	46	85	52	85
DGDG, DGMG	149	292	119	297	115	265	115	247
APE, ALPE	80	137	92	156	75	135	66	139
Diacyl PL	42	78	59	114	99	197	203	398
Lyso PL	11	57	24	103	36	125	85	177
Total	916	1227	1725	2246	2250	2676	3647	4091
FNL/FPol	1.66	—	3.93	—	5.06	—	6.00	—

[a] From Morrison et al. (1982)

wheat lipolysis is faster and is associated with proliferation of molds and a marked loss of baking quality (Daftary et al., 1970a, 1970b).

Lipolysis is always faster after milling due to disruption of cells and better mixing of lipid substrates with enzymes, and it is faster in wholemeal flour than in white flour (Morrison, 1978a, 1988a). Wheat bran has a very active lipase (TG acyl hydrolase) that can raise the FFA content of the free lipids to 28% in wholemeal, stored at c.14% moisture and 20°C, within 20 weeks (Barnes and Lowy, 1986), and in one particularly poor sample this level was reached in 4 weeks (T. Galliard, personal communication). For comparison in low-grade flours (12.5-13.5% moisture) stored at 15°C for six months or at 37°C for three months FFA contents reached 15-22% of total lipids, equivalent to 21-29% of free lipids, and in high grade flours stored at 15°C for six months or 25°C for four months they reached 10-14% of total lipids, equivalent to 15-21% of free lipids (Clayton and Morrison, 1972).

In white flours there is non-specific stepwise deacylation of all glycerolipids to FFA, partial glycerides and even some fully deacylated residues (Clayton and Morrison, 1972). The enzyme hydrolyzing TG and its partial glycerides is probably a true lipase which has no effect on PoL, and the enzyme hydrolyzing all types of PoL is a polar lipid acyl hydrolase which does not hydrolyze TG (Morrison, 1988a). There is also slight oxidation of lipids, possibly due to lipoxygenase, in flours stored at 12 ± 2°C for five years (Warwick et al., 1979).

FUNCTIONALITY OF LIPIDS IN BREADMAKING

Nearly all of the effects of lipids in breadmaking are due to the non-starch lipid fraction. Some are substrates for lipoxygenases, and most (if not all) lipids interact physically with dough proteins directly or indirectly in ways that modify dough rheology, gas retention and baking performance. The lipids inside the starch granules are probably inert in dough, but they undoubtedly moderate the gelatinization and swelling of the starch granules during baking, and subsequent crumb firming and staling (Morrison and Milligan, 1982; Kim and Hill, 1984a, 1984b; Soulaka and Morrison, 1985b). However, starch lipids are not one of the major variable factors that affect quality.

(a) Lipoxygenase

There is negligible hydrolysis of lipids in doughs, and all biochemical changes are initiated by the enzyme lipoxygenase (LOX) which requires atmospheric oxygen and linoleic (L) or linolenic (Ln) acids as substrates. The enzyme in wheat, optimum pH 6-6.5, is type I (LOX-1) and it converts the initial peroxy radicals into relatively stable hydroperoxides with little further breakdown to volatile fragments, and negligible co-oxidation and bleaching of carotenoids, or co-oxidation of other substrates (Grosch et al. 1976; Grosch, 1986). The hydroperoxides are then largely converted by LOX, hydroperoxide isomerase and other enzymes, and possibly also by non-enzymic mechanisms, into hydroxyepoxy acids, trihydroxy acids and other products (Gardner, 1988).

Substrates for wheat LOX are L and Ln as free acids, and to a lesser extent as monoglycerides (Mann and Morrison, 1974; Morrison and Panpaprai, 1975), and stereospecificity is high-giving 85% 9-D-LOOH and 15% 13-L-LOOH (Graveland, 1973a; Grosch et al., 1976; Nicolas et al. 1982). Of the three LOX isoenzymes in wheat germ, A and B give over 85% 9-LOOH while isoenzyme C gives over 83% 13-LOOH (Galliard, 1983a). According to Grosch et al. (1976) the enzyme gives 85% 9-LnOOH and 15% 13-LnOOH from linolenate, but Graveland (1973b) claims that peroxidation is exclusively at the 9-position and 9-LnOOH does not accumulate. Wheat LOX activity is obviously dependent on the extent of prior lipolysis in the flour, and on the availability of oxygen in doughs or batters. In white flour doughs lipid oxidations are very incomplete and end after 5-10 min of mixing (Laignelet and Dumas, 1984). In suspensions of whole-meal with high FFA content, the system becomes anaerobic within a few seconds of mixing because there is insufficient dissolved oxygen and replenishment from air bubbles is rate limiting (Galliard, 1983b; Galliard et al., 1987).

Bread doughs may be supplemented with enzyme-active flours from soya and other legumes which are rich sources of type II LOX (soya type I LOX has optimum activity at pH 8.5-9.0, but none in dough at pH 5-5.5; Mann and Morrison, 1975). Soya LOX-2 oxidizes free L and Ln acids and all esterified forms found in wheat lipids (Morrison and Panpaprai, 1975), producing similar amounts of 9-LOOH and 13-LOOH (Grosch et al., 1976). Peroxidation is accompanied by high levels of co-oxidation reactions that produce volatile degradation products, bleaching of carotenoids and modification of gluten proteins. High levels of horse bean (Vicia faba) LOX in doughs mixed at high speed cause off-flavors (Nicolas and Drapron, 1983).

The endogenous LOX-1 in white flour has little effect on baking quality compared with soya LOX-2 (Kieffer and Grosch, 1980; Kieffer et al., 1982 Galliard, 1983b; Grosch, 1986). In bread doughs soya LOX-2 selectively releases bound TG (see below), promotes carotenoid bleaching, increases dough mixing tolerance, and improves dough rheological properties (Mann and Morrison, 1975; Frazier 1979; Hoseney et al., 1980; Kieffer and Grosch, 1980; Faubion and Hoseney, 1981; Nguyen-Brem et al., 1983, Nicolas and Drapron, 1983). The precise mechanisms whereby LOX-2 exerts these effects have been the subject of much debate (Morrison, 1976, 1978a, 1988a; Faubion and Hoseney, 1981; Grosch, 1986).

(b) Free and Bound Lipids

Most effects of NL and PoL in dough are due to their physical properties in solution or at interfaces, or to lipoprotein-type complexes stabilized by polar or by hydrophobic interactions. When flour is hydrated in a work-free system some free lipids become bound, and on anaerobic mixing the majority of the free lipids become bound but some are released irreversibly into the free state if oxygen is admitted to the mixer (Daniels, 1975; Frazier,

1979). The lipid that is released is TG, and nearly all of the GL and PL remain bound (Mann and Morrison, 1974). Soya LOX-2 promotes release of some bound TG, and this is thought to be desirable for breadmaking quality (Frazier, 1979). A protein fraction, ligolin, that binds much of the TG in the liquid state (i.e. natural wheat TG or the low-melting component of any added fat) has been partially characterized (Frazier and Daniels, 1986). Most of the high-melting component of added fat is always free in dough.

Flour lipid extraction and reconstitution experiments have shown that TG have a negative effect on loaf volume; PoL at very low levels are also detrimental, but otherwise are beneficial and when reconstituted at above-normal levels they give increased loaf volume (Hoseney and Finney, 1971; MacRitchie, 1981, 1983). Synthetic GL and surfactants with similar hydrophile-lipophile balance (HLB) values are also most effective, and permit additions of non-wheat protein supplements with minimal loss of bread quality (Chung and Pomeranz, 1977).

Models for hydrophilic and hydrophobic interactions between PoL or surfactants and proteins in dough have been proposed (Chung, 1986). However, there is an alternative view that PoL form mesomorphic dispersions in the aqueous phase, and that it is the surfactant properties of these dispersions that improve gas retention in dough and hence give better loaf volume (Larsson, 1986). Recently, evidence has been obtained using NMR and TEM which indicates that PL are not in classical membrane-type lipoprotein complexes, but are more likely in mesomorphic dispersions (Marion et al., 1987).

It is now fully accepted that prolamin proteins, particularly the high molecular weight glutenins, have the greatest influence on the loaf volume potential of wheat. There is evidence that lipids associate with particular protein fractions (Bushuk, 1986) but it is extremely difficult to exclude the possibility that at least some of these associations are artifacts. GL and some PL complex with sulfur-rich globulins to form hexane-soluble lipopurothionins (Lasztity, 1984; Morrison, 1988a), but there is no evidence that they have any effect on baking quality (Hoseney et al., 1970; Patey et al., 1976).

Early work showed that the free PoL were more beneficial to baking quality than the bound PoL (Hoseney and Finney, 1971) largely because they have higher proportions of GL. This led several groups to study correlations between FGL, FPL and FPoL in wheat or flour and loaf volume. Best correlations were obtained when the NL were taken into account, and the negative correlation between the ratio of FNL/FPoL and loaf volume is now widely used. Table V shows some impressively significant correlations for US, Canadian, Greek and Hungarian wheats, especially when the data are normalized to a standard flour protein content. However, others have failed to establish correlations in British (Bell et al., 1987), French (Berger, 1983), Australian (F. MacRitchie, personal communication) and New Zealand wheats (Larsen et al., 1986), possibly reflecting the high degree of selection by breeders in developing the varieties that were used in these particular studies.

(c) Genetic Control of Free Polar Lipids

The chromosomal locations of the structural genes for the wheat prolamins (gliadins and glutenins) are now known, but there can be no such direct links for the lipids whose structures are determined by many enzymes (proteins) and hence by many genes. However, quantitative differences in free GL appear to be determined by genes in the group 5 chromosomes (Morrison, 1988a; Morrison et al., 1984b, 1988), probably through some regulatory mechanism. The most effective gene appears to be in the short arm of chromosome 5D. Aneuploid lines lacking this chromosome arm, and tetraploid durum wheats which do not have the D-genome, are very hard and have much less free PoL than normal hexaploid wheats (A, B, D genomes) or aneuploids that have the full 5D chromosome (Morrison et al., 1984b, 1988).

Experiments with ditelosomic lines, intervarietal and interspecies group 5 chromosome substitution lines, and with recombinant lines have shown unequivocally that genes in chromosome 5D have

TABLE V

Linear Correlation Coefficients (r) Between Ratio of Free Nonpolar/Polar Lipids and Loaf Volume

Wheat or flour	Samples	r	r(corr.)[a]	Authors
US HRW, wheat	23	-0.901	-0.925	Chung et al. (1982)
US HRW, flour	23	-0.907	-0.904	Chung et al. (1982)
Canadian HRS, flour	5	-0.945		Zawitowska et al. (1984)
Canadian HRS, flour	25	-0.944[b]	-0.973[b]	Bekes et al. (1986)
Canadian HRS, flour	26	-0.791[b]	-0.961[b]	
Canadian HRS, flour	26	-0.637	-0.976[b]	
Canadian HRS, grain	25	-0.895	-0.910	
Greek winter, flour	10	-0.957[c]	-0.977[c]	Matsoukas and
	10	-0.924[d]	-0.925[d]	Morrison (in preparation)
Hungarian, flour	8[e]		-0.940	Karpati et al., (1988)

[a] Loaf volumes normalized to standard protein content. [b] Same wheats but significant site differences, hence data not combined. [c] Long-fermentation process. [d] Activated (chemical) dough development process. [e] Wheats grown at seven sites.

a major effect on levels of FPoL and on loaf volume potential (Morrison et al., 1988). It appears that there may be two genes whose effects are normally opposed, so that one increases FPoL while the other decreases FPoL, although not necessarily to the same extent. Wheats with different levels of FPoL will have allelic variation in these genes.

Interestingly, the gene on the short arm of chromosome 5D is indistinguishable from a major gene that regulates endosperm softness (Morrison et al., 1988), and possibly also quantities of the starch granule surface protein 'friabilin' (Greenwell and Schofield, 1986; Schofield and Greenwell, 1987). This raises many questions as to the various causes of differences in FPoL and loaf volume. There is no doubt that FPoL are directly beneficial, but they could also be fortuitous markers for other unidentified factors which are equally important (Morrison et al., 1988).

LITERATURE CITED

BARNES, P.J., and LOWY, G.D.A. 1986. The effect on baking quality of interaction between milling fractions during storage of wholemeal flour. J. Cereal Sci. 4:225.

BEKES, F., ZAWITOWSKA, U., ZILLMAN, R.R., and BUSHUK, W. 1986. Relationship between lipid content and composition and loaf volume of twenty-six common spring wheats. Cereal Chem. 63:327.

BELL, B.M., DANIELS, D.G.H., FEARN, T., and STEWART, B.A. 1987. Lipid composition, baking quality and other characteristics of wheat varieties grown in the UK. J. Cereal Sci. 5:277.

BERGER, M. 1983. Soft wheat lipids. 2. Composition of free and bound lipids in flour from eight varieties of French soft winter wheat. Sci. Aliment. 3:181.

BUSHUK, W. 1986. Protein-lipid and protein carbohydrate interactions in flour-water mixtures. Pages 147-154 in: Chemistry and Physics of Baking. J.M.V. Blanshard, P.J. Frazier, and T. Galliard, eds. R. Soc. Chem., London.

CLAYTON, T.A., and MORRISON, W.R. 1972. Changes in flour lipids during the storage of wheat flour.

J. Sci. Food Agric. 23:721.
CHUNG, O.K. 1986. Lipid-protein interactions in wheat flour, dough, gluten, and protein fractions. Cereal Foods World 31:242.
CHUNG, O.K., and POMERANZ, Y. 1977. Wheat flour lipids, shortening and surfactants: a three way contribution to breadmaking. Bakers Dig. 51(5):32.
CHUNG, O.K., POMERANZ, Y., and FINNEY, K.F. 1982. Relation of polar lipid content to mixing requirement and loaf volume potential of hard red winter wheat flour. Cereal Chem. 59:14.
DAFTARY, R.D., POMERANZ, Y., HOSENEY, R.C., SHOGREN, M.D., and FINNEY, K.F. 1970b. Changes in wheat flour damaged by molds during storage. Effect in breadmaking. J. Agric. Food Chem. 18:617.
DAFTARY, R.D., POMERANZ, Y., and SAUER, D.B. 1970a. Changes in wheat flour damaged by mold during storage. Effect on lipid, lipoprotein and protein. J. Agric. Food Chem. 18:613.
DANIELS, N.W.R. 1975. Some effects of water in wheat flour doughs. Pages 573-583 in: Water Relations of Foods. R.B. Duckworth, ed. Academic Press, London.
FAUBION, J., and HOSENEY, R.C. 1981. Lipoxygenase. Its biochemistry and role in breadmaking. Cereal Chem. 58:175.
FRAZIER, P.J. 1979. Lipoxygenase action and lipid binding in breadmaking. Bakers Dig. 53(6):8.
Frazier, P.J., and Daniels, N.W.R. 1986. Protein-lipid interactions in bread dough. Pages 299-326 in: Interactions of Food Components. G.G. Birch and M.G. Lindley, eds. Elsevier Applied Science, London.
GALLIARD, T. 1983a. Enzymic oxidation of wheat flour lipids. Pages 419-424 in: Developments in Food Science. Vol. 5A: Progress in Cereal Chemistry and Technology. J. Holas, ed. Elsevier, Amsterdam.
GALLIARD, T. 1983b. Enzymic degradation of cereal lipids. Pages 111-148 in: Lipids in Cereal Technology. P.J. Barnes, ed. Academic Press, London.
GALLIARD, T., TAIT, S.P.C., and GALLACHER, D.M. 1987. Rapid enzymic peroxidation of polyunsaturated fatty acids on hydration of wheat milling products. Pages 413-415 in: Metabolism, Structure and Function of Plant Lipids. P.J. Stumpf, J.B. Mudd, and W.D. Nef,

eds. Plenum Press, New York.

GARDNER, H.W. 1988. Lipoxygenase pathways in cereals. Adv. Cereal Sci. Technol. 9:161.

GRAVELAND, A. 1973a. Analysis of lipoxygenase nonvolatile reaction products of linoleic acid in aqueous cereal suspensions by urea extraction and gas chromatography. Lipids 8:599.

GRAVELAND, A. 1973b. Enzymic oxidation of linolenic acid in aqueous wheat flour suspensions. Lipids 8:606.

GREENWELL, P., and SCHOFIELD, J.D. 1986. A starch granule protein associated with endosperm softness in wheat. Cereal Chem. 63:379.

GROSCH, W. 1986. Redox systems in dough. Pages 155-169 in: Chemistry and Physics of Baking. J.M.V. Blanshard, P.J. Frazier, and T. Galliard, eds. R. Soc. Chem., London.

GROSCH, W., LASKAWY, G., and WEBER, F. 1976. Formation of volatile carbonyl compounds and co-oxidation of β-carotene by lipoxygenase from wheat, potato, flax and beans. J. Agric. Food Chem. 24:456.

HARGIN, K.D., and MORRISON, W.R. 1980. The distribution of acyl lipids in the germ, aleurone, starch and non-starch endosperm of four wheat varieties. J. Sci. Food Agric. 31:877.

HARGIN, K.D., MORRISON, W.R., and FULCHER, R.G. 1980. Triglyceride deposits in the starchy endosperm of wheat. Cereal Chem. 57:320.

HOSENEY, R.C., and FINNEY, K.F. 1971. Functional (breadmaking) and biochemical properties of wheat flour components. XI. A review. Baker's Dig. 45(4):30.

HOSENEY, R.C., POMERANZ, Y., and FINNEY, K.F. 1970. Functional (breadmaking) and biochemical properties of wheat flour components. VII. Petroleum ether-soluble lipoproteins of wheat flour. Cereal Chem. 47:153.

HOSENEY, R.C., RAO, H., FAUBION, J. and SIDHU, J.S. 1980. Mixograph studies. IV. The mechanism by which lipoxygenase increases mixing tolerance. Cereal Chem. 57:163.

KARPATI, E.M., BEKES, F., LASZTITY, R., ORSI, F., SMIED, I., and MOSONYI, A. 1988. Wheat protein-lipid interactions and their role in baking technology. Poster, Cereals '88. ICC Symp., Lausanne, Switzerland (30 May-30 June).

KIEFFER, R., and GROSCH, W. 1980. Improvement of the baking properties of wheat flour by type II lipoxygenase from soybeans. Z. Lebensm. Unters. Forsch. 170:258.

KIEFFER, R., MATHEIS, G., BELITZ, H.-D., and GROSCH, W. 1982. Occurrence of lipoxygenase, catalase and peroxidase in wheat flours with different baking performances. Z. Lebensm. Unters. Forsch. 175:5.

KIM, H.O., and HILL, R.D. 1984a. Modification of wheat flour dough characteristics by cycloheptaamylose. Cereal Chem. 61:406.

KIM, H.O., and HILL, R.D. 1984b. Physical characteristics of wheat starch gelatinization in the presence of cycloheptaamylose. Cereal Chem. 61:432.

LAIGNELET, B., and DUMAS, C. 1984. Lipid oxidations and the distribution of oxidised lipids during mixing of dough made from bread wheat flour. Lebensm. Wiss. Technol. 17:226.

LARSEN, N.G., BARUCH, D.W., and HUMPHREY-TAYLOR, V.J. 1986. Lipid quality factors in breeding New Zealand wheats. Agron. Soc. N.Z. special publ. 50:278.

LARSSON, K. 1986. Functionality of wheat lipids in relation to gluten gel formation. Pages 62-74 in: Chemistry and Physics of Baking. J.M.V. Blanshard, P.J. Frazier and T. Galliard, eds. R. Soc. Chem., London.

LASZTITY, R. 1984. Pages 79-87 in: The Chemistry of Cereal Proteins. CRC Press, Boca Raton, FL.

MacRITCHIE, F. 1981. Flour lipids: Theoretical aspects and functional properties. Cereal Chem. 58:156.

MacRITCHIE, F. 1983. Role of lipids in baking. Pages 165-188 in: Lipids in Cereal Technology. P.J. Barnes, ed. Academic Press, London.

MANN, D.L., and MORRISON, W.R. 1974. Changes in wheat lipids during mixing and resting of flour-water doughs. J. Sci. Food Agric. 25:1109.

MANN, D.L., and MORRISON, W.R. 1975. Effects of ingredients on the oxidation of linoleic acid by lipoxygenase in bread doughs. J. Sci. Food Agric. 26:493.

MARION, D., LeROUX, C., AKOKA, C., TELLIER, S., and GALLANT, D. 1987. Lipid-protein interactions in wheat gluten: a phosphorus nuclear magnetic resonance spectroscopy and freeze-fracture electron microscopy study. J. Cereal Sci. 5:101.
MIRBAHAR, R.B., and LAIDMAN, D.L. 1982. Gibberelic acid-stimulated α-amylase secretion and phospholipid metabolism in wheat aleurone tissues. Biochem. J. 208:93.
MORRISON, W.R. 1976. Lipids in flour, dough and bread. Baker's Dig. 50(4):29.
MORRISON, W.R. 1978a. Cereal lipids. Adv. Cereal Sci. Technol. 2:221.
MORRISON, W.R. 1978b. The stability of wheat starch lipids in untreated and chlorine-treated cake flours. J. Sci. Food Agric. 29:365.
MORRISON, W.R. 1978c. Wheat lipid composition. Cereal Chem. 55:548.
MORRISON, W.R. 1979. Lipids in wheat and their importance in wheat products. Pages 313-335 in: Recent Advances in the Biochemistry of Cereals, D.L. Laidman and R.G. Wyn-Jones, eds. Academic Press, London.
MORRISON, W.R. 1981. Starch lipids: a reappraisal. Starch/Staerke 33:408.
MORRISON, W.R. 1983. Acyl lipids in cereals. Pages 11-32 in: Lipids in Cereal Technology. P.J. Barnes, ed. Academic Press, London.
MORRISON, W.R. 1984. Plant lipids. Pages 247-260 in: Research in Food Science and Nutrition, Vol. 5. Food Science and Technology: Present Status and Future Direction. J.V. McLoughlin and B.M. McKenna, eds., Boole Press, Dublin.
MORRISON, W.R. 1988a. Lipids. Pages 373-439 in Wheat: Chemistry and Technology, 3rd. edn., Vol. 1. Y. Pomeranz, ed. Am. Assoc. Cereal Chem., St. Paul, MN.
MORRISON, W.R. 1988b. Lipids in cereal starches: a review. J. Cereal Sci. 8:1.
MORRISON, W.R., and BARNES, P.J. 1983. Distribution of wheat acyl lipids and tocols in flour millstreams. Pages 149-163 in: Lipids in Cereal Technology. P.J. Barnes, ed. Academic Press, London.

MORRISON, W.R., and COVENTRY, A.M. 1985. Extraction of lipids from cereal starches with hot aqueous alcohols. Starch/Staerke 37:83.
MORRISON, W.R., COVENTRY, A.M., and BARNES, P.J. 1982. The distribution of acyl lipids and tocopherols in flour millstreams. J. Sci. Food Agric. 33:925.
MORRISON, W.R., and GADAN, H. 1987. The amylose and lipid contents of starch granules in developing wheat endosperm. J. Cereal Sci. 5:263.
MORRISON, W.R., and HARGIN, K.D. 1981. Distribution of soft wheat kernel lipids into flour milling fractions. J. Sci. Food Agric. 32:579.
MORRISON, W.R., LAW, C.N., WYLIE, L.J., COVENTRY, A.M., and SEEKINGS, J. 1988. The effect of group 5 chromosomes on the free polar lipids and breadmaking quality of wheat. J. Cereal Sci., in press.
MORRISON, W.R., MANN, D.L., WONG, S., and COVENTRY, A.M. 1975. Selective extraction and quantitative analysis of non-starch and starch lipids from wheat flour. J. Sci. Food Agric. 26:507.
MORRISON, W.R., and MILLIGAN, T.P. 1982. Lipids in maize starches. Pages 1-18 in: Maize: Recent Progress in Chemistry and Technology. G.E. Inglett, ed. Academic Press, New York.
MORRISON, W.R., MILLIGAN, T.P., and AZUDIN, M.N. 1984a. A relationship between the amylose and lipid contents of starches from diploid cereals. J. Cereal Sci. 2:257.
MORRISON, W.R., and PANPAPRAI, R. 1975. Oxidation of free and esterified linoleic and linolenic acids in bread doughs by wheat and soya lipoxygenases. J. Sci. Food Agric. 26:1225.
MORRISON, W.R., WYLIE, L.J., and LAW, C.N. 1984b. The effect of group 5 chromosomes on the free and total galactosylglycerides in wheat endosperm. J. Cereal Sci. 2:145.
MORRISON, W.R., TAN, S.L., and HARGIN, K.D. 1980. Methods for the quantitative analysis of lipids in cereal grains and similar tissues. J. Sci. Food Agric. 31:329.
NICOLAS, J., AUTRAN, M., and DRAPRON, R. 1982. Purification and some properties of wheat germ lipoxygenase. J. Sci. Food Agric. 33:365.

NICOLAS, J., and DRAPRON, R. 1983. Lipoxygenase and some related enzymes in breadmaking. Pages 213-236 in: Lipids in Cereal Technology. P.J. Barnes, ed. Academic Press, London.

NGUYEN-BREM, P.T., KIEFFER, R., and GROSCH, W. 1983. Influence of additives during dough mixing on the molecular weight and fatty acid composition of wheat gluten fractions. Getreide Mehl Brot 37:35.

PATEY, A.L., SHEARER, G., and WARWICK, M. 1976. Wheat albumin and globulin proteins: purothionin levels of stored flour. J. Sci. Food Agric. 27:688.

SOULAKA, A.B., and MORRISON, W.R. 1985a. The amylose and lipid contents, dimensions, and gelatinisation characteristics of some wheat starches and their A- and B-granule fractions. J. Sci. Food Agric. 36:709.

SOULAKA, A.B., and MORRISON, W.R. 1985b. The bread baking quality of six wheat starches differing in composition and physical properties. J. Sci. Food Agric. 36:719.

SCHOFIELD, J.D., and GREENWELL, P. 1987. Wheat starch granule proteins and their technological significance. Pages 407-420 in: Cereals in a European Context. I.D. Morton, ed. Ellis Horwood, Chichester, UK.

WARWICK, M.J., FARRINGTON, W.H.H., and SHEARER, G. 1979. Changes in total fatty acids and individual lipid classes on prolonged storage of wheat flour. J. Sci. Food Agric. 30:1131.

ZAWITOWSKA, U., BEKES, F., and BUSHUK, W. 1984. Intercultivar variations in lipid content, composition and distribution and their relation to baking quality. Cereal Chem. 61:527.

FUNCTIONAL SIGNIFICANCE OF WHEAT LIPIDS

Okkyung Kim Chung
USDA-Agricultural Research Service
U.S. Grain Marketing Research Laboratory
1515 College Avenue
Manhattan, KS 66502

INTRODUCTION

Wheat lipids, a minor constituent, play major roles in wheat production, storage, processing, products, nutrition, and consumer acceptance of finished goods. Quantitative and qualitative differences in lipids in various structural parts of grains are responsible for multifaceted functions. In germination nonpolar lipids (NL, see Table I for definitions of abbreviations) are energy sources and polar lipids (PoL) (glycolipids plus phospholipids, GL + PL) are structural components of cellular membranes. The literature on wheat lipids up to 1970 was reviewed by Mecham (1971) and Pomeranz (1971), on cereal lipids from 1969 to 1976 by Morrison (1978a), and more recently on wheat lipids by Morrison (1988), and on the functionality of wheat flour lipids in breadmaking by Chung et al (1978), Chung and Pomeranz (1977, 1981), Daniels et al (1971), Frazier et al (1979), MacRitchie (1981), Morrison (1976), and Pomeranz (1988). This review mainly covers the role of lipids in wheat products and their end-use properties.

TABLE I
Abbreviations Used in This Chapter

Grouping	Abbreviations	Definitions
Lipid Term:	BL	Bound lipids
	FL	Free lipids
	GL	Glycolipids
	HL	Hydrolysate lipids
	NL	Nonpolar lipids
	NSL	Nonstarch lipids
	NSTL	Nonstarch total lipids
	PL	Phospholipids
	PoL	Polar lipids
	SL	Starch lipids
	SSL	Starch surface lipids
	TL	Total lipids
Lipid Class:	DG	Diglycerides
	DGDG	Digalactosyldiglycerides
	FA	Fatty acids
	FFA	Free fatty acids
	MG	Monoglycerides
	MGDG	Monogalactosyldiglycerides
	SE	Steryl esters
	TG	Triglycerides
Others:	Et_2O	Diethyl ether
	G	Gluten(s)
	HMW	High molecular weight
	HPLC	High performance liquid chromatography
	HRS	Hard red spring
	HRW	Hard red winter
	LV	Loaf volume
	MC	Moisture content
	MT	Mixing time
	MW	Molecular weight
	PetE	Petroleum ether
	RI	Refractive index
	UV	Ultraviolet
	WA	Water absorption
	WS	Water-soluble
	WSB	Water-saturated *n*-butanol

RECENT PROGRESS IN WHEAT LIPID METHODOLOGY

Based on extraction methods including extractants, commonly used categories are hydrolysate lipids (HL), free lipids (FL) extractable with hexane, PetE or Et_2O, bound lipids (BL) extractable with very polar solvents such as WSB or other aqueous alcohol mixtures after FL-extraction, and total lipids (TL = FL + BL) extractable with polar solvents (Morrison, 1988).

Since Acker and co-workers (Acker, 1974; Acker and Becker, 1971; Acker and Schmitz, 1967; Becker and Acker, 1971) investigated the extraction conditions for starch lipids (SL), wheat lipids are classified, based on their locations, into three categories, i.e. nonstarch lipids (NSL), starch lipids (SL), and starch surface lipids (SSL). NSL are present in wheat fractions other than starch granules and extractable at ambient temperatures. SL are bound to starch, present inside starch granules, and extractable most efficiently by a mixture of 1-propanol or 2-propanol with water (3:1, by volume) under N_2 at 100°C (Morrison and Coventry, 1985). The third category, SSL, are portions of NSL which become firmly absorbed onto or into starch granules during the preparation of pure starch. Thus, TL could be FL + BL or NSL + SL. In this review, nonstarch total lipids will be denoted as NSTL.

NL are eluted from silicic acid columns with $CHCl_3$ or migrate on thin-layer plates with similar solvents, and PoL refers to all other GL and PL. After NL elution, acetone elutes GL, and then MeOH elutes PL. If the acetone-elution is bypassed, MeOH elutes PoL after the NL-elution.

Complete procedures for the quantitative analysis of lipids (including SL) in cereal grains have been described by Morrison et al (1980). Although WSB is the most efficient extractant for wheat endosperm NSL, solvents based on $CHCl_3$-MeOH mixtures are, in general, best for dry gluten (Barnes, 1983), aleurone, germ, and bran mixtures (Morrison et al, 1980).

Several researchers extracted lipids from wheat or flour to demonstrate the role of flour lipids in breadmaking. In such cases, extracting SL is not feasible, because defatted starch would not only be complexed with butanol, but would also be gelatinized. Although WSB is the most efficient extractant for wheat lipids, it is not used when the defatted flour is to be used for baking. Extracting conditions are critical for NSL, because the quantity and composition of extracted lipids depend on the extraction method (regular vs vacuum Soxhlet, Soxhlet vs shaker) (Chung et al, 1977a, 1977b), solubility parameter values of extractants (Chung et al, 1980a, 1984) and MC of flour samples (Chung et al, 1978, 1984).

Some extraction conditions (e.g. solvent and temperature) affect the extractability of NL relatively little but significantly affect the extractability of PoL. Chung et al (1980a) established conditions of lipid extraction that differentiate among HRW wheat flour that vary in breadmaking characteristics. Six solvents were compared. The ratio of NL to PoL extracted with PetE or Skellysolve B differentiated the five flours according to LV potential (Fig. 1, PL denotes polar lipids in Figs. 1 and 2). The slope was largest for PetE-extraction and smallest for WSB-extraction.

The most recent fractionation techniques include HPLC of wheat lipids. They include: FA content of lipids by derivatization with *p*-bromophenacyl esters to be detected at 254 nm with a UV detector (Tweeten and Wetzel, 1979); DGDG and MGDG contents in PoL by a variable wavelength detector at 206 nm (Tweeten, 1979); or by an RI detector (Tweeten et al, 1981); GL as benzoyl derivatives by an UV detector at 254 nm (Walker, 1988); and NSTL by a mass spectrometer (Christie and Morrison, 1988).

WHEAT LIPID COMPOSITION AND DISTRIBUTION

A wheat kernel weighs 30 to 45 mg (db) and consists of 2.5-3.0% germ, 4-10% aleurone, 7-9% pericarp and 79-85% starchy endosperm containing

17-19% nonstarch fractions and 60-68% starch fractions (Hargin and Morrison, 1980).

Lipids are only 3-4% of the wheat kernel. Lipids are unevenly distributed in wheat structural parts. About 25-30% of wheat lipids are present in germ, 22-33% in aleurone, 4% in pericarp, and the remaining 40-50% are in starchy endosperm fraction having 20-31% NSL and 16-22% SL (Chung and Pomeranz, 1981;

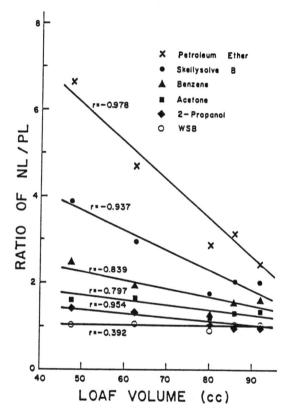

Fig. 1. Loaf volume of bread baked with HRW wheat flours versus the ratio of nonpolar lipids (NL) to polar lipids (PL) extracted with petroleum ether, Skellysolve B, benzene, acetone, and 2-propanol by a Soxhlet method and with water-saturated n-butanol (WSB) by a Stein mill. (Reprinted, with permission, from Chung et al, 1982)

Hargin and Morrison, 1980; Morrison, 1988). Lipids in structural parts differ in composition: germ lipids are 79-85% NL, 0-4% GL, and 13-17% PL; aleurone lipids are 72-83% NL, 2-10% GL, and 14-18% PL; NSL are 33-47% NL, 20-35% GL, and 22-35% PL; and SL are 4-6% NL, 1-7% GL, and 89-94% PL. Thus, whole kernel lipids are 44-57% NL, 8-14% GL, and 31-42% PL. About 80% of wheat grain GL are distributed mainly in starchy endosperm, especially nonstarch part and 50-60% of wheat PL are in starch granules.

Both wheat germ and aleurone fractions are rich in lipids: 25-30% of germ weight and 9-11% of aleurone weight (Morrison, 1988). Those fractions are relatively easily separated from milled flour as by-products due to their high-lipid contents. Lipids are partially transferred from germ and aleurone in the form of tissue fragments and as oil adhering to the flour particle surface.

ROLE OF LIPIDS IN REDUCTION OF DUSTINESS

Chung et al (1980b) studied flour dustiness by counting dust particles with an automatic particle counter and by sifting through U.S. standard sieves of a sonic sifter. Shawnee flour was defatted by Skellysolve B and the extracted lipids were fractionated to NL, GL, and PL. Unfractionated lipids, their fractions, mineral oil or soybean oil were added to the nondefatted or defatted flour. The dust index of both untreated and defatted flours, as determined by the particle counter, decreased significantly with an increase in levels of any of the additives: log value of the dust index was linearly related to the level of additives ($r = -0.967$ to -0.998). Mid point of sample (untreated or defatted) particle size, determined by the sonic sifter, was linearly related to the level of additives ($r = 0.907$ to 0.991). Defatting increased the dust index by a factor of up to 100 times. Therefore, if there were no lipids in flour by nature, the grain industry would have faced more serious problems with grain dust explosions.

LIPIDS IN STORED WHEAT AND FLOUR

Biochemical components are changing during storage of wheat and flour: those changes depend on storage conditions such as temperature, MC, storage time, etc. It has been well recognized that an increase in fat acidity during the early stages of deterioration is more rapidly detected than changes in the other wheat flour constituents. Consequently, it has been suggested that fat acidity be used as a measure of grain damage. Unfortunately, there are limitations on using fat acidity as the sole criterion of damage in storage, especially for grain stored for short periods. If fat acidity is to be used to assess and predict storability, fat acidity and the baking quality of stored wheat or flour must be highly correlated; this is not always the case.

Chung and Pomeranz (1981) correlated the lipid content (data from Warwick et al, 1979) to LV of breads (data from Bell et al, 1979) of wheat flours (12.2 to 13.6% MC) stored at 12 ± 2°C for five years. LV was linearly related to the amount of TL, of NL, and of PoL in flour stored for different periods at 12°C. There was no significant change in LV for an increase in FFA from about 50 to 288 mg or for a decrease in TG from 630 to 372 mg/100 g flour for the first 18 months. LV decreased linearly, however, with an increase in FFA from 288 to 450 mg or with a decrease in TG from 372 to 119 mg in 100 g flour which had been stored for 18 to 54 months. Changes in individual lipid (especially major) classes during early (18 months) storage of flours with a low moisture at 12°C were more rapid than damage that could be recognized by the baking test.

Lipid components may be the proper indicators to detect incipient deterioration in stored grain. However, to predict baking quality of stored grain or flour, a further study is needed to use some other parameters such as the NL/PoL ratio of FL but not of BL or TL.

LIPIDS IN DIFFERENT WHEAT CLASSES AND VARIETIES

Shollenberger et al (1949) reported, as early as 1949, that FL were a varietal characteristic and Fisher et al (1964, 1966) demonstrated varietal, seasonal, and environmental effects on the quantity and quality NSTL. Subsequently, there has been little success in differentiating between wheat classes (e.g. HRS vs HRW) on the basis of NSTL content or composition.

There are, however, significant differences in lipid quantity and composition between hexaploid and tetraploid wheats. A summary of data reported by three groups (Hargin and Morrison, 1980; Lin et al, 1974a; Pomeranz et al, 1966a, 1966b) shows that durum wheats, their flours, or hand-dissected starchy endosperm, contain more NSTL than HRS wheat lipids.

Hargin and Morrison (1980) showed differences in lipid composition in hand-dissected endosperm in four wheats. Whereas NSL in the endosperm of the English soft wheat (Atou) and the U.S. amber durum wheat (Edmore) contained significantly less PoL than in that of the English hard winter wheat (Flinor) or Waldron (HRS), NL contents were about the same for the four wheats. Atou and Edmore contained much more SL than Flinor and Waldron. No generalizations can be made based on data from a single durum wheat and further investigations on SL of different wheat classes and/or varieties are recommended.

Davis et al (1980) evaluated NSL content, tocopherol content, and FA composition in 290 wheat samples from eight classes and subclasses. Durum wheats had the highest lipid contents but the lowest tocopherol contents. Red wheats contained more lipids than white wheats but white wheats, generally, contained more tocopherols (especially δ-tocopherols) than red wheats. White wheat lipids were, in general, poorer in oleic acid but richer in linoleic acid than red wheat lipids. Durum wheat lipids were richer in palmitic acid and poorer in linoleic and linolenic acids than lipids in wheats from the other classes.

High levels of SE could be used as evidence of

contamination of durum semolina with hexaploid farina (Artaud et al, 1979; Hsieh et al, 1981; Laignelet, 1983; Morrison, 1978a), because when only steryl palmitate or saturated SE are measured, durum wheat semolinas contain 0-1.5 mg per 100 g, while most hexaploid wheat flours contain 3.0-57.6 mg per 100 g.

LIPIDS IN RELATION TO LV-POTENTIAL

As shown in Fig. 1, PetE was the best extractant for lipids to differentiate among five HRW wheat flours according to LV potential. Chung et al (1982) have extended that preliminary study to HRW wheats grown in the Great Plains of the U.S. and their straight grade flours. The correlation between LV and NL/PoL ratio was highly significant for both wheat and flour, whether on the as-received or constant protein basis (Fig. 2). A significant linear relationship was also found between LV and the amount of PoL or of lipid-galactose content in FL occurring naturally in wheat or its milled flour. The highly significant correlations point to the potential usefulness of PoL, NL/PoL, and lipid-galactose content for estimating LV potential of HRW wheat flours.

Similarly, high proportions of free PoL (GL + PL) or low NL/PoL ratios were highly correlated with LV potential of Canadian HRS wheats (Bekes et al, 1986; Zawistowska et al, 1984) and of the intervarietal and interspecies chromosome substitution lines (Morrison et al, 1988).

Marston and MacRitchie (1985) observed differences in lipid content and composition of Australian baker's flours but no significant relationship between lipid characteristics and breadmaking quality. The results with six varieties grown in the U.K. at three sites in 1983 appeared to confirm reports of significant correlations between LV and free PoL contents in wheat samples from the USA and Canada. However, with 15 winter varieties and six spring varieties grown at three sites in 1984, no significant correlations were found (Bell et al, 1987). European wheats studied by O. K. Chung

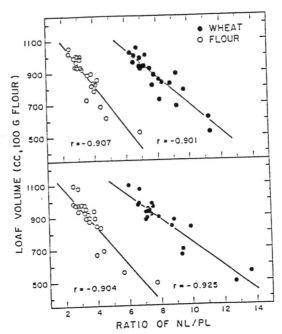

Fig. 2. Relation between loaf volume (LV) of bread baked from 100 g of flour and the ratio of nonpolar lipids (NL) to polar lipids (PL) of wheat or flour. Top. LV and NL/PL ratio on the as-received basis of protein content. Bottom. LV and NL/PL ratio corrected to a constant protein basis. The corrected ratio of NL to PL was obtained from NL content (as received) divided by PL content corrected to 12% protein for flour or 13% for wheat. (Reprinted, with permission, from Chung et al, 1982)

and Y. Pomeranz (unpublished data, 1980-82) indicated no significant correlation between lipid content and composition and LV potential. Although the reasons for differences between North American wheats and Australian or European wheats are unknown at the present time (Pomeranz, 1988), further research could offer a possible new route to wheat quality improvement through the genetic regulation of free PoL (Morrison, 1988), because deletion of the short arm of chromosome 5A raised the levels of free DGDG, whereas eliminating the short arm of chromosome 5D greatly reduced the levels of free and total MGDG and DGDG (Morrison et al, 1984).

Morrison et al (1988) showed an interesting relationship between grain hardness and free GL, which suggests that both are related through a gene or genes in the group 5 chromosomes. There were inverse relationships between wheat hardness (expressed as grinding time) and the amount of DGDG, MGDG, or PL in FL. However, the time to grind was positively related to the NL/PoL ratio of FL.

LIPIDS IN WHEAT PRODUCTS BY WET PROCESSING

Gluten (G)-washing yields 5-6% (flour wt, db) of WS fraction (20-26% protein content), 75-80% starch (prime + tailing, 2:1 ratio), and 14-19% crude G (76-85% protein content). Thus, 7-11% of total flour protein is in the WS fraction, 2-4% in tailing starch, approximately 1% in prime starch, and 80-90% in G (Chung, 1986).

The distribution of proteins in the various fractions depends on the quantity and quality of flour proteins, the overall flour quality, and the presence or absence of native flour FL (Chung et al, 1987).

When G is washed from defatted flour (only FL are removed), the G yield is comparable to or slightly higher than that from nondefatted regular flour, provided that proper care is taken during washing. When G is washed from regular flour containing native FL, slightly more protein is distributed in the WS fraction and both starch fractions. The protein content of G is higher for the G obtained from good-quality breadmaking flour than that from poor-quality flour, because poor-quality G is less cohesive and less tenacious and thus, contains more starch entrapped in the G mass (Chung et al, 1987).

Bekes et al (1983) washed G from flours defatted by hexane, butanol, and WSB which extracted lipids in increasing amounts, i.e. 0.78% (FL), 1.04%, and 1.52% (NSTL). G yield and protein content in G preparation differed: 11% G-yield (78.5% protein content) was from control flour; 12.3% G-yield (81.7% protein), 12% yield (82.4% protein), and 10.8% yield (86.6% protein) were from flours defatted by hexane, butanol, and WSB, respectively. Thus, protein recovery (% of flour protein) was highest in G from flour from which mainly FL were removed. If BL were also removed from flour, G yield decreased, and protein recovery lowered. However, functionality of those G preparations was not tested in breadmaking.

Commercially available vital G face difficulties in storing for prolonged periods at room temperature because of their high lipid contents. On the laboratory scale, G were prepared from two flours with widely varying breadmaking characteristics (Chung et al, 1987). The G processing variables included various MT and FL removal from flours prior to G-washing. Those G preparations were added to bread formulation at the 2% level (200 mg per 10 g flour on 14% mb). In general, the LV increases and especially improvement in crumb grain were greater with the addition of G prepared from regular flours than from defatted flours. Therefore, vital G should be prepared from regular flours to take advantage of lipid-protein interactions taking place during G-washing in spite of storage difficulty.

In addition, the separation of tailing from prime starch was difficult when G was washed from the optimally mixed dough from PetE-defatted flour: the yield of prime starch decreased from 45-49% to 39-42% and that of tailing starch increased from 29-30% to 37% by defatting the flour prior to G-washing (Chung et al, 1987). Such decreases in prime starch yield would be economically undesirable.

EFFECTS OF LIPID REMOVAL ON PROTEINS

Simmonds and Wrigley (1972) found that much less protein was extracted with 6 M urea from G than from storage proteins defatted by $CHCl_3$-benzene mixtures. Reconstitution experiments, involving wetting of storage proteins in the presence of added NSL, suggested that the reduced G solubility resulted from lipid-protein association during dough formation. Charbonnier (1973), however, showed that protein solubility in 55% EtOH decreased when flour was defatted with $EtOH-Et_2O-H_2O$ (2:2:1, by vol), mainly due to a decrease in glutenin solubility. Thus, available evidence indicates that lipid can suppress or enhance the solubility of flour or its isolated proteins (Chung, 1986).

Protein extractability in flour decreased curvilinearly with the increase in lipid removal (Chung et al, 1979). The decreased protein solubilities of defatted flours were, to a large extent, restored when the defatted flours were reconstituted with their respective extracted NL or total (NL + PoL) lipids but not with PoL alone. Protein solubility, as well as breadmaking quality, was almost completely restored when lipids extracted with Skellysolve B were returned to the defatted flour but not when 2-propanol extracted lipids were recombined with their flour residue.

Lipid removal changes the ratio of glutenin to gliadin proteins: this ratio decreased in proteins from defatted flours (Chung and Pomeranz, 1978), whereas it increased in G obtained from defatted flours containing lipids in a decreasing order (Bekes et al, 1983). The seemingly contradictory conclusions may result from many factors, such as different defatting procedures including extractants, different fractionation procedures of gliadin and glutenin proteins, and/or the opposite effects due to true differences in proteins from either flour or G.

The effects of lipid removal on disaggregating HMW gliadin (Bekes et al, 1983) or lower MW glutenin (Zawistowska et al, 1985) fractions were reported. Both fractions were subfraction I of gliadins and glutenins by gel filtration chromatography. Reconstituting study demonstrated the aggregating role of lipids.

ROLE OF LIPIDS IN END-USE PROPERTIES

Effects on Breadmaking

The phenomenon of lipid binding may not be fully

understood on a molecular level. However, lipid binding, on macromolecular levels, has been evidenced for a long period of time and its significant effects have been well recognized by cereal chemists. Binding of FL to proteins during dough mixing or G formation is the most important first stage of breadmaking. Lipids are generally thought to be inactive during the fermentation stage and then become actively involved in multiple interactions with other constituents including starch during the oven stage.

The other functional role of flour lipids is their polyunsaturated FA being substrates of lipoxygenase to increase dough mixing tolerance and improve rheological properties, internal structures of bread, and bread flavor (Faubion and Hoseney, 1981; Frazier, 1979; Hoseney et al, 1980).

The defatting and reconstituting studies conducted during the last 20-25 years point to the following conclusions: (a) defatting flour increases WA and MT requirements; (b) it reduces LV, impairs crumb grain and accelerates crumb firming rate; (c) the degree of lipid removal from flour by varying lipid extraction methods decides the bread characteristics of defatted or reconstituted flours, i.e. reconstitution restores functionality of flours with only nonpolar solvent treatment; the BL removal impairs bread quality to an extent that adding back the extracted lipids cannot fully restore it; (d) functionality of FL classes differ widely, i.e. NL have no improving effects (sometimes detrimental) whereas PoL have improving effects on breadmaking; (e) among NL, FFA are detrimental, and the effects of TG, DG, MG depend on shortening in the bake formulations whereas among PoL, GL, especially DGDG, are the most efficient bread improver; and (f) native FL are essential to fully exert the shortening responses to all characteristics of finished breads.

Recent research documented quantitative effects of lipids on several breadmaking parameters. The amount of lipids in a bake formulation can be varied in several ways: (a) by adding different amounts of flour lipids to the defatted flour (Chung et al,

1980c; Lin et al, 1974b, MacRitchie, 1976, 1977); (b) by defatting flour differentially by varying solvents and temperatures so the amounts of residual lipids that remain differ (Chung et al, 1980d); (c) by flour lipid quantities and qualities changing under different storage conditions (Warwick et al, 1979; Bell et al, 1979); (d) by inherent lipid quantities and qualities differing in wheat varieties and in milled flours (Bekes et al, 1986; Chung et al, 1980a, 1982); and (e) by adding different levels of shortening to defatted, partially-reconstituted, or nondefatted flours (Bell and Fisher, 1977; Chung et al, 1981).

Effects on Cookie Quality

Cookie diameter is the major parameter determining the cookie quality as LV is the main one in bread quality. Despite the high levels of fat added to the system, the role of wheat lipids is not masked (Table II). FL content was generally proportional to flour protein content. The reduction in cookie diameter by defatting differed widely among flours and were, in general, related to the amount of FL removal.

Flour lipid removal reduced cookie diameter and cookie top-grain scores (Clements, 1980; Clements and Donelson, 1981; Cole et al, 1960; Kissell et al, 1971). In addition, removal of lipids impaired the internal structure of cookies (Clements, 1980; Yamazaki and Donelson, 1976). In reconstitution studies, the original diameter was restored fully with the unfractionated FL and PoL but partially with NL (Clements and Donelson, 1981; Cole et al, 1960). However, the cookie grains were restored fully by adding the unfractionated FL (NL + PoL) but partially by adding NL or PoL. Yamazaki and Donelson (1976) found that it was necessary to remove lipids first, then separate the defatted flour into four components for complete restoration of functionality of the reconstituted flour, because interaction of FL with tailings and WS had an adverse effect on the internal structure of cookies, and this effect was intensified

TABLE II
Effects of Free Lipid Removal on Cookie Diameter

Flour Type	Protein (%)	FL (%)	Cookie Diameter (cm) Control	Defatted	ΔD^a (cm)
Avon	7.3	0.81	17.7	17.0	-0.7
Yorkstar	7.6	1.00	17.8	17.7	-0.1
Thorne	8.4	0.91	17.8	16.9	-0.9
Purkof	8.9	0.89	16.8	15.9	-0.9
Blackhawk	9.9	0.93	18.1	16.8	-1.3
Arthur	12.7	1.20	17.7	15.6	-2.1

[a] Calculated data from Clements (1980) for Yorkstar and Arthur flours and Kissell et al (1971) for the other flours. ΔD is difference in cookie diameters of defatted and control (nondefatted) flours.

when gluten was added as an interactant. Therefore, native flour lipids contribute significantly to the overall cookie quality.

Effects on Cake Quality

Montzheimer (1931) reported that Cl_2-treatment of flour effectively avoided structural collapse in high-ratio cakes. Since then, numerous researchers have investigated the effect of chlorination on physical and chemical changes in flour or flour constituents, mainly nonlipid constituents. Chlorinating flour reduces the degree of unsaturation in lipids (Morrison, 1978b) and increases the oil-binding (Shuey et al, 1963; Seguchi and Matsuki, 1977).

The effects of chlorination and FL on cake properties are shown in Table III. FL removal reduced cake volume (Johnson et al, 1979; Kissell et al, 1979; Spies and Kirleis, 1978) and other quality characteristics. Reconstituting with the extracted FL generally restored the original cake quality.

TABLE III
Effects of Free Lipids on Cake Baking Properties[a]

Flour:	Cl_2-Treatment	Volume (cc)	Cake Characteristics		
			Grain[b]	Contour	Collapse
Nondefatted:	Yes	525	F/C	Round	No
	No	445	T/O	Flat	Yes
Defatted:	Yes	500	T/O	Flat	No
	No	458	C/O	Flat	No
Reconstituted:	Yes	536	F/O	Round	No
	No	440	T/O	Flat	Yes

[a] Selected data from Johnson et al (1979).
[b] F/C = fine/close, T/O = thick/open, C/O = coarse/open.

Effects of lipid removal were more pronounced with Cl_2-treated flour than untreated flour (Table III). Johnson et al (1979) showed that adding acetone-soluble fraction (81%) of FL to unchlorinated defatted flour increased cake volume from 445 to 563 cc and produced cake with a round contour and no collapse, but had little effect on the crumb grain. Adding acetone-insoluble lipids (19%) of FL to unchlorinated defatted flour decreased volume to 410 cc and caused excessive collapse (Johnson et al, 1979).

Based on interchange combinations of lipids extracted from flours treated with different levels of Cl_2, Kissell et al (1979) confirmed the importance of FL *in situ* at the time of chlorination. Spies and Kirleis (1978) reported, however, that chlorinated defatted flours reconstituted with unchlorinated lipids produced larger cakes with poorer textures than chlorinated defatted flours reconstituted with chlorinated lipids. Johnson et al (1979) reported that lipids both from chlorinated and unchlorinated flours restored baking properties of the chlorinated defatted flour. Therefore, the beneficial effect of Cl_2-treatment must be due to its interaction with flour component(s) other than lipids (Johnson et al, 1979; Spies and Kirleis, 1978). Johnson and Hoseney (1979) also found the necessity of defatting flour prior to fractionating other components by wet processing in reconstituting studies to completely restore the volume and grain of the cake. The beneficial effect of removing FL from flour before a wet-fractionation might be due to a lesser G development during fractionation, so the cake crumb would not be too dense.

Although FL play important functional roles in both cookie and cake qualities, their improving mechanisms in soft wheat products must differ from that in breadmaking. Lipid binding in dough or G formation is critically important in breadmaking, whereas lipid-protein interactions in G formation should be avoided in cookie and cake making.

A decrease in cake volume from 472 to 380 cc and excessively fine cells resulted from removing FL by Et_2O and from 472 to 335 cc from NSTL removal by WSB (Seguchi and Matsuki, 1977). Complete restoration of volume and cell structure was obtained by adding 0.2% (flour wt) lipids (unfractionated lipids extracted by either Et_2O or WSB and WSB-extracted PoL) to the Et_2O-extracted flour, but not by WSB-extracted NL (even at the 0.6% level). However, the pan-cake baking quality of the WSB-extracted flour was irreversibly damaged by the WSB extraction; adding the WSB-extracted lipids (1.36%) back to the WSB-defatted flour could not restore the pan-cake baking quality as was previously demonstrated in breadmaking.

Effects on Spaghetti, Noodles, and Other Extrudates

The significance of the endogenous semolina lipids was studied by extracting NSTL with WSB and proteins from raw spaghetti (Dahle and Muenchow, 1968). They found that removing NSTL resulted in increased stickiness after cooking whereas removing both lipids and proteins resulted in greater leaching of amylose into the cooking water and decreased firmness of the cooked pasta. They concluded that the proteins were essential for retention of high cooking quality but that NSTL improved the functional properties of proteins and minimized the secondary cooking effects such as stickiness.

The firmness of cooked pasta decreased slightly from 3.6 to 3.3 when semolina had been defatted with PetE, but increased considerably from 3.6 to 5.0 when the WS fraction had been extracted from semolina (Laignelet, 1983). Defatting the semolina which already had the WS fraction removed had little effect on firmness and decreased the extent of swelling in the cooked pasta. Although Laignelet (1983) presumed that the functional properties of FL could be related to the presence of WS components, it may be due to FL binding to G during the fractionation of WS components by wet-processing.

Extraction of FL from semolina resulted in a higher pasta WA and in loss of yellow color in spaghetti (Lin et al, 1974b). Addition of 0.6% NL to the PetE-defatted semolina restored the spaghetti color. NL and MGDG increased WA of the nondefatted semolina and the firmness of spaghetti while DGDG and PL decreased these factors. Unlike Dahle and Muenchow (1968), Lin et al (1974b) did not consider cooking losses and change in stickiness, and concluded that neither PoL nor NL were factors of great importance in cooking quality. However, Matsuo et al (1986) reported that removal of durum semolina FL (mainly NL) increased surface stickiness, the surface carbohydrate content, and cooking loss, whereas enrichment with the extracted FL fully restored them.

Less lipids were extracted from pasta than from semolina (Fabriani et al, 1968) because about 90% of FL in semolina became bound during commercial pasta processing, especially during the drying stage (Barnes et al, 1981). Thus, the functional significance of semolina lipids must lie in their binding ability during pasta production.

The effect of lipids on Japanese "hand-stretched" noodles was investigated by Niihara et al (1982), who reported that FFA produced during storage contributed to the texture of cooked noodles by inhibiting the swelling of starch granules and affecting the viscosity of gelatinized starch after cooking. Rho et al (1986) reported that FL removal decreased surface firmness from 52.7 to 42.2 g/mm, but increased the cutting stress, breaking stress, color, and cooking loss of Oriental dry noodles. Quality of the noodles from original wheat flour was fully restored by adding FL back to the defatted flour. Generally, NL were most effective in restoring the surface firmness of cooked noodles, and PoL increased the breaking stress of dry noodles more than NL.

The role of lipids in the extrusion cooking of wheat starch and flour was studied (Faubion and Hoseney, 1982). PetE-extracted 0.2% FL from starch and 1.0% FL from medium-protein flour. Removal of FL from starch had little effect on the extrudate properties. Defatted flour gave extrudates with increased expansion and increased textural strength. Adding 1% flour FL back to the defatted flour or to wheat starch resulted in decreased extrudate expansion and texture (shearing and breaking strengths).

LITERATURE CITED

ACKER, L. 1974. Cereal lipids: Their composition and significance. Getreide Mehl Brot 28:181-187.

ACKER, L., and BECKER, G. 1971. New research on the lipids of cereal starches. II. Lipids of various types of starch and their binding to amylose. Staerke 23:419-424.

ACKER, L., and SCHMITZ, M. J. 1967. The lipids of wheat starch. III. The remaining lipids of wheat starch and the lipids of other types of starch. Staerke 19:275-280.

ARTAUD, J., IATRIDES, M. C., and ESTIENNE, J. 1979. Application of high-pressure liquid chromatography to the determination of soft wheat in pastas. Ann. Falsif. Expert. Chim. 72:153-157.

BARNES, P. J. 1983. Nonsaponifiable lipids in cereals. Pages 33-55 in: Lipids in Cereal Technology. P. J. Barnes, ed. Academic Press, London.

BARNES, P. J., DAY, K. W., and SCHOFIELD, J. D. 1981. Commercial pasta manufacture: Changes in lipid binding during processing of durum wheat semolina. Z. Lebensm. Unters. Forsch. 172:373-376.

BECKER, G., and ACKER, L. 1971. New research on lipids of cereal starches. I. Introduction and Methodology. Staerke 23:339-343.

BEKES, F., ZAWISTOWSKA, U., and BUSHUK, W. 1983. Protein-lipid complexes in the gliadin fraction. Cereal Chem. 60:371-378.

BEKES, F., ZAWISTOWSKA, U., ZILLMAN, R. R., and BUSHUK, W. 1986. Relationship between lipid content and composition and loaf volume of twenty-six common spring wheats. Cereal Chem. 63:327-331.

BELL, B. M., CHAMBERLAIN, N., COLLINS, T. H., DANIELS, D. G. H., and FISHER, N. 1979. The composition, rheological properties and breadmaking behavior of stored flours. J. Sci. Food Agric. 30:1111-1122.

BELL, B. M., DANIELS, D. G. H., FEARN, T., and STEWART, B. A. 1987. Lipid composition, baking quality and other characteristics of wheat varieties grown in the UK. J. Cereal Sci. 5:277-286.

BELL, B. M., and FISHER, N. 1977. The binding of model shortenings during mixing of mechanically developed bread doughs from fresh and stored flours. J. Am. Oil Chem. Soc. 54:479-483.

CHARBONNIER, L. 1973. Studies of the alcohol-soluble proteins of wheat flour. Biochimie 55:1217-1225.

CHRISTIE, W. W., and MORRISON, W. R. 1988. Separation of complex lipids of cereals by high performance liquid chromatography with mass detection. J. Chromatogr. 436:510-513.

CHUNG, K. H., and POMERANZ, Y. 1978. Acid-soluble proteins of wheat flours. I. Effects of delipidation on protein extraction. Cereal Chem. 55:230-243.

CHUNG, O. K. 1986. Lipid-protein interactions in wheat flour, dough, gluten, and protein fractions. Cereal Foods World 31:242-244, 246-247, 249-252, 254-256.

CHUNG, O. K., and POMERANZ, Y. 1977. Wheat flour lipids, shortening, and surfactants - A three way contribution to breadmaking. Baker's Dig. 51(5):32-34, 36-38, 40 42-44, 153.

CHUNG, O. K., and POMERANZ, Y. 1981. Recent research on wheat lipids. Baker's Dig. 55(5):38-50, 55, 96, 97.

CHUNG, O. K., and POMERANZ, Y. 1980-82. Unpublished data.

CHUNG, O. K., POMERANZ, Y., FINNEY, K. F., HUBBARD, J. D., and SHOGREN, M. D. 1977a. Defatted and reconstituted wheat flours. I. Effects of solvent and soxhlet types on functional (breadmaking) properties. Cereal Chem. 54:454-465.

CHUNG, O. K., POMERANZ, Y., FINNEY, K. F., and SHOGREN, M. D. 1977b. Defatted and reconstituted wheat flours. II. Effects of solvent type and extracting conditions on flours varying in breadmaking quality. Cereal Chem. 54:484-495.

CHUNG, O. K., POMERANZ, Y., and FINNEY, K. F. 1978. Wheat flour lipids in breadmaking. Cereal Chem. 55:598-618.

CHUNG, O. K., POMERANZ, Y., HWANG, E. C., and DIKEMAN, E. 1979. Defatted and reconstituted wheat flours. IV. Effects of flour lipids on protein extractability from flours that vary in breadmaking quality. Cereal Chem. 56:220-226.

CHUNG, O. K., POMERANZ, Y., JACOBS, R. M., and HOWARD, B. G. 1980a. Lipid extraction conditions to differentiate among hard red winter wheat that vary in breadmaking. J. Food Sci. 45:1168-1174.

CHUNG, O. K., POMERANZ, Y., and MARTIN, C. R. 1980b. Role of native wheat flour lipids or oil additives in reduction of flour dustiness. Cereal Foods World 25:523-524.

CHUNG, O. K., POMERANZ, Y., FINNEY, K. F., SHOGREN, M. D., and CARVILLE, D. 1980c. Defatted and reconstituted wheat flours. V. Bread-making response to shortening of flour differentially defatted by varying solvent and temperature. Cereal Chem. 57:106-110.

CHUNG, O. K., POMERANZ, Y., SHOGREN, M. D., FINNEY, K. F., and HOWARD, B. G. 1980d. Defatted and re-constituted wheat flours. VI. Response to shortening addition and lipid removal in flours that vary in bread-making quality. Cereal Chem. 57:111-117.

CHUNG, O. K., SHOGREN, M. D., POMERANZ, Y., and FINNEY, K. F. 1981. Defatted and reconstituted wheat flours. VII. The effects of 0-12% shortening (flour basis) in breadmaking. Cereal Chem. 58:69-73.

CHUNG, O. K., POMERANZ, Y., and FINNEY, K. F. 1982. Relation of polar lipid content to mixing requirement and loaf volume potential of hard red winter wheat flour. Cereal Chem. 59:14-20.

CHUNG, O. K., POMERANZ, Y., and JACOBS, R. M. 1984. Solvent solubility parameter and flour moisture effects on lipid extractability. J. Am. Oil Chem. Soc. 61:793-797.

CHUNG, O. K., AL-OBAIDY, K. A., and HUBBARD, J. D. 1987. Wheat glutens: Effects of processing variables and flour quality on their enhancing characteristics in breadmaking. Vol. I-2. Pages 36-47 in: Symposium Proc., International Symposium on New Technology of Vegetable Proteins, Oils and Starch Processing, CICCST, China Association for Science and Technology, Beijing, China.

CLEMENTS, R. L. 1980. Note on the effect of removal of free flour lipids on the internal structure of cookies as observed by a resin-embedding method. Cereal Chem. 57:445-446.

CLEMENTS, R. L., and DONELSON, J. R. 1981. Functionality of specific flour lipids in cookies. Cereal Chem. 58:204-206.

COLE, E. W., MECHAM, D. K., and PENCE, J. W. 1960. Effects of flour lipids and some lipid derivatives on cookie-baking characteristics of lipid-free flours. Cereal Chem. 37:109-121.

DAHLE, L. K., and MUENCHOW, H. L. 1968. Some effects of solvent extraction on cooking characteristics of spaghetti. Cereal Chem. 45:464-468.

DANIELS, N. W., FRAZIER, P. J., and WOOD, P. S. 1971. Flour lipids and dough development. Baker's Dig. 45(4): 20-25, 28.

DAVIS, K. R., LITTENEKER, N., LE TOURNEAU, D., CAIN, R. F., PETERS, L. J., and MC GINNIS, J. 1980. Evaluation of the nutrient composition of wheat. I. Lipid constituents. Cereal Chem. 57:178-184.

FABRIANI, G., LINTAS, C., and QUAGLIA, G. B. 1968. Chemistry of lipids in processing and technology of pasta products. Cereal Chem. 45:454-463.

FAUBION, J. M., and HOSENEY, R. C. 1981. Lipoxygenase: Its biochemistry and role in breadmaking. Cereal Chem. 58:175-180.

FAUBION, J. M., and HOSENEY, R. C. 1982. High-temperature short-time extrusion cooking of wheat starch and flour. II. Effects of protein and lipid on extrudate properties. Cereal Chem. 59:533-537.

FISHER, N., BROUGHTON, M. E., PEEL, D. J., and BENNETT, R. 1964. The lipids of wheat. II. Lipids of flours from single wheat varieties of widely varying baking quality. J. Sci. Food Agric. 15:325-341.

FISHER, N., BELL, B. M., RAWLINGS, C. E. B., and BENNETT, R. 1966. The lipids of wheat. III. Further studies of the lipids of flours from single wheat varieties of widely varying baking quality. J. Sci. Food Agric. 17:370-382.

FRAZIER, P. J. 1979. Lipoxygenase action and lipid binding in breadmaking. Baker's Dig. 53(6):8-10, 12, 13, 16, 18, 20, 29.

FRAZIER, P. J., BRIMBLECOMBE, F. A., DANIELS, N. W. R., and EGGITT, P. W. R. 1979. Better bread from weaker wheats - A rheological attack. Getreide Mehl Brot 33:268-271.

HARGIN, K. D., and MORRISON, W. R. 1980. The distribution of acyl lipids in the germ, aleurone, starch, and non-starch endosperm of four wheat varieties. J. Sci. Food Agric. 31:877-888.

HOSENEY, R. C., RAO, H., FAUBION, J., SIDHU, J. S. 1980. Mixograph studies. IV. The mechanism by which lipoxygenase increases mixing tolerance. Cereal Chem. 57:163-166.

HSIEH, C. C., WATSON, C. A., and MC DONALD, C. E. 1981. Direct gas chromatographic estimation of saturated steryl esters and acylglycerols in wheat endosperm. Cereal Chem. 58:106-110.

JOHNSON, A. C., and HOSENEY, R. C. 1979. Chlorine treatment of cake flours. III. Fractionation and reconstitution techniques for Cl_2-treated and untreated flours. Cereal Chem. 56:443-445.

JOHNSON, A. C., HOSENEY, R. C., and VARRIANO-MARSTON, E. 1979. Chlorine treatment of cake flours. I. Effect of lipids. Cereal Chem. 56:333-335.

KISSELL, L. T., POMERANZ, Y., and YAMAZAKI, W. T. 1971. Effects of flour lipids on cookie quality. Cereal Chem. 48:655-662.

KISSELL, L. T., DONELSON, J. R., and CLEMENTS, R. L. 1979. Functionality in white layer cake of lipids from untreated and chlorinated patent flours. I. Effects of free lipids. Cereal Chem. 56:11-14.

LAIGNELET, B. 1983. Lipids in pasta and pasta processing. Pages 269-286 in: Lipids in Cereal Technology. P. J. Barnes, ed. Academic Press, London.

LIN, M. J. Y., YOUNGS, V. L., and D'APPOLONIA, B. L. 1974a. Hard red spring and durum wheat polar lipids. I. Isolation and quantitative determinations. Cereal Chem. 51:17-33.

LIN, M. J. Y., D'APPOLONIA, B. L., and YOUNGS, V. L. 1974b. Hard red spring and durum wheat polar lipids. II. Effect on quality of bread and pasta products. Cereal Chem. 51:34-45.

MAC RITCHIE, F. 1976. The liquid phase of dough and its role in baking. Cereal Chem. 53:318-326.
MAC RITCHIE, F. 1977. Flour lipids and their effects in baking. J. Sci. Food Agric. 28:53-58.
MAC RITCHIE, F. 1981. Flour lipids: Theoretical aspects and functional properties. Cereal Chem. 58:156-158.
MARSTON, P., and MAC RITCHIE, F. 1985. Lipids: Effects on the breadmaking qualities of Australian flours. Food Technol. Aust. 37:362-365.
MATSUO, R. R., DEXTER, J. E., BOUDREAU, A., and DAUN, J. K. 1986. The role of lipids in determining spaghetti cooking quality. Cereal Chem. 63:484-489.
MECHAM, D. K. 1971. Lipids. Pages 393-451 in: Wheat: Chemistry and Technology, 2nd Ed. Y. Pomeranz, ed. Am. Assoc. Cereal Chem., St. Paul, MN.
MONTZHEIMER, J. W. 1931. A study of methods for testing cake flour. Cereal Chem. 8:510-517.
MORRISON, W. R. 1976. Lipids in flour, dough, and bread. Baker's Dig. 50(4):29-34, 36, 47-49.
MORRISON, W. R. 1978a. Cereal lipids. Pages 221-348 in: Advances in Cereal Science and Technology, Vol. II. Y. Pomeranz, ed. Am. Assoc. Cereal Chem., St. Paul, MN.
MORRISON, W. R. 1978b. The stability of wheat starch lipids in untreated and chlorine-treated cake flours. J. Sci. Food Agric. 29:365-371.
MORRISON, W. R. 1988. Lipids. Pages 373-439 in: Wheat: Chemistry and Technology, 3rd Ed. Y. Pomeranz. ed. Am. Assoc. Cereal Chem., St. Paul, MN.
MORRISON, W. R., and COVENTRY, A. M. 1985. Extraction of lipids from cereal starches with hot aqueous alcohols. Staerke 37:83-87.
MORRISON, W. R., TAN, S. L., and HARGIN, K. D. 1980. Methods for the quantitative analysis of lipids in cereal grains and similar tissues. J. Sci. Food Agric. 31:329-340.
MORRISON, W. R., WYLIE, L. J., and LAW, C. N. 1984. The effect of group 5 chromosomes on the free and total galactosyl diglycerides in wheat endosperm.

J. Cereal Sci. 2:145-152.
MORRISON, W. R., LAW, C. N., WYLIE, L. J., COVENTRY, A. M., and SEEKINGS, J. 1988. The effect of group 5 chromosomes on the free polar lipids and breadmaking quality of wheat. J. Cereal Sci. (In press)
NIIHARA, R., NISHIDA, Y., and YONEZAWA, D. 1982. Role of fatty acids produced in the storage process of tenobe-somen (a kind of noodle) called "Yaku". Pages 973-978 in: Progress in Cereal Chemistry and Technology. Proc. World Cereal Bread Congr. 7th Prague. J. Holas and J. Kratochvil, eds. Elsevier, Amsterdam.
POMERANZ, Y. 1971. Composition and functionality of wheat flour components. Pages 585-674 in: Wheat: Chemistry and Technology, 2nd Ed. Y. Pomeranz, ed. Am. Assoc. Cereal Chem., St. Paul, MN.
POMERANZ, Y. 1988. Composition and functionality of wheat flour components. Vol. II. Pages 219-370 in: Wheat: Chemistry and Technology, 3rd Ed. Y. Pomeranz, ed. Am. Assoc. Cereal Chem., St. Paul, MN.
POMERANZ, Y., CHUNG, O. K., and ROBINSON, R. J. 1966a. Lipids in wheat from various classes and varieties. J. Am. Oil Chem. Soc. 43:511-514.
POMERANZ, Y., CHUNG, O. K., and ROBINSON, R. J. 1966b. Lipid composition of wheat flours varying widely in breadmaking potentials. J. Am. Oil Chem. Soc. 43:45-58.
RHO, K. L., CHUNG, O. K., and SEIB, P. A. 1986. Functional properties of wheat flour gluten and lipids in Oriental dry noodles. Cereal Foods World 31:607-608.
SEGUCHI, M., and MATSUKI, J. 1977. Studies of pancake baking. II. Effects of lipids on pan-cake qualities. Cereal Chem. 54:918-926.
SHUEY, W. C., RASK, O. S., and RAMSTAD, P. E. 1963. Measuring the oil-binding characteristics of flour. Cereal Chem. 40:71-77.
SHOLLENBERGER, J. H., CURTIS, J. J., JAEGER, C. M., EARLE, F. R., and BAYLES, B. B. 1949. The chemical composition of various wheats and factors influencing their composition. Tech. Bull. No. 995, U.S. Dept. Agric., Washington, D.C.

SIMMONDS, D. H., and WRIGLEY, C. W. 1972. The effect of lipid on the solubility and molecular weight range of wheat gluten and storage proteins. Cereal Chem. 49:317-323.

SPIES, R. D., and KIRLEIS, A. W. 1978. Effect of free flour lipids on cake-baking potential. Cereal Chem. 55:699-704.

TWEETEN, T. N. 1979. Analytical high performance liquid chromatography of digalactosyl diglyceride in wheat flour lipid fractions. Ph.D. Dissertation, Kansas State University, Manhattan, KS.

TWEETEN, T. N., and WETZEL, D. L. 1979. High performance liquid chromatographic analysis of fatty acid derivatives from grain and feed extracts. Cereal Chem. 56:398-402.

TWEETEN, T. N., WETZEL, D. L., and CHUNG, O. K. 1981. Physicochemical characterization of galactosyldiglycerides and their quantitation in wheat flour lipids by high performance liquid chromatography. J. Am. Oil Chem. Soc. 58:664-672.

WALKER, G. C. 1988. Determination of flour glycolipids as their benzoyl derivatives by high-performance liquid chromatography with ultraviolet detection. Cereal Chem. 65:433-435.

WARWICK, M. J., FARRINGTON, W. H. H., and SHEARER, G. 1979. Changes in total fatty acids and individual lipid classes on prolonged storage of wheat flour. J. Sci. Food Agric. 30:1131-1138.

YAMAZAKI, W. T., and DONELSON, J. R. 1976. Effects of interactions among flour lipids, other flour fractions, and water on cookie quality. Cereal Chem. 53:998-1004.

ZAWISTOWSKA, U., BEKES, F., and BUSHUK, W. 1984. Intercultivar variations in lipid content, composition and distribution and their relation to baking quality. Cereal Chem. 61:527-531.

ZAWISTOWSKA, U., BEKES, F., and BUSHUK, W. 1985. Involvement of carbohydrates and lipids in aggregation of glutenin proteins. Cereal Chem. 62:340-345.

21

THE USE OF WHEAT FLOURS IN EXTRUSION COOKING

Robin C.E. Guy

Flour Milling and Baking Research Association
Chorleywood, Hertfordshire, U.K.

INTRODUCTION

Modern extrusion cooking processes were developed from simple single screw forming processes for pasta and breakfast cereals. To-day there are several forms of extrusion cooking in use in industry (Harper 1981), which range from simple heating processes to those which also incorporate high levels of mechanical energy input and the use of powerful kneading effects. All such extrusion processes have variable reaction times but these are generally quite small in the range 20 to 500s. Consequently extrusion cooking may be classified as a form of high temperature/ short time (HTST) processing.

The extrusion cooking processes are used in the manufacture of foodstuffs because of their efficiency in comparison with older methods involving batch processing in multistage operations. It has been shown (Ziminski, 1981) that the continuous extrusion process for breakfast cereal flakes saves costs in the following areas:
  (a) energy
  (b) manpower
  (c) factory space
  (d) raw materials
  (e) capital equipment

For roller drying operations the energy savings are more dramatic since the extruders operate at low moisture levels (~25%) compared with the high levels (75-80%) used on roller dryers.

Even in areas of high moisture such as processes for intermediate moisture pet foods and texturised proteins, extrusion cooking processes may offer handling efficiencies and control which permit unique treatments to be carried out that are impossible using batch processing technologies. Extrusion cooking processes are used commercially to manufacture products for the biscuit, breakfast cereal, breadings, snackfoods and food ingredients (pregelled flours, maltodextrins, texturised proteins, etc.) sectors of the food market and for many forms of pet foods and animal feedstuffs.

Wheat flours, starches and proteins are used in most of these sectors and form the major ingredients in cereal-based products in the U.K.

Research studies at FMBRA were undertaken to evaluate the performance of wheat flour and other major cereal types in extrusion cooking processes to assist in raw material selection and to provide information for modelling studies of such processes.

In order to understand the range of possibilities for cereal processing which might be available in such a complex multivariate process it was thought necessary to examine the fundamental changes occurring in relation to the native biopolymer systems present in the raw materials, building on earlier studies with maize starch by Colonna et al (1983).

Preliminary studies with wheat flours were made over a range of process conditions to assess the changes to starch, protein and bran components. Observations over a wide range of process conditions with hard and soft milling wheat flours and other cereals such as maize and rice revealed (Guy and Horne, 1988) that a general mechanism existed for all these starch-rich systems in the moisture range 10 to 25%, (wet weight basis). Further studies with wheat flours in the higher moisture region 25 to 40% (wwb) have been carried out to study this region and in particular to examine additional features related to the free water in the system.

In this paper are reviewed the basic mechanisms for development of cereals in relation to the 10 to 40% (wwb) moisture range which covers almost all commercial processes. The major processing factors related to the development mechanisms are highlighted to show their variation across the moisture range. Finally the particular features of wheat flour which may influence its choice for use in such processes are discussed and compared with other cereals.

MECHANISMS FOR THE DEVELOPMENT OF CEREALS

Observation of the processing pattern of cereal flours and grits in starch-rich formulations on a standard screw configuration (Fig.1) for a Baker Perkins MPF50D twin screw extruder revealed a common sequence of changes to the raw materials over a wide range of machine settings. This processing sequence was found to be similar at all moisture levels in the range 10 to 40% (wwb) for wheat flour and most other cereals. If the mechanism for development is considered in detail there are small differences at moistures >30% (wwb). Therefore the mechanism for the low moisture region will be described in the first instance and the differences at higher moisture levels discussed in the following section.

Mechanism For The Development Of Cereals In The Low Moisture Range, 10-25% (wwb)

The sequence of changes to the raw materials observed by the use of a 'dead stop' shutdown technique were as illustrated on the macroscopic scale in Fig.2. They involved:
1. Compression of powders by the action of the filling screws to a compacted mass (density ~1.0g/ml).

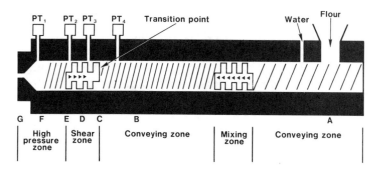

Fig.1 Standard screw configuration used for comparison of materials

2. An initial kneading of the compressed powder at pressures of 5 to 20 atmospheres, causing some breakdown of large particles and rapid heat input to the powder by frictional and mechanical effects.
3. A rapid temperature rise of 10 to 40°C within a short distance (2 to 20mm) which raises the mass temperature sufficiently to melt all the crystalline structures within the starch granules.
4. Release of the softened starch granules from the wedge protein and their deformation by the kneading action of the screws under forces of compression, elongation and shear. This caused changes in the microscopic appearance of the raw materials as air was excluded from the system and the flour was transformed into a plastic or viscoelastic fluid mass (density, 1.2 to 1.45g/ml).
5. Under further intensive kneading by the screw elements the starch polymers are dispersed from their native aggregates to form a continuous phase. The loss of starch aggregates causes a sharp reduction in fluid viscosity as indicated by lower die pressures. Within the continuous starch phase the proteins are macerated to small globules or rods (<100 m) and appear as discontinuities, together with the residual granules and the bran platelets in the starch phase.
6. In the final pumping section at the die the viscoelastic fluids tend to lose all their entrapped air (densities, 1.4-1.5g/ml) and may be extruded as fluids or expanded foams depending on their exit temperatures.
7. At temperatures >100°C bubbles of steam are nucleated in the fluid and are retained by the continuous starch phase. Little further change occurs to the biopolymer structures during this expansion process.

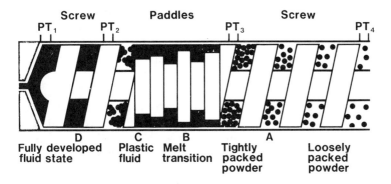

Fig.2. A view of the development of wheat flour in the cooking zone.

Variations In The Mechanism Of Development Of Cereals At High Moisture Levels (25 to 40%, wwb)

Several factors vary with moisture content most notably the dilution and lubrication of particles at higher moisture levels and changes in both the critical melting temperatures and the stability of the starch granules in the softened state to dispersal by kneading forces.

It has been shown (including Donovan, 1979; Burt and Russell, 1983) that the temperature values for starch granules' melting points increase with decreasing moisture levels. Most workers have used static techniques such as Differential Scanning Calorimetry or light microscopy in sealed cells. No reports have been made to cover the effects of the mechanical kneading of the granules during the heating process.

In limited studies with wheat flour at 20 and 30% moisture levels (wwb) the loss of birefingence was monitored at constant screw speed (300 rpm) and feed rate (0.8kg/min) while raising the barrel temperature very slowly. The observed melting points were at ~120 and 130°C, respectively. Both values are lower than the reported measurements by the DSC or light microscopy techniques and indicate some influence from the mechanical effects. These would be greater at the 20% than at the 30% moisture level.

The second important change in the character of the starch granules at high moisture levels is due to swelling and diffusion within the polymer aggregates as free water becomes available. These effects appear to weaken the stability to dispersal of the aggregates as in gelatinisation processes in excess moisture.

The cereal proteins are denatured and macerated into small pieces of fairly similar shapes, globules and rods, at all moisture levels. At high moisture levels the outlines of the fragments appear less regular under the light microscope.

PROCESSING VARIABLES AND MOISTURE CONTENT

The most important process inputs are those related to the heat input required to raise the mass temperatures to exceed the melting points of the starch granules and also to the mechanical effects acting on the softened granules. Raw material characteristics and the moisture content in the mix play very important roles in determining these inputs during cereal processing.

For example, the specific mechanical energy (SME) input, measured from the motor torque, which gives a measure of the applied kneading forces and makes an important contribution to the heat input, is markedly affected by both factors. In Fig.3 the SME inputs for soft wheat, rice and maize are shown for the moisture range 15 to 25% and for wheat flour up to 40%. SME inputs falls sharply as the particles are diluted at low moisture levels, 15 to 20%, but levels off as the starch granules begin to swell at 25 to 30% before falling slightly from 35 to 40% moisture. Maize and rice, which contain particles with harder endosperm textures than the soft wheats, give significantly higher SME inputs at low moisture levels.

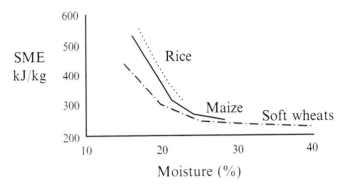

Fig. 3. SME inputs for cereal flours (300rpm, screw speed, 0.8kg.hr^{-1} using twin circular dies, 4mm diameter).

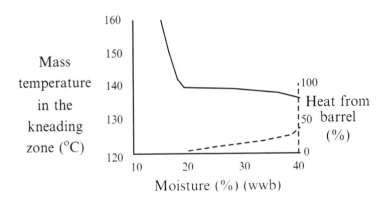

Fig.4. Mass temperature values in the shear zone for wheat flour processed as in Fig.3.

The graph of mass temperature of soft wheat flour shows a similar trend to the SME input (Fig.4). However the temperatures obtained at higher moistures than 25% represent an increasing heat input from the barrel heating systems as indicated in Fig. 4. The flour samples shown in these graphs are completely free of any birefringent starch granules indicating that the temperatures shown in Fig. 4 exceeded the critical melting points of the granules. An attempt was also made to estimate the extent to which the granules had been dispersed using a combination of light microscopy and a paste viscosity technique similar to that used by Launay and Lisch (1983) with 9.1% w/w flour pastes.

The graph shown in Fig.5 represents the estimates of the amount of completely dispersed starch polymer. It can be seen that the value for this particular form of starch decreases to a minimum at 20-30% moisture as the kneading forces are reduced but increases once more at high moisture levels as swelling and diffusion processes probably weaken the forces binding the polymer aggregates together.

RAW MATERIAL CHARACTERISTICS OF WHEAT FLOURS

All the major cereals and many starch-rich materials such as potato have been used in extrusion processes. The choice of any particular material may be related to its costs, physical performance, flavour, appearance, nutritional value, etc. in the products.

Generally wheat flour is quite widely used because of its low costs and versatile performance.

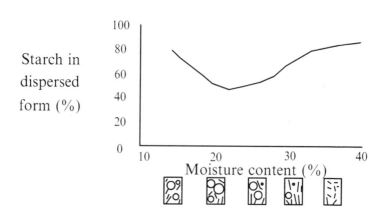

Fig.5. The extent of starch dispersal for wheat flour processed as in Fig. 3.

The performance may be judged in terms of the general mechanism of development and particular requirements for the physical state of starch and its concentration to create the structure of a product. Wheat flours are milled from a wide variety of wheats and may be obtained with varying properties and compositions. However, the starch granules tend to be of a similar form and composition. The range of powder properties, particularly the variation from a soft to hard endosperm textures permits a flour blend to be produced to match process requirements at optimal costs related to both materials and energy inputs. In addition, modified forms of flour or separated components may be used to give further variations as shown by Faubion et al. (1982). Hard wheats may be used to create high SME inputs as found with maize or rice flours (Guy and Horne, 1988) for processes involving short barrel extruders or where extensive degradation of the starch is required. Separated wheat starch may also be used for these purposes but in other situations the use of soft milling wheats can reduce SME inputs and provide a wider processing window thus helping to reduce possible product variations and retain nutritional qualities.

In high moisture conditions or systems in which the flour particles are diluted to <70% of the mix the soft wheat flours or starch offer better opportunities to manipulate the physical form of the starch. The more cohesive pieces of protein and starch in maize or hard wheat may fail to disperse in such relatively low shear conditions.

Wheat flours tend to be richer in proteins than maize or rice. These act as diluents and lubricants in the kneading processes. The lipid content is fairly high but much of this material is bound within the biopolymer aggregates and does not act as a lubricant. However, wheat flour is fairly sensitive to added lipids and emulsifiers. Distilled monoglyceride at 0.25 to 0.50% exerts a protective effect on wheat starch granules while having little effect on those of maize or rice.

An important feature of the use of cereals in biscuits and snacks is the flavour and appearance. Most wheat flours products have a bland, faintly cereal flavour which forms the basis for flavour creation. However this is a very complex field as may be illustrated by comparison of base raw materials and their interactions with added flavours (Anon., 1985) or by the possibilities available through the Maillard type reactions at high temperatures (Buckholz, 1988). The appearance of these products varies with expansion, air cell numbers, etc, from tan to an off white. Pure wheat starch forms white products and the yellowish tan colour is introduced in the flour.

In the creation of textures and flavours from cereal extrudates, wheat flours form an excellent base material and are compatible with other cereals, proteins, etc.

ACKNOWLEDGEMENTS

This work forms part of a research project sponsored by the UK Ministry of Agriculture, Fisheries and Food. The results of the research are the property of the Ministry of Agriculture, Fisheries and Food and are Crown Copyright.

LITERATURE CITED

ANON. 1985. Endurance tests for flavours. Food Eng. Intern. 10:35-40.

BUCKHOLZ, L. L. 1988. The role of the Maillard technology in flavouring food products. Cereal Foods World 33:547-551.

BURT, D. J. and RUSSELL, P. L. 1983. Gelatinisation of low water content wheat starch-water mixtures. Starke 35:354-360.

COLONNA, P., MELCION, J. P., VERGNES, B., and MERCIER, C. 1983. Flour mixing and residence time distribution of maize in a twin screw extruder. J. Cereal Sci. 1:115-125.

DONOVAN, J. W. 1979. Phase transition of starch-water systems. Biopolymers 18:263-275.

GUY, R. C. E. and HORNE, A. W. 1988. Extrusion and coextrusion of cereals. Pages 331-267 in: Food Structure-Its Creation and Evaluation. J.M.V. Blanshard and J.R. Mitchell, eds. Butterworth Press: London and Boston.

HARPER, J. M. 1981. History of food extrusion. Pages 3-5 in: Extrusion of Foods, ed. 2 Vol. 1. CRC Press: Boca Raton, Florida.

LAUNAY, B. and LISCH, J. M. 1983. Twin screw extrusion cooking of starches and flow behaviour of starch pasties. J. Food Eng. 2:259-280.

FAUBION, J. M., HOSENEY, R. C. and SEIB, P. A. 1982. Functionality of grain components in extrusion. Cereal Foods World. 27:212-216.

ZIMINSKI, R. D. 1981. Economics of twin screw processing of foods. Ann. Mtg. Inst. Am. Chem. Eng., New Orleans, Werner Pfeiderer Corp, Ramsey, NJ.

22

THE TWIN-SCREW EXTRUSION COOKER AS A VERSATILE TOOL FOR WHEAT PROCESSING

Pekka Linko

Helsinki University of Technology
Laboratory of Biotechnology and Food Engineering
SF-02150 Espoo, Finland

INTRODUCTION

Extrusion cooking technology has been applied in the food and animal feed industries for a long time, but only recently extrusion cooking has become a tool for other industrial processing of biological material (Linko, P. et al 1981; Mercier et al 1989). Such applications are largely based on biopolymer modifications such as the gelatinization and complex formation of starches, protein denaturation, enzymic hydrolysis, and a number of chemical reactions. Typically, extrusion cooking is a high-temperature short-time (HTST-) process, in which the material is plasticized under a relatively high temperature, pressure, and shear, before it is extruded through a die assembly to ambient pressure and temperature (Olkku et al 1980). The recent approach is to treat an extrusion cooker as a special type of a continuous bioreactor, in which the biological material is processed at a relatively short residence time. Thus in the actual reaction zone(s) in front of the die element and/or any reversed pitch screw elements and other restriction devices the residence time has been shown to be only of the order of a few seconds, and in a typical process no more than two minutes (Olkku et al 1980). In the food and animal feed processing this results in a high nutritional quality with little losses in vitamins and other nutrients (Guzman-Tello and Cheftel 1987; Linko, P. et al 1981), and significant destruction of microorganisms (Bouveresse et al 1982; Gry et al 1984) and of certain toxic compounds (Fenwick et al 1986; Gréhaigne et al 1983).

An extrusion process is economically viable for most applications, allowing widely different products to be obtained with the same central processing unit simply by adjusting the screw design and processing conditions. Consequently, the product quality is also a function of the extruder design, and for example the degree of cook of cereal starches can be conveniently chosen from between relatively raw to well cooked. The wear of the screws and of the barrel lining may quite dramatically affect the results (Miller 1984). Such effects have been little investigated, and we have recently initiated a systematic study of related problems. For example we have observed that for a crisp-bread type product even slight wear could result in several tens of °C increase in the mass temperature before the die element under the same set processing conditions such as the barrel temperature. Such effects of the increased friction, back flow, and residence time further illustrate the importance of the monitoring the actual mass temperature, as has been emphasized before (Linko, P. 1982). Although both single- and twin-screw extruders can be used for applications, there are also marked differences between the two, which cannot be dealt with in detail here. As a typical example, the residence time distribution in a single-screw extruder is usually larger, allowing under- or over-cooked particles to pass through with a greater ease than in a twin-screw extruder.

The raw material composition may also markedly influence the results. For example only small additions of salt, or lipids and emulsifiers may significantly affect the degree of starch gelatinization and, thus, the degree of cook. Consequently, for example results obtained with barley, corn (maize), oats, rice, and rye, and with various blends, cannot necessarily be applied as such to the extrusion cooking of wheat. Further, in many applications involving cereal grain feed water content, together with the mass temperature is one of the most important parameters affecting product quality. The product water activity which is of particular importance in certain intermediate moisture applications, can in many cases be conveniently controlled by the raw material composition and extrusion processing conditions (Linko, P. et al 1985; Vainionpää et al 1984).

Novel uses of extrusion cooking in bread baking have been recently reviewed elsewhere (Linko, P. et al 1984a). Conventional food and animal feed applications shall not be further treated in this paper. The paper discusses a twin-screw extrusion cooker as a bioreactor against the background given above, for potential non-food industrial uses as applied to wheat processing.

THERMOMECHANICAL PRETREATMENT FOR BIOTECHNICAL PROCESSING

For most biotechnical applications based on cereal grain, starch gelatinization is necessary. The water activity of $a_w \sim 1$ required for the gelatinization of cereal starches at ambient pressure can be achieved in extrusion cooking at feed moisture levels less than 20%. The gelatinization under such conditions can be clearly demonstrated, for example, by X-ray diffraction patterns, paste consistency profiles, and scanning electron microscopy (Linko, P. 1982; Linko, Y.-Y. et al 1980).

Mercier and Feillet (1975) first demonstrated that cold water soluble starches can be obtained by twin-screw extrusion cooking without observable maltodextrin formation. A few years later Y.-Y. Linko et al (1979, 1980) reported the direct saccharification of extrusion cooked starch by glucoamylase with excellent results, although the very high paste viscosities of such extrudates limited the solids content in subsequent enzymic processing to relatively low levels. Further, such starches retrograde rapidly unless kept at above 60°C until further processing.

During the extrusion cooking of starchy materials the degree of cook increases with a decrease in feed moisture and an increase in mass temperature. However, even a small quantity of monoglycerides and some other minor ingredients markedly inhibit gelatinization. Suzuki (1970) reported a product DP ∿ 200-300 in thermomechanical liquefaction of starch under conditions not unlike those prevailing during extrusion cooking, and similar results have been obtained by Reinikainen et al (1986) in HTST-extrusion of wheat starch. Small molecular weight oligosaccharides were not detected, and the DE-value was less than 2 after extrusion. Such extrudates of very low DE-value are especially suitable for enzymic high maltose syrup production (Linko, Y.-Y. et al 1983).

A systematic study on the effects of acid or alkali treatment of wheat starch in a twin-screw extruder has also been reported (Kervinen et al 1985). However, no significant differences on subsequent saccharification with glucoamylase were observed as compared with starch extrusion alone.

BEHAVIOR OF ENZYMES DURING EXTRUSION COOKING

Because of the HTST-nature of the twin-screw extrusion cooking process the effects of the treatment on a number of enzymes is of interest. No doubt, any residual enzyme activities in processed foods and animal feed may have a marked influence on product shelf-life and quality. Cheftel (1989) has recently extensively reviewed the inactivation of enzymes during extrusion cooking as it relates to food stability. However, the behavior of enzymes during extrusion cooking is also of interest if the extruder is to be used as an enzyme bioreactor.

The early work suggested nearly complete inactivation of enzymes during extrusion cooking (Linko, P. et al 1981), and it is clear that under severe processing conditions enzymes are inactivated (Linko, P. et al 1980; Linko, Y.-Y. et al 1980; Lorenz and Jansen 1980; Nierle et al 1980). The controlled inactivation of α-amylase can be used to advantage in the utilization of sprout damaged wheat in extrusion baking (Linko, P. et al 1984a, c; Mattson et al 1984), and in the production of fur animal feed (Strandberg et al 1988).

The first report on only partial inactivation of an enzyme, peroxidase, during extrusion cooking is that by Gardner et al (1969). An initial moisture content of 20-35% and a temperature of 110-149°C was observed sufficient for complete inactivation. Quite recently residual peroxidase activities have been reported in rice bran extruded with a Brady single-screw extruder (Randall et al 1985; Sayre et al 1985). The interest in the early work was directed to the complete inactivation of enzymes. Further, insufficient available data makes the evaluation and comparisons of the results difficult. For example, the temperature reported was usually given as the barrel temperature, if defined at all. The potential of an extrusion cooker as a bioreactor was first realized in the late 1970s when it was observed that a number of cereal enzymes, including α-amylase, remained active after extrusion cooking under relatively mild processing conditions (Linko, P. et al 1978, 1980; Linko, Y.-Y. et al 1980). Most of this work was carried out with wheat and rye using a Clextral (formerly Creusot-Loire) BC 45 twin-screw extruder, although some later work was also done with a Werner & Pfleiderer C 58 twin-screw machine, and also with barley and oats. The highest residual enzyme activities were obtained at a relatively high initial moisture level, low mass temperature, low shear, and a short residence time with little backmixing. These results have been later confirmed with several enzymes by Fretzdorff and Seiler (1987), who used a Werner & Pfleiderer C 58 twin-screw extruder.

Although in food and in animal feed application HTST-processes such as extrusion cooking are usually designed for complete enzyme inactivation under conditions minimizing nutrient losses, the possibility to control an extrusion cooking process for a given objective so that the inactivation of a given enzyme is minimized has given rise to investigations on enzyme biocatalysis during and after extrusion cooking.

EXTRUSION COOKING FOR ENZYMIC BIOCONVERSIONS

Enzymic Liquefaction

Thermostable α-amylase can be used to the advantage in the extrusion liquefaction of starches (Linko Y.-Y. et al 1980). although cereal starches can be liquefied thermomechanically alone (Linko, Y.-Y. et al 1979; Suzuki 1970), the initiation of the enzymic hydrolysis already during extrusion cooking by the addition of a suitable quantity of thermostable α-amylase to the feed stream markedly decreases the problems otherwise arising from the excessive paste viscosity of the extrudate, and thus allows enzymic down-stream processing even at 45% solids level with ease (Hakulin et al 1983; Linko, P. et al 1983). The effect is quite dramatic, although with a short (500-600 mm) barrel extruder the DE-value remains low immediately after extrusion. First after the Termamyl α-amylase dosage was increased to 30 ml/kg d.s. a DE $\sim$ 5 was obtained. However, significantly higher DE-values can be obtained during extrusion cooking with α-amylase with a long barrel twin-screw extruder such as the 1222.5 mm barrel Werner & Pfleiderer C 58 used by Hakulin et al (1983), and the 1000 mm barrel Clextral BC 45 used by Chouvel et al (1983). Such enzymic pretreatment can be used either for subsequent saccharification or as a step in the fractionation of wheat and other cereal grain into their components.

The rapid inactivation of Termamyl α-amylase after extrusion cooking is difficult, although it does inactivate in hot acetate. Once initiated during the extrusion cooking the enzymic starch hydrolysis thus continues rapidly (Reinikainen et al 1986). This has to be also taken into account in subsequent biotechnical processing in order to arrive at the optimum DE-values for each subsequent step. Reinikainen et al (1986) also clearly demonstrated that enzymic hydrolysis of starch takes place already during extrusion cooking regardless of the low DE-value obtained.

The extruder may be connected directly either to a tubular reactor or to a scraped surface heat exchanger (SSHE), instead of subsequent batch processing. Hakkarainen et al (1985) obtained with wheat starch DE-values between 10 and 27 using a Contherm SSHE at 100-120°C at about 7 min residence time, which agrees well with the DE $\sim$ 20 reported by Chouvel et al (1983) with corn starch extrusion processed at 125°C, 50% initial moisture, and a residence time of 2-8 min. If the hydrolysis is allowed to continue at about 80°C after the extrusion cooking with the short barrel Clextral BC 45 extruder (mass temperature 137°C, initial moisture 55%, feed rate 12 kg/h d.s., screw speed 75 rev/min, 50 mm reversed pitch screw elements next to the dies, Termamyl dosage 30 ml/kg d.s.), a DE $\sim$ 17 with wheat starch was obtained at a total residence time of about 10 min. No attempt was made to optimize the conditions after the extrusion. Further, the enzyme dosage has a marked effect on the rate of hydrolysis, as can be expected. Under similar conditions, a DE $\sim$ 20 was obtained in about 2 h with 5 ml/kg d.s. Termamyl. For many applications, however, this is quite satisfactory.

Because of the short residence time in extrusion cooking, Termamyl α-amylase was reasonably stable up to about 150°C (Linko, P. 1982). For the initiation of the enzymic liquefaction during the extrusion cooking, the initial moisture content has to be above 40%, with an optimum at 55-60%. The high water content makes it necessary to set the barrel temperature at a markedly higher level than the desired mass temperature at the die plate, a further proof of the fact that the barrel temperature alone may tell little about the actual process conditions. When the enzyme dosage is increased the mass temperature decreases owing to decreased friction, thus requiring an increase in the set barrel temperature. It also follows that the optimal barrel temperature for enzymic liquefaction is significantly higher than that for the extrusion pretreatment without α-amylase. Finally it should be noted that the use of a long barrel extruder allows continuous processing by optimizing different stages separately (Hakulin et al 1983). For example, the water content and mass temperature can first be adjusted for starch gelatinization, with a portion of the α-amylase added at this stage. Next, with continuing gelatinization, the rest of the enzyme and water are added, and the temperature adjusted, if necessary, for the efficient initiation of enzymic liquefaction.

Saccharification

Glucose syrup

The saccharification with glucoamylase is influenced both by the extrusion cooking conditions and by the initial DE-value reached before the saccharification (Hakulin et al 1983; Linko, P. et al 1983). For wheat starch extrusion processed with the short barrel CLextral extruder the highest DE 96-98 (DX 94-96) after 24 h saccharification was obtained when starch was first extrusion cooked in the presence of Termamyl α-amylase at mass temperature of 125-145°C, initial moisture 60%, and with a 30% solids level in saccharification. However, because of the presence of α-amylase during extrusion cooking a DE 93 in 24 h was obtained even at 45% solids, when without α-amylase subsequent batch saccharification at higher than 10% solids was difficult. The DE-value obtained after 10 h of saccharification clearly increased with an increase in the extrusion feed moisture level. The results obtained are in a good agreement with those reported by Chouvel et al (1983) with corn starch.

The long barrel Werner & Pfleiderer C 58 extruder allowed even the initiation of saccharification during the extrusion processing. For a 3-stage process, 75 mm reversed pitch screw elements were placed at 470 mm distance to initiate gelatinization under optimal moisture and temperature. Three short mixing elements were placed at 590 mm, 835 mm, and 1080 mm distance, respectively. Starch was preferably added to the extruder as a premixed slurry. When the mass temperature during the initiation of gelatinization and liquefaction was 124°C, and the temperature of the two last barrel sections was decreased to 64°C for the addition of glucoamylase in a buffer to bring the pH down to 4.5 and the total water content to 70%, the total feed rate was 72 kg/h with the total α-amylase dosage 0.9% (w/w d.s.), and the saccharification was allowed to continue at 60°C after the extrusion, a DE

94 was reached in 5 h and DE 97 in about 10 h. Consequently, the total saccharification time was reduced from the conventional of about 2 days to about 10 h.

High maltose syrup

The starch pretreatment requirements for high maltose syrup production differ markedly from those for glucose syrup. Unlike in the latter case, the initial DE-value before saccharification with β-amylase and pullulanase should be as low as possible for sufficiently liquefied starch in order to avoid excessive maltotriose formation which would reduce yield and quality (Takasaki and Yamanobe 1981). Inasmuch as extrusion cooking could yield liquefied starches of low DE-value both in the presence and absence of thermostable α-amylase, extrusion pretreatment appeared to be very suitable for high maltose syrup production. The lowest DE-values on liquefaction and the highest maltose content of 87.5% on saccharification were obtained when starch was thermomechanically treated without any α-amylase (Linko, Y.-Y. et al 1983). Again, the absence of α-amylase during extrusion cooking made batch saccharification at solids levels above 10% difficult. It is, however, possible to optimize the process for somewhat lower maltose yield, but markedly higher solids level during saccharification, if α-amylase is used during the extrusion stage.

BIOTECHNOLOGY OF EXTRUSION PROCESSED WHEAT

Syrups produced form extrusion pretreated cereal based materials such as wheat or wheat starch can be used as substrate in many biotechnical processes. Especially, if a relatively low DE syrup can be used as a substrate, continuous extrusion cooking pretreatment could be an attractive alternative. Consequently, ethanol fermentation was chosen as the example process inasmuch as yeast can most effectively produce ethanol during simultaneous saccharification and fermentation. Although a number of reports have been published on the extrusion pretreatment of cereal based materials for ethanol fermentation, few contain actual data on ethanol.

P. Linko and Y.-Y. Linko (1982) demonstrated with a twin-screw extruder, and Wenger et al (1981) and Ben-Gera et al (1983) with a single-screw extruder that extrusion cooked cereal grain present no special problems in subsequent ethanol fermentation. Ethanol yield both from high moisture and steam cooked grain is about equal, with the energy balance favoring extrusion cooking. We have later shown that roller milled whole barley, wheat, and oats, and various cereal starches all can be used as a raw material with only minor changes in processing conditions. Clextral BC 45 twin-screw extruder was exclusively used in these experiments at feed rates ranging from 20 to 60 kg/h, total water content from 10 to 65%, mass temperature from 105 to 160°C, while screw speed was kept constant at 150 rev/min. Ethanol fermentations were carried out from material extruded either with or without thermostable α-amylase, with varying glucoamylase dosage, with Saccharomyces cerevisiae yeast and Zymomonas mobilis bacterium, in both cases with free and immobilized cells.

The α-amylase treatment during extrusion cooking has little influence on ethanol fermentation carried out in the simultaneous saccharification and fermentation mode, if other conditions are comparable (Linko, P. et al 1983). Instead, the glucoamylase level had a marked effect both on the fermentation rate and on ethanol yield. Good results were obtained if the yeast was added after about one hour of saccharification at about DE 58, and at all solids' levels investigated best results were obtained by simultaneous saccharification and fermentation. The fermentation progressed without glucose inhibition, and solids' levels of 35% or even higher could be handled without difficulty. An increase in the initial quantity of yeast, and the use of Z. mobilis instead of yeast both markedly increased the initial rate of fermentation. The results obtained agree well with those reported by Chay et al (1984) on extrusion liquefied corn starch, and corn and wheat flour, fermented both with yeast and with Z. mobilis.

CONCLUSIONS

Wheat is, no doubt, excellent raw material for extrusion cooking. Extrusion cooking is a versatile, continuous process, which can be used in a large variety of applications. In addition to more conventional food and animal feed applications it provides, for example, means for the use of high α-amylase sprout damaged grain in flat-bread and snack production. More interestingly extrusion cooker can be used as a bioreactor in cereal grain and starch processing, for example, to glucose and high maltose syrups, and to ethanol. Alongside with the increasing interest in biotechnical applications, the extrusion cooker also has been used as a chemical reactor for a number of pilot and industrial scale applications. A good example is the production of modified starches by continuous extrusion cooking. It is likely that other novel applications will be seen in the near future.

ACKNOWLEDGEMENT

The author is grateful to the Ministry of Trade and Industry (Finland) for the generous financial support which made this work possible as a contribution to the European scientific and technological cooperation project COST 91 bis. Thanks are also due to all collaborators who carried out the innovative experiments, with special thanks to my wife Dr. Yu-Yen Linko who initially came up with the idea of applying extrusion cooking in ethanol production.

LITERATURE CITED

BEN-GERA, I., ROKEY, G. J., and SMITH, O. B. 1983. Extrusion cooking of grains for ethanol production. J. Food Eng. 2:177-188.

BOUVERESSE, J. A., CERF, O., GUILBERT, S., and CHEFTEL, J. C. 1982. Influence of extrusion cooking on the thermal destruction of Bacillus stearothermophilus spores in a starch-protein-sucrose mix. Lebensm.-Wiss.-Technol. 15:135-138.

CHAY, P. B., CHOUVEL, H., CHEFTEL, J. C., GHOMMIDH, G., and NAVARRO, J. M. 1984. Extrusion-

hydrolyzed starch and flours as fermentation substrates for ethanol production. Lebensm. -Wiss.-Technol. 17:257-267.
CHEFTEL, J. C. 1989. Extrusion cooking and food safety. In: Extrusion Cooking. C. Mercier, P. Linko, and J.M. Harper, eds. Am. Assoc. Cereal Chem., St. Paul, MN (in press).
CHOUVEL, H., CHAY, P. B., and CHEFTEL, J. C. 1983. Enzymatic hydrolysis of starch and cereal flours at intermediate moisture contents in a continuous extrusion reactor. Lebensm.-Wiss.-Technol. 16:346-353.
FENWICK, G. R., SPINKS, E. A., WILKINSON, A. P., HEANEY, R. K., and LEGOY, M. A. 1986. Effects of processing on antinutrient content of rapeseed. J. Sci. Food Agric. 37:735-741.
FRETZDORFF, B., and SEILER, K. 1987. The effects of twin-screw extrusion cooking on cereal enzymes. J. Cereal Sci. 5:73-82.
GARDNER, H. W., INGLETT, G. E., and ANDERSON, R. A. 1969. Inactivation of peroxidase as a function of corn processing. Cereal Chem. 46:626-634.
GRÉHAIGNE, B., CHOUVEL, H., PINA, M., GRAILLE, J., and CHEFTEL, J. C. 1983. Extrusion cooking of aflatoxin-containing peanut meal with and without addition of ammonium hydroxide. Lebensm.-Wiss. -Technol. 16:317-322.
GRY, P., HOLM, F., and KRISTENSEN, K. H. 1984. Degermination of spices in an extruder. Pages 185-188 In: Thermal Processing and Quality of Foods. P. Zeuthen, J.C. Cheftel, C. Eriksson, M. Jul., H. Leniger, P. Linko, G. Varela, and G. Vos, eds. Elsevier Applied Science Publishers, London.
GUZMAN-TEOLLO, R., and CHEFTEL, J. C. 1987. Thiamine destruction during extrusion cooking as an indicator for the intensity of thermal processing. Intern. J. Food Sci. Technol. 22:549-565.
HAKKARAINEN, L., LINKO, P., and OLKKU, J. 1985. State vector model for Contherm scraped surface heat exchanger used as an enzyme reactor in wheat starch conversions. J. Food Eng. 4:135-153.

HAKULIN, S., LINKO, Y.-Y., LINKO, P., SEILER, K., and SEIBEL, W. 1983. Enzymatic conversion of starch in twin-screw HTST-extruder. Starch 35:411-414.
KERVINEN, R., SUORTTI, T., OLKKU, J., and LINKO, P. 1985. The effects of acid and alkali on wheat starch extrusion cooking. Lebensm.-Wiss. -Technol. 18:52-59.
LINKO, P. 1982. HTST-(High-Temperature-Short-Time-) Extruder als Biochemischer Reaktor. Getreide, Mehl Brot 36:326-332.
LINKO, P., ANTILA, J., LINKO, Y.-Y., and MATTSON, C. 1984a. Extrusion cooking in bread baking. Pages Q1-Q12 In: Proceedings of the International Symposium on Advances in Baking Science and Technology. Kansas State University, Manhattan, KS.
LINKO, P., ANTILA, J., and OLKKU, J. 1978. Retention of amylolytic activity in HTST-extrusion cooking. Kemia-Kemi 5(1):691.
LINKO, P., COLONNA, P., and MERCIER, C. 1981. High-temperature short-time extrusion cooking. Adv. Cereal Sci. Technol. 4:145-235.
LINKO, P., HAKULIN, S., and LINKO, Y.-Y. 1983. Extrusion cooking of barley starch for the production of glucose syrup and ethanol. J. Cereal Sci. 1:275-284.
LINKO, P., KERVINEN, R., KARPPINEN, R., RAUTALINNA, E.-K., and VAINIONPÄÄ, J. 1985. Extrusion cooking for cereal-based intermediate-moisture products. Pages 465-479 In: Properties of Water in Foods.
LINKO, P., and LINKO, Y.-Y. 1982. Continuous ethanol fermentation by immobilized biocatalysts. Enzyme Eng. 6:335-342.
LINKO, P., MATTSON, C., LINKO, Y.-Y., and ANTILA, J. 1984c. Production of flat bread by continuous extrusion cooking from high α-amylase flours. J. Cereal Sci. 2:43-51.
LINKO, P., OLKKU, J., ANTILA, J., and ROSENBERG, K. 1980. Reduktion der Enzymaktivität während der Hochtemperatur-Kurzzeiterhitzung beim Extrudieren. Getreide, Mehl Brot 34(3):78-81.
LINKO, Y.-Y., LINDROOS, A., and LINKO, P. 1979. Soluble and immobilized enzyme technology in

bioconversion of barley starch. Enzyme Microb. Technol. 1:273-278.

LINKO, Y.-Y., MÄKELÄ, H., and LINKO, P. 1983. A novel process for high maltose syrup production from barley starch. Ann. New York Acad. Sci. 413:352-354.

LINKO, Y.-Y., VUORINEN, H., OLKKU, J., and LINKO, P. 1980. The effect of HTST-extrusion on retention of cereal α-amylase activity and on enzymatic hydrolysis of barley starch. Pages 210-223 In: Food Process Engineering, Vol. 2, Enzyme Engineering in Food Processing. P. Linko and J. Larinkari, eds. Elsevier Applied Science Publishers, London.

LORENZ, K., and JANSEN, G. R. 1980. Nutrient stability of full-fat soy flour and corn-soy blends produced by low-cost extrusion. Cereal Foods World 25:161-162, 171.

MATTSON, C., ANTILA, J., LINKO, Y.-Y., and LINKO, P. 1984. Extrusion cooking of high α-amylase flour for baking. Pages 262-266 In: Thermal Processing and Quality of Foods. P. Zeuthen, J.C. Cheftel, C Eriksson, M. Jul, H. Leniger, P. Linko, G. Varela, and G. Vos, eds. Elsevier Applied Science Publishers, London.

MERCIER, C., and FEILLET, P. 1975. Modification of carbohydrate components by extrusion-cooking of cereal products. Cereal Chem. 52:283-297.

MERCIER, C., LINKO, P., and HARPER, J. M., eds. 1989. Extrusion Cooking. Am. Assoc. Cereal Chem., St. Paul, MN.

MILLER, R. C. 1984. Effect of wear on twin-screw extruder performance. Food Technol. 38(2):58-61.

NIERLE, W., ELBAYA, A. E., SEILER, K., FRETZDORFF, B., and WOLFF, J. 1980. Veränderungen der Getreideinhaltstoffe während der Extrusion mit einem Doppelschneckenextruder. Getreide, Mehl Brot 34:73-78.

OLKKU, J., ANTILA, J., HEIKKINEN, J., and LINKO, P. 1980. Residence time distribution in a twin-screw extruder. Pages 791-794 In: Food Process Engineering, Vol.1, Food Processing Systems. P. Linko, Y. Mälkki, J. Olkku, and J.

Larinkari, eds. Elsevier Applied Science Publishers, London.

RANDALL, J. M., SAYRE, R. N., SCHULTZ, W. G., FONG, R. Y., MOSSMAN, A. P., TRIBELHORN, R. E., and SAUNDERS, R. M. 1985. Rice bran stabilization by extrusion cooking for extraction of edible oil. J. Food Sci. 50:361-364.

REINIKAINEN, P., SUORTTI, T., OLKKU, J., MÄLKKI, Y., and LINKO, P. 1986. Extrusion cooking in enzymatic liquefaction of wheat starch. Starch 38:20-26.

SAYRE, R. M., NAYYAR, D. K., and SAUNDERS, R. M. 1985. Extraction and refining of edible oil from extrusion stabilized rice bran. J. Am. Oil. Chem. Soc. 62:1040-1043.

STRANDBERG, T., KERVINEN, R., and LINKO, P. 1988. Extrusion cooking of sprout damaged wheat. J. Cereal Sci. 8:111-123.

SUZUKI, S. 1970. Novel industrial processes for enzymic conversion of starch. Pages 484-490 In: Proc. SOS/70, Third International Congress of Food Science and Technology, IFT, Washington, DC.

TAKASAKI, Y., and YAMANOBE, T. 1981. Production of maltose by pullulanase and β-amylase. Pages 73-78 In: Enzymes in Food Processing. C.G. Birch, N. Blakebrough and K.J. Palmer, eds. Elsevier Applied Science Publishers, London.

VAINIONPÄÄ, J., KARPPINEN, R., RAUTALINNA, E., KERVINEN, R., and LINKO, P. 1984. Adsorptive and desorptive modes in extrusion of cereal based intermediate moisture foods. Pages 701-707 In: Engineering and Food, Vol. 2, Processing applications. B.M. McKenna, ed. Elsevier Applied Science Publishers, London.

WENGER, L. G., ROKEY, G. J., BEN-GERA, I., and SMITH, O. B. 1981. Enzymatic conversion of high moisture shear extruded and gelatinized grain material. U.S. Patent 4,286,058.

23

EXTRUSION COOKING OF WHEAT PRODUCTS

B. van Lengerich

Nabisco Brands, Inc., Fair Lawn, NJ 07410

F. Meuser and W. Pfaller

Technical University Berlin, Berlin, West Germany

INTRODUCTION

Extrusion cooking of wheat products has become a considerably important process technique in the food industry. In the area of snacks, breakfast cereals and flat breads, extrusion is already a well established technology and product varieties are constantly increasing (Ben-Gera 1982, Scales 1982). Wheat based raw materials are usually extruded in mixtures with sugars, proteins, fats and dietary fiber to achieve specific sensory and nutritional final product properties.

Under the conditions of cooking extrusion the raw material mixtures show a rather complex reaction behavior. Depending upon process conditions and ingredients used, specific mechanical and thermal energy is introduced which can lead to final products of a most diverse nature.

It has been observed that the properties of the wheat based raw materials affect the extrusion process and influence the quality attributes of the final product (Faubion et al 1982a, Faubion et al 1982b, Kim et al 1980). The required optimization of the process and products could in the past only be achieved through empirical changes of the formula and/or the process variables. It was shown to be extremely difficult to specifically influence the reaction behavior of the mass in the extruder through changes in formulation or process conditions in order to compensate for variations in the wheat flour quality.

A systematic investigation of both process and raw material specific interrelationships, would require an extremely large experimental effort. It was, therefore, recommended to utilize statistically proven methods for the design and analysis of extrusion experiments (van Lengerich 1984). Under the assumption that influencing variables can be differentiated from influenced parameters, statistical experimental designs reduce significantly the number of tests to be conducted without compromising the general validity of the results.

The aim of this study was to investigate the influence of raw material related properties of milled wheat products on their reaction behavior during extrusion cooking and the influence on final extrudate attributes. To accomplish this, it was necessary to characterize and identify wheat flour quality related properties which are directly related to their extrusion behavior. The overall relationships between extrusion parameters, raw material properties, energy input and extrudate attributes are described by means of a previously developed system analytical model. This will be outlined in detail for the extrusion of wheat starches.

MATERIALS AND METHODS

System Analytical Model

In previous work a system analytical model was introduced to describe the extrusion cooking process of starches and starch containing raw materials (van Lengerich 1984, Meuser et al 1982, Wiedmann et al 1984). The model (Figure 1) distinguishes among independent process parameters, system parameters and target parameters. The process parameters correspond to operational variables of the extruder and/or raw material variations. These parameters govern the specific thermal and mechanical energy input during extrusion. Both, thermal and mechanical energy affect the microscopic and molecular structure of the starch

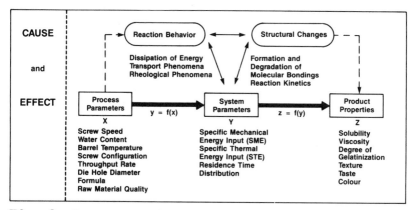

Fig. 1. System analytical model for the extrusion of starch containing food materials

which in turn, causes a change in the final product attributes.

The capability to introduce energy into a product during extrusion depends to a large extent upon the mass and heat transport phenomena in general, but specifically upon the rheological behavior of the plasticized mass in the extruder barrel. The influence of extrusion parameters on the rheological behavior of wheat starch within an extruder was already characterized and described in previous work (van Lengerich 1984).

The principal validity of the system analytical model was verified by carrying out extrusion experiments with wheat starch according to statistical experimental designs to first describe functional relationships between process variables and system parameters and secondly, between system parameters and final product attributes. It should be emphasized that statistical design and analysis of experiments can be used as a powerful tool to describe trends and changes in complex relationships. However, regression equations may not describe a relationship itself on a mechanistic basis.

Figure 2 shows as an example the effect of

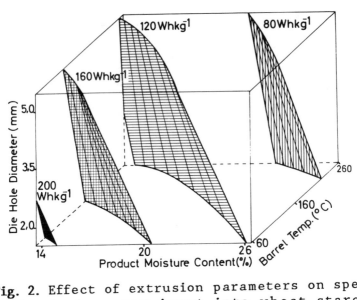

Fig. 2. Effect of extrusion parameters on specific mechanical energy input into wheat starch

moisture content, barrel temperature and die hole diameter on the specific mechanical energy input (SME) during wheat starch extrusion. The illustration is a result of a polynomial regression equation generated by using the measured data from extrusion tests which were conducted according to a fractional factorial design. The Figure shows that a reduction in the level of all variables causes an increase in the mechanical energy. The slight undulations of the isoplanes also show that the relationships were not linear but influenced by non-linear and interactive components. It also becomes obvious that practically an infinite number of parameter combinations of variable levels are possible to achieve a definite SME. On the other hand, each extrusion parameter could become a limiting factor if a definite minimal energy input would be required. Thus, it was not possible for example to exceed 200 Wh kg^{-1} in the experimental range when the moisture content of the wheat starch was above 15%. In this case, other parameters such as screw configuration or screw speed would have to

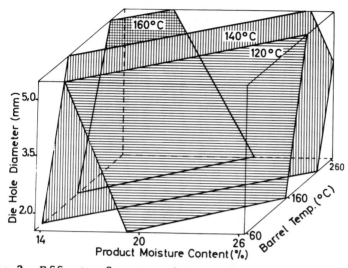

Fig.3. Effect of extrusion parameters on product temperature of wheat starch

be considered in order to attain an increase in mechanical energy.

It was noted on the basis of measured and calculated results, which described the effect of extrusion parameters on the product temperature (PT) that - as expected - high product temperatures could be reached through lowering the cross-sectional die area, the moisture content and/or increasing the barrel temperature (Figure 3).

The description of the relationships between extrusion parameters and energy input with statistical methods is related to the individual extruder used. However, if a laboratory or pilot plant analysis is conducted, the trends of the observed changes are similar in a production size extruder and the experimental effort for process or product optimization can be significantly reduced (Wiedmann et al 1984).

All further relationships described in the system analytical model are independent of extruder size and machine manufacturer. It is obvious that the key system variable to generate a specific product by extrusion is the energy history a product is

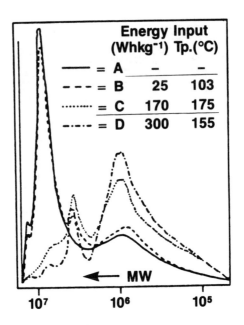

Fig. 4. Influence of energy introduction on the relative molecular weight distribution of extruded wheat starch

exposed to during the process. Primarily, the amount of specific mechanical energy causes a more or less significant change in the molecular structure of a given material as shown in Figure 4. The chromatogram of native wheat starch has two pronounced peaks which can be assigned to amylopectin and amylose.

The amylopectin is characterized by the larger peak, the amylose by the smaller peak. A very small energy input already showed a measurable change in the molecular weight distribution of the extruded starch. In addition, a new peak appeared between the two molecular sizes, which indicated the formation of a new intermediate fraction. If the energy input was increased, the results were a more severe decrease in the amylopectin portion, an enlargement of the middle peak as well as an enlargement of the amylose peak. Thus, it could be inferred from these results that primary valence bonds of starch are broken open in each case by the extrusion process. In particular, the amylopectin molecules were mostly affected by the molecular breakdown; they are split into smaller fragments, which clearly differ from one another in their size. However, the smaller molecular parts had the same size as the amylose molecules and thus caused an increase in the amylose peak in the chromatograms.

To further differentiate the influence of the forms of energy input on the molecular weight distribution, it was studied to what extent the molecular breakdown was influenced by the change of the reaction behavior of the starch. For that study, starches were extruded with different barrel temperatures and different product water contents (Samples C and D in Figure 4). Higher barrel temperatures lead to a lower structural viscosity of the plasticized mass in the extruder because of an increase in the mass temperature; the resulting decrease in the viscosity was indicated by a lower SME. The results shown here indicate that, in spite of higher mass temperatures, (Sample C in Figure 4), the starches extruded with less SME had less molecular breakdown and thus confirmed the assumption that under extrusion conditions, the amount of energy dissipation expressed as SME exerts the decisive influence on the breakdown of the primary valence bonds, whereas the mass temperature plays only a subordinate role in this case.

By plotting the average molecular weights versus SME, the reaction course of the molecular breakdown could be estimated (Figure 5). Up to an SME of 200 Wh/kg^{-1} the decrease in the average molecular weight can generally be described by the indicated exponential function. The course of the curve shows that in this region of energy input, the molecular breakdown due to a constant increase in SME depends on the remaining concentration of the average molecular weight. Above an SME of 200 Wh/kg^{-1} the fit of the determined curve is not as good, but the measurement point for the SME of 350 Wh/kg^{-1} indicates that a further molecular breakdown is no longer to be expected.

Breakdown of primary valency bonds leads to a

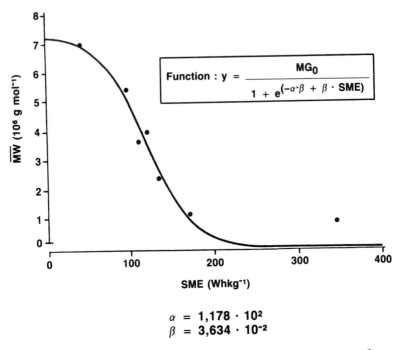

Fig. 5. Influence of SME on the medium molecular weight of extruded wheat starch

change in solubility and viscosity behavior. The values determined for these target dimensions were plotted versus the average molecular weight. It can be seen from Figure 6 that, as expected, the solubility of the extrudates increased with decreasing molecular weight and moved toward a final value which was about 70%.

The relationship between the molecular weight and the hydration behavior (Figure 7) showed that in the first phase of the molecular breakdown, which had been caused by a small energy input, (right side of the diagram in Figure 7), the swelling capacity and the structural viscosity increase at first and become drastically smaller with decreasing molecular size after passing through a maximum value. In the first section of the curve, the extruded wheat starches showed a greater hydration capacity with increasing energy input, although the solubility values increased at the same time. In addition,

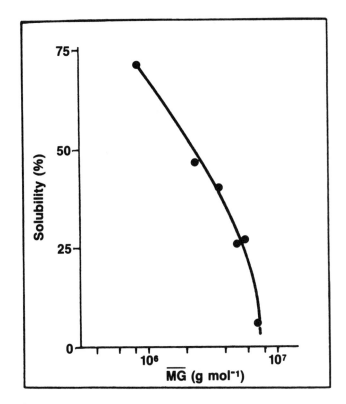

Fig. 6. Influence of molecular degradation on the solubility of extruded wheat starch

intact starch granules could be detected up to an SME of abut 50 Wh/kg^{-1}. This shows that, under the influence of a small energy input, wheat starch must be subject to an extremely non-homogeneous strain under extrusion conditions, and after the extrusion process, exists as a substance in which partially deformed starch granules, partially gelatinized starch and a starch mass that is structurally completely decomposed occur side by side. Only an increase in the thermal and mechanical energy input shifts this composition in favor of an increasing structural breakdown in the super molecular and molecular region and leads to a decrease in the swelling capacity and to a loss in viscosity.

Based on the results shown thus far, it was

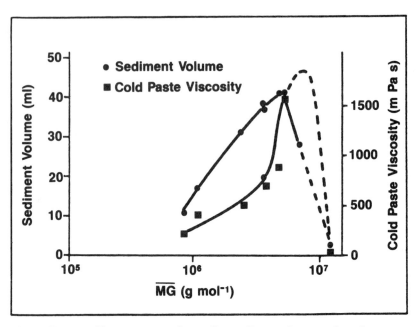

Fig. 7. Influence of molecular degradation on hydration behavior (sediment volume and cold paste viscosity) of extruded wheat starch

concluded that wheat starch structure can be influenced within wide limits by the input of thermal and mechanical energy. Characterization of the relationship between the structural and functional properties indicated that solubility, apparent viscosity of cold pastes as well as swelling behavior are closely related to the degree of cleavage of the primary and secondary valency bonds of starch and thus were used as suitable indicators to characterize structural changes in the molecular and super molecular region of the starch.

Thus, it should also be possible to derive a response relationship between the energy input into the starch, on the one hand, and the functional properties, on the other hand. This response relationship was derived by means of a second polynomial regression.

Figure 8 shows the influence of the energy input on the solubility of the extrudates. The regression equation predicts the solubility values in the experimental region studied for any random

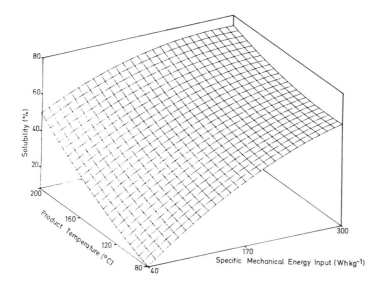

Fig. 8. Effect of energy input on solubility of extruded wheat starch

combination of the forms of energy input. This relationship was also independent of machine size and was valid for single and twin screw extruders (van Lengerich 1984).

It follows that increasing the product temperature and SME basically lead to a higher solubility. However, the higher the SME, the less is the influence of the product temperature. As expected, the highest solubility values can be achieved with a high input of thermal and mechanical energy at the same time; however, in the experimental region, maximum values of about 75% are not exceeded.

In this study the system analytical model will be used also to describe these relationships by extruding differently milled wheat flours of uniform quality as well as various wheat varieties uniformly milled.

RAW MATERIALS

For the first test phase of this study, a uniform batch of wheat (GDR, Harvest 1982) was milled into

TABLE I
Analytical Characterization of Wheat Flours for Test Phase I

CHARACTERIZATION		WHEAT FLOUR TYPE					
		405	550	812	1050	1600	1700 f
Water Content	(%)	13,3	13,6	12,7	12,4	11,8	14,3
Ash Content	(% DS)	0,46	0,57	0,89	1,11	1,59	1,77
Protein Content	(% DS)	11,8	13,1	14,9	15,9	16,8	14,3
Starch Content	(% DS)	80,9	79,3	76,0	71,4	67,5	64,7
Fat Content	(% DS)	0,9	1,2	1,9	2,4	3,4	1,6
Hagberg Falling Number	(s)	446	446	438	428	425	385

TABLE II
Variations of Wheat Flour Qualities Used in Test Phase II

CHARACTERIZATION			LEVEL		
		#	LOW	HIGH	MEDIUM
Starch Content	(% DS)	a	74,1	76,9	75,5
Viscosity W/O HgCl$_2$	(BU)	b	120	1650	885
Viscosity with HgCl$_2$	(BU)	c	1880	2700	2290
Hagberg Falling Number	(s)	d	172	462	317
Corrected Water Binding Capacity	(% DS)	e	70,3	87,4	78,8
Soluble Fraction of Flour	(% DS)	f	2,8	4,0	3,4
Protein Content	(% DS)	g	11,3	13,7	12,5
Soluble Protein	(% DS)	h	1,3	2,3	1,8
Soluble Starch	(% DS)	i	1,0	2,0	1,5
Soluble Pentosanes	(% DS)	k	0,3	0,6	0,45
Fat Content	(% DS)	l	1,0	1,4	1,2
Kernel Hardness	(kNm)	m	14,1	23,8	19,0
Milling Time	(s)	n	10,8	21,9	16,4
Milling Energy	(Wh)	o	57,4	93,6	75,5
Particle Size Distribution					
>125 µm	(%)	p	9,9	18,7	14,3
>100 µm	(%)	r	11,7	22,5	17,4
> 80 µm	(%)	s	10,6	45,8	27,9
< 80 µm	(%)	t	24,0	73,5	48,8

flours with various ash-contents according to the standard flour types in West Germany (Table I). For the second test phase, various wheats, which differed in their chemical composition and quality characteristics, were milled under constant conditions into a medium light flour (type 550 = 0.55% ash) with a Buhler pilot-size mill. Analytical results are shown in Table II.

Analytical Characterization of the Milled Wheat Products and Extrudates

The chemical composition of the milled wheat products were analyzed according to standardized methods (ICC 1982, Richter et al 1968). Specifically, water (ICC-Method 10), ash (ICC-Method 104), protein, fat and starch (DIN 10300) were determined. The amylolytic activity was characterized using the Hagberg Falling Number (ICC-Method 107). The flours produced from wheat of various qualities were additionally characterized by determining the amount of soluble carbohydrates, proteins and pentosans (Abdel-Gawad 1982).

The hardness of the kernels and the particle size distribution of the flours were determined by previously described methods (Meuser 1987). Viscograms were recorded with and without 0.1% $HgCl_2$ to determine the maximum viscosity and temperature of gelatinization with and without the effect of active flour enzymes.

The extrudates were analyzed after equilibrating for 18 hours at 40°C and ground with a high speed hammer mill. The apparent viscosities of the cold pastes were determined using a Haake Rotovisco RV3 at shear rates of 150 s^{-1} and concentrations of 10% at 20°C. Solubility was measured by stirring a 1% extrudate-water suspension for 30 minutes at 20%, filtering (S+S, 595, 110 mm), washing (50 ml distilled water) and drying of the residue at 130°C for 90 minutes. Radial Expansion Index was determined by using the ratio between extrudate diameter and die diameter (n=10). Total expansion was determined from the specific weight of the extrudates using seed displacement method.

Extrusion Tests

The extrusion tests (Pfaller 1987) were conducted using a co-rotating twin screw extruder (Werner & Pfleiderer, Continua 37, l/d=12). The experimental plan was designed as a fractionated factorial Box-Wilson design. The relationships between extrusion parameters, energy input and final product properties were calculated using second order polynomial regression equations. (Minitab Reference Manual 1982).

In Test Phase I, six different flours (Table 1) which were milled from uniform batches of wheat were processed under various extrusion conditions. Each extrusion parameter was varied at 5 levels according to Table 3. In Test Phase II, extrusion tests were conducted with flours from 12 different wheat varieties, each milled under constant conditions. The extrusion parameters were varied at 3 levels according to Table IV. Test Phase II enabled us to detect not only the influence of extruded parameters but also the influence of raw material quality on reaction behavior and final product attributes. The die opening diameter was held constant (2x4 mm) and the screw contained 3 consecutive 90° staggered reverse pitch elements in the compression zone.

The flours were gravimetrically fed via a twin screw loss-in-weight feeder (K-TRON LWF 2-20). Total product water content was adjusted through water injection at a distance of 10 l/d from exit.

TABLE III
Experimental Range of Extrusion Parameters for Test Phase I
(Various Wheat Flours of Uniform Quality)

EXTRUSION PARAMETERS		LEVEL				
		-2	-1	0	+1	+2
M = Throughput Rate	(kg h^{-1})	10,3	12,5	14,7	16,8	19,1
N = Screw Speed	(min^{-1})	140	180	220	260	300
D = Die Hole Diameter	(mm)	2	3	4	5	6
W = Product Water Content	(%)	15,0	17,5	20,0	22,5	25,0
T = Barrel Temperature	(°C)	80	120	160	200	240
S = Screw Configuration	(*)	1	2	3	4	5

* Number of reverse pitch elements in the compression zone.

TABLE IV
Experimental Range of Extrusion Parameters for
Test Phase II (Uniform Wheat Flours
Varying in Qualities)

EXTRUSION PARAMETERS		LEVEL		
		-1	0	+1
M = Throughput Rate	(kg h⁻¹)	9,0	13,5	18,0
N = Screw Speed	(min⁻¹)	80	180	280
W = Product Water Content	(%)	15,0	20,0	25,0
T = Barrel Temperature	(°C)	100	160	220

The barrel temperature was adjusted through two individual thermo oil heaters. Product temperature was measured at the die head immediately before product exit. The specific mechanical energy introduction (SME) was determined using the following formula:

$$\text{SME (Wh/Kg)} = \frac{\text{Torque (Nm) x Angular velocity (s}^{-1})}{\text{Product Mass Flow (kg/h)}}$$

RESULTS AND DISCUSSION

Influence of Extrusion Parameters and Raw Material Quality Characteristics on Energy Introduction
The relationship between investigated extrusion parameters and the specific mechanical and thermal energy introduction was characterized by second order polynomial regression equations generated from the observed and measured data. Interactions between the parameters were neglected because of their small statistical significance. Variations in flour type were considered in the equation and the

degree of milling (F) was related to the following values:

T.405 = -2
T.550 = -1.538
T.812 = -0.732
T.1050 = 0
T.1600 = 1.692
T.1700 = 2

The remaining variables and levels are adjusted according to Table III. The resulting equations which express the influence of the extrusion parameters on the energy input are as follows:

PT (°C) = 162.12 + 1.72N - 4.72D - 5.5W + 15.5T

+ 2.51S - 0.51F

+ 0.55D^2 + 1.24W^2 + 0.84S^2

r^2 = 0.98 , P > 99.9%

SME(Wh/kg) = 81.0 - 3.49M + 5.42N - 9.42D - 13.06W

- 25.94T + 11.85S - 3.6F

+ 3.38D^2 + 2.43W^2 + 2.6T^2 + 1.6F^2

r^2 = 0.95 , P > 99.9%

The equations show that under extrusion conditions where all parameters are adjusted at their 0-level, all terms (except for the constant value) become 0 and a SME of 81 Wh/kg and a PT of 162.1°C results. This is the center point of the experimental design. If, for example, the screw speed (N) is increased from the 0 to the +1 level, the resulting SME is 81.0 + 5.42 = 86.42 Wh/kg, and the PT raises to 162.1 + 1.7 = 163.8°C. In this case, no quadratic effects influence the relationship and a raise in screw speed at first changes the mechanical energy, and the resulting frictional heat then raises the product temperature.

If the product water content (W) is decreased from its middle point to the -1 level, and all remaining parameters are at their 0-level, the resulting SME is:

$$81.0 - (-13.06) + 2.43 = 96.5 \text{ Wh/kg}$$

and the product temperature is:

$$162.12 - (-5.5) + 1.24 = 168.9°C$$

This example shows that the quadratic effects of the change in water content increase the linear effect of the water content on SME and PT.

Decrease of the water content increases the shear stress of the plasticized mass in the extruder, which results in a higher torque and consequently increases the SME. The increase from 81.0 to 96.5 Wh/kg also causes a simultaneous increase of the frictional heat which raises the product temperature from 162.12 to 168.9°C. The equations can be used to determine the SME and PT of each theoretical point within the entire test range. They also can be used to estimate the set point and combination of extrusion parameters, if a definite combination of SME and PT is required.

For a better illustration of the effect of all extrusion parameters on SME and PT, the equations were used here to generate isoplanes for SME and PT values within the experimental range. These are shown in Figures 9 and 10.

Figure 9 shows the influence of die hole diameter, water content and barrel temperature on the specific mechanical energy introduction (SME) into wheat flour (T.550). The illustration represents the test range investigated at a constant screw speed and a constant product mass flow. The location and slope of the isoplanes show that similar to wheat starch, the largest SME values always result (A) when the product is extruded at low water contents (17.5%), low barrel temperatures (120°C) and through small dies (3 mm). The SME

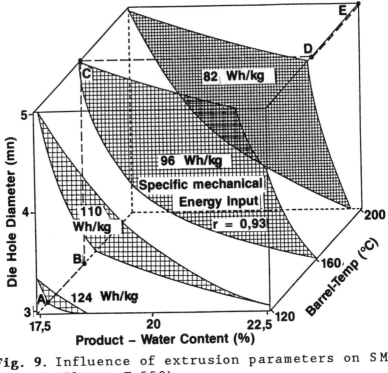

Fig. 9. Influence of extrusion parameters on SME (Wheat Flour, T.550)

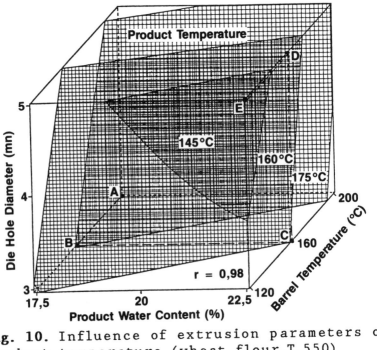

Fig. 10. Influence of extrusion parameters on product temperature (wheat flour, T.550)

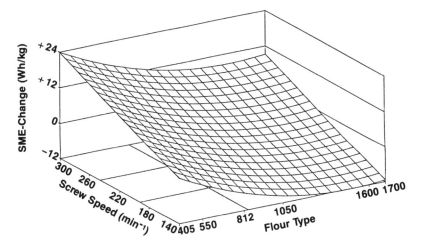

Fig. 11. Influence of screw speed and flour type on SME-change (Wheat Flour)

decreased from 125 to 113 Wh/kg when only the barrel temperature was raised from 120°C to 160°C (B). An increase of die hole diameter from 3 to 5 mm resulted in a SME-decrease from 113 Wh/kg to 96 Wh/kg (C). A following increase of the product moisture content by 5% decreased the SME further from 96 Wh/kg to 76 Wh/kg (D). Lowest SME values were achieved, when finally the barrel temperature was increased to 200°C (E).

The influence of the remaining extrusion parameters (screw speed and degree of milling) on the SME-change is shown in Figure 11.

These tests were conducted at the medium level of product water content (20%), die opening diameter (4 mm) and barrel-temperature (160°C). The results show that an increase in the degree of milling of the flours consistently decreased SME. The slightly bent slope of the planes shows that the change in SME during extrusion of the wheat products is characterized through a non-linear relationship between the degree of milling and the SME-change. SME decreased by 15 Wh/kg when the degree of milling changed from its lowest to its highest value at a constant screw speed of 300 RPM. This change in SME could be compensated through adjustment of other extrusion parameters. It was possible, for example, to increase SME back to 85 Wh/kg by using a different wheat flour, type 1700 instead of type 405, and by an increase of the screw speed from 160 to 270 RPM.

The influence of raw material quality characteristics and extrusion parameters on SME was also determined and is shown in Figure 12. SME was determined at the test condition of the middle point of each experimental design conducted for all investigated wheat flours of different qualities. Under those extrusion conditions, no significant SME-change takes place as a result of raw material quality variations. This indicates that the raw material quality parameters, in general, do not

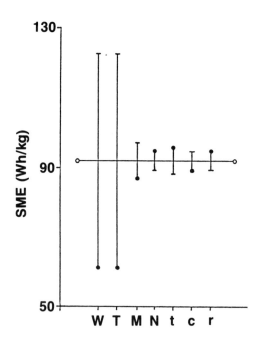

W = Product Moisture Content (%)
T = Barrel Temperature (°C)
M = Mass Flow (kg/h)
N = Screw Speed (min⁻¹)
t = Particle Size < 80 μm (%)
C = Viscosity with $HgCL_2$ (BU)
r = Partical Size > 100 μm (%)

Fig. 12. Influence of extrusion parameters and raw material quality characteristics on SME

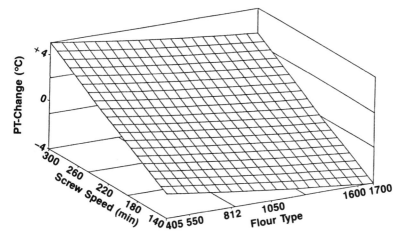

Fig. 13. Influence of screw speed and flour type on PT-change

affect significantly the specific mechanical energy introduction. The largest influence on SME was caused in these tests through changes in product water content, barrel temperature, and product mass flow. An increase of these extrusion variables to their upper level decreased SME, whereas, an increase of the screw speed increased SME.

Figure 10 shows the influence of die opening diameter, product water content, and barrel temperature on the product temperature during the extrusion of wheat flour, type 550.

The isotherms indicate that an increase in product water content and die opening diameter resulted in a decrease in the product temperature.

An increase in the barrel temperature resulted, as expected, in an increase in the product temperature. The highest thermal energy introduction was consistently achieved when the product moisture was adjusted to 17.5%, the die diameter to 3mm, and the barrel temperature to 200°C. Under these conditions the product temperature was 200°C (A). The product temperature decreased to 175°C when the barrel temperature was decreased from 200°C to 160°C at constant product water content and die diameter (A-B). An increase of the product water content from 17.5 to 22.5% at a barrel temperature of 160°C

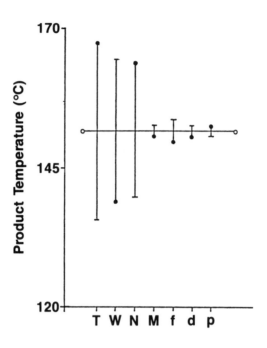

W = Product Moisture Content (%) f = Soluble Part of Flour (%)
T = Barrel Temperature (°C) d = HFN (s)
M = Mass Flow (kg/h) p = Partical Size >125 μm (%)
N = Screw Speed (min^{-1})

Fig. 14. Influence of extrusion parameters and raw material characteristics on product temperature.

and a die diameter of 3 mm resulted in a further reduction of the product temperature from 175°C to 160°C (B-C). The product temperature was reduced to 150°C when the die diameter was increased from 3 to 5 mm at a product water content of 22.5% and a barrel temperature of 160°C (C-D). A further decrease of the barrel temperature from 160°C to 120°C resulted in a product temperature of 135°C (D-E).

The influence of the degree of milling and the screw speed on product temperature during wheat flour extrusion is shown in Figure 13. The result can be interpreted to show that all values in Figure 10 can be corrected by the value of the temperature change, which results from a change in screw speed and flour type. The shape of the plane shows that a

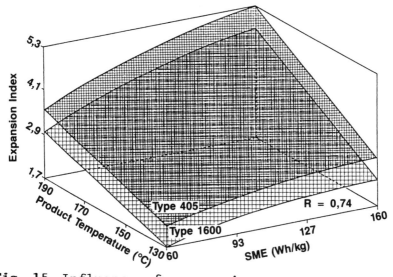

Fig. 15. Influence of energy input on expansion index of extruded wheat flours

screw speed change has a much larger influence on the product temperature than a change in flour type. For example, the product temperature was decreased only by 2°C when a flour of type 1700 was extruded instead of a flour of type 405. The same change in product temperature could be achieved through lowering the screw speed by only 46 RPM.

Figure 14 shows that the quality characteristics of different wheat flours do not cause a significant change in the resulting product temperature during extrusion. Moreover, only the extrusion parameters such as barrel temperature, water content and screw speed cause a significant product temperature change. The influence of raw materials specific quality parameters on the product temperature was very small and statistically not significant. This is shown by the patterns in Figure 14 which indicate the direction and the magnitude of the product temperature change in the experimental test range when the parameters change from the -1 to the +1 level in the experimental design. The influence of a simultaneous change of several process parameters on the product temperature can be calculated through addition of the patterns in Figure 14.

Influence of Energy Introduction and Raw Material Quality Characteristics on Functional Properties of the Extrudates

The results generated thus far show that raw materials' specific quality parameters in most cases exert a relatively small effect on the energy input. The influence on SME and product temperature was superimposed by the influence of the extrusion parameters. Only the degree of milling (ash content - starch content) of the flours significantly influenced the SME; the product temperature changed little when various flours were used. The other extrusion parameters, however, caused significant changes in the thermal and mechanical energy input, which in turn influenced the functional properties of the extrudates.

Influence of Energy Introduction and Raw Material Quality Characteristics on the Expansion Behavior of the Extrudates

Figure 15 shows the influence of the energy input and the degree of milling of wheat flour on the radial expansion index (EI) of the extrudates.

The Figure shows as an example the influence of SME and product temperature on the EI of two extruded flours (Type 405 and 1600). The results show that an expansion index of 5.3 was reached when a white wheat flour was extruded at an SME of 160 Wh/kg and a product temperature of 190°C. Radial expansion decreased with increase in flour extraction. The lowest value of the expansion index resulted when wheat flour was extruded at lowest product temperature, lowest SME, and high flour extraction. An increase of SME consistently increased the expansion. The same was observed for the change in the product temperature.

The influence of extrusion parameters and raw material quality characteristics on the expansion index of wheat flour is shown in Figure 16. The expansion index of the extrudates decreased with decrease in barrel temperature and product water content. An increase in screw speed, however,

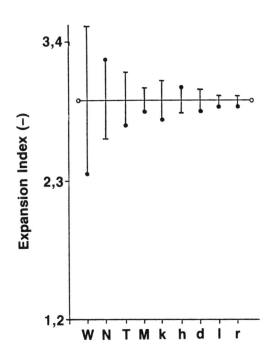

W = Product Moisture Content (%)
T = Barrel Temperature (°C)
M = Mass Flow (kg/h)
N = Screw Speed (min⁻¹)
k = Soluble Pentosanes (% DS)
h = Soluble Proteins (% DS)
d = HFN (s)
l = Fat Content (% DS)
r = Partical Size > 100 μm (%)

Fig. 16. Influence of extrusion parameters and raw material characteristics on expansion index

increased the radial expansion. The amounts of soluble pentosans and soluble proteins influence expansion. Those effects, however, were of small significance. The radial expansion index of the wheat flour extrudates was highly inversely correlated with the specific weight of the extrudates.

The specific weight can be interpreted as a target parameter to describe total expansion of a product considering both radial and longitudinal expansion. Figure 17 shows the influence of SME and product temperature on the specific weight. Higher energy introduction consistently lowered specific weight and the effect of an increase in product

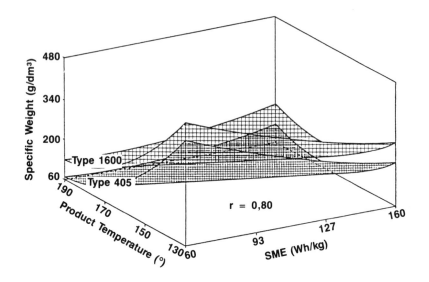

Fig. 17. Influence of energy input on specific weight of extruded wheat flours

temperature was stronger than in SME. Also, an increased amount of starchy endosperm particles in the flour (lower degree of milling) increased expansion values. The high bran content in flour probably decreases expansion by two effects. First, it has a thinning effect with respect to the amount of starch. Bran itself is considered as an expansion inert substance. With higher amounts of bran, the reduced amount of starch decreases expansion. In addition, bran particles may cause a mechanical disruption of growing air bubbles and lower expansion and specific weight.

Figure 18 shows how specific weights of extrudates are affected by extrusion parameters and wheat flour quality characteristics. The most significant effect on total expansion results from the product moisture content. A decrease in moisture content by 1% raises the expansion significantly and causes a decrease in specific weight of 24 g/dm^3. This strong influence is mainly due to a primary effect of the product moisture on SME and the rheological behavior of the

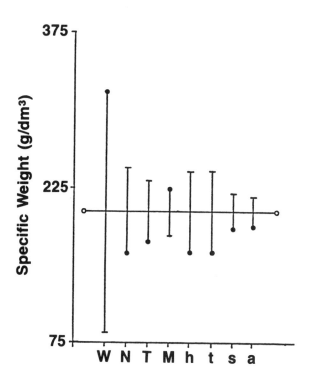

W = Product Moisture Content (%)
T = Barrel Temperature (°C)
M = Mass Flow (kg/h)
N = Screw Speed (min⁻¹)
h = Sol. Protein (% DS)
t = Partical Size <80 μm (%)
s = Partical Size <100 μm (%)
a = Starch Content (% DS)

Fig. 18. Influence of extrusion parameters and raw material characteristics on specific weight

plasticized mass in the extruder. Lowering the product moisture content resulted in a higher SME and in a higher shear stress of the mass at the end of the screw. The higher shear stress is indicated by an increase in pressure drop (van Lengerich 1984), which affects final expansion. Screw speed, barrel temperature, and product mass flow also affected the total expansion via the mechanical and thermal energy input.

A statistically significant correlation was determined between the amount of soluble proteins and the extent of total expansion. Higher levels of soluble proteins increased expansion. On an average, the specific weight was lowered by 40 g/dm^3 when the amount of soluble protein increased by 0.5% DS. The effect of soluble proteins on expansion might be due to an influence on the gas holding capacity of the plasticized mass during bubble growth. Another important factor affecting the total expansion of wheat flour is particle size and particle size distribution. The pattern for particle sizes (s and t) indicates the following effect: specific weight decreases on an average by 12.5 g/dm^3 when the amount of particles which are lower than 100 microns increases by 5%. In other words, the more fine particles the flour contains, the more intense the expansion will be. A similar tendency was found for the starch content of the flours: increase of starch by 1% decreases specific weight by 11 g/dm^3.

Influence of the Energy Input and Raw Material Quality Characteristics on Cold Water Solubility of Wheat Flour Extrudates

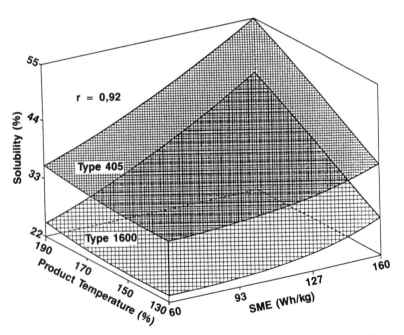

Fig. 19. Influence of energy input on solubility of extruded wheat flours

Cold water solubility and cold paste viscosity are important product properties with respect to the application of ground extrudates as half-products or intermediate-products. Cold water solubility, for instance, is used to characterize the chewing behavior of direct expanded products as well as for determination of the functional quality of intermediate-products such as pregelatinized flours. The hydration behavior of extruded starch-containing raw materials is usually determined through measurement of the apparent viscosity of cold pastes. The cold paste viscosity further relates to the amount of mechanical and thermal energy introduced into the mass during extrusion and indirectly describes

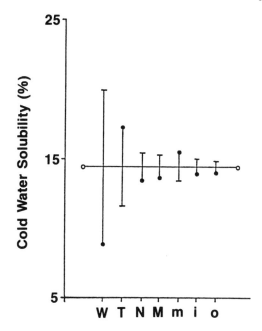

W = Product Moisture Content (%)　　m = Kearnel Hardness (k Nm)
T = Barrel Temperature (°C)　　　　　i = Soluble Starch (% DS)
M = Mass Flow (kg/h)　　　　　　　　o = Milling Energy (Wh)
N = Screw Speed (min⁻¹)

Fig. 20. Influence of extrusion parameters and raw material quality on cold water solubility of extrudates from wheat flour

changes in the molecular structure of the starch as described earlier. It was, therefore, of importance to investigate the influence of energy input and degree of milling of wheat flours on cold water solubility and cold paste viscosity of the extrudates. Figure 19 shows that cold water solubility of the extrudates decreased when degree of milling increased. A higher solubility was achieved in the experimental test range when the mechanical energy and the product temperatures were at their highest levels. Under these conditions, a cold water solubility of 55% was achieved when wheat flour of Type 405 was extruded. The uneven slope of the planes also shows an interaction between product temperature and mechanical energy input with respect to the influence on solubility: at low SME values, a distinct increase of product temperature caused a small increase of cold water solubility; at high SME values, the same product temperature change caused a significantly larger increase of cold water solubility. On the other hand, a definite increase of SME at low product temperatures caused a smaller increase in cold water solubility than at high product temperatures.

Under the extrusion conditions of the middle point test in the experimental design, variations in wheat flour quality did not significantly change the cold water solubility of the extrudates (Figure 20). The only significant influence was exerted by a change in product water content, which primarily changes SME, and therefore the product temperature. The resulting effect is a change in molecular breakdown of starch, which in turn, affects cold water solubility. Solubility of wheat flour extrudates increased from 14.4% to 20% when product water content decreased from 25% to 15%. Barrel temperature had a relatively small effect on cold water solubility: increase in barrel temperature by 60°C increased extrudate solubility by 2.8%.

Influence of Energy Input and Raw Material Quality Characteristics on Cold Paste Viscosity of Wheat Flour Extrudates

Structural changes on a molecular level, i.e.,

cleavage of primary and secondary bonds can be monitored through viscosity of the cold pastes of the extrudates.

Figure 21 shows that cold paste viscosity of the extrudates from wheat flours of high degree of milling (T.1600) is consistently lower than cold paste viscosity of extrudates from wheat flour of low degree of milling (T.405). Under the conditions of the center point test of the experimental design, cold paste viscosity of the wheat flour extrudates was 460 m Pas. A change of screw speed and product mass flow caused wide changes in cold paste viscosity: increase of the screw speed from 180 to 280 RPM increased cold paste viscosity from 460 to 665 m Pas. The same increase was observed by decreasing the product mass flow from 15 to 10 kg/hr. Increasing the product water content from 20% to 25% changed cold paste viscosity from 460 to

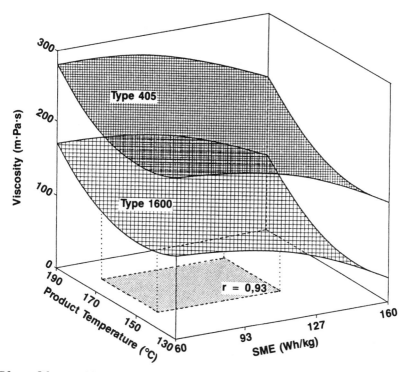

Fig. 21. Influence of energy input on cold paste viscosity from extruded wheat flours

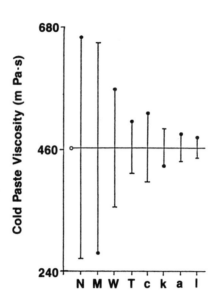

W = Product Moisture Content (%) c = Viscosity with Hg Cl_2 addition (BU)
T = Barrel Temperature (°C) k = Sol. Pentosanes (% DS)
M = Mass Flow (kg/h) a = Starch Content (% DS)
N = Screw Speed (min^{-1}) l = Fat Content (% DS)

Fig. 22. Influence of extrusion parameters and raw material quality on cold paste viscosity of extrudates from wheat flour

568 m Pas (Figure 22).

Those regression calculations showed that cold paste viscosity of the wheat flour extrudates is also affected by the gel-visosity of the wheat flour before extrusion (measured in the presence of added $HgCl_2$), the amount of water soluble pentosans, and starch content of the wheat flours. The largest change in cold paste viscosity of extrudates was caused by the wheat flour gel-viscosity (with $HgCl_2$ addition). Higher values for this raw material parameter led to a significant increase in the cold paste viscosity. Starch, the largest component of a flour, is primarily responsible for the development of cold paste viscosity. The figure shows, that even under the conditions of uniformly milled flours, a variation in starch content affects directly extrudate viscosity. A 1% increase in starch flour increases cold paste viscosity by 18 m Pas.

SUMMARY

Wheat of uniform quality was milled to six flours with various ash contents. Each flour was extruded and extrusion parameters were systematically varied according to a statistical experimental design. Mechanical and thermal energy input during extrusion and resulting final product attributes, such as specific weight, radial expansion, cold paste viscosity, and cold water solubility were measured. The relationships between extrusion conditions, energy input and final product attributes were characterized using polynomial regression analysis. The input of mechanical energy under similar extrusion conditions decreased by 14 Wh/kg if whole wheat flour (T.1700) was used, compared to a white flour (T.405). This effect caused a slight decrease in frictional heat which in turn decreased the product temperature at the die by 2°C. When the energy input was adjusted to a constant level, the extrudates of flours with higher ash contents were more dense, had less cold paste viscosity and a lower water solubility.

The influence of wheat quality on the extrusion behavior and final product attributes was investigated using 12 different wheat varieties. Each of the wheats was milled to a standard flour (T.550). The amount of water soluble proteins in the wheat flours influenced density and hydration behavior of final extrudates, whereas the starch content of the wheat flours directly influenced cold paste viscosity.

LITERATURE CITED

ABDEL-GAWAD, A.S. 1982. Isolierung und Charakterisierung von Pentosanfraktionen aus verschiedenen Weizensorten. Disseration, D 83, FB 13, Nr. 132, TU Berlin.

BEN-GERA, I. 1982. Markt und Nährwert von Extruderprodukten in den USA. Getreide Mehl und Brot 36: 248-251.

FAUBION, J.M. and HOSENEY, R.C. 1982a. High-temperature-short-time extrusion cooking of wheat starch and flour. I. Effect of moisture and flour type on extrudate properties. Cereal Chem. 59: 529-533.

FAUBION, J. M. and HOSENEY, R.C. 1982b. High-temperature-short-time extrusion cooking of wheat starch and flour. II. Effect of protein and lipid on extrudate properties. Cereal Chem. 59: 533-537.

FAUBION, J. M., HOSENEY, R. C. and SEIB, P. A. 1982. Functionality of grain components in extrusion. Cereal Foods World. 27: 212-216.

INTERNATIONALE GESELLSCHAFT FUER GETREIDECHEMIE (ICC): Standard-Methoden.- Detmold: Schäfer (1982).

KIM, J.C. and ROTTIER, W. 1980. Modification of aestivum wheat semolina by extrusion. Cereal Foods World 25: 62-66.

MEUSER, 1987. Untersuchung der Wirkung charakteristischer Eigenschaften von Getreidemahlerzeugnissen auf deren Kochextrusionsverhalten und die resultierenden Endprodukteigenschaften. AIF-Forschungsvorhaben Nr. 5989, Technische Universität Berlin, Berlin Institut für Lebensmitteltechnologie, Getreidetechnologie.

MEUSER F., VAN LENGERICH, B. and KÖHLER, F. 1982. Einfluß der Extrusionsparameter auf funktionelle Eigenschaften von Weizenstärke. Starch/Stärke. 34: 366-372.

MINITAB REFERENCE MANUAL, 1982 Edition. Minitab Project, Statistics Department, The Pennsylvania State University Pa 16802, USA.

PFALLER, W. 1987. Einfluß der Zusammensetzung von Getreidemahlerzeugnissen und ausgewählten Rezepturen auf das Kochextrusionverhalten und die Extrudateigenschaften. Disseration, D83, FB 13, Nr. 220, TU Berlin.

RICHTER, M., AUGUSTAT, S. and SCHIERBAUM, F. 1968. Ausgewählte Methoden der Stärkechemie. Wissenschaftliche Verlagsanstalt mbH, Stuttgart.

SAS/GRAPH USER'S GUIDE: Graph, 1981 Edition. SAS Institute Inc. Gary, North Carolina 27511, USA.

SCALES, H. 1982. The U.S. snack food market. Cereal Foods World 27: 203-205.

VAN LENGERICH, B. 1984. Entwicklung und Anwendung eines rechnerunterstützten systemanalytischen Modells zur Extrusion von Stärken und stärkehaltigen Rohstoffen. Dissertation D83, FB 13, Nr.165 TU Berlin.

WIEDMANN, W. M., VAN LENGERICH, B., HIRSCH, A.R. and PIEL, A. 1984. Enfluß von Extrusionsparametern auf die physikalischen Eigenschaften kochextrudierter Mehle mit unterschiedlichen Ausmahlungsgraden. 35 Tagung für Getreidechemie, Detmold.

24

PROCESSING AND UTILIZATION OF AIR-CLASSIFIED WHEAT FLOUR FRACTIONS

Frank W. Sosulski

Department of Crop Science and Plant Ecology
University of Saskatchewan
Saskatoon, SK S7N 0W0

Darek M. Nowakowski

Multifoods Inc.
Scarborough, ON M1P 2Z7

INTRODUCTION

The basic processes of flour milling involve grinding, grading, purifying and sifting on the rollermill, purifier and plansifter (Panter and Brennan, 1982). Modern mills are substantially more efficient and reliable, with greater throughput, than mills developed in the late nineteenth century, but the technological developments have remained consistent with traditional processes.

AIR CLASSIFICATION - EARLY STUDIES

The discovery by millers that the finest flour particles contain more protein has led to a novel milling process that involves separation of flours into subsieve size classes. Jones et al. (1959) air classified English wheat flours into segments of endosperm cells, detached starch granules with some adhering protein and small wedges of protein in association with small starch granules. The intermediate starchy fraction was substantially depleted in protein content while the finest fraction had twice the protein content in the parent flour. Gracza (1959) separated a soft red winter (SRW) wheat flour at 7.8% protein into seven progressively-larger cut sizes on an air classifier. The fraction with the smallest particles contained 22.2% protein and the intermediate-sized fraction had only 2.5-3.3% protein (Table 1). Other indices such as ash, fat, diastatic activity, MacMichael viscosity and water absorption increased with protein content while bulk density increased with particle size. The degree of protein shift into the fine fraction was calculated to be 27.7% as compared to only 5.4% for a hard red spring (HRS) wheat flour when air classified by the same procedure (Gracza 1960).

Bode et al. (1964) separated SRW and hard red winter (HRW) wheat flours at four consecutively-wider settings of the air classifier. The positive protein shift indices, in percent, were calculated by the Gracza (1959) formula:

TABLE I
Characteristic Indices of Soft Wheat Flour and Its Seven Subsieve-Size Fractions Obtained By Air Classification at Increasing Critical Cut Size, 14 mb[a]

Cut Size (μm)	MPS[b] (μm)	Yield (%)	Protein (%)	Ash (%)	Fat (%)
Parent	12	–	7.8	0.3	0.7
< 12	4	5	22.2	0.4	1.8
12-16	4	7	20.0	0.4	1.6
16-19	8	10	10.0	0.3	0.9
19-23	14	22	3.1	0.3	0.4
24-30	16	15	2.5	0.3	0.3
31-36	17	13	3.3	0.3	0.4
> 37	23	28	8.6	0.3	0.8

[a] Source: Gracza (1959); used by permission.
[b] MPS = mean particle size.

$$\frac{\text{cut protein \% − parent protein \% × cut yield \%}}{\text{parent protein \%}}$$

and found to be 26.7 and 5.2%, respectively, for the soft and hard wheats. The high protein fraction from the SRW flour improved loaf volumes of a control flour to a greater degree than that of the HRW flour.

Regrinding the roller-milled flours on impact mills, such as the pinned disc grinder, was effective in disintegrating the starch-protein clusters and separating the starch granules from adhering protein (Jones et al. 1959). Particle size reduction into the fine and medium range (0-35 μm) was only partially successful in hard wheat flours but regrinding yielded 68% of medium and 20% of fine fraction for a soft wheat flour (Table II). The protein shift into the fine fraction was 25.3%. Wichser (1958) and Kent (1965) obtained similar results with soft and hard wheat flours.

Therefore, the technique of impact milling and air classification is used primarily in areas where only soft wheats are available, especially where mills specialize in the production of flours for the pastry and confectionery trade. At 5% residual protein content, the starchy intermediate fraction has not been a suitable substitute for refined wheat starch in industrial applications.

TABLE II
Yields and Protein Contents of Air-Classified Fractions of Soft Wheat Before and After Pinned Disc Grinding, 14% mb[a]

Flour fraction	Unground		Ground	
	Yield (%)	Protein (%)	Yield (%)	Protein (%)
Initial flour	100	8.3	100	8.3
Fine	9	15.9	20	18.8
Intermediate	34	3.9	68	5.7
Coarse	57	9.7	12	8.9

[a]Source: Jones et al. (1959); used by permission.

STARCH DAMAGE

Rigorous impact milling of hard wheat to achieve greater particle size reduction, and improve yields of HPF, has resulted in substantial breakage of starch granules themselves (Jones, 1940; Kent, 1966). Excessive starch damage increases maltose number and has deleterious effects on baking quality. Also, fractured starch granules would pass into the fine fraction during air classification and reduce its protein content.

However, a certain degree of starch damage is necessary for adequate water absorption and gas production in breadmaking processes. Kent (1966) reported starch damage values in hard wheat flours of less than 10%. But modern milling systems based on high capacity and a reduction in the number of grinding passages has brought about a general decrease in the damage to starch granules during the milling process (Panter and Brennan, 1982).

Tipples and Kilborn (1968) found that pin milling of HRS wheat flours increased starch damage, baking absorption and bread yield in short-time breadmaking systems. There was little advantage in pin milling the HRS flours intended for the long bulk fermentation baking systems, but the reduction in particle size had no adverse effects on these bread characteristics. Similarly, Dexter et al. (1985) milled five flours to four starch damage levels and found they had no adverse effects on pan breads prepared by the remix procedure. However, hearth bread quality deteriorated as starch damage increased due to the specific flour water absorption requirements of the formulation.

The trends to mechanical high speed breadmaking processes in large commercial bakeries and short-time chemical dough development in small bakeries has led to a preference for flour with consistently higher levels of starch damage. Thus, the tendency has been to mill to higher starch damage by adjusting the grinding pressures at certain stages of the milling process. Nowakowski et al. (1986) found 16% starch damage in a commercial first middling (1M) flour (Table III). The fifth break (5B) flour from the

TABLE III
Distribution of Flour Particle Size for Roller-,
Pin- and Attrition-Milled 5B and 1M Flours, % of
Flour Volume[a]

Particle Diameter (μm)	Roller Mill 5B (%)	Roller Mill 1M (%)	Pin Mill(2X) 5B (%)	Pin Mill(2X) 1M (%)	Attrition Mill 5B (%)	Attrition Mill 1M (%)
0- 15	0	0	21	17	28	39
16- 30	3	3	52	56	62	56
31- 45	7	4	22	22	10	5
46- 60	12	5	2	5	0	0
61- 75	17	8	3	0	0	0
76- 90	17	11	0	0	0	0
91-105	14	11	0	0	0	0
105	30	58	0	0	0	0
DR[b]	10	7	95	95	100	100
SD[b]	11	16	20	31	18	27

Sources: Nowakowski et al. (1986) and Sosulski et al. (1988); used by permission.
DR = degree of reduction, SD = starch damage.

same mill had only 11% starch damage due to the less intensive grinding action applied to this low quality flour.

Nowakowski et al. (1986) observed that pin milling caused considerable shattering and cracking of starch grains with radial fissures appearing in many of the larger granules. Small starch granules appeared resistant to damage. SEM-micrographs of attrition-milled flours showed a different pattern of starch damage (Fig. 1). Damaged granules tended to be intact but exhibited extensive surface abrasions, deep pitting and exfoliation of the granule surface. The implications of the differences in type of starch damage for pin- and attrition-milled flours has not been evaluated.

AIR-CLASSIFIED HARD WHEAT FRACTIONS

Wichser (1958) obtained three fractions by two-stage air classification of a HRW wheat flour and

Fig. 1. Types of starch damage in attrition-milled first middling flour where exfoliation (a) and pitting (b,d) of granules predominated over granule disintegration (c). Magnification x 2716. (Reprinted by permission from Nowakowski et al. 1986).

found the HPF and coarse fraction were satisfactory bread flours while the intermediate cut was low in protein content and suitable for angelfood and layer cakes. Bean et al. (1969a) blended 20-30% of HPF from five HRW cultivars with low-protein base flours and found that loaf volumes and farinograph curves of the blends were strongly influenced by the characteristics of the HPF parent flours. HPF from strong gluten cultivars enhanced loaf volumes of the blends, and weak gluten cultivars gave LPF that were suitable for cookie and cake applications (Bean et al. 1969b).

Hayashi et al. (1976) processed two middling and one break flour from three HRS wheat varieties by fine grinding and air classification. A marked increase in starch damage and concentration of broken starch granules into the fine fraction was noted. In general, the HPF gave high loaf volumes but the crumb structures were inferior to those obtained with the coarse fraction. The intermediate fractions were satisfactory for cakes but poor for bread and cookies. MacArthur and D'Appolonia (1976, 1977) found that HPF from similar wheat varieties contained primarily small starch granules, as well as 18.7-26.1% protein (14% mb), and much of the lipid, pentosans, ash, reducing and nonreducing sugars from the original flours. The high starch damage values and water binding capacities in HPF also indicated the presence of damaged starch.

Dick et al. (1977) air classified several pin-milled flour streams from two HRS wheat varieties. The yields of HPF were 9.2 and 13.3% which represented protein shifts of 5.7 and 9.4%, respectively. Rheological properties were correlated with protein contents of blends of the air-classified fractions, and there was a strong varietal effect. In addition, reconstituted blends of air-classified fractions did not demonstrate a clear superiority in breadmaking quality (Dick et al. 1979). Best results were obtained by blending air-classified fractions from an acceptable bread variety with a base flour from the same variety.

PARTICLE SIZE DISTRIBUTIONS

The areas of irregular flour particles visualized at high magnification under a microscope have been quantitated by image analysis and the data converted to spheres of equivalent mass (Sosulski et al. 1988). The number of particles and their volumes were used to compute the flour volumes occupied by particles within specific particle diameter ranges from 0 to 150 μm (Table III). Flour particles greater than 100 μm diameter constituted over 40% of the flour volume of 5B flour and nearly 60% of 1M flour. Only 7-10% of these rollermilled flours were <45 μm diameter and potentially free protein or starch particles. A single pass of these flours through the pin mill reduced about 80% of the particles to <45 μm, defined as the 'degree of reduction' in the Nowakowski et al. (1987) study. Two passes (2X) through the pin mill achieved 95% reduction (Table III) while attrition milling resulted in 100% level of reduction. The lower percentages of starch damage in the attrition milling experiment suggests that further investigations could be fruitful in obtaining particle size reduction at acceptable levels of starch damage.

OPTIMIZED AIR-CLASSIFICATION CUTS

The pin- and attrition-milled 5B and 1M flours were passed through an Alpine Mikroplex 132 MP spiral air classifier at five progressively-wider vane settings to give six fractions of increasing particle size (Nowakowski et al. 1987; Sosulski et al. 1988). Each flour was initially classified at a narrow vane of 12 to separate the finest flour particles (Table IV). The coarse residue was reclassified at vane setting 16, and then at 20, 26 and 34, so that the sum of all cuts plus the >34 coarse residue constituted the whole flour.

For the attrition-milled 5B and 1M flours, the VS12 cut separated a HPF containing 30.7% protein that constituted a positive protein shift of 15.7 relative to the parent 5B flour (Table IV). This fraction was also depleted in starch but contained 2.0% ash and 4.0% lipid. Since the coarse fraction at VS34 also exhibited a positive protein shift, the optimum air classifier cuts for 5B flour were established at VS12 and VS34 (Table V). The HPF at VS12 could serve as a replacement for gluten supplements in weak gluten doughs and the second HPF at VS34 with low ash but uniform, intermediate

438

TABLE IV
Composition of Consecutive Air-Classification Cuts on Two Attrition-Milled Flours in %, 14% mb[a]

Vane Setting	Fraction Yield	Fraction Composition			Protein Shift
		Protein	Starch	Ash	
Fifth Break Flour					
Flour	100	19.8	58.6	1.1	-
12	28	30.7	41.3	2.0	15.7
16	13	17.5	56.0	1.0	-1.4
20	5	14.5	66.0	0.8	-1.3
26	22	11.8	68.5	0.5	-9.4
34	22	14.3	69.4	0.5	-6.1
>34	10	24.8	58.2	0.5	2.6
First Middling Flour					
Flour	100	12.1	71.0	0.4	-
12	34	21.6	55.8	0.6	26.7
16	10	10.7	61.7	0.4	- 1.1
20	4	8.4	75.8	0.4	- 1.3
26	24	5.6	77.7	0.3	-13.1
34	20	6.3	79.0	0.3	- 9.3
>34	8	9.3	72.9	0.3	- 1.9

[a] From Sosulski et al. (1988); used by permission.

particle size could be used to upgrade low protein patent flours intended for household use. The medium protein fraction (MPF) could be blended off into baker's patent flour due to its relative low ash level (Nowakowski et al. 1987). Thus, there was potential for utilizing all products from the air-classified 5B flour as added-value protein supplements or blending flours.

An 1M flour represents one of the most valuable products from a HRS mill and would not likely be value added through further processing by fractionation on the air classifier. However, it was of interest to determine the potential degree of protein displacement in this flour from the center of

TABLE V
Projected Yield and Composition of Fractions From Two-Stage Classification of Attrition-Milled Flours in %, 14% mb[a]

Vane Setting	Fraction Yield	Fraction Composition			
		Protein	Starch	Ash	Lipid
Fifth Break Flour					
12 fines	28	30.7	41.3	2.0	4.0
34 fines	62	13.9	66.0	0.6	1.6
34 coarse	10	24.8	58.2	0.5	1.4
First Middling Flour					
12 fines	34	21.6	55.8	0.6	1.1
20 fines	14	10.0	66.0	0.4	0.7
20 coarse	52	6.4	77.4	0.3	0.4

[a] From Sosulski et al. (1988); used by permission.

the kernel whereas 5B originated from the outer endosperm. The degree of protein shift of 26.7% at VS12 was 70% greater than was obtained in 5B (Table IV) and was comparable to the protein shift in soft wheats (Gracza, 1960). In addition, there were low protein fractions obtained at VS26 and VS34 that contained about 78.5% starch (91.3% starch on a dry basis). Due to the high proportion of these fractions, 52% in the optimized system proposed in Table V, it is unlikely that the products of air-classified 1M could be marketed economically relative to the cost of the initial flour.

BAKING QUALITY OF HPF

Five market classes of flour were supplemented with an additional two percentage units of protein from gluten, HPF/5B and HPF/1M in bread dough formulations containing 15 ppm bromate (Nowakowski et al. 1987). There were wide differences in positive responses among the five flours but, on average, HPF/5B gave comparable increases in loaf volume to those obtained with vital gluten (Table VI). The results with HPF/1M indicated a poorer response in the lower protein base flours. Nowakowski et al. (1987) also determined the effects of blending the MPF from 5B flour into patent and bakers' patent flours. The replacement of up to 9% MPF in HRS patent flour had no adverse effects on any bread characteristic. For bakers' patent flour, 7% replacement had no visible effects on the straight dough breads (Table VII). At higher levels of replacement, MPF reduced loaf volumes by about 5% in 'no bromate' formulations but the effect was minimal when 15 ppm bromate was added.

CONCLUSIONS

While protein shifts of over 25% are common during air classification of remilled soft wheat flours, the separation efficiency of protein into the fine fraction of air-classified hard wheat flours has usually ranged from 5-10%. The recent trend to higher starch damage in bread wheat flours may permit more intensive milling of hard wheats to increase the degree of reduction to achieve greater separation

TABLE VI
Increases in Loaf Volume After Supplementation With Two Percentage Units of Protein from Gluten and HPF in %[a]

Market Class of flour	Gluten	HPF/5B	HPF/1M
Soft Wheat	9.2	6.7	3.3
Prairie Spring	8.3	11.0	4.8
HRS Patent	12.5	9.6	5.1
HRS Bakers'	4.0	6.1	7.4
Triticale	11.9	15.2	15.2
Average	9.2	9.7	7.2

[a] From Nowakowski et al. (1987); used by permission.

TABLE VII
Influence of MPF from Air-Classified 5B Flour
on Bread Volume of Bakers' Patent Flour[a]

Flour Blend with MPF/5B	Protein Content (%,14%mb)	Volume Change in %	
		0 Bromate	15 ppm Bromate
Control	13.7	0.0	5.4
+ 7.0%	13.7	0.0	6.7
+11.0%	13.7	-4.9	2.4
+17.5%	13.7	-5.5	-2.4

[a]From Nowakowski et al. (1987); used by permission.

efficiency. Twice pin-milled and attrition-milled HRS break flours provided HPF with over 30% protein at protein shifts of nearly 16%. These HPF from HRS wheats appeared to be satisfactory replacements for vital gluten in bread formulations and the by-product MPF could be value-added by blending off into patent streams. These studies demonstrate the need for further research on mechanisms for regrinding hard wheat flours with less starch damage so that more efficient separations can be achieved with a single cut on the air classifier.

LITERATURE CITED

BEAN, M.M., ERMAN. E., and MECHAM, D.K. 1969a.
  Baking characteristics of high-protein fractions from air-classified Kansas hard red winter wheats. Cereal Chem. 47:27.
BEAN, M.M., ERMAN, E., and MECHAM, D.K. 1969b.
  Baking characteristics of low-protein fractions from air-classified Kansas hard red winter wheats. Cereal Chem. 47:35.
BODE, C.E., KISSELL, L.T., HEIZER, H.K. and B.D.
  MARSHALL. 1964. Air-classification of a sof and a hard wheat flour. Cereal Sci. Today 9 432.

DEXTER, J.E., PRESTON, K.R., TWEED, A.R., KILBORN, R.H., and TIPPLES, K.H. 1985. Relationship of flour starch damage and flour protein to the quality of Brazilian-style hearth bread and remix pan bread produced from hard red spring wheat. Cereal Foods World 30: 511.

DICK, J.W., SHUEY, W.C., and BANASIK, O.J. 1977. Adjustment of rheological properties of flours by fine grinding and air classification. Cereal Chem. 54:246.

DICK, J.W., SHUEY, W.C., and BANASIK, O.J. 1979. Bread-making quality of air-classified hard red spring wheat manipulated flour blends. Cereal Chem. 56: 480.

GRACZA, R. 1959. The subsieve-size fractions of a soft white flour produced by air classification. Cereal Chem. 36:465.

GRACZA, R. 1960. The subsieve-size fractions of a hard red spring wheat flour produced by air classification. Cereal Chem. 37: 579.

HAYASHI, M., D'APPOLONIA, B.I., and SHUEY, W.C. 1976. Baking studies on the pin-milled and air-classified flour from hard red spring wheat varieties. Cereal Chem. 53:525.

JONES, C.R. 1940. The production of mechanically damaged starch in milling as a governing factor in the diastatic activity of flour. Cereal Chem. 17:133.

JONES, C.R., HALTON, P., and STEVENS, D.J. 1959. The separation of flour into fractions of different protein contents by means of air classification. J. Biochem. Microbiol. Technol. Eng. 1:77.

KENT, N.L. 1965. Effect of moisture content of wheat and flour on endosperm breakdown and protein displacement. Cereal Chem. 42:125.

KENT, N.L. 1966. Technology of Cereals with Special Reference to Wheat. Pergamon Press, London.

MACARTHUR, L.A. and D'APPOLONIA, B.L. 1976. The carbohydrates of various pin-milled and air-classified flour streams. I. Sugar analyses. Cereal Chem. 53:916.

MACARTHUR, L.A., and D'APPOLONIA, B.L. 1977. The carbohydrates of various pin-milled and air-classified flour streams. II. Starch and pentosans. Cereal Chem. 54: 669.

NOWAKOWSKI, D., SOSULSKI, F.W. and HOOVER, R. 1986. The effect of pin and attrition milling on starch damage in hard wheat flours. Starch/Starke 38:253.

NOWAKOWSKI, D., SOSULSKI, F.W. and REICHERT, R.D. 1987. Air classification of pin-milled break and middling flours from hard red spring wheat. Cereal Chem. 64: 363.

PANTER, A. and BRENNAN, P. 1982. Flour milling technology. in Grains & Oilseeds - Handling, Marketing, Processing. 3rd Ed. Canadian International Grains Institute, Winnipeg, MB.

SOSULSKI, F.W., NOWAKOWSKI, D.M. and REICHERT, R.D. 1988. Effect of attrition milling on air classification properties of hard wheat flours. Starch/Starke 40: 100.

TIPPLES, K.H. and KILBORN, R.H. 1968. Effect of pin-milling on the baking quality of flour in various breadmaking methods. Cereal Sci. Today 13:331.

WICHSER, F.W. 1958. Baking properties of air-classified flour fractions. Cereal Sci. Today 3:123.

25

A SHORT MILLING PROCESS FOR WHEAT

Lone E. Gram and Lars Munck
Department of Biotechnology
Carlsberg Research Laboratory
10 Gamle Carlsberg Vej
DK-2500 Valby, Copenhagen
Denmark

Michael P. Andersen
United Milling Systems A/S
8 Gamle Carlsberg Vej
DK-2500 Valby, Copenhagen
Denmark

INTRODUCTION

Milling of wheat was started by the hand-driven stone mill and was developed into the typical wind or water mill displaying circular grinding elements with central feeding. In the last century, metal roller mills were introduced implying a successive removal of the endosperm from bran by 10-40 passages, each step separating the fine flour from the bran by sifting. A modern wheat mill displays an impressive machinery with large series of roller mills, sifters, purifiers etc. bound together with an elaborate and energy-consuming pneumatic transportation system.

The question arises, is it possible to simplify this system by introducing a mill which is easier to maintain than the roller mill and which in one step can remove a large portion of fine flour (endosperm) while still obtaining a reasonable separation between bran and flour as expressed in colour and ash analyses. We have at our department at Carlsberg Research

Center in cooperation with United Milling Systems A/S (UMS), a fully owned subsidiary of Carlsberg A/S, developed (Gram et al., 1988) such a mill and a short milling process for wheat which we will describe in the following together with some preliminary tests.

The Disc Mill

Like the old stone mills, the disc mill consists of a stationary and a rotary disc (Fig. 1). Unlike in stone mills, these discs are made of steel and mounted vertically. The rotary disc has a velocity of 3600 RPM, which means that the speed is 350 km/h in the periphery of the 50 cm discs. The grinding surface consists of 24 special corrugated hard metal segments (Fig. 1 B) mounted in the periphery of each disc. This means that each segment covers an angle of 15°. The elements can be easily changed when needed without elaborate tools. The capacity of the mill is about 1 ton of wheat per hour.

The wheat grain is led to the mill through the centre of the stationary disc, and by the centrifugal force and the crossing angle between the grinding elements it is thrown towards the periphery of the discs where the grain is ground. After one passage through the disc mill followed by sifting, the products from a Danish soft wheat will be:

50-55% flour
35-40% middlings
10-12% bran

In a conventional wheat mill, each break roll releases only a few percentages of flour.

When the disc mill is combined with 2-3 passages of roller mills, a complete and extremely short milling system (Fig. 2) is obtained. After a simple break system based on disc mills, the middlings are cleaned in a purifier and milled on smooth roller mills. The flour yield will constitute 78-80%. If a very high flour yield is wanted, the bran fraction (CB 1) is divided into a heavy and a light fraction in an airsifter. Milling of the heavy bran fraction

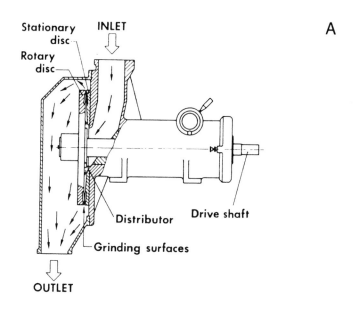

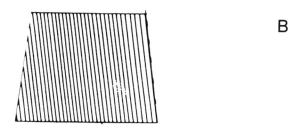

Fig. 1. Cross section of the UMS disc mill MHA 600 (A) with a disc section (B) (Andersen, 1987).

on a corrugated roller mill extends the flour yield to 80-82%. Other end-products are: coarse bran about 8% and fine bran 10-12%.

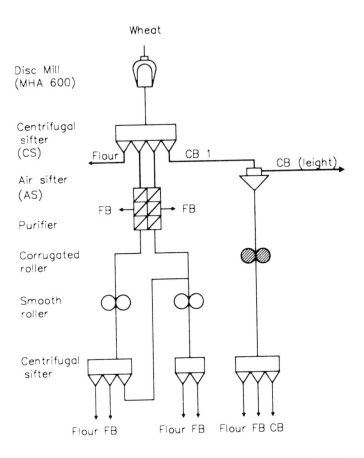

Fig. 2. Milling diagram for the UMS short disc milling system; FB = fine bran, CB = coarse bran.

MATERIALS AND METHODS

Wheat Samples

Two Danish grown varieties (Kraka and Solid winter wheats) and four Swedish wheat samples from Kalmar and Örebro from unknown spring and winter wheat varieties with widely different protein levels were used for the test together with one extremely hard Canadian spring wheat sample.

Milling

The wheat samples were milled in our UMS pilot wheat mill (capacity 1 ton/h), following the milling diagram described, and on a Bühler laboratory roller mill (MLU 205, capacity 5 kg/h) for comparison.

Methods of Analysis

Moisture, protein (N x 6.25), ash, fat, fibre, and flour colour (wet Agtron, green mode) were obtained by standard AACC methods (1983). Starch was measured by enzymatic degradation to glucose (Haastrup Pedersen, 1987). Damaged starch was determined as described by Hallgren (1985). The germ content of raw materials and flours was measured fluorometrically as described by Barnes and Jørgensen (1987). The particle size of the flours was determined by sifting on an Alpine laboratory air-jet sifter (100 LS). The sifting time was 3 min on the 32 μm screen and 1 min on 180 μm screen.

Baking Test

The method and formula were based on our standard procedure for test baking with wheat and composite flour (Hallgren, 1985); (300 g flour was used).

Gluten Content

The content of wet gluten in the flours was determined after washing in a Glutomatic 2100 (Falling Number Co., Sweden). The content of dry gluten was measured after drying the gluten ball in a Glutork 20-20 dryer for 5 min at 150 $^{\circ}$C.

RESULTS

The detailed milling data for three of the wheat samples are summarized in Table I. The flour yield from the disc mill of the three extreme Scandinavian wheats was highest for the Danish Kraka wheat (56%) and lowest for the Swedish Kalmar wheat (47.5%).

TABLE I
Milling Yields Examplified with Three Wheat Varieties on UMS Short Milling System and the Laboratory Roller Mill

Milling System and Product	Yield (%) from Wheat Variety		
	Örebro	Kraka	Kalmar
UMS short disc milling system			
Disc mill	54.0	56.0	47.5
Smooth roller	23.1	23.6	29.1
Corrugated roller	2.5	1.8	2.5
Total flour extraction	79.6	81.4	79.1
Fine bran	14.6	11.3	13.3
Coarse bran	5.8	7.3	7.6
Laboratory roller mill			
Total flour extraction	74.3	77.5	72.1
Fine bran	6.2	4.5	6.1
Coarse bran	19.5	18.0	21.8

The three roller milling steps equalize some of the differences in flour extraction, so that there is only about 2% difference in flour extraction of Kraka and Kalmar wheat. Milling on the laboratory roller mill led to flour extractions 3.7 to 7.0% lower than from the UMS wheat milling system.

The chemical data of the raw materials are shown in Table II and of flours in Table III together with total extraction rates. There are large differences between the wheats especially with respect to protein and colour. When comparing the flours from the two mills, it is important to keep in mind the differences in flour extraction. A higher content of ash, protein, fat, and fibre plus lower starch content in flours from the UMS short milling system is expected, since the extraction rate is higher. Because most of the ash, fat, and fibre is concentrated in aleurone, germ, and pericarp, respec-

TABLE II
Proximate Analyses of Raw Materials
(on dry matter basis)

VARIETY	ASH (%)	PROTEIN (N x 5.7, %)	STARCH (%)	FIBRE (%)	FAT (%)	COLOUR
KRAKA	1.67	13.2	71.6	2.5	1.8	37
SOLID	1.48	14.8	66.0	3.2	1.8	16
KALMAR SPRING	1.68	16.3	67.4	3.1	2.1	30
KALMAR WINTER	1.79	12.8	67.4	2.5	2.1	25
ÖREBRO SPRING	1.60	11.7	71.2	3.0	2.3	53
ÖREBRO WINTER	1.80	10.9	67.2	2.9	2.2	40
CANADIAN	2.01	17.6	62.5	2.8	1.9	6

TABLE III
Comparison of Flour from 7 Wheat Cultivars, Ground in the UMS Wheat Disc Mill Process and in a Laboratory Roller Mill
(on dry matter basis)

VARIETY	MILL TYPE	YIELD (%)	ASH (%)	PROTEIN (N x 5.7, %)	STARCH (%)	FIBRE (%)	FAT (%)	DAMAGED STARCH (%)	GERM (%)	COLOUR AGTRON
Kraka	Disc	81.4	0.75	12.6	80.7	0.7	1.5	9.9	30.5	58
	Roller	77.5	0.55	12.1	80.8	0.8	1.3	9.0	17.7	75
Solid	Disc	81.1	0.72	13.9	77.0	0.5	1.4	12.6	35.8	39
	Roller	75.8	0.58	13.4	79.6	0.5	1.2	11.1	15.9	67
Kalmar spring	Disc	79.1	0.75	16.0	77.6	0.8	1.6	12.5	23.7	47
	Roller	72.1	0.48	15.0	78.9	0.6	1.3	10.3	11.7	79
Kalmar winter	Disc	79.2	0.85	12.6	80.0	0.6	1.6	11.2	26.2	49
	Roller	71.9	0.61	12.0	81.9	0.4	1.3	9.6	15.7	76
Örebro spring	Disc	77.7	0.88	11.3	80.3	0.8	1.8	11.9	26.9	70
	Roller	75.3	0.57	10.8	80.9	0.5	1.5	11.4	16.1	89
Örebro winter	Disc	79.6	0.79	10.0	80.2	0.9	1.7	11.3	30.3	57
	Roller	74.3	0.63	9.6	82.6	0.5	1.3	9.9	15.2	85
Canadian	Disc	77.5	0.88	17.0	75.3	0.7	1.5	10.8	32.0	35
	Roller	73.5	0.65	16.5	78.7	0.4	1.1	10.0	17.5	59

tively, the flour analyses indicate that flours from the UMS mill contain more of these botanical components especially germ and aleurone. This means that they contain relatively less endosperm, which is also confirmed by the starch figures. The differences in starch content of flours between the two milling systems are, however, rather small especially when milling soft wheat varieties and considering the higher extraction rates obtained for the disc mill system.

The slightly higher protein figures for the UMS-flours are probably caused by higher contents of germ and/or aleurone, which are both higher in protein than the endosperm. The results from the germ analysis (Table III) reveal that the UMS-flours contain about twice the amount of germ than the flours from the laboratory roller mill, which is an improvement from a nutritional point of view. The fluorometric germ test estimates germ tissue not germ oil.

Measuring the colour of the flour is another way to estimate the amount of dark bran particles in the white endosperm flour. In Table III, it is seen that the UMS-flours are darker than the flours from the laboratory roller mill, which is natural due to the lower extraction rate of the laboratory roller mill. The difference in colour between the milling systems is not greater than is the variation among wheat varieties.

The screening analysis showed that the UMS short milling system produces a finer granulated flour than the laboratory roller mill (Fig. 3). This can also explain the higher amount of damaged starch found in some (but not all) of the UMS-flours under the present running conditions.

The baking tests showed only small differences between extreme flours tested (Örebro, Kraka and Kalmar) from the two mills (Table IV). Bread baked from the UMS-flours was 2.9-5.5% smaller than bread from the flours from the laboratory roller mill. These small differences were expected because of the greater flour extraction from the UMS short milling system. Visual evaluation of bread crumb porosity showed no difference between the two mills. The

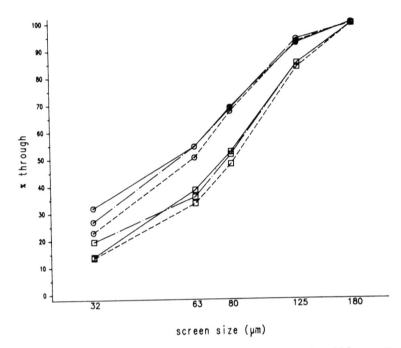

Fig. 3. Screening curves of flours from 2 mills. O = UMS short disc milling system, □ = Laboratory roller mill. Wheat varieties: —— Örebro, --- Kraka, -·-· Kalmar

TABLE IV
Results from Baking Test and Gluten Determination

Wheat Variety	Mill	Bread Volume (ml)	Bread Weight (g)	Spec. Volume (ml/g)	Crumb Porosity[a]	Gluten Wet (% DM)	Dry (% DM)
Örebro	Disc	236.9	83.9	2.86	7	23.7	8.3
	Roller	243.8	83.2	2.93	7	22.9	8.0
Kraka	Disc	245.0	82.3	2.98	7	32.5	11.6
	Roller	258.8	82.2	3.15	7	31.2	11.2
Kalmar	Disc	243.8	83.9	2.91	6	36.7	13.5
	Roller	256.3	83.6	3.07	6	36.1	13.0

a) Dahlman Scale

flours from the UMS short milling system showed a little higher yield of wet and dry gluten, but the differences were very small.

It is thus evident that by using the UMS disc mill technology it is possible to reduce drastically the number of passages in milling wheat to different types of flour while retaining acceptable baking quality and saving capital costs, maintenance and energy.

The new disc mill may also be installed as a prebreak-system in an existing conventional wheat mill, thus enlarging capacity by 20 to 30% (without any noticeable changes in the performance of the final baking flour).

LITERATURE CITED

AMERICAN ASSOCIATION OF CEREAL CHEMISTS. 1983. Approved Methods of the AACC, 8th ed., The Association, St. Paul, Minn., U.S.A.

ANDERSEN, M. 1987. Method of milling and apparatus for carrying out the method. U.S. Patent No. 4,667,888.

BARNES P.J., and JØRGENSEN, K.G. 1987. Fluorimetric measurement of wheat germ in milling products. II. Sodium tetraborate reagent. J. Cereal Sci. 5:149.

GRAM, L.E., ANDERSEN, M., and MUNCK, L. 1988. Fractionation of wheat. In Cereal Science and Technology. Proc. 23. Nordic Cereal Congress, August 17-20, 1987, Copenhagen, Denmark (L. Munck, ed.) The Danish Cereal Society, c/o Carlsberg Research Center, Dept. of Biotechnology, Valby, Denmark, pp. 391-399.

HALLGREN, L. 1985. Physical and structural properties of cereals, sorghum in particular, in relation to methods and product use. Thesis No. 96, ATV, Copenhagen, Denmark.

HAASTRUP PEDERSEN, L. 1987. Development of screening methods for evaluation of starch structure and synthesis in barley. Thesis No. 131, ATV, Copenhagen, Denmark.

26

WHEAT FRACTIONATION AND UTILIZATION

Yrjö Mälkki, Jaana Sorvaniemi,
Olavi Myllymäki, Jari Peuhkuri
and Erkki Pessa

Technical Research Centre of Finland
Food Research Laboratory
SF-02150 Espoo, Finland

INTRODUCTION

Fractionation of wheat has been proposed by many researchers as an option for marketing of surplus wheat. The most important market for gluten is in improving baking properties of soft and low-protein wheats. Starch has been marketed for similar uses as starches from other sources. One possibility for improving the economy of these processes would be to use low-priced raw material, another to find special functional properties and uses for the fractions to improve their competitiveness. The purpose of this study has been to compare the quality of glutens isolated from samples of several wheat varieties and with varying quality, to evaluate their applicability as raw material for gluten manufacture, and to compare the properties of wheat starch with starches from other sources in papermaking and as a gelling agent.

TABLE I
Varieties for Isolation of Glutens and
Analytical Properties of the Flours.

Variety /Flour	Harvest year	Protein, N x 5.7 %	Ash %	Falling no.	Wet gluten[a] %
1. Tähti	85	13.3	0.40	472	30.1
2. Luja	86	13.3	0.37	65	27.2
3. Reno	86	12.1	0.47	371	38.2
4. Ruso	85	12.1	1.01	134	23.4
5. Kadett	86	10.6	0.43	174	29.0
6. Tapio	83	9.6	0.40	371	32.0
Finn.flour	87	10.9	0.66	284	33.6
Dutch flour	87	12.1	0.48	321	38.2

[a] According to AACC Method 38-11, calculated on 15 % flour moisture basis.

PROPERTIES OF GLUTENS FROM VARIOUS SOURCES

Materials and Methods

Gluten was isolated by manual washing on a laboratory scale from samples of six spring wheat varieties cultivated in Finland. Analytical properties of the flours from which the isolation was made are presented in Table I. In addition, two commercial glutens were tested, one of domestic (Raisio Group, Finland), another of Dutch (Cargill) origin.

Test bakings were performed using one domestic and one Dutch commercial wheat flour, and adjusting the protein content of the flour to 13 % or 16 % level by using the gluten to be tested. Loaf volumes were measured one hour after the baking, texture of the crumb using an Instron Universal Testing Machine after 1 or 2 days at room temperature.

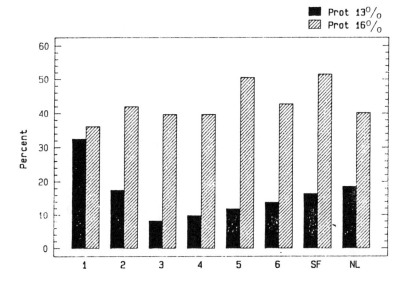

Fig.1. Effect of glutens on volume increase of breads from Finnish flour. Glutens 1 to 6 from varieties acc. to Table I, SF = Finnish commercial gluten, NL = Dutch commercial gluten. Dough protein levels adjusted with glutens to 13 or 16 per cent.

Effects on Loaf Volume

In test bakings where the various glutens were mixed with Finnish wheat flour to obtain a 13 % total protein level (Figure 1), the average level of volume increase was the same as when using commercial glutens. Cv. Tähti having the highest protein content in flour and the highest falling number gave clearly the highest volume increase, whereas with cv. Ruso low volume increase coincided with an unusually high ash and low wet gluten content of the flour. At the higher protein level, the differences in volume increase from adding glutens of different varieties and commercial glutens were small, and no correlation with analytical quality criteria of the flours was noticeable.

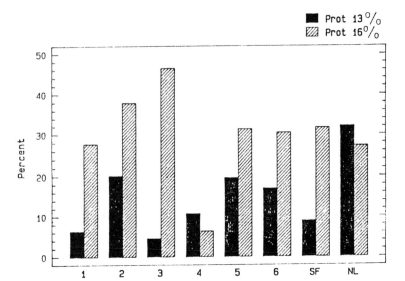

Fig. 2. Effect of glutens on bread volume increase using Dutch flour. Gluten additions as in Fig. 1.

In baking tests with the Dutch flour (Fig. 2), differences in volume increase were more pronounced at both protein levels, although due to the higher protein content of the flour, the additions of glutens were small at the lower level. The highest volume increase was obtained with the commercial Dutch gluten, followed by the cv. Luja, which had a high flour protein content but an exceptionally low falling number. At both levels of added gluten, there was seemingly no correlation between the total protein content of the variety and volume increasing effect of the gluten. As in the baking tests with Finnish flour, cv. Ruso which was high on ash and low in wet gluten gave the lowest volume increases. The great variation in the effects of the different varieties and the smaller effect of the commercial Finnish gluten indicate possible incompatibilities of the indigenous proteins of the flour and those of added gluten preparations, which is probable in the case of cv. Ruso with a high ash content.

Two varieties with low falling number, Luja and Kadett, did not exhibit weaker effects on volume increase than average; in the case of cv. Luja the effect was even higher for Dutch flour. As expected, there was no significant correlation between the falling number and quality or breakdown of proteins. The results indicate, that some wheats rejected on the basis of a low falling number, could be used for gluten manufacture, except in cases of high flour ash, and provided that starch hydrolysis does not affect its end use properties.

Effects on Staling

Although the increase in firmness is in general inversely proportional to the volume, some deviations

TABLE II
Young's Module of the Crumb After Storage at Room Temperature

Gluten Source cf. Table I	Young's module, $Nm^{-2} \times 10^{-4}$							
	Finnish flour				Dutch flour			
	1 day		2 days		1 day		2 days	
	13%	16%	13%	16%	13%	16%	13%	16%
Tähti	5.1	4.5	9.1	5.6	5.4	3.6	4.6	7.3
Luja	5.7	3.5	9.5	4.5	2.6	2.9	4.3	3.3
Reno	7.0	3.0	9.8	6.5	3.9	2.2	6.5	4.8
Ruso	7.2	3.8	10.2	5.1	4.7	3.3	5.5	5.2
Kadett	5.7	3.5	12.9	5.4	4.3	2.9	7.5	3.7
Tapio	6.1	4.3	9.7	5.5	4.9	2.6	6.2	4.6
SF commercial	6.2	4.2	10.3	5.1	5.5	4.2	7.3	5.1
NL commercial	4.5	3.7	7.2	6.1	2.5	2.2	2.8	4.8
Control	9.7		14.5		6.0		9.6	

from this correlation can be observed from the results (Table II). In the test bakings using both Finnish and Dutch flour, the Dutch commercial gluten yielded the softest crumb and the slowest staling at the lower protein level, although the volume increase in bakings with Finnish flour was not exceptionally high. Conversely, gluten from cv. Ruso yielded breads of average softness and staling rate despite the low loaf volumes. Again, no correlations can be found with the analytical quality criteria of the flours. Similarly, no significant correlation was established between farinograph data and gluten quality.

STARCH IN PAPERMAKING

Native starches are seldom used in papermaking due to their high viscosities and relatively weak retention of starch and filler substances in the pulp fiber network. Modification is required, and the general approach is to select the cheapest starch. Reducing viscosity by hydrolysis is difficult to control, and derivatization increases costs. Thus minimally modified starch is preferred.

Wheat starch yields dispersions low in viscosity after cooking and cooling and can bind filler particles at a low concentration, giving a higher ash content than other starches at low concentrations (Figure 3). Since the fiber network becomes less dense, this results in a lower tensile strength at low starch concentrations. With an increase in concentration, the binding properties of the starch compensate for this effect, and with 1 % starch, a higher tensile strength can be obtained with wheat than with other starches having the same degree of substitution after cationization (Figure 4). In our experiments, this concentration for a wheat starch with a substitution degree of 0.03 % was also optimum for surface smoothness. At higher concentrations

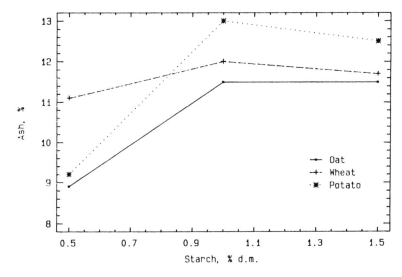

Fig. 3. Effect of cationized (degree of substitution 0.03 %) starches on the binding of filler

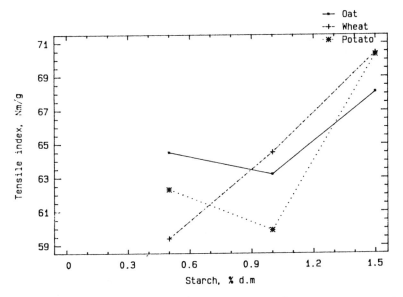

Fig. 4. Effect of cationized starches on tensile strength of paper.

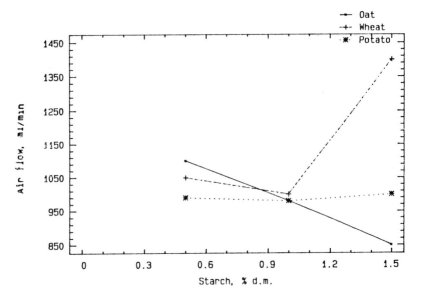

Fig. 5. Effect of cationized starches on the roughness of paper surface.

the roughness of the surface and also the porosity of the paper increased, this effect being dissimilar to the effects of potato and oat starches. (Figure 5).

STARCH FOR GELATINIZATION

For gelatinization, e.g. in industrially prepared desserts and in the candy industry, derivatives of waxy starches are the usual choice. Native wheat starch is too viscous for most of the applications, and its sensory properties are not optimal. For replacing other, more expensive gelling agents such as vegetable gums, pectin, agar and gelatin, there would be market for a new starch-based product, provided the rheological properties could be tailored specifically to give sufficiently rapid gelatinization and a reduced adhesiveness, compared to the present preparations.

TABLE III
Textural Characteristics of Gels of Starch Hydrolysates[a].

Starch Source	Amylase	Penetration Energy	Adhesion Energy
Wheat	Bacterial	64	19
Wheat	Fungal	50	12
Potato	Bacterial	120	34
Potato	Fungal	140	11

[a]Concentration: 15 %. Figures related to the energy during penetration and backstroke.

We have used enzymatic hydrolysis to reduce viscosity to a level applicable in the present processes for soft candy products, and compared the behavior of wheat and potato starches under these conditions. We used bacterial amylase BAN and fungal amylase Fungamyl, of Novo, Copenhagen. The objective was to achieve a fluidity value between 30 and 50.

Despite of similar fluidity values, the two enzymes and the two starches tested gave gels of differing properties (Table III). The gels of wheat starch hydrolysates were less hard than those of potato starch. The adhesiveness of hydrolysates with fungal amylase was smaller than those of hydrolysates with bacterial amylase.

When these hydrolysates were used as gelling substances in making soft candies, properties presented in Table IV were obtained. As in plain starch gels, the wheat starch hydrolysate candies were softer than those made of potato starch hydrolysates. However, their adhesiveness and especially the ratio of penetration to adhesion energy was unfavorable as compared to the candies made with potato starch hydrolysates.

TABLE IV

Textural Characteristics of Soft Candies Made from Hydrolysates described in Table III[a].

Starch Source	Amylase	Penetration Energy	Adhesion Energy	Ratio (Penetr/Adh.)
Wheat	Bacterial	4.3	3.4	1.3
Wheat	Fungal	11	11	1.0
Potato	Bacterial	45	24	1.8
Potato	Fungal	23	10	2.3

[a]Starch hydrolysate concentration 10 % of dry matter.

Attempts were also made to modify wheat starch by cooking extrusion in a twin-screw extruder, at moisture contents from 20 to 30 %, and end temperatures from 160 to 200 °C. Under these conditions and with no enzymes added, fluidity of the mixtures was not significant.

CONCLUSIONS

Of the main fractions obtained from industrial wheat fractionation, only gluten has unique properties which offer distinct advantages over similar fractions from other raw materials. The competitiveness of this industry can be improved by using wheat which has poor baking properties. At least in a part of wheat lots rejected from food use on the basis of low falling number, gluten is still intact.

Differences in the structure and physical properties of wheat starch, as compared to other starches, cannot at present be commercially exploited and further research is needed towards improvements in processing, modification and applications of wheat starch.

27

A STUDY OF THE FACTORS AFFECTING THE SEPARATION OF WHEAT FLOUR INTO STARCH AND GLUTEN

R.J. Hamer, P.L. Weegels, J.P. Marseille and M. Kelfkens

TNO Cereals, Flour and Bread Institute, Wageningen, the Netherlands

INTRODUCTION

Wheat produced in the northwestern part of Europe is highly variable in quality characters. This situation calls for methods to determine the quality of wheat. The milling industry applies a number of such methods. In addition the potential baking quality of most varieties is known, enabling the milling industry to make reliable selections.

The wheat starch industry is also urged to select wheat for processing. For this purpose it has adopted the methods used to test dough and bread-making properties. There is, however, no theoretical basis to correlate baking quality to processing quality. As a result the industry's demand towards wheat in terms of quality specifications is broad and of a limited value. The use of a small scale separator, simulating the separation process, could overcome this problem.

Godon and coworkers (1983) developed a pilot plant for starch-gluten separation according to the Martin process. In this process, gluten is prepared from a dough by intensive washing and removal of the starch. The pilot plant of Godon has been demonstrated to be useful in evaluating the processing properties of different wheat varieties. However, the Martin process differs greatly from the process nowadays in general use by the European wheat starch industry. This is the 'Batter' process where starch and gluten are separated from a dilute wheat slurry by hydrocyclones or decanters.

For this reason we decided to build a laboratory system for starch/gluten separation based on the ba ter process. A schematic outline of the system given in Figure 1.

Separation System

Firstly, a dough is prepared from five kg flour. This dough is diluted to a slurry of 50 lite while circulating through a monopump in the therm statted separation tank. At this stage small glut agglomerates are formed. Then the slurry is pump onto a stack of vibrating sieves. The top sieve has mesh of 400 μm with the following sieves decreasing mesh from 250 to 30 μm.

Depending on the speed of coagulation, gluten retained on the 400 and 250 μm sieves. Bran particl are retained on the 125 and 90 μm sieve; hemicellulo on the 50 and 31 μm sieve. Starch and solubles pa the stack of sieves. With a disc-nozzle separator t starch is concentrated and separated from the proce sing water containing the solubles. The technical c tails of the separating system were published els where (Weegels et al 1988). The system provides info mation on gluten coagulation, on yields and purity gluten and starch and on losses of protein and oth soluble matter in the waste water.

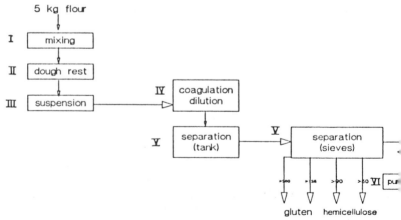

Figure 1: Schematic outline of the separation syste

TABLE I

Characteristics of the Tested Varieties

Variety	Baking Quality	Protein Content (% d.m.)*	Kernel Hardness	Damaged Starch (%)
Taurus	poor	10.1	soft	4.69
Kraka	good	12.8	hard	6.88
Minaret	fair	9.1	soft	6.15
Ralle	very good	11.9	hard	8.16
Camp Remy	good	12.1	hard	6.78
Tombola	poor	11.8	soft	4.75
Kanzler	good	11.2	soft	5.42
Obelisk	fair	11.3	hard	9.49
Okapi	moderate	12.6	soft	8.37
Granta	fair	10.2	hard	10.87

* N x 5.7, in the flour

Processing Properties of Ten Wheat Varieties

The system was used to assess the processing quality of ten wheat samples of different varieties and baking qualities. In Table I are shown some characteristics of the tested varieties. From the characteristics in Table I it is clear that wheat varieties were selected of both good and poor bread-making quality. The variety Camp Remy can be considered as a reference. This variety is well-known for its good processing properties and is preferentially used by the wheat starch industry. Camp Remy was grown in France. All other varieties were grown in the Netherlands. Table II gives the results obtained with these varieties in the processing system.

It is clear that Camp Remy gives a good gluten yield. For the other varieties gluten yields vary from 11 to 15.7 % with the exception of Minaret, which apparently has very poor processing properties. With this variety gluten yield is as low as 1.2 %. These results demonstrate the potential of the separating system to evaluate the economics of

TABLE II

Results Obtained for Ten Wheat Varieties

Variety	Total Gluten (%)*	Starch (%)*	Solubles (%)*
Taurus	12.14	79.20	7.71
Kraka	14.02	77.93	7.08
Minaret	1.21	82.76	7.27
Ralle	13.83	78.56	8.78
Camp Remy	14.36	69.24	15.87
Tombola	12.74	78.86	8.56
Kanzler	13.75	78.16	8.85
Obelisk	11.02	79.43	12.14
Okapi	15.70	76.48	8.91
Granta	12.27	79.06	9.15

* Results are expressed as percentage of flour.

wheat varieties in terms of gluten yield.
The yield of starch is also an important feature. Crude starch yields range from 69.2 % for Camp Remy to 82.8 % for Minaret. Apparently poor coagulation favours a high recovery of starch. With Camp Remy the recovery is surprisingly low and presumably due to loss of starch in the solubles fraction.
The amount of material recovered in the solubles fraction must also be considered to be of importance. The possibilities of recirculation of processing water and the costs of waste water treatment are important features of the industrial process. Again, large differences are observed. Generally 7 to 9 % of material was recovered in the processing water. Obelisk and Camp Remy, however, yielded more solubles: 12.1 % and 15.9 %, respectively. As stated before, the solubles of Camp Remy contain a high amount of starch. With Obelisk this value is due to a high amount of soluble proteins in the flour.
These different parameters must be carefully evaluated to conclude about the suitability and

economic feasibility of certain wheat varieties for starch/gluten processing. Our system reliably provides this information.

Coagulation Properties

The separation system can also provide information on an important processing related characteristic of the flour, its coagulation properties. Our system provides this information with the coagulation index. The coagulation index is calculated from the ratio of gluten on the 400 μm sieve and the total gluten recovered. Therefore the index gives an indication of the rate of coagulation. In Figure 2 the coagulation index is presented for the ten varieties tested.

The varieties Granta, Okapi, Obelisk and Camp Remy have an index ranging from 89 to 98 representing nearly complete recovery of the gluten on the 400 μm sieve. The varieties Tombola and Kanzler have an index of about 24 indicating a much slower coagulation. This demonstrates that the coagulation index is a sensitive parameter giving information on an important processing property.

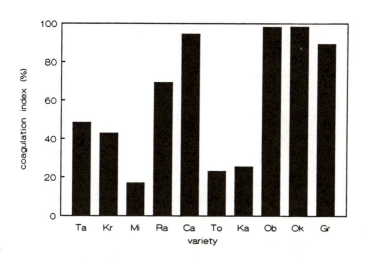

Figure 2: Coagulation index of ten wheat varieties.

TABLE III

Relationship Between Baking Quality
and Processing Quality

Variety	Baking Quality	Protein Recovery (%)*	Coagulation Index
Taurus	poor	81.2	48.5
Kraka	good	82.0	42.9
Minaret	fair	5.9	17.1
Ralle	very good	82.4	69.4
Camp Remy	good	90.2	94.6
Tombola	poor	79.8	23.2
Kanzler	good	87.0	25.5
Obelisk	fair	71.7	98.2
Okapi	moderate	92.3	98.4
Granta	fair	90.5	89.4

* % of flour protein recovered as gluten

Effect of Protein Content and Flour Particle Size

This leaves us with the question which parameters measured in the flour do correlate with processing properties. We found a good correlation between protein content of the flour and gluten yield ($r = 0.79$). This indicates that the poor processing properties of Minaret are likely to be due to the low protein content of the Minaret flour

Relation Between Bread-making Quality and Processing Properties

Our results also corroborate the hypothesis that there is no relation between the bread-making quality of wheat and its starch/gluten processing properties. This is illustrated in Table III where bread-making quality is compared with gluten yield and coagulation index. From this table it is clear that varieties with poor bread-making quality like Taurus and Tombola still give a good protein recovery and have acceptable coagulation properties.

(9.1 %). An interesting observation is that the protein content of the flour does not correlate with the coagulation index. This parameter only seems to be related with the extent of starch damage in the flour (r = 0.77) indicating an effect of grain hardness and milling conditions. Upon further investigation we found a good correlation between the coagulation index and the particle size distribution of the flour. This is shown in Figure 3 where we graphed the coagulation index against the amount of flour passing a 45 μm sieve.

The correlation was high (r = 0.92) indicating the important effect of milling and grain hardness on the coagulation properties. Although particle size is usually related to grain hardness, figure 3 shows that wheats, considered to be soft wheats, sometimes may yield coarse flours.

Effect of Year on Processing Properties

A matter of particular importance is the year to year variation. We started a study on this mat-

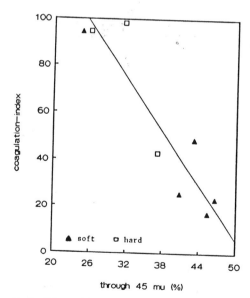

Figure 3: Relation between coagulation index and flour particle size

ter by measuring the processing properties of three wheat varieties grown in 1986 and 1987. Although this study is still in progress, some initial results on this matter are presented here. The characteristics of these varieties are shown in Table IV. Three varieties were tested: Camp Remy, Okapi and Granta. Table IV shows a clear year effect on protein content and Falling Number, typical indicators of growing and harvesting conditions. The differences in the amount of starch damage and ash content indicate differences in the effect of milling.

In Table V the results of the separation are presented. From these results it is clear that with all three varieties the recovery of protein in gluten is significantly higher for wheat grown in 1987 than in 1986. The differences observed with the variety Okapi are striking. Gluten recovery and coagulation index are 52 % and 24, respectively, for the 1986 samples. In 1987, values of 92 % and 98 were obtained. Although this may result from a higher content of protein in the flour, such a relation is not conclusive. The coagulation indexes of Okapi over the two years agree reasonably well with the relation between coagulation index and flour particle size. These are, however, limited and preliminary results. More experimental data are needed to explain the important effect of har-

TABLE IV

Characteristics of Flours from Two Years

Variety	Year	Protein Content (% dmb)	Falling Number (sec)	Damaged Starch (% dmb)	Ash Content (% dmb)
Camp Remy	1986	10.6	396	8.76	0.53
	1987	12.1	212	6.78	0.43
Okapi	1986	10.5	345	5.86	0.53
	1987	12.6	290	8.37	0.46
Granta	1986	10.4	361	14.38	0.53
	1987	10.2	209	10.87	0.61

TABLE V

Processing Quality of Three Varieties from Two Years

Variety	Year	Protein Recovery (% dmb)	Coagulation Index	Starch (% dmb)	Solubles (% dmb)
Camp Remy	1986	74.3	88.8	78.34	2.68
	1987	90.2	94.6	69.24	15.87
Okapi	1986	51.7	24.2	75.55	6.08
	1987	92.3	98.4	76.48	8.91
Granta	1986	79.4	94.6	73.33	10.25
	1987	90.5	89.4	79.06	9.15

TABLE VI

Effect of Cellulolytic Enzymes on the Processing Properties of Wheat

Enzyme Added	Total Gluten (%)	Starch Flour (%)	Solubles (% dmb)	Loaf Volume (% increase)
Flour 1, control	0.6	75.0	12.4	6.5
A	4.8	72.1	12.5	6.9
B	6.5	73.2	15.3	9.6
C	10.4	62.5	24.7	10.3
Flour 2, control	9.2	77.4	8.2	10.0
A	9.2	76.0	7.7	7.2
B	9.2	78.9	6.4	14.2
C	8.5	78.9	6.4	10.2

* Yield of total gluten and starch are expressed as % of flour. A, B, C represent the different cellulolytic enzyme mixtures used.

vesting conditions on starch/gluten processing properties.

Effect of Cellulolytic Enzymes on Processing Properties

We also studied the effect of cellulolytic enzymes on the processing properties of wheat. The enzymes used were almost completely devoid of amylolytic and proteolytic activities and were added at a concentration of 0.01 %, flour basis, directly before processing. A summary of the results is shown in Table VI.

Addition of cellulolytic enzymes resulted in an enormous increase in the gluten yield of a wheat variety with very poor processing properties. Addition of the same enzymes to a variety with good processing properties had no effect. The baking quality of the gluten thus produced, however, in nearly all cases was improved. This indicates that cellulolytic enzymes are powerful tools to improve the processing properties of poor quality wheats and yield a better product.

CONCLUSIONS

In conclusion, we have developed a new system for the evaluation of the processing properties of wheat for use in the wheat starch industry. This system provides a powerful tool not only for the evaluation of wheat varieties but also for systematic studies to reveal analytical parameters related to starch/gluten processing properties. Our study has led to the following conclusions:

- gluten yield and processing properties are not related to bread-making quality;
- flour protein content correlates positively with gluten yield;
- the coagulation index does not correlate with protein content but strongly correlates with flour particle size. This indicates that milling is an important factor.

Furthermore we have observed that:

- year-to-year effects can greatly influence processing properties. This is in part due to variations in protein content and flour particle size;
- processing properties can be improved using cellulolytic enzymes.
- addition of cellulolytic enzymes can also lead to an improved baking quality of the gluten.

LITERATURE CITED

GODON, B., LEBLANC, M., and POPINEAU, Y. 1983. A small scale device for wheat gluten separation. Qual. Plant Foods Hum. Nutr. 33: 161-168.

WEEGELS, P.L., MARSEILLE, J.P., and HAMER, R.J. 1988. Small scale separation of wheat flour in starch and gluten. Starch 40: 342-346.

28

DEVELOPMENTS IN THE EXTRACTION OF STARCH AND GLUTEN FROM WHEAT FLOUR AND WHEAT KERNELS

F. Meuser and F. Althoff
Technische Universität Berlin
Bundesrepublik Deutschland, 1000 Berlin 65

and H. Huster
Westfalia Separator AG
Bundesrepublik Deutschland, 4720 Oelde

ABSTRACT

Wheat starch and gluten are obtained industrially either by washing them out of wheat flour doughs in extraction vessels or by centrifugal separation of wheat flour dispersions with decanters or hydrocyclones. In the doughs and dispersions the gluten proteins form 3-dimensional and elastic meshlike agglomerates from which the starch granules can be extracted. The agglomeration of the gluten is caused either by kneading or by homogenizing the flour-water mixtures. The high tendency of gluten proteins to agglomerate has hindered, up till now, the use of a wet milling process to extract starch and gluten directly from wheat kernels. The processes developed for such extractions cause, during the wet milling step and the separation of the endosperm from the fibres and germ, a partial agglomeration of the gluten. This becomes mixed together with the fibres from which it can then no longer be separated. Trials on a laboratory scale have shown that the gluten can then be extracted when the steeped wheat grains are carefully milled using a corundum disk mill and the endosperm released is immediately sieved and diluted with water to form a slurry. By applying this procedure gluten agglomeration did not occur. A relatively high yield of gluten could be obtained after concentrating the slurry to a dough like mass in which the gluten was agglomerated by kneading.

INTRODUCTION

One effect of the European Community (EC) agriculture policy is that wheat is now preferred to corn as a raw material for starch extraction. This has led to a considerable expansion in wheat starch and gluten production which has, in turn, necessitated the expansion of existing wheat starch factories and the building of new ones. One result of this development is the requirement for advanced processing technologies capable of economically processing the raw material, hitherto only wheat flour. Some of these aspects will be dealt with in detail within the framework of this paper.

The majority of European wheat starch factories have a production capacity of only 50-200 tonnes/day of wheat flour. This is relatively low compared to European corn starch factories which can process up to 3000 tonnes/day of corn. The size of corn starch factories compared to all other starch factories, including potato starch factories, is to be explained by the fact that the composition of corn makes it an ideal raw material which can be processed all year round in a "bottled-up" processing system. The initial dry substance is recovered almost completely in the form of marketable end products so that the process employed causes minimal environmental problems in comparison with starch recovery processes from other raw materials.

A previous disadvantage for the corn starch factories in the EC was that the effectiveness of the process depended on using corn of a certain quality which is difficult to grow in Central Europe with the result that corn starch factories in the Federal Republic of Germany (F.R.G.) for example, were forced to import the raw material from the USA. However, this problem seems to have been solved by the application of new process techniques (Meuser et al. 1985), which enable comparable results to be obtained with corn grown in Western and Southern Europe. This is worth mentioning here because of the potential negative impact on the preference given to wheat as a raw material for starch extraction in the EC. In the light of the above, it is clear that wheat will only be used for starch production on an increasing scale if wheat starch factories can operate as economically as corn starch factories.

A look at the products recoverable from the two raw materials shows that the profitability of both processes is dependent on the yield of the various valuable constituents (Fig. 1). These are starch and germs in the case of corn and A-starch and vital gluten in the case of wheat. With regard to the profitability, it is interesting to note that the competitiveness of wheat starch factories as opposed to corn starch factories was largely guaranteed by the demand of the baking industry for vital gluten which consistently fetched a good price. However, this market, as with other established markets for vital gluten, has only a limited absorption capacity which appears to be

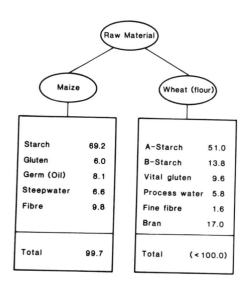

Fig. 1. Material balance for maize and wheat starch plants calculated on a dry matter basis.

exhausted by the existing production capacity. This would also appear to be indicated by the recent sharp drop in prices for vital gluten.

From this, it can be concluded that the raw materials, corn and wheat, for starch extraction will in future be in a dynamic competitive relationship which can be brought into equilibrium in dependence on the raw material costs and the profits attainable on the recovered products. This equilibrium can be influenced in favour of one or the other raw material by process technological measures. In this respect, the prime objective of wheat starch factories must be to increase the yield of A-starch and to realize profitable recovery of the substances dissolved in the process water. Both objectives should be combined with a simplification of the process technology to facilitate possible increases in the production capacities of the individual factories.

The existing possibilities for achieving these objectives and new approaches to the solution of these problems are the subject of the following exposition. The state-of-the-art forms the starting point for the special task of using laboratory tests to establish criteria and the basis for a process for the direct extraction of starch and gluten from wheat kernels.

STATE-OF-THE-ART

In the F.R.G. three concepts for the raw material supply to wheat starch factories are practiced (Fig. 2). They are primarily distinguished by the method of grinding wheat to flours used in the starch factories as raw materials. In the factories, the flour is wet-processed to agglomerate the gluten which is then separated from the starch utilizing various techniques.

Concepts for raw material supply

The most widely employed concept for raw material supply is to purchase wheat flour with an ash content of approx. 0.8% d.m. from mills which produce flours with different ash contents. The starch factories can acquire flours with this ash content at a favourable price since the mills are able to attain higher

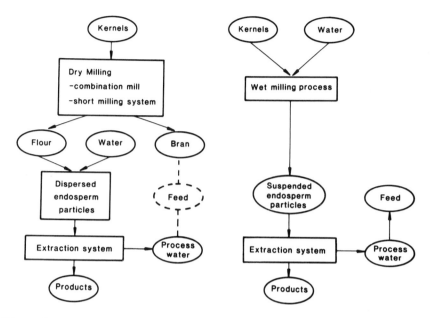

Fig. 2. Proposed wet milling system for the extraction of starch and vital gluten from wheat compared to conventional processes.

profits on their main markets with the remaining flours with a relatively lower ash content.

This advantage is also exploited by a second concept, whereby flour with an ash content of 0.8% d.m. is diverted as a partial stream of the flour production. This has the additional advantage of savings in flour transportation costs and the milling costs can be accounted internally.

This concept further offers the possibility of transferring the soluble constituents in the process water to the bran by means of an evaporation and drying process. This is not yet done owing to the high thermal energy requirement which would also necessitate the installation of evaporators. In one factory, the process water containing approx. 3% soluble substances from the flour is directly used as fodder by sending it to farmers. The money obtained in this way covers the transportation costs.

Factories working according to the first concept have not yet found as satisfactory a way of disposing of their process water. They, therefore, irrigate the process water as effluent, send it to municipal clarification plants or ferment it anaerobically (Witt 1985; Witt and Kröner 1988). The hitherto least expensive solution to the effluent problem was irrigation. In practice, this is becoming increasingly problematic due to legislation governing pollution control. A further disadvantage is that the caloric value of the dissolved dry substance is almost completely lost. Part of the caloric value can be utilized in anaerobic fermentation by taking advantage of the combustion heat of the methane obtained. However, at the present time, it is not possible to speculate as to whether the process costs can be recovered by utilizing the methane (Witt and Kröner 1988).

The third concept is to combine a mill with a starch factory (Dahlberg 1978). The mill can grind the highest possible flour yield from the wheat without having to comply with the legal regulations (BGBL. 20.10.1981) governing the flour trade in the F.R.G. and with the object of attaining the highest starch and gluten yield. For this system it is assumed to be advantageous that the raw materials can be selected specifically for suitability for starch and gluten extraction, the milling costs need not show an independent profit margin, the flour transportation costs are saved and the process water can be concentrated by evaporation and then be dried with the bran.

By this means, one wheat starch factory attains a dry substance yield which is comparable with that of a corn starch factory. The costs, however, are appreciably higher when related to the recovery of the dissolved substances in the process water; the solid content in the process water amounts to less than 50% of that in corn steepwater. Approximately three times more water must be evaporated to concentrate the process water of this wheat starch factory to the normal solids content normally found in concentrated corn steepwater.

From the above elaborations, it would seem logical to process the whole wheat, as unmilled grains as is the case of corn. Indeed, this has already been attempted (Langford and Slotter 1944; Galle et al. 1976). The main reason for the failure of these trials was that the wet extraction process was not capable of producing gluten of sufficient quantity or quality. The advantage of such a processing method was that the total yield of starch and gluten could be recovered in one process line. This could not be technically realized at that time because the modern techniques of gluten agglomeration and separation from the starch had not been developed.

Techniques for starch/gluten separation

The latest developments in processing techniques in wheat starch factories are characterized by two basic technological advances: agglomeration of the gluten with homogenizers (Westfalia Separator AG 1987) and centrifugal separation of the gluten from the starch using decanters (Kerkkonen et al. 1979) or hydrocyclones (Verbene and Zwitserloot 1978). These developments have led to the demise of the most important, long-standing process, the Martin process, whereby the starch is separated from the gluten by washing out dough formed from flour and water in

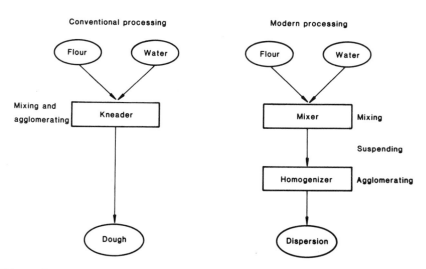

Fig. 3. Procedures used to agglomerate the gluten in wheat flour.

washing installations. The dough is formed in a dough kneader by mixing flour with water, whereby the gluten is agglomerated by kneading (Fig. 3).

The agglomeration of the gluten is the basic precondition for its separability from the starch by washing. However, the effectiveness of the washing process does not require a doughy mass, as demonstrated by the Batter process (Anderson et al. 1958) in which a soft dough is fragmented in a centrifugal pump after adding water with no significant losses of gluten in the subsequent washing out via screens. These findings where of fundamental importance regarding centrifugal separation techniques for the development of new agglomeration processes, since the centrifugal separation in decanters and hydrocyclones requires that the mass to be separated is of a pumpable consistency.

One possible method could therefore be to convert the dough into a pumpable dispersion. However, this technique need not be adopted as the gluten can also be agglomerated in flour-water suspensions with homogenizers. The agglomeration takes place due to friction of the mass in the machine's valve. The extent of the agglomeration depends on the attainable degree of friction in the gluten which can, inter alia, be influenced by the concentration of the suspension.

Normally, part of the flour is thoroughly mixed with water in a continuous mixer for a few seconds. The discharging suspension is pumped to the homogenizer where, due to its working principle, most of the necessary agglomeration takes place. The mass leaves the machine as a dispersion which is diluted with water prior to centrifugal separation of the gluten from the starch (Fig. 4).

Centrifugal separation produces a different distribution of the mass constituents. Whereas the washing process produces a pure gluten fraction and the total starch together with the insoluble pentosans forms a second fraction suspended in the process water, with centrifugal separation a relatively pure A-starch fraction is obtained and a second fraction in which the gluten, B-starch and pentosans are concentrated.

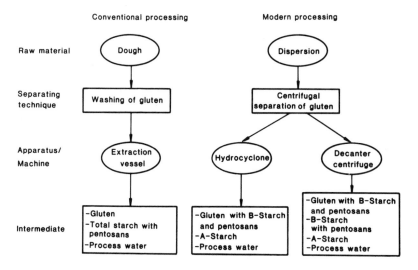

Fig. 4. Processes used for the separation of flour components from dough or flour-water dispersion.

The main advantage of centrifugal separation, as opposed to washing, is that the main fraction, the A-starch, is separated out of the diluted dispersion in a matter of seconds, whereas it takes up to half an hour with the washing process. This shortening of the retention time of the mass in the separation system in combination with the small solids and water volume in the system per unit of time (approx. 20 times smaller than in the Martin process) is extremely advantageous with regard to the process water flow and its pH.

Recent advances in the field of centrifugal separation technology now make it possible to separate the suitably diluted dispersion into three solids-containing fractions. This is especially advantageous with regard to the necessary refining of the gluten, since in this way solid and dissolved substances are removed from the fraction containing the gluten. The splitting of a dispersion with a decanter operating according to this principle is illustrated in Figure 5 (Witt 1988).

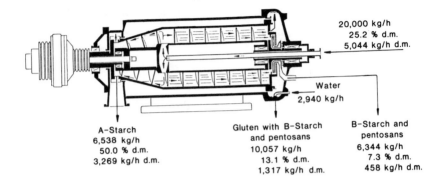

Fig. 5. Centrifugal decantation of a flour-water dispersion into three phases.

Figure 6 is a simplified flow diagram of the distribution of the dry substance from the flour in a starch factory running this process with a production capacity of 5000 kg flour/hr. For the sake of simplicity, the process water flow is not depicted in detail.

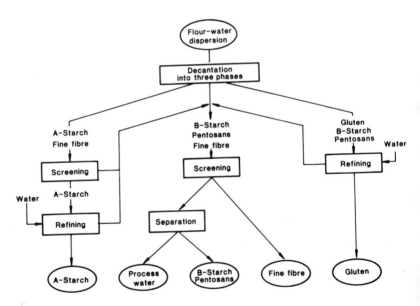

Fig. 6. Separation and refining of the components of a homogenized flour-water dispersion.

In this context, it should be mentioned that the process water obtained after separation of the B-starch and pentosans contains most of the soluble flour constituents. The process water is used in this factory partly to dilute the dispersion before decanting. By this means, a constant concentration of dissolved substances of 3.0-3.3% can be adjusted in the process water at a constant process water volume. The product specifications do not permit a higher concentration so that the process water volume to be treated as effluent is precisely defined.

An appreciable increase in the process water concentration, in order to reduce the effluent volume, would require that the flour dry substance be subjected to a completely different preparation which would alter the installation concept accordingly. A logical approach would be to extract the soluble substances before agglomeration of the gluten. This technique seems to be realizable since, with homogenizers, the gluten can be agglomerated in flour-water suspensions obtained by concentrating highly diluted flour-water suspensions. This has been shown by laboratory-scale experiments.

Recovery of starch and gluten from wheat kernels on a laboratory scale

The laboratory tests were concerned with assessing the steeping characteristics of wheat kernels and wheat flour products, the suitability of various wet milling techniques, and the screenability of the flour fractions. We describe only those tests which are of interest for converting the project idea into an industrial scale process.

MATERIAL AND METHODS

<u>Raw material</u>. Soft-milling wheat grown in Germany was used as raw material. The wheat was analyzed using established methods (AfG 1978) to determine its starch, protein, fat, ash and water contents. The intermediate products from the wheat processing were, if necessary, analyzed using the same methods in order to compile mass balances on the basis of the measured

values and the masses of the intermediate and end products.

Steeping tests. The wheat was steeped in tap water in glass beakers at 323 K under 100 kPa and 750 kPa in a 3 l pressure vessel in batches of 250 g for 3 and 6 hours. The water absorption after 3 and 6 hours was assessed by determining the water content of the kernels after pouring off the steepwater. In addition, the steepwater dry substance was determined by freeze-drying the total steepwater volume. The water content of the steepwater dry substance was assumed to be 5%. The steepwater dry substance established in this way was related to the dry substance of the wheat kernels and expressed as a percentage.

Milling and screening tests. Water was added to the steeped wheat kernels, which were ground in a corundum disk mill (Fryma, type: KMR-2). The volume of added water was such that the suspension contained a mass concentration of 2.5% after milling. The suspension was screened over a test screen (mesh width: 160 μm, diameter: 20 cm) without further addition of water. The oversize was dried overnight in a forced-air drying oven at 333 K. The water, protein and starch content of the dried and weighed mass were then determined. The results were used to calculate the distribution of protein and starch in the milled fractions.

Tests for starch and gluten recovery. Wheat grains steeped without pressure at 313 K were ground in a corundum disk mill so that the suspended endosperm could be screened over two series of connected test screens (1st screen: mesh width: 500 μm, 2nd screen: mesh width: 160 μm; diameter of both screens: 20 cm) after further addition of water. The undersize from the second screen was concentrated with a laboratory centrifuge (Hereaus, type: Junior 15000). The clear phase discharge of the centrifuge was recirculated for milling the wheat kernels and screening the milled material. The ratio of milled material dry substance and process water was approx. 1:10. Milling the grains, screening the fibres and concentrating the undersize took approximately 30 minutes.

After the fibres had been screened off, the overflow from the centrifuge was collected in a glass beaker. The fibres were dried overnight at 333 K in the forced-air oven and then weighed. The residue in the chamber of the centrifuge rotor which contained the steeped and largely agglomerated gluten was carefully washed out by hand on a test screen (mesh width: 160 µm; diameter 20 cm). The washed out gluten was freeze-dried and weighed. The overflow collected in the glass beaker from the preceding screening was used as wash water. The wash water from the gluten washing was again recirculated via the centrifuge, whereby the test screen undersize was sent across a further test screen (mesh width: 63 µm; diameter: 20 cm) which served as a refining screen. The oversize from this screen was collected as fine fibre. The fine fibre was dried overnight at 333 K in the forced-air oven and weighed.

It took about 30 minutes to wash out all the gluten. The solids in the permeate from the refining screen were separated out in the rotor chambers of the centrifuge. The total overflow volume was collected and its mass determined by weighing. An aliquot was freeze-dried to determine the mass proportion and the composition of the soluble matter from the milled raw material.

The sediment was removed from the rotor chamber and diluted with tap water at a mass ratio of 1:1. The suspension was then sedimented in a test-tube centrifuge (Hereaus, type: Junior II). Two layers of solids were formed and were covered by a liquid phase. The liquid top phase was poured into a glass beaker. The two layers of solids were separated with a spatula into A- and B-starch. The B-starch was combined with the liquid top phase, freeze-dried, and weighed. The A-starch was dried overnight at 333 K in the forced-air drying oven and weighed.

RESULTS AND DISCUSSION

Wheat from steeping, when processed as for corn in the HD system (Meuser et al. 1987) was hydrated uniformly in 6 hours: this steeping period being independent of the pressure. Within this period, wheat attained a water content of 45%, whereby approx. 1% of its dry substance was leached out into the steep water (Table I). Thus it can be concluded whereby approx. 1%

TABLE I
Water Absorption of Wheat Kernels and Leaching out of Soluble Matter into Steepwater: Dependence on Steeping Conditions

Steeping Conditions			Water Content of Steeped Kernels (%)	Solubilized Matter (% of total dry matter)
Pressure (kPa)	Temperature (K)	Time (h)		
100	323	3	40.5	0.7
100	323	6	45.5	1.1
750	323	3	41.0	0.7
750	323	6	45.0	0.9

of its dry substance was leached out into the steep water (Table I). Thus it can be concluded that steeping of wheat on an industrial scale can likewise be carried out without pressure. The construction of a continuously operating steeping installation would therefore be considerably less complicated than that required for the HD process for corn.

Wet milling of the wheat grains and wet screening of the milled material produced 86.5% endosperm in the

TABLE II
Distribution of Kernel Dry Matter between the Fractions Resulting from Dry and Wet Milling of Wheat

Process	Distribution of Dry Matter (%)	
	Bran/Fibre	Flour/Endosperm Suspension
Combination mill	17.0[a]	83.0
Short milling system	15.0[b]	85.0
Wet milling system	13.5	86.5

[a]Assumption: Flour yield 83% with 0.8% ash d.m.
[b]Dahlberg 1978.

extraction stage (Table II). This means that a mass 3.5% greater than that achievable in the combination milling of wheat (83%) was available for extraction of starch and gluten. The yield obtained was even 1.5% higher than that attainable (85%) with the short milling system (Dahlberg 1978). It can be assumed that the fibre fraction to be dried in the wet milling process would be correspondingly smaller.

Wet screening of the fibres resulted in 97.5% of the starch contained in the kernels being transferred to the extraction stage (Table III). This is ample proof that the fibre fraction can be easily separated from the endosperm, a necessary pre-condition for a high starch yield. Calculation by difference based on a mean starch content of 18% in conventional bran, shows that wet milling techniques could in this respect offer slight advantages over dry milling of wheat.

Even more notable are the results shown in Table IV for the ratio of A- and B-starch. The mass ratio related to the starch content of the two fractions was 82.1 to 9.6 (8.5 to 1). This was significantly higher than the mass ratio normally obtained in German starch factories. This ratio is, for example, 4.6 to 1 in a factory depicted diagramatically in Figure 6. The much better ratio achieved on a laboratory scale was due to the process water flow which, with modifications,

TABLE III
Distribution of the Starch Content between the Fractions Resulting from Dry and Wet Milling of Wheat

Process	Distribution of Starch Content (% of total starch content)	
	Bran/Fibre	Flour/Endosperm Suspension
Combination mill	4.5[a]	95.5
Short milling system	3.9[a]	96.1
Wet milling system	2.5	97.5

[a]Assumption: Starch content of bran = 18% d.m.

TABLE IV
Extraction Rate of Starch from Wheat Kernels and its Distribution between A-Starch and B-Starch Fractions

Recovery of the Kernel Dry Matter (%)	Extraction Rate[a] of Starch (% of total starch content) Starch	Extraction Rate of A- and B-Starch (% of total starch content) A-Starch	B-Starch
94.9	87.0	77.9	9.1
100.0	91.7	82.1	9.6

[a]The extraction rate of starch was calculated as the ratio of the weight of A- and B-starch fractions to the analytically determined total starch content. In this calculation the purity of the starch fractions was not taken into consideration.

could be employed just as successfully on an industrial scale.

The A-starch was of high purity (Table V); its protein content was only 0.2%. In contrast, the protein content of the B-starch was too high (17.4%). This was attributable to the washing process. The manual washing on the screen had such an abrasive effect on the soft gluten that gluten passed through the

TABLE V
Gross Composition Starch Extracted from Wheat by Wet Milling

Component	A-Starch (% d.m.)	B-Starch (% d.m.)
Protein	0.2	17.4
Fat[a]	0.6	
Ash	0.24	
Starch		62.3

[a]Extracted with petroleum ether after hydrochloric acid treatment.

TABLE VI
Recovery of Protein in the Fractions from Wet Milling of Wheat

Fractions	Distribution of Protein Recovered (% of total protein)
Fibre >500μm	13.6
Fibre <500μm >160μm	6.4
Fine fibre <160μm > 63μm	2.1
Gluten	42.1
A-Starch	0.5
B-Starch	13.3
Process water dry matter	16.4
Total	94.4

screen. This problem would not be encountered on an industrial scale as the gluten would be better agglomerated after concentration of the undersize obtained from the fibre screens with the homogenizer. Moreover, centrifugal separation of the fractions in the decanter would be more gentle and effective than that attainable in this laboratory experiment.

The recovery of protein in the gluten was only 42.1% of the total in all intermediate products (Table VI). This is a very low recovery in the gluten in comparison with recoveries in wheat starch factories where a yield of 70% is obtained, resulting in a gluten yield of about 10% related to the kernel dry matter. However, it could be increased relatively easily due to the expected better separation of the gluten from the B-starch on an industrial scale. A further increase in the gluten yield would not be possible if part of the gluten passed into the fibre fraction. This unintended mixing in the laboratory experiment was a result of agglomeration of the gluten during the wet milling and wet screening steps. On an industrial scale, this could be largely avoided by better overall control of the process.

The latter is indicated by the fact that adjusting the suspension from a solids' concentration of 10%

TABLE VII
Protein Content of the Fibre Fraction after Sieving of Wet Milled Wheat Suspended in Water

Concentration of of the Suspended Mass (% d.m.)	Protein Content of the Fibre Fraction (% of total protein content of wheat)
10.0	20.0
2.5	13.2

to 2.5% in the laboratory experiment led to a significant reduction in protein associated with the fibre fraction (Table VII). The calculated value was reduced from 20.0% to 13.2% related to the protein mass present in the whole kernel. The protein content of the fibre fraction, which was used to compute this figure was 14.6%. It, therefore, corresponded to the protein content of the outer kernel layers. Consequently, the fibre fraction was free of gluten.

The test results clearly show that, in order to achieve complete separation of the fibres from the endosperm, it is essential to prevent agglomeration of the gluten during wet milling the grains and wet screening the fibres. On an industrial scale, the concentration of the suspension must be selected so that a minimal process stream is handled in the decanting stage for concentrating the undersize.

A dry substance yield of 95% was attained in the laboratory experiment. The protein yield of 94.4% was comparably high. The analytically, determined dry substance yield was extrapolated to 100% to compare the yields of individual masses with the yields in other industrial-scale processes. In Table VIII, the calculated values are compared with the mass yields actually obtained in a starch factory, typical of the short milling process on the basis of published results. In all three cases, the figures are related to the dry substance of wheat as raw material, for comparison across all process stages.

This comparison makes it clear that the advantage of the application of the wet milling process on an

TABLE VIII
Distribution of Wheat Kernel Dry Matter Components in Starch Processing Systems

Fraction	Distribution of Wheat Kernel Components[a] (% of d.m.)		
	A[b]	B[c]	C[d]
Bran/Fibre	17.0	15.0	13.5
Fine fibre	1.6	*	1.4
Gluten	9.6	10.2	8.4
A-Starch	51.0	53.5	58.2
B-Starch	13.8	14.4	10.6
Process water	5.8	6.9	7.9

[a] Based on the assumption of 100% dry matter recovery.
[b] Combination mill, mixer, homogenizer, decanter centrifuge.
Assumption: Flour yield 83% with 0.8% ash d.m..
[c] Source: Dahlberg 1978. B: Short milling system, mixer, homogenizer, decanter centrifuge.
* Assumption: Fine fibre fraction included in B-Starch fraction.
[d] Laboratory scale experiment. Wet milling system, sieve, centrifugal concentration.

industrial scale would be primarily in the potentionally higher A-starch yield. Together with the realizable high concentration of dissolved substances in the process water, this could significantly improve the economic efficiency of the process, particularly in comparison with the "combination mill" starch factory on the same site. However, this would require a higher gluten yield than achieved in the laboratory experiment. The optimization of the process could then lead to the yield relations shown in Table IX.

The higher transfer of soluble substances into the process water compared to the dry milling process need not be considered a disadvantage as the soluble substances in the process water can be concentrated up to 6%. This fraction could be recovered economically with the fibre fraction by a suitable evaporation and drying process.

TABLE IX
Calculated Distribution of Wheat Kernel Dry Matter Components between Fractions from a Newly-Developed Wet Milling System

Fractions	Distribution of Wheat Kernel Components[a] (% of d.m.)	
	Laboratory Scale -Status Quo-	Industrial Scale -Prediction-
Fibre	13.5	12.7
Fine fibre	1.4	1.4
Gluten	8.4	10.4
A-Starch	58.2	58.2
B-Starch	10.6	7.5
Process water	7.9	9.7

[a] Based on the assumption of 100% dry matter recovery.

LITERATURE CITED

ANDERSON, R.A., PFEIFER, V.F., and LANCASTER, E.B. 1958. Continuous batter process for separating gluten from wheat flour. Cereal Chem. 35:449.

ARBEITSGEMEINSCHAFT FÜR GETREIDEFORSCHUNG. 1978. Standard-Methoden für Getreide, Mehl und Brot. 6th ed., Verlag Moritz Schäfer, Detmold (F.R.G.).

BUNDESGESTZBLATT I. vom 20.10.1981. P. 1039: 17th Durchführungsverordnung zum Getreidegesetz, Bonn (F.R.G.).

DAHLBERG, B. I. 1978. A new process for the industrial production of wheat starch and wheat gluten. Stärke 30(1):8.

GALLE, E.L., KOLOSKO, J.F., and MAYOU, J.L. 1976. Verfahren zur Hydrobehandlung von Weizen. German patent application, DT 2607829 Al. Date of application 26.02.1976.

KERKKONEN, H.K., LAINE, K.M., ALANEN, M. A., and RENNES, H.V. 1979. Verfahren zur Abtrennung von Gluten aus Weizenmehl. German patent specification DE 2345129 C3. Date of application 07.09.73. Date of issue of patent letter 15.02.79.

LANGFORD, C.T., and SLOTTER, R.C. 1944. Wheat starch production, a new method. Ind. Eng. Chem. 36:404.
MEUSER, F., GERNMAN, H., and HUSTER, H. 1985. The use of a high-pressure disintegration technique for the extraction of starch from corn. Page 161 in: Progress in Biotechnology, Vol 1: New Approaches to Research on Cereal Carbohydrates. R. D. Hill and L. Munck, eds. Elsevier, Amsterdam.
MEUSER, F., WITTIG, J., HUSTER, H., and HOLLEY, W. 1987. Recent developments in the extraction of starch from various raw materials. Page 285 in: Cereals in a European Context. I.D. Morton, ed. VCH Publishers, Weinheim, New York.
VERBERNE, P., and ZWITSERLOOT, W. 1978. A new hydrocyclone process for the separation of starch and gluten from wheat flour. Stärke 30(10):337.
WESTFALIA SEPARATOR AG. 1987. Information on Starch Production, No. 4. Oelde (F.R.G.).
WITT, W. 1985. Verfahrenstechnischer und wirtschaftlicher Vergleich von Methoden zur Reinigung des Abwassers einer Weizenstärkefabrik unter besonderer Berücksichtigung der anaeroben-aeroben Reinigung. Thesis, Technische Universität Berlin D 83.
WITT, W. 1988. Kröner-Stärke Ibbenbüren (F.R.G.).
WITT, W., and KRONER, H. 1988. Beseitigung von Weizenstärkeindustrieabwässern-Ein Wirtschaftlichkeitsvergleich -. Stärke 40(4):139.

29

THE ENGINEERING OF A MODERN WHEAT STARCH PROCESS

Derek J. Barr

Barr & Murphy Overseas Ltd
BPP House, 142-154 Uxbridge Road
London W12 8AA, England

HISTORY

Twenty years ago the processing of wheat flour into starch and gluten was a small scale industry and the process plant involved was often homemade.

Fundamentally the Martin process, originally developed in the late 19th century, was used for separation of starch and gluten in the following way. The flour was mixed with water to make a dough and this was washed in a variety of mixing troughs with very large amounts of water to get a clean starch-free gluten and a dilute starch milk which was then concentrated and separated into 'A' and 'B' starch.

Although this equipment could be purchased commercially, at least half of the gluten and starch processors built their own systems based upon principles that were publicly available. The common feature of all these processes was that they used at least ten tons of water per ton of flour, giving rise to high levels of effluent. They also needed frequent shut-downs for cleaning.

The industry was small with most plants processing 3-5 tons/hour of flour. The high price of gluten was the reason for their existence, with the starch being treated as a by-product.

Ten to twelve years ago all this changed. The larger gluten producers, particularly in Australia, saw a need for more efficient processing and reduction of waste water, and larger processing units ranging from 6 to 10 tons/hour of flour.

PROCESS DEVELOPMENTS

Starting in the early 1970's several new process development initiatives were taken (Fig. 1) (Fig. 2). These varied widely from developments of the Martin process to improve the engineering and quality of the production, to radically new developments such as a whole wheat process which was designed to avoid flour milling completely by steeping of whole wheat. The research lasted about five years until one or two of the processes became commercially acceptable and the others were withdrawn from active use. The whole wheat processes were not successful, principally because the separation of fiber from the gluten proved difficult, and also they involved drying of all the bran which was an additional major energy expense that flour based processes could avoid. The flour based processes again fell into two categories, those using more orthodox washing techniques for separating and concentrating the starch and those using decanters for gluten and starch separation. The traditional processes, modernized by the use of hydrocyclones proved more compact and less expensive than the decanter processes, and full decanter processes were commercially installed in only two or three locations, with limited success. The hydrocyclone process uses a mixture of agglomeration, hydrocyclone separation and centrifugal separation techniques in combination, for maximum ease and reliability of operation.

THE B & M HYDROCYCLONE PROCESS

In the last ten years the B & M hydrocyclone process has been the principal process installed in the wheat industry throughout the World. This has been a period of continuous development, and considerable technical advances have been made in the engineering of the process to maximize the yields and improve the product qualities. At the same time the continuity and reliability of operation has been improved to a level undreamed of ten years previously

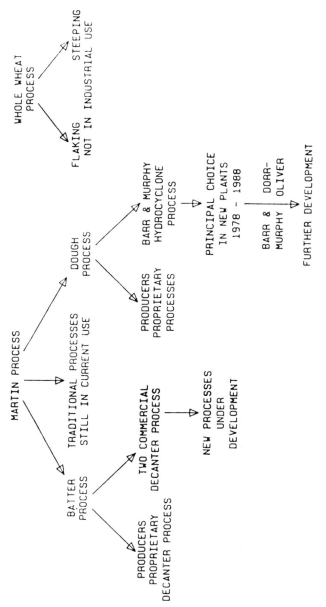

Fig. 1. Process developments.

The process shown in the diagram (Fig. 3) starts with wheat flour, frequently produced in a special mill to give a processing quality with high

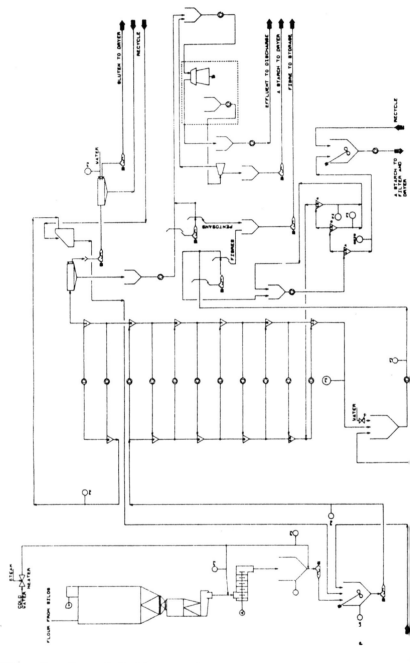

Fig. 2. P and I diagram for wet process.

extraction. The wheat flour is mixed with a controlled amount of water into a dough, which after gluten development is passed to a dilution and agglomeration tank. There it is diluted with further addition of water, and the tank is agitated to allow gluten to form. The starch/gluten mixture with gluten formed within the starch milk is pumped to a set of hydrocyclones where the gluten is separated and the starch is progressively refined. From the underflow of the hydrocyclones the 'A' starch is extracted in pure form. The gluten and 'B' starch leave the top of the hydrocyclones together and the gluten is separated in a special rotary screen. The remaining starch milk is screened to remove fine fibers and then separated into two fractions, one of which is a second 'A' starch fraction, the other of which is the 'B' starch. All three principal products, gluten, 'A' starch and 'B' starch, are normally dried. The coarse and fine fibers are taken wet as a separate stream and the remaining solubles are contained in the effluent which can either be treated by evaporation or by some other acceptable form of treatment.

This is a simplified description of the process that has now been accepted in the wheat starch industry worldwide. The process is adaptable and reliable and fulfills the need for larger sizes of plant in the range from 10 to 20 tons/hour of flour input.

ENGINEERING OBJECTIVES

In the engineering of a wheat starch process the designer has to analyze carefully the objectives of the customer in building a plant, and use them as the basis for the design. These can be summarized as follows.

Products

The first objective is the quality and use of the products, namely the gluten and starch, and the consideration of the uses available for 'B' starch and any by-products. Effluent disposal may also be a problem to be examined. Product qualities, yields and

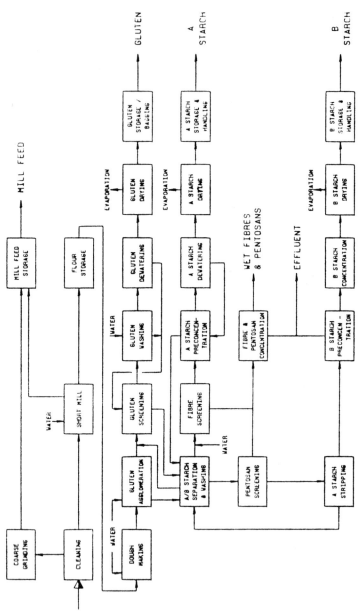

Fig. 3. Barr & Murphy wheat separation process.

water consumption are all closely interrelated and t
some extent can be independent of the type of process
For example if the highest possible 'A' starch qualit

s required a higher water usage is needed than if a standard quality is acceptable. In the wheat starch market a protein content of 'A' starch of between 0.30 and 0.33% is normally acceptable. However it is possible through careful operation of the plant to achieve much lower protein levels if required.

Raw Materials

The second consideration is the type of wheat that is available or that is to be used. Frequently production considerations require a specific quality of wheat for processing rather than a lesser quality which may be locally available. This is particularly true in Europe where French wheat has been adopted as the standard for most wheat process plants due to its higher protein content and better processing characteristic compared to other European wheats.

Cost

The third consideration is cost. Cost involves many aspects, the most important being the capital cost and the operating cost of the plant. The process must produce the right products of the right qualities from the raw material chosen, at an economic operating cost and at a justifiable capital cost.

This involves using fully developed technology to the latest designs with a clear understanding of the operating requirements of the user of the plant. The right balance must be achieved between capital and running costs, high and medium technology levels, the country of operation, the type of operators available, and the capital budget constraints. Although this may be seen as a commercial exercise it is fundamental to the engineering of the plant and to the way the plant is designed and built.

Operability

The final consideration is operability. While this term may not be easy to define, this will often be the user's measure of satisfaction with the plant. The process must be flexible, and able to cope with changes in the raw material which naturally occur between seasons and between different raw material sources, without affecting the final product qualities or the products that the plant produces. A well engineered process does not require much operator attention for minor product changes. Good system design as well as control design will ensure that changes between batches of wheat or flour do not affect the smooth operation of the plant.

This criterion of operability is the truest qualitative test of the engineering of the plant. A new process prototype will always be more difficult to operate than a plant whose design has been refined over several years and numerous installations. In essence a well developed process which is intrinsically simple will always be more easy to operate, have less downtime, and be more satisfactory in its overall performance than a newer process which is in its final development stages.

EXPERIENCE AND DEVELOPMENTS

The hydrocyclone process has undergone steady development over the last ten years to become flexible and operable. As the requirements of the industry have become more sophisticated, have required larger installations, much higher degrees of automation and much tighter control on product qualities, the engineering development of the process has optimized it to enable the industry to maximize its return. This is strengthened by combining the engineering expertise and know-how of two equipment suppliers involved in the hydrocyclone wheat process thereby providing a complete and fully co-ordinated service to the wheat starch industry.

The approach draws together technological resources available to fulfil the four engineering objectives and made possible to build the principal process equipment in the plant. This ensures a depth of design knowledge that can tailor a process to give products matched to the user's markets and needs.

30

PRODUCTION OF WHEAT STARCH AND GLUTEN: HISTORICAL REVIEW AND DEVELOPMENT INTO A NEW APPROACH

W. R. M. Zwitserloot

Westfalia Separator Nederland B.V.
Cuijk, The Netherlands

The production of vital wheat gluten has been taking place on a commercial scale only since the first dryer for vital gluten was developed in Australia in the 1930's.

Production facilities were marked by several factors:
1. Raw material was wheat flour.
2. Production capacities were limited, maximum a few tons per hour of flour processing because of several limiting factors such as: size of the equipment especially the gluten washers, limited markets, production of effluent (up to 15 m^3/ton flour) in spite of the fact that there was no requirement to treat the effluent in most countries.
3. The separation process of the components always depended on the gluten properties: a dough or batter was formed, the gluten appeared in its rubbery, chewing gum like structure, from which the starch granules were separated by large amounts of water.

Mainly two processes were used: the Martin process based on washing of a stiff dough and the Batter process based on separation of gluten and starch from a thin batter.
This was the situation until the mid 1970's.

In the meantime the corn and potato starc industry had developed enormously, mainly in Europ and the US.

Wheat starch production until those days staye behind and was never more than a few percent of cor and potato starch production.

In the mid 70's effluent problems became a threa to the industry and new "low water" processes wer developed in order to survive:
a) <u>Improved Martin or Batter process</u> working wit effluent recycle wherever possible. This reduce water consumption from 15:1 to 5-7:1,
b) <u>The hydrocyclone process</u>, based on separation o gluten and starch out of a diluted slurry i hydrocyclones reduced water consumption to 4-5:1 and the
c) <u>Decanter based processes</u> achieved a 4:1 rati (US Patent 3.951.938).

All these had the disadvantage of being sensitiv to flour quality (especially using the soft Europea flours) and achieved lower yields, mainly in protei recovery (as gluten) and A starch.

Because of the relatively high effluent recycl bacteriological growth increased and resulted in mor losses of solubles. A drop in pH caused equilibriu breakdown, lead to product losses and viscosit increases, and resulted in loss of small granule starch to B starch due to clarification effects. Bu the "killer", the effluent, became a manageabl problem.

RAW MATERIAL

While the raw material for starch and glute production continued to be wheat flour, the industr slowly started to think of using whole wheat. Durin the 70's numerous technologies were invented (an patented) to use whole wheat as raw material. Some o them were:

<u>Staley USA:</u> grinding whole wheat with hammer mill followed by the classical Martin process (BRD Paten 24.54.236).

disadvantage: large amounts of wet fiber to be dried and poor quality (color, fiber, ash and fat content) gluten.

Armarco USA had a process based on flaking wheat kernels, further processed by the Martin technology (US Patent 3.790.553). This system had the same disadvantages as the Staley process.

Subsequently, different technologies were developed, such as steeping of whole wheat in HCL (US Patent 3.851.085), foam separation of gluten, and separation of gluten by chemicals. None of these became viable.

Another approach to using whole wheat was the development of the short mill by Buhler (BRD Patent 6.42.628). The short mill is based on conventional milling technology but with fewer streams; it produces a somewhat coarser flour with less damaged starch giving better A:B starch ratio. A higher flour extraction was satisfactory since there was no need to produce a perfect flour; an ash content of 0.80-0.85% as compared to the previously required 0.5-0.6%) posed no problem. It enabled the starch producer to select a wheat and produce a series of flours. The famous saying of those days was: "the problems of the flour mill always end up in the starch plant". I can confirm this situation based on my own experience. Whenever there was a processing problem in "my" starch plant, a telephone call to the flour mill had the required effect. Of course, the miller claimed ignorance that anything had gone wrong, but somehow with the next truckloads our problem was largely over.

The short mill still had one disadvantage: its price! In many European countries there is a milling overcapacity and the short mill is not economical, even where it has been further simplified by Buhler and other companies. Thus, the United Milling Systems use disc mills in the first break and claim that this leads to a better quality flour (lower starch damage), as confirmed by laboratory analyses.

HD PROCESS (High Pressure) FOR STARCH/GLUTEN SEPARATION

In the last two decades the equipment suppliers and the starch industry improved the mentioned above processes and made them reasonably stable and reliable; in addition, a lot of work was carried out to reduce even further effluent production. However, minimum water consumption reached in the Martin, Batter, Raisio and Weipro processes has been about 4-5:1.

Further reduction resulted in irreversible process instabilities, which made it impossible to separate gluten and starch.

Here is where the HD process comes in (Fig. 1). The high pressure process is in principle the decanter-based Weipro process mentioned earlier. It does not make use of the flour and gluten dough properties for separation of gluten and starch, but is based on the gravity force for separation.

The process was developed by the Technical University of Berlin, West Germany together with Westfalia Separator AG (Witt and Kröner 1988) and works as follows.

Flour and water are mixed for as short a time as possible, 30 seconds is sufficient, to a lump free slurry. The slurry is transferred to a three-phase decanter via a high pressure homogenizer, where, at pressures of up to 100 bar, the high shear forces in the homogenizer cause a very sharp split of A starch (phase 1) and gluten plus B starch (phase 2) and effluent + pentosans (phase 3) (Fig. 2). The slurry after the homogenizer can be simply separated in the three-phase decanter. The three phases can be purified, as required, using conventional technology. Water consumption is reduced to well below 3:1 resulting in a 1.5 to 1.8:1 effluent. Additional advantages are that as a result of the very sharp split the A starch yield is increased by up to 10% compared to the Martin or the Hydrocyclone process.

No chemicals are required to control the pH because of the holding time of seconds in the process between flour and products. Cleaning in place is required only after over 3 weeks of continuous

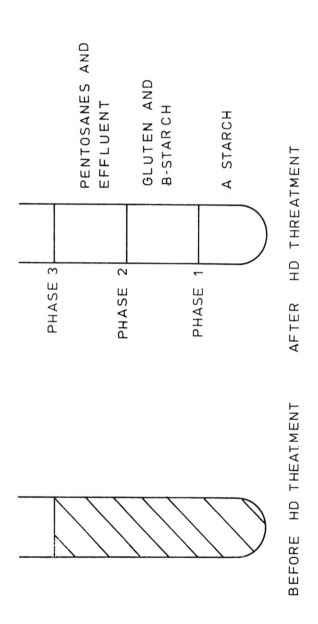

Fig. 1. Westfalia H. D. Process.

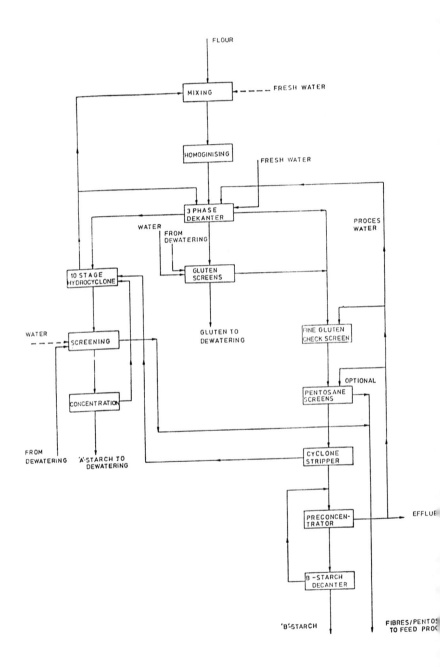

Fig. 2. Westfalia Separator.

production and does not affect adversely yields or product qualities.

Finally, a very important factor (especially in countries where soft wheats are used) is the fact that flour quality per se is not as important as in the other processes since the HD process is based on gravity separation.

Some flours that break down in the Martin or the hydrocyclone process can be used in the HD process. Of course, if the gluten-forming capacity is destroyed, no gluten can be produced. Still, the sensitivity to gluten quality exhibited in other processes is removed largely in the HD process.

The HD process itself was not developed primarily for wheat starch and gluten production. The invention originated with replacing the rasps by the HD pump in potato starch production and was followed by application of HD in corn starch production, where the major breakthrough was made. In corn starch production, maize is steeped with SO_2 during 45-60 hours. This step was replaced by high pressure steeping and grinding, without SO_2, within 3 hr. The process has many advantages; details are available upon request.

Presently processes for production of starch from peas and beans are being tested. In addition, production of starch and gluten from whole wheat, is underway; it is summarized later in the paper.

The dream of a multipurpose plant is becoming more of a reality as the principles of the HD process can be used for a variety of raw materials.

CURRENT STATUS OF WHOLE WHEAT PROCESSING

At the beginning of this paper, the statement was made that the older wheat gluten plants were limited in size. This has changed too in the early 80's. Because of the development of the new process systems and several political reasons, most of the limiting factors in the gluten industry were eliminated and plant sizes increased from about 5 t/hr to well over 10 t/hr of flour handling. Presently, in Europe two plants of 20 t/hr and one of up to 30 t/hr of flour processing are in operation.

This resulted in the pressure to increase markets for the products of the gluten and starch industry. In Europe this problem was solved in a simple manner, EEC policy forced the starch industry to use wheat instead of "imported" corn for the production of glucose. This led to an increase of flour handling from 350,000 t in 1980 to well over 1 million t in 1988, by simply replacing part of the corn starch.

Australian glucose producers used wheat starch for many years and, not knowing better, had no great difficulties with it. I can assure you, however, that in Europe (where historically potato and corn starch were the basis for glucose production) the introduction of wheat was responsible for a lot of grey hair. Wheat starch is a difficult starch to produce glucose from. Last year the fantastic increase in wheat starch production in Europe and, as a consequence, the increase in gluten production resulted in a sharp fall in gluten prices.

One has to wonder whether gluten is not becoming a common commodity, compared by some to corn gluten or potato protein. The consequence is a low price just a little above that of cattle feed (proteins); the substantial research on new applications for gluten, notwithstanding. Under those circumstances, the demand for high quality viable gluten will be much reduced and this will reopen the door for whole wheat processes.

As mentioned before, using whole wheat invariably led to gluten quality problems (color, fat, ash, etc).

At present, however, two systems are being tested (more or less successfully) in plants that have HD as a basis:
1. a very simple flour mill based on the use of ultra rotor mills for grinding wheat followed by simple 1-stage screening yielding up to 85% flour extraction of a low quality from the flour millers' point of view. This flour can be separated well into acceptable starch and gluten of reasonable quality.

   An additional advantage is that the ratio of A to B starch is increased (less damaged starch, less B).

2. The use of whole wheat in a process similar to production of corn starch; it involves simple steeping and homogenizing followed by a 3-phase separation. The process is described in the paper by Meuser et al, elsewhere in proceedings of this symposium.

As stated previously, the ultimate objective is a multipurpose plant, where various raw materials can be processed. Wheat, corn, and peas can be processed interchangeably, depending on their availability and price. Processing corn grits is the easiest.

Equipment to process the various products (e.g. a wheat gluten dryer adapted to handle by-products from the other raw materials) is available. I am convinced that the first multipurpose plants will be operational within the next few years.

B STARCH AND EFFLUENT

One of the difficulties in production of wheat starch and gluten is the B starch. Starch producers basically like to produce one grade of starch: A. In Europe B starch has been always a product with few applications. It was, however, cheaper to isolate the B starch than to dispose of it with the effluent.

Applications for native, ungelatinized B starch include: cattle feed, some minor uses as core binder in the foundry industry, and corrugated boards. Pregelatinized B starch is used as milk replacer in calf feeds, as core binder, and several other minor products. Some high quality B starch is used in human foods, but only in limited amounts and at low prices.

We have, therefore, together with our friends of Latenstein Zetmeel B.V. (Netherlands), developed a new process to separate the B starch into two fractions:
  - small starch granules of high purity, and
  - a feed component.

The process is similar to a new rice starch process (again, multipurpose!): a B starch slurry or rice flour slurry is high-pressure treated, after enzyme addition, and purified by fine screens, separators, and decanters.

It is a simple, straightforward technology and produces up to 75% of high value small-granule starch with a purity similar to that of regular A starch. The starch has good properties for further modification into tailormade starches, e.g. for specialty paper.

Forecasts in Europe are that this modified, small granule starch will yield better prices than A starch. It can be added, of course, to A starch thus increasing A starch yield and reducing B starch yield (70% A and only 5% B are achievable).

You can imagine the impact of this development: A starch yield out of wheat approached maize starch yield!

Finally, a few words about wheat starch effluent treatment since effluent has been a major problem of the industry.

When a new effluent legislation came into force in the Netherlands in 1970, there was no available process, except for the conventional and economically not feasible aerobic technology. Probably more efforts have been made to find the appropriate process for effluent treatment than to find new processes for starch and gluten production.

This resulted in various systems.

a. Evaporation

Wheat starch effluent concentrate is a good component of swine rations and can be supplied in wet form to large swine feedlots. Presently, the effluent of wheat starch plants in the Netherlands (total production up to 25 m^3/hr concentrate) is fed to swine. The whole Dutch swine industry is geared to use the concentrate.

Evaporation, however, does not solve the problem entirely. A residual condensate has at least 500 ppm COD. Since in many countries the effluent should be below 25 ppm, further treatment is necessary. Some plants use an anaerobic, others an aerobic post-treatment, others pay for the pollution.

b. Anaerobic treatment

Production of methane gas from the solubles in the effluent is operational in several plants, mainly where there is no swine production. It is a matter of cost comparison; while evaporation requires a high energy input, anaerobic treatment produces energy.

The investment is comparable to that of evaporation (Witt and Kröner 1988); when energy prices are high, the methane gas produced in anaerobic treatment may be more valuable. When energy prices are low and if there is an outlet for the concentrate, an evaporator may be more valuable.

If in addition to wheat starch and gluten, glucose and/or modified starches are produced, anaerobic treatment is more advantageous since the total water load from the starch facility is much higher than the starch/gluten effluent above.

The recent, novel developments in the wheat starch and gluten industry were directed not only towards solving old and new problems but also towards opening new approaches and frontiers in industrial wheat utilization.

New ways had to be and are being found. The explosive increase in capacity depressed drastically the prices of the products and affected the whole economic structure of the wheat gluten and starch industry. New uses for wheat gluten and starch must be found to maintain not merely the acceptability of the starch and vitality of the gluten but also the vitality of the whole industry.

LITERATURE CITED

WITT, W., and KRONER, H. 1988. Beseitigung von Weizenstärkeindustrieabwässern - Ein Wirtschaftlichkeitsvergleich -. Stärke 40(4):139.

31

PROCESS FOR THE INDUSTRIAL PRODUCTION OF WHEAT STARCH FROM WHOLE WHEAT

W. Kempf
Bundesforschungsanstalt für Getreide- und Kartoffel-
verarbeitung Detmold
Bundesrepublik Deutschland, 4930 Detmold

und C. Röhrmann
Fachhochschule Lippe
Bundesrepublik Deutschland, 4920 Lemgo

ABSTRACT

Of a total of fifteen different processes for the industrial production of wheat starch, six are described in the present paper. The fifteen processes can be grouped according to whether the raw material used is whole wheat or wheat flour and whether the separation technique used is mechanical, chemical, or fermentative. They can also be grouped according to whether the gluten obtained as a by-product is vital or devitalized. The classification used in the present case is that based on the raw material used. Six of the fifteen production processes use whole wheat as the raw material. The paper ends with a description of a modified Longford-Slotter process in which unmilled wheat was steeped in the laboratory for 15 hours at 37 °C with the addition of 0.2 % of sulfur dioxide. With this process it was possible to recover 54 % of starch having a protein content of 0.4 %, corresponding to an efficiency of just under 81 %, with a single discontinuous centrifugation and a single tabling operation.

INTRODUCTION

Many different processes are available for the industrial production of wheat starch, in contrast with the situation in the potato starch and corn starch industries. However, it is extremely difficult in many cases to decide whether the various processes really do embody new technologies, or whether they are merely technical improvements, or possibly even differ only in the use of equipment and machinery from different manufacturers.

CLASSIFICATION

A comprehensive and careful literature study has shown that if the above distinction is taken into account, the total number of different processes for the industrial production of wheat starch is fifteen (Table I), six of which will be discussed below.

There are basically three criteria that could be used to classify these fifteen processes. The first is the raw material used (Table II).

Another possible classification is based on whether the starch/gluten separation is predominantly mechanical, chemical, or fermentative (Table III).

Finally, the processes can also be classified according to whether the gluten obtained as a by-product is vital or devitalized (Table IV).

PROCESSES

The first of these three possibilities appears to be the most significant. In the following discussion, therefore, the production processes will be classified according to whether whole wheat or wheat flour is used as the raw material.

Whole wheat as the raw material

Whole wheat is used as the raw material in six of the fifteen processes for the industrial production of wheat starch (Rao, 1979). Only these six will be discussed here (Table V).

These six processes all involve the same basic opperations, i.e. steeping and milling the wheat on

TABLE I
Processes for the industrial production of wheat starch - chronological order

Alsatin	(USA)
Halle	(D)
Alsace	(F)
Martin	(F)
Fesca	(D)
Batter	(USA)
Dimler	(USA)
Longford-Slotter	(USA)
Phillips-Sallans	(USA?)
Alfa-Laval-Raisio	(S/SF)
Verberne-Zwitserloot	(NL)
Weipro	(D)
Far-Mar-Co	(USA)
Pillsbury	(USA)
Tenstar	(GB)

TABLE II
Classification of processes for the industrial production of wheat starch - raw materials

Whole Wheat	Wheat Flour
Alsatin	Martin
Halle	Fesca
Alsace	Batter
Longford-Slotter	Dimler
Far-Mar-Co	Phillips-Sallans
Pillsbury	Alfa-Laval-Raisio
	Verberne-Zwitserloot
	Weipro
	Tenstar

TABLE III
Classification of processes for the industrial production of wheat starch - separation

Mechanical	Chemical	Fermentative
Alsatin	Dimler	Halle
Alsace	Longford-	
Martin	Slotter	
Fesca	Phillips-	
Batter	Sallans	
Alfa-Laval-Raisio	Far-Mar-Co	
Verberne-Zwitserloot	Pillsbury	
Weipro		
Tenstar		

the one hand and washing out, purifying, and drying the starch on the other. Gluten recovery, in contrast, is only of minor importance. In comparison with processes using wheat flour as the raw material, the use of whole wheat has the important processing advantage that wheat having characteristic physical and chemical properties is always available and ready for use. The possibility of supplyside bottlenecks is avoided, since there is no dependence on the milling industry. Other advantages are the substantially higher yields of first-grade starch, as a result of reduced damage to the starch granules, and an improvement in the vitality of the gluten in some cases, since the separation of the endosperm protein is almost complete.

Alsatin process (Kerr, 1950, Radley, 1953). The Alsatin process, which was developed in the USA, is based on the observation that steeping the wheat in warm water not only makes grinding considerably easier, but also leads to higher starch yields, with a simultaneous improvement in quality (Fig. 1).

The steeping with water is carried out at 30 to

TABLE IV
Classification of processes for the industrial
production of wheat starch - gluten

Vital	Devitalized
Alsatin	Halle
Martin	Alsace
Fesca	Dimler
Batter	Longford-Slotter
Alfa-Laval-Raisio	Phillips-Sallans
Verberne-Zwitserloot	Pillsbury
Weipro	
Far-Mar-Co	
Tenstar	

TABLE V
Classification of processes for the industrial
production of wheat starch - whole wheat as raw
material

Alsatin	Longford-Slotter
Halle	Far-Mar-Co
Alsace	Pillsbury

35 °C and takes between 24 and 48 hours. Continuous renewal of the steep water helps to wash out soluble material, and at the same time prevents the acidity from becoming too high. The steeping is followed by coarse grinding between either millstones or fluted rollers. The resulting dough is thoroughly kneaded in extractors consisting of a finely perforated with rotating mixing elements, and the starch/gluten suspension is simultaneously washed with a continuous supply of fresh water from a spray, with separation of

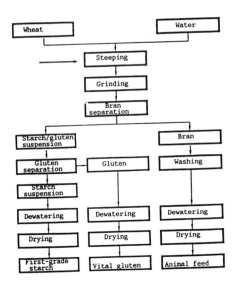

Fig. 1. Processes for the industrial production of wheat starch - Alsatin process

the bran. The gluten is then separated by screening, and the remaining starch suspension is converted by separation, concentration, and drying stages into first-grade starch. Starch yields of about 45 % are obtained by this process. A further 10 to 15 % of second-grade starch can be obtained by re-use of the residue from the first extraction stage. The final residue consists of germs and hull fractions, and is used as animal feed. Since separation of the gluten presents considerable difficulty and rarely gives yields of more than 5 %, it is dispensed with in most cases.

In a variant of the American Alsatin process, the grinding is combined with the washing. The starch is washed out as the swollen wheat is ground, and the starch slurry is then screened to separate gluten and bran. The screen residue is reground to allow an additional starch separation. The resulting second-grade starch is refined further. To swell and dissolve the gluten, acetic acid or ammonia is added during steeping, or an intermediate fermentation stage is used. This allows a much better separation of the gluten from the starch, as well as improved washing and sedimentation.

Halle process (Singer, 1937, Kerr, 1950, Crafts-Lighty et al., 1980.) The Alsatin process described above obviously led on to the so-called Halle process, which was in general use at one time. In Germany it was found mainly in and around the Halle/Saale region, hence the name by which it is known (Fig. 2).

The main difference from the Alsatin process is that the Halle process includes a fermentation stage.

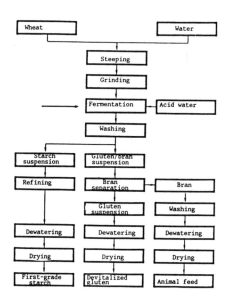

Fig. 2. Processes for the industrial production of wheat starch - Halle process

The cleaned wheat is first steeped, an operation that is vitally important to the starch yield, and whose duration depends on the type of wheat; the fermentation may then be carried out either during the steeping process, immediately after grinding, or during both of these stages. Regardless of the variant adopted, the fermentation, which is initiated by the addition of so-called acid water, takes about a week and proceeds at a temperature of about 25 C. Apart from small quantities of alcohol, the early stages of fermentation lead mainly to the formation of organic acids with the aid of acid-forming bacteria. The end of fermentation is marked by the start of putrefaction, which is accompanied by the appearance of a strong, unpleasant odor due to the decomposition of proteins. This odor signals the end of fermentation and also of the desired action on the gluten, which should be sufficiently degraded by the fermentation to make its separation from the starch as easy as possible. When fermentation is complete and the acid water has been removed, the starch is washed out in rotating perforated drums. The resulting starch slurry leaves the drum through openings in the wall, while the gluten and the hull fractions remain inside the drum. The starch slurry can then be purified either by settling and screening in stirred tubs or by centrifugation. The refining and concentration stages are followed by drying. Starch yields of 60 % can be achieved by the Halle process, with low protein contents. An undesirable side effect of this process, however, is the unsatisfactory smell of the starch. The bran is washed, dewatered, and dried and then marketed as animal feed.

Though the Halle process was used for many centuries until the French apothecary Martin developed the production of starch from wheat flour in 1835, it has now been totally replaced by other processes, since the destruction of the gluten meant that the recovery of marketable gluten was impossible. The profitability of a wheat starch factory using the Halle process was thus severely limited by the loss of the gluten, which would have commanded high prices on the market (Parow, 1928, Singer, 1937, Kerr, 1950,

Radley, 1953, and Knight, 1965).

<u>Alsace process</u> (Parow, 1928). The so-called Alsace process, which was developed in France, and is sometimes also known as the Hungarian process, is derived from the Halle Process, which predominated in Germany. In the Alsace process, the wheat is steeped for several days, either as whole grain or in a coarsely ground form during this operation the gluten changes into a plastic mass that is easily separated from the starch. The swollen wheat is coarsely ground, and the starch is then washed out as starch slurry in rotating perforated drums and collected in tubs equipped with stirrers. First-grade starch and possibly also second-grade starch are obtained after repeated settling of the starch slurry followed by screening, centrifugation, dewatering, and drying.

The residue remaining when the starch has been washed out is washed in a perforated drum with pins inside it to separate the gluten, which collects inside the drum, from the hulls and germs, which are carried away through openings in the wall of the drum. The gluten is then allowed to ferment spontaneously for about three to four days, with the result that it becomes viscous and turns yellowish. It is then spread out on greased metal sheets and dried either on racks or in drying tunnels at 50 to 60 C.

Whereas the gluten that had been decomposed by fermentation in the Halle process was practically unusable even as cattle feed, the gluten from the Alsace process made very good cattle feed. However, the yield was very low, since its coherent structure was severely damaged by the long steeping process, and a large proportion of the gluten was lost together with the wash water. Whereas the uneconomical Halle process with its foul-smelling waste water fell totally into disuse in time, a series of substantial improvements and technological modifications were made to the Alsace process before it was gradually ousted by the Martin process developed in 1835 (Parow, 1928, Finger, 1937, Kerr, 1950, Radley, 1953, Horan, 1955, Kempf, 1955, Maurer, 1956, Katsuya and Ishiwatari, 1962, Katsuya and Kojima, 1962, Knight, 1965, Finley

et al., 1973, Schäfer, 1975, Kröner, 1975, Meuser and Smolnik, 1976, Skogman, 1976) (Fig. 3).
Longford-Slotter process (Parow, 1928, Singer, 1937, Longford and Slotter, 1944, 1945, Kerr, 1950, Radley, 1953, Horan, 1955, Kempf, 1955, Maurer, 1956, Iwanjuk and Purisman, 1959, Katsuya and Ishiwatari, 1962, Katsuya and Kojima, 1962, Ludewig, et al., 1963, Knight,1965, Schäfer, 1975, Kröner, 1975,). The Longford-Slotter process, which was developed in the USA, derives directly from the American Alsatin process. The main difference is that the wheat is steeped in the presence of a relatively high concentration of sulfur dioxide (Fig. 4).

The unmilled wheat is steeped for 24 h at 38 °C. The added sulfur dioxide concentration varies between 0.3 and 0.5 %. The sulfurous acid formed on addition of the sulfur dioxide opens up the starch/gluten matrix sufficiently to eliminate any difficulty in the

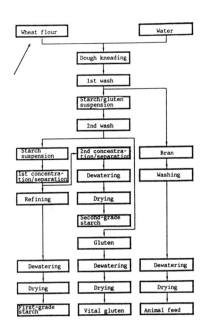

Fig. 3. Process for the industrial production of whe starch - Martin process

wet grinding stage, which is preferably carried out in stone centrifugal mills. The chemical treatment of the gluten avoids difficulties in the subsequent screening, such as deposition of the starch together with the gluten. The residue remaining after screening is again ground and screened. The gluten obtained after washing, dewatering in filter presses, and drying contains about 30 % of protein and has a relatively high starch content; because of the denaturing that has taken place, moreover, it is no longer suitable for applications requiring native gluten. The starch yields are between 55 and 60 %, with a protein content of 0.2 %.

<u>Far-Mar-Co process</u> (Nishimura and Takaoka, 1960, Rao, 1979). It is not certain whether the Far-Mar-Co process was also developed in the USA and also derives from the Alsatin process.

The first production process to be developed in

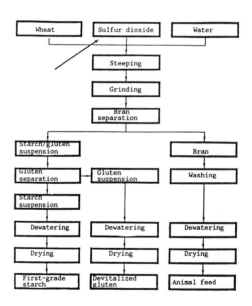

Fig. 4. Processes for the industrial production of wheat starch - Longford-Slotter process

recent times, it also uses unmilled wheat as its raw material (Fig. 5).

The first step in the process is to wet the wheat with water. The wheat is then flaked by passage between rollers, and the flakes are continuously worked with water into a thick dough. It is particularly important here to avoid any excess of water. For this reason, the hydration is the decisive, even critical phase of the entire process, since its extent determines the effectiveness of the separation of the wheat flakes. Subsequent separation into a gluten fraction and a starch/bran/germ fraction is effected by washing with high-pressure sprays. Further screening separates the starch/bran/germ fraction into a starch suspension and a mixture of bran and germs. The starch suspension obtained from this second separation is subjected to separation, concentration, and drying to produce first-grade starch.

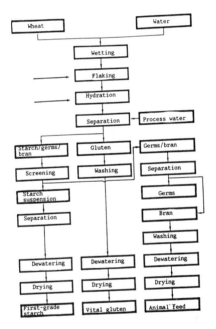

Fig. 5. Processes for the industrial production of wheat starch - Far-Mar-Co process

The remaining starch and bran fractions are
moved from the gluten in a separate wash, and the
cess moisture is eliminated in the subsequent
watering and drying operations. The final product is
vital gluten having a protein content of between
-80 %. The germs and the bran may be either
parated or sold together as animal feed.

Pillsbury process (Hunwick, 1980). Attempts to
ply corn wet grinding technology to wheat led to the
llsbury process, which was developed in 1975, again
the USA. In this process, unmilled wheat is steeped
r 16 hours with warm water and acid and with a
npathogenic bacterial culture. The purpose of the
cterial culture (Lactobacillus) is to prevent the
read of pathogenic microorganisms that can normally
velop during the process. This steeping operation is
llowed by wet grinding or further steeping to
arantee an endosperm fraction having a very specific
rticle size for the subsequent screening operation.
rms and bran are retained and removed. The endosperm
dewatered in settlers and dried to form a floury
oduct. Agglomeration of gluten in the endosperm can
held in check if the pH is kept in the acidic
nge. The objective of the Pillsbury process is not
obtain pure starch, but to produce high-quality
eat flour. However, starch/gluten separation may be
fected in a further operation.

Wheat flour as the raw material

Wheat flour is used as the raw material in nine
the fifteen processes for the industrial production
wheat starch (Table VI) (Parow, 1928, Singer, 1937,
rr, 1950, Radley, 1953, Horan, 1955, Kempf, 1955,
urer, 1956, Ludewig, 1956, Anderson et al. 1960,
tsuya and Ishiwatari, 1962, Katsuya and Kojima,
62, Knight, 1965, Plaven, 1965, Knight, 1969, CPC
ternational, Inc., 1969, Finley et al., 1973,
häfer, 1975, Kröner, 1975, Meuser and Smolnik, 1976,
ogman, 1976, Alfa-Laval Process Group, 1978,
hlberg, 1978, Rao, 1979, Wu and Stringfellow, 1980,
afts-Lighty et al., 1980, Westfalia Separator AG,).
would unfortunately lie outside the scope of the

TABLE VI
Classification of processes for the industrial production of wheat starch – wheat flour as the raw material

Martin	Alfa-Laval-Raisio
Fesca	Verberne-Zwitserloot
Batter	Weipro
Dimler	Tenstar
Phillips-Sallans	

present paper to enumerate even the distinguishing features of these nine processes.

YIELDS

The relatively extensive specialist literature contains only very occasional information on the yields of first-grade and second-grade starch and of vital and devitalized gluten, and it is not clear from the few figures available whether they are expressed as % of air-dried product or % of dry solids, and whether they are based on commercial wheat flour or on wheat flour dry solids. Despite these difficulties, an attempt has been made to compare all the known data on the yields of first-grade and second-grade starch and of vital and devitalized gluten (Table VII).

In its essential features, the Longford-Slotter modification of the Alsatin process corresponds closely to the industrial processes for the production of corn starch, since before the wheat is actually processed it is steeped with the addition of sulfur dioxide, and is then ground and sieved to remove the hulls, though without prior separation of the germs. The hull fraction is washed, dried, and added to animal feed. The gluten is separated from the starch/gluten suspension by tabling and horizontal separation, and drying then yields a mixture of dried gluten and second-grade starch. The gluten-free starch slurry is refined again, thickened, and dried to obtain total starch; however, this consists essentially of first-grade starch (Longford and Slotter, 1944, 1945, Kerr, 1950, Radley, 1953, Iwanjuk and Purisman, 1959, Ludewig et al., 1963, Knight, 1965, Verberne and Switserloot, 1978) (Fig. 6).

As a result of a number of preliminary tests in the laboratory, it was found that steeping for 15 h at 37 °C with the addition of 0.2 % of sulfur dioxide gave the optimum results, and these conditions were accordingly adopted as standard. The swollen wheat was ground in a Starmix, and the hull fraction was separated in a vibratory sieve. The subsequent starch-gluten separation was carried out discontinuously or continuously in laboratory centrifuges and tables. To summarize the results, all single stage processes were basically found to be excellent (Pelshenke and Lindemann, 1954, Lindemann, 1955, Kempf and Tegge, 1961, 1962) (Table VIII).

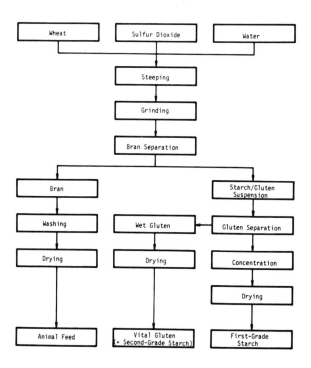

Fig. 6. Flow diagram of the Longford-Slotter process

TABLE VII

Processes for the industrial production of wheat starch and their yields

Process	Raw material	Starch yield	Gluten yield
Alsatin	Whole wheat	45 % first-grade 10-15 % second-grade	5 % vital
Halle	Whole Wheat	60 % first-grade (< 0.5 %)	-
Alsace	Whole wheat	-	-
Longford-Slotter	Whole wheat	55-60 % first-grade (0.2 %)	Devitalized (30 %)
Far-Mar-Co	Whole wheat	-	-
Pillsbury	Whole wheat	-	-
Martin	Wheat flour	45-55 % first-grade, 10-20 % second-grade	10-15 % vital (70-80 %)
Fesca	Wheat flour	-	-
Batter	Wheat flour	90 % total	Vital (80 %)
Dimler	Wheat flour	70-80 % total	-
Phillips-Sallans	Wheat flour	--	Devitalized (70-80 %)
Alfa-Laval-Raisio	Wheat flour	60-70 % first-grade, 12-16 % second-grade	9-10 % (77-80 %)
Verberne-Zwitserloot	Wheat flour	-	-
Weipro	Wheat flour	-	-
Tenstar	Wheat flour	-	-

In the Longford-Slotter modification of the Alsatin process, 54.0 % of starch with a protein content of 0.42 %, corresponding to an efficiency of 80.8 %, was obtained in the laboratory after a single discontinuous centrifugal separation followed by single tabling operation. At the same time, 22.3 % of a mixture of dry gluten and second-grade starch ntaining 36.6 % of protein was also obtained (Kempf, 1977).

Following the laboratory investigations, the Longford-Slotter modification of the Alsatin process was tested for the production of wheat starch from soft wheat in an Austrian corn starch factory on the Bodensee (Kempf and Bode, 1977) (Table IX).

This trial was carried out with 20 t of soft wheat having a moisture content of about 14 % and a

TABLE VIII
Laboratory yields from the Longford-Slotter process

Raw starch/wheat	66.8 %
Yield/first-grade starch	54.0 %
Efficiency/first-grade starch	80.8 %
Protein/first-grade starch	0.42 %
Yield/gluten mixture	22.3 %
Protein/gluten mixture	36.6 %

TABLE IX
Industrial yields from the Longford-Slotter process

Soft wheat	
Quantity	20 t
Moisture	14 %
Protein	10 %
Steeping	
Time	12 h
Temperature	39 °C
Sulfur dioxide	0.2 %
Water absorption	37 % after 4 h
	55 % after 12 h
Steep water	3.2 °Bé
Yield	
Total starch	73 %
Protein	0.6-0.8 %

protein content of about 10 %. The soft wheat wa steeped for 12 h at 39 °C with the addition of 0.2 of sulfur dioxide. The water absorption during th steeping operation was 37 % after 4 h and 55 % afte 12 h. The steep water had a concentration of 3.2 °Bé It was evaporated in a falling film evaporator, but n information is available on the yield of steep wate solids.

The yield of total starch, which was dewatered i an overflow centrifuge, was 73 % of dry solids, bu this contained 0.6 to 0.8 % of protein. The tota starch was not separated into first-grade an second-grade starch, but such a separation seem possible. The total starch was then saccharified b enzyme-modified acid hydrolysis; this did not presen any problems apart from considerable filtratio difficulties. The crystal dextrose obtained from th total starch was of excellent quality, and wa indistinguishable from that produced in the norma manner by enzyme-modified acid hydrolysis of cor starch. No gluten fraction was obtained, since th gluten was dried together with the bran and hulls t produce an animal feed. The fine fibers in particula had very high gluten contents. No information i available on the yields of gluten, bran, and hulls The same applies to the consumption of fresh water an the production of waste water.

No processing difficulties of any kind wer encountered at any stage, so that the results of thi trial on the industrial production of wheat starc from soft wheat by the Longford-Slotter modificatio of the Alsatin process can be described as very good

CONCLUSION

A comparison of the various processes for th industrial production of wheat starch from whole whea or from wheat flour led to the conclusion that lik most things in life, they each have their advantage and disadvantages. In comparison with processes usin wheat flour as the raw material, all those based o unmilled wheat have the advantage that they avoid th need for dry milling, which is energy-intensive an costly. Against this they have the disadvantage tha hen unmilled wheat is used, the separation of the gluten presents considerable difficulties, and accordingly gives low yields. Moreover, only devitalized gluten is obtainable in most cases.

However, the importance of this fact should not be over-estimated, since as well as the dry milling of the wheat, it is also possible to dispense with the separation of the second-grade starch, and sales of vital gluten cannot be increased indefinitely in any case. Even though current technology and the market situation are such that vital gluten accounts for only % of the total yield but 30 % of total sales in the case of the Martin process, which is the only process used in the Federal German Republic, it should not be forgotten that unlimited gluten output could not be absorbed by the protein market, which is already glutted with potato, milk, and soy protein (Kempf, 1977).

LITERATURE CITED

ALFA-LAVAL PROCESS GROUP DE LAVAL SEPARATOR COMPANY INC. 1978. Milling & Baking News.

ANDERSON, R. A., PFEIFER, V. F., LANCASTER, E. B., VOJNOVICH, C., and GRIFFIN, E. L. 1960. Cereal Chem. 37(2):180-188.

APC INTERNATIONAL, INC. 1969. DOS 19 16 597.

RAFTS-LIGHTY, A. L., BEECH, G. A., and EALDEN, T. N. 1980. J. Sci. Food Agric. 31(3):299-307.

AHLBERG, B. I. 1978. Starch/Stärke 30(1):8-12.

AHLBERG, B. I. 1978. Private communication.

FINLEY, J. W., GAUGER, M. A., and FELLERS, D. A. 1973. Cereal Chem. 50(4):465-474.

ORAN, F. F. 1955. North Western Miller 1:3a.

UNWICK, R. J. 1980. Food Technology in Australia 32(8):463.

WANJUK, M. J., and PURISMAN, J. I. 1959. Ssacharnaja Promyshlennosst 33(8):62-66.

ATSUYA, N., and ISHIWATARI, Z. 1962. J. Technol. Soc. Starch 10(1):8-13.

ATSUYA, N., and KOJIMA, K. 1962. J. Technol. Soc. Starch 10(1):13-19.

KEMPF, W. 1955. Starch/Stärke 7(5):123-127.

KEMPF, W. 1977. Starch/Stärke 29(9):307-315.

KEMPF, W., and BODE, H. 1977. Unpublished results
KEMPF, W., and TEGGE, G. 1961. Starch/Stärk 13(10):363-368.
KEMPF, W., and TEGGE, G. 1962. Getreide und Mehl 12(5):51-54.
KERR, W. R. 1950. Chemistry and Industry of Starch Academic Press, New York.
KNIGHT, J. W. 1965. Wheat Starch and Gluten Leonard Hill, Ltd., London.
KNIGHT, J. W. 1969. The Starch Industry. Pergamo Press, Ltd., Cambridge, England.
KRONER, H. 1975. DOS 25 03 787.
LINDEMANN, E. 1955. Der Züchter 25(4):129-132
LONGFORD, C. T., and SLOTTER, R. L. 1944. Int. Eng Chem. 36:404.
LONGFORD, C. T., and SLOTTER, R. L. 1945. U.S Patent 2:368, 668.
LUDEWIG, H. 1956. East German Patent 19 679
LUDEWIG, H., GOEDECKE, G., and STEPHANI, R. 1963 East German Patent 34 813.
MAURER, W. 1956. Starch / Stärke 8(1):6-12
MEUSER, F., and SMOLNIK, H.-D. 1976. Starch/Stärk 28(12):421-425.
NISHIMURA, A., and TAKAOKA, K. 1960. J. Ferm. Techn 38:15.
PAROW, E. 1928. Handbuch der Stärkefabrikation Verlag Paul Parey, Berlin.
PELSHENKE, P. F., and LINDEMANN, E. 1954 Starch/Stärke 6(8):177-182.
PLAVEN, E. 1965. DAS 152 67 381.
RADLEY, J. A. 1953. Starch and Its Derivatives Vol. 2, Chapman & Hall, Ltd., London.
RAO, G. V. 1979. Cereal Foods World 24(8):334-335
SCHAFER, R. 1975. Starch/Stärke 27(8):257-262
SINGER, W. 1937. Mühle 74(2):34.
SKOGMAN, H. 1976. Starch/Stärke 28(8):278-282
WESTFALIA SEPARATOR AG. Westfalia-Magazin, Issue No 4.
WU, Y. V., and STRINGFELLOW, A. C. 1980. J. Foo Sci. 45(5):1383-1386.
VERBERNE, P., and ZWITSERLOOT, W. 1978. Starch Stärke 30(10):337-338.

32

NEW INDUSTRIAL APPLICATIONS OF CHEMICALLY MODIFIED WHEAT GLUTEN

W. Kempf and W. Bergthaller
Institut für Stärke- und Kartoffeltechnologie
Bundesforschungsanstalt für Getreide und
Kartoffelverarbeitung
Bundesrepublik Deutschland, 4930 Detmold

and B. Pelech
Research and Development Center
Feldmühle AG
Bundesrepublik Deutschland, 4060 Viersen

ABSTRACT

Research on opening new application fields for by-products of renewable resources has demonstrated that chemically modified wheat gluten is well suited for use as a co-binder in the paper-coating industry. In comparison with modified soy protein isolate, that is considered a standard co-binder in industry, wheat gluten is equal (and in some cases superior) in important technical properties such as binding power, wet-picking resistance, whiteness, activation of optical brighteners, printability, and water retention capacity.

According to information from a German paper manufacturer, up to 6,000 t of modified wheat gluten can be processed on one large paper-coating machine alone. In 1982 about 12,000 t of soy protein isolate, which could be replaced by chemically modified wheat gluten, were used in the Federal Republic of Germany as a co-binder in paper- and card-coating mixtures. This is equivalent to a demand for about 145,000 t of wheat, which requires a cultivation area of about 23.000 ha at a present average yield of 6.3 t/ha, and corresponds to 1 % of the total wheat cultivation area in the Federal Republic of Germany.

Preliminary studies on the use of natural gluten, chemically modified gluten, and the two gluten fractions gliadin and glutenin as fillers in urea-formaldehyde resin glues have shown that corresponding glue batches with natural gluten and gliadin additions have very good properties with respect to the dry-binding strenght and warm water strength of glued beech veneers.

Finally, attempts to replace plasticized and extruded collagen by plasticized vital gluten as a component in synthetic edible sausage skins also have been very successful. Sausage skins of very high quality were produced from mixtures with the addition of 5-20 % wheat gluten. They were assessed as acceptable in preparation and consumption tests, and as fully equivalent to collagen.

INTRODUCTION

In 1979, 1.5 million t of agricultural products were converted into 650,000 t starch, derivatives, and saccharification products by the starch industry in the Federal Republic of Germany. Referred to as regular commercial starch, maize represented by far the largest proportion of the source materials at 71 %, followed by potatoes at 18 %, and wheat at 11 %. In the meantime, however, these ratios have changed considerably since the production of potato starch and of wheat starch in particular have increased considerably at the expense of maize starch. In 1985, a total of 2 million t of agricultural products was used as raw material in the starch industry, of which maize made up 50 %, potatoes 28 %, and wheat 22 %.

According to production figures in 1986, 7 plants in the German wheat starch industry converted 340,000 t wheat into starch and starch products, producing 18,000 t of vital gluten as a by-product in the process. At its annual general meeting on December 2nd, 1988 in Cologne, the president of the Association of the Starch Industry in Bonn (Fachverband der Stärkeindustrie e.V. in Bonn) stated that the high proceeds from vital gluten make up about 70 % of the price of the source material, but that on the one hand the position would change dramatically if increased production from new investments came to bear fully on the market and on the other hand, the sales potential could not be stretched any further, if for example no more foreign wheat was to be substituted. In the meantime, it is precisely this dramatic change in the situation which has occured (Hees 1988).

In the EEC in 1987 about 1.5 million t of wheat was converted into starch and starch products, which accounts for 24 % of the usage of source materials based on grain. This production capacity probably will increase considerably in the near future. When one considers that vital gluten only makes up about 9-11 % of the wheat flour, but accounts for 30-35 % of the turnover, the inevitable conclusion is that the wheat starch industry can only exist in the long or medium term if sales of vital gluten are guaranteed at acceptable prices.

There are only two possibilities to counteract the disastrous collapse in the gluten market, which has started and made its effects felt. One of those two possibilities involves developing techniques in which wheat gluten is no longer produced at all as the by-product vital gluten and, therefore, no longer has to be marketed. This would require the process to be much simpler and considerably less expensive in order to offset or even compensate for the loss of revenue from vital gluten. The second possibility is based on opening up completely new areas of application and, therefore hitherto unknown, sales possibilities for wheat gluten. In view of the completely saturated food and feedstuffs market, the only yields that can be considered are chemical and industrial ones. This avenue has been followed successfully by carrying out the research projects described in detail in what follows. Intensive research has opened promising potential uses in the paper and wood materials industry, and in the production of edible synthetic sausage skins (Kempf and Pelech 1986).

STATEMENT OF THE PROBLEM

High-grade special papers for printing are achieving a high growth rate in the current commodity world. Such types of paper are coated and polished. In coating the raw paper, the applied pigments require binders for anchoring them to each other and to the raw paper, the so-called coating mixtures employ natural binders as co-binders in addition to the prevailing synthetic polymers made from fossil raw materials (Silvernail and Bain 1961, Strauss 1975, Kotte 1978). The natural binders, some of which are plant proteins, exert favourable regulatory effects on some important properties of coating mixtures, e.g. the flow properties and the water retention capacity. Often, however, they cannot fully exert their specific effects until they have undergone targeted chemical digestion and modification (Garey 1987, Coco 1983, Coco 1984, Whalen-Shaw 1984).

Wheat gluten potentially represents a valuable raw material in the chemical-technical field. Its characteristic property of forming a viscous to highly elastic mass, depending on the degree of crosslinking of the protein chains (Friedmann 1978, Fujimaki et al. 1980), has to be modified to suit the specific requirements of the application. As far as use as a binder in the paper industry is concerned, the aim of chemical digestion and/or modification of the gluten protein is to produce residue-free solubility and to improve technical properties. Those include ability to be incorporated into paper coating mixtures, viscosity regulation, and binding power of the pigments and their microbiological stability, while achieving the greatest possible product yield. As a standard, one must refer to technical properties of modified soy protein isolates, which are obtained with yields between 65 and 80 % and have been used widely for this purpose. Also, the suitability of

eat gluten as a filler and formaldehyde scavenger
urea-formaldehyde resin glues is to be tested, as
ll as its use in synthetic edible sausage skins.

DISCUSSION OF RESULTS

Co-binders in paper coating mixtures

The first problem to be overcome in the poten-
ial use of chemically modified wheat gluten as a
-binder in paper-coating mixtures is the selection

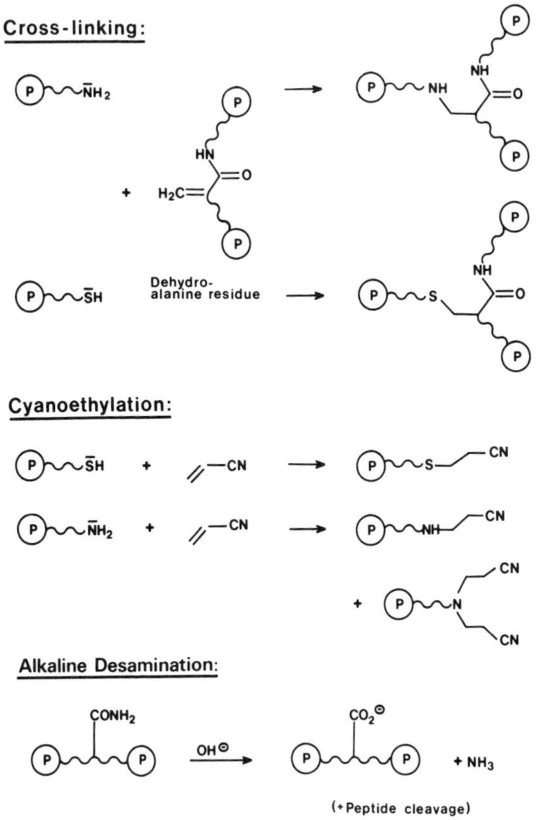

Fig. 1. Main reactions in cross-linking, cyanoethylation, and alkaline deamination of wheat gluten

of the best possible residue-free dissolution tech
nique and the inhibition of amino and sulfhydry
group cross-linking in the wheat gluten, which pre
vents dissolution in alkali Modification with acry
lonitrile (Coffman 1951) with simultaneous side-chai
de amination in a strongly alkaline medium is particu
larly effective (Fig. 1). The effect of reactio
time, acrylonitrile concentration, gluten extrac
tion, and consumption of sodium hydroxide in th
side-chain deamination of wheat gluten was investiga
ted in detail for this modification process. Furthe
optimization of the experimental parameters was no
possible.

For this purpose, vital wheat gluten was dissol
ved in an aqueous sodium hydroxide solution, mixe
with acrylonitrile in the strongly alkaline region

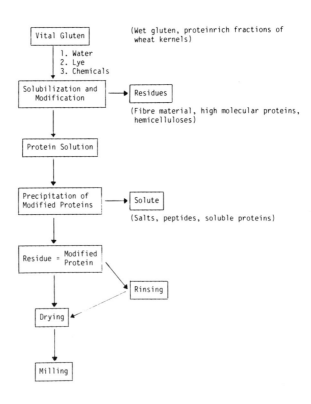

Fig. 2. Flow diagram of wheat gluten modification

and stirred for varying lengths of time. After completion of the modification the excess acrylonitrile was distilled off, the solution was freed from impurities, and the protein was precipitated in the mildly acid region at pH 4.6 and finally gently air-dried after separation and purification. The dried material was then ground finely for further use in the manufacture of paper coating mixtures (Figs. 2 and 3).

Specifically, the addition of 5 (CN1/2) and 10 (CN) weight % acrylonitrile, gluten dry substance basis, was used to modify wheat gluten, the 10 % addition already guaranteeing a reagent excess. To investigate the effect of wheat lipids on the properties of the modified gluten, butanol-extracted and therefore lipid-free gluten (CNBu) was used for the modification experiments with the 10 % addition (Coffman 1951, Mecham and Mohammad 1955). Finally, gluten was also modified without the addition of reagent (CN*) under the conditions of cyanoethylation, in which essentially only a side-chain deamina-

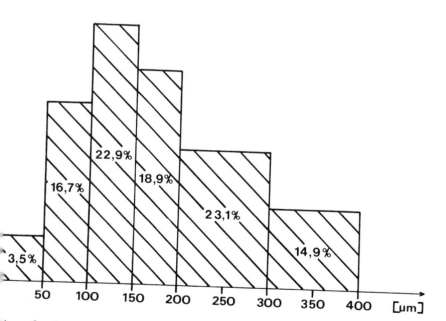

Fig. 3. Particle size distribution of ground cyanoethylated wheat gluten

tion was carried out. This led to a rapid and large increase in viscosity and considerably impaired the application possibilities.

Under the conditions of cyanoethylation there was significant splitting off of side-chain amide groups (deamination, Fig. 4). Of the sodium hydroxyde used for alkaline digestion of the gluten, more than 50 % was required for deamination. The results of analyses of the amino acid composition of the gluten modified with acrylonitrile revealed that the amino acid lysine, which is preferentially involved in the cross-linking characteristic of wheat gluten, is modified relatively quickly; this inhibits or even prevents the disadvantageous cross-linking of the protein molecules.

The yields of cyanoethylated wheat gluten decreased with increasing reaction time and averaged about 85 % of the starting material (Fig. 5). In lipid-free gluten the yields decreased to about

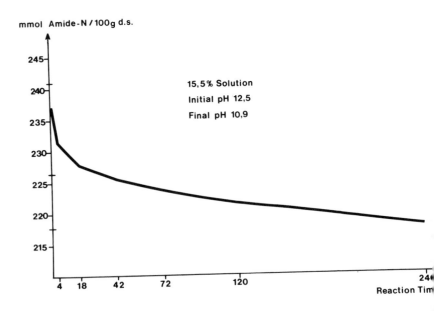

Fig. 4. Decrease in amide nitrogen content in the alkaline deamination of wheat gluten

2 %. The losses of gluten were mainly attributable to non-precipitable protein fractions. In contrast, the yield decreased sharply during modification under conditions of side-chain desamination, i.e. without the addition of acrylonitrile. Formation of insoluble residues was responsible mainly for this decrease in yield. In addition to the modification with acrylonitrile, preliminary acylation experiments using succinic anhydride were conducted (Grant 1973, Schwenke 1978, Barber and Warthesen 1982). As a result of blocking of the free amino groups of the protein, a loosening of the gluten structure improved solubility in both the neutral and the alkaline ranges. After modification and the removal of insoluble residues the protein was precipitated and processed as in the cyanoethylation. Relatively high yields between 85% and 90 % were obtained.

As a prerequisite to satisfactory application in the paper industry, the next studied factor was solubility of modified wheat gluten in comparison with vital gluten. However, chemicals and other auxi-

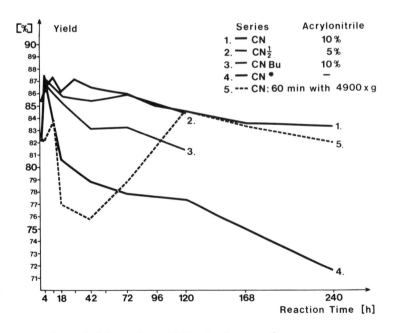

Fig. 5. Yields of modified wheat gluten

liary agents used in the paper-coating industry, were not included. To monitor the dissolution behaviour, 15 weight % protein (dry matter basis) in each case was placed in a sodium hydroxide solution or in a dilute ammonia solution at pH 12. Natural gluten showed incomplete dissolution under those conditions.

deaminated non-cyanoethylated gluten showed only slightly better behaviour. In contrast, the solubility was highly satisfactory in the 10 % dispersions, even though the insoluble residues and the deficient stability due to the formation of gel structures proved to be troublesome. On the other hand, the solubility of cyanoethylated gluten was good in every respect with residues up to a maximum of 8 % and solution times between 0.5 and 8 hrs at room temperature (Table I). The viscosity of the solutions remained stable if modification was adequate, but an after-thickening effect was observed within the first 18 hrs. In lipid-free samples the viscosity increase

TABLE I

Solubility of Cyanoethylated Wheat Gluten (in % of protein dry substance) in Sodium Hydroxyde Solution[a] and Ammonia Solution[b] on the Basis of the Insoluble Residue (centrifuging, 60 min, 4900 x g, 15 % solutions**)

Reaction time (h)	Residue (% dry substance)				
	CN^a	CN^b	$CN1/2^a$	$CNBu^a$	$CN*^a$
4	1.4	4.8	1.2	1.7	54
18	4.5	3.6	8.4	1.7	32
42	1.1	1.1	3.7	1.3	10
72	1.3	2.3	4.0	1.9	7
120	1.7	1.9	3.1	1.8	12

** For CN* 10 % solutions
[a] Sodium hydroxide solution (pH 12, room temperature, 3 hrs)
[b] Ammonia solution (10 weight % NH_3 (25 %) referred to protein dry substance, 55 °C, 30 min).

extended until gel formation. Gluten modified with succinic anhydride decomposed within a short time owing to microbial infection.

To test the technical properties of modified wheat gluten, hand coating experiments (among others) were carried out with model coating mixtures (Table II). The proportion of co-binder was increased by a factor of 1.6 in the model coating mixtures, compared to those used in industrial practice. In testing the pick resistance of the coated paper samples, only the cyanoethylated gluten with long modification times was superior to the standard. The binding power showed a tendency to increase with increasing modification time (Fig. 6). Lipid-free cyanoethylated gluten yielded the best overall results (Fig. 7). In contrast, whiteness, gloss, and absorbency of the paper showed no clear trend. The water retention capacity of the model coating mixtures with cyanoethylated gluten surpassed the values of the standard although there was no distinct trend. In all cases, however, it was clearly greater than with vital gluten or alkaline deaminated gluten.

TABLE II
Composition of Model Coating Mixtures for Hand Coater Experiments

100	parts coating clay (whiteness 85.5 ± 0.7, 80 % < 2 μm)
8	parts protein (15 % solution)
5	parts synthetic dispersion binder (50 %)
0.3	parts dispersion agent (Polysalt)
0.5	parts optical brightener

Solids content:	52.5–55.0 %, pH 10.5
Application:	12 g/m^2 on one side (wire side) with hand coater on wood-free body paper (68 g/m^2)
Pigment slurry:	66.5–67.5 % dry substance, pH 7.5.

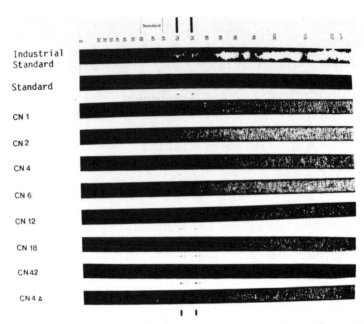

Fig. 6. IGT pick tests for determination of coating strength of hand-coated papers with modified wheat gluten as co-binders

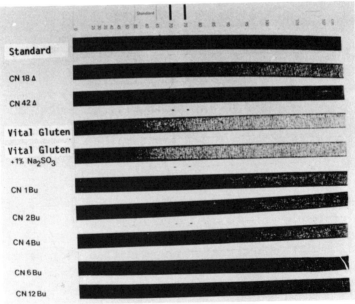

Fig. 7. IGT pick tests for determination of coating strength of hand-coated papers with vital wheat gluten and modified wheat gluten as co-binders

The largely stable viscosity of the model coating mixtures with cyanoethylated wheat gluten was below the values of the standards. Lipid-free cyanoethylated gluten reached higher viscosity values in each case. When vital gluten and alkaline deaminated gluten were used the viscosities increased extremely strongly if model coating mixtures were not diluted.

Using the results from the hand coating experiments as a starting point, coating experiments were carried out for selected modified wheat gluten under standardized conditions using the coating equipment of a pilot plant. In those experiments account was also taken of the degree of agreement between the coating mixture viscosity and the requirements. The composition and the application of the coating mixture as well as the quality of the raw paper were suited to the conditions of offset printing. Apart from wet-reinforcers and dispersion agents no auxiliary substances were used (Table III).

The viscosity of the coating mixtures decreased with increasing modification time of the gluten (Fig. 3), while the water retention capacity increased

TABLE III
Composition of Model Coating Mixtures for Pilot Plant Experiments

60	parts coating kaolin (whiteness 85.5 ± 0.5, particle size $82\pm2\,\%<2\,\mu m$)
40	parts calcium carbonate
8	parts synthetic dispersion binder (50 %)
5	parts protein (17.5 % solution)
0.4	parts dispersion agent (Polysalt)
0.5	parts wet-reinforcer

Solids content:	min. 56.8 %, pH 9-9.5
Application:	about 12 g/m^2 at 110 m/min (on one side, with doctor knife on wood-free body paper (60 g/m^2)
Drying temperature:	160 °C

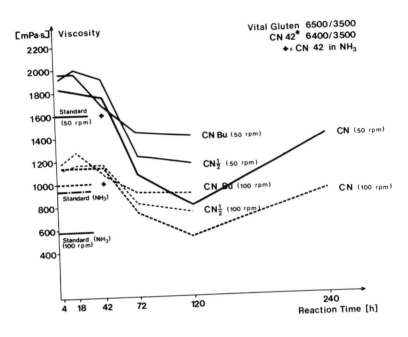

Fig. 8. Viscosities of model coating mixtures for pilot plant experiments

correspondingly (Fig. 9). The lipid content of the gluten before modification played no role in this respect. The resistance of the coating mixture to microbial degradation was so great as a result of the preservative effect of the wet-reinforcer that succinylated gluten could be used likewise, with success, in preliminary experiments.

To evaluate suitability for offset printing, the speckle of the lighter blue areas and of the individual colours was assessed. No clear trend was established, although some gluten modified over a short reaction time showed less speckle. The absorption behaviour of a blue test colour in the offset test was assessed as considerably more favourable for modified gluten than for the reference standard. The use of ammonia as solvent gave the most uniform absorption.

As far as the coated papers are concerned, the gloss and whiteness studies revealed no uniform trend

wards improvement in the quality by addition of
dified gluten as co-binders. In the wet-picking
st, which is one of the deciding factors for
fset printability, the suitability of the cyanoethyted gluten was good to very good, although longer
dification times were undesirable. In this case
o, ammonia was best as a solvent for the gluten.

The pick resistance of the machine-coated papers
th cyanoethylated wheat gluten as co-binders ineased with increasing reaction time. However, a
early positive effect of the acrylonitrile concenation was only discernible at reaction times above
h. Lipid-free cyanoethylated gluten once again
elded the best pick resistance.

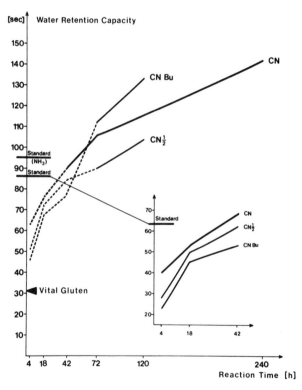

g. 9. Water retention capacity of model coating mixres with cyanoethylated wheat gluten as co-binder

Succinylated wheat gluten also yielded in‐
teresting results. Taking into consideration all the
other co-binder properties investigated, starting
points for further advantageous developments finally
emerged with this type of product (Figs. 10 and 11).

Fillers in urea-formaldehyde resin glues

We evaluated possible applications of vital glu‐
ten, cyanoethylated wheat gluten, and the two gluten
fractions gliadin and glutenin as fillers in urea-
formaldehyde resin glues. Interest was centered on
dry-binding power, warm water strength, and wood
covering of the glue joint after a shearing test of
glued beech veneer. In the latter test a larger wood
covering corresponds to better gluing. Batches of
glue supplemented with vital gluten and gliadin gave

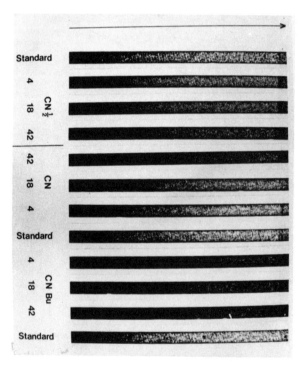

Fig. 10. IGT pick tests for determination of coating
strength of machine-coated papers with cyanoethylated
wheat gluten as co-binders

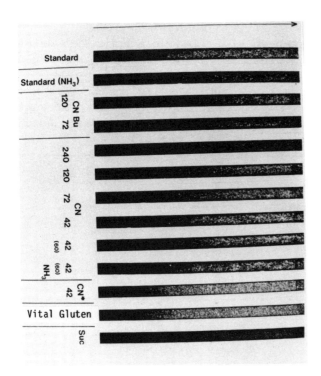

Fig. 11. IGT pick tests for determination of coating strength of machine-coated papers with modified wheat gluten as co-binders

very encouraging results (Table IV). Incorporation and the adjustment of viscosity of vital gluten, however, presented difficulties.

As far as thickening of the glue solution and ropiness are concerned, the gluten fractions gliadin and glutenin and possibly also cyanoethylated gluten deserve special attention from the point of view of process technology. Special attention was also paid for topical reasons to the formaldehyde given off from glued plywood boards. When the sides are open, formaldehyde given off could be significantly reduced by the addition of natural gluten (Table V).

TABLE IV
Binding Strength of Filled Urea-Formaldehyde Resin Glues in Plywood Bonding

Filler	Dry-binding strength		Warm water strength	
	MPA	Wood covering (%)	MPA	Wood covering (%)
None	3.06	90	2.31	10
Vital gluten	3.78	90	2.74	30
Gliadin	3.51	85	2.85	35
Gliadin (delipidized)	3.61	100	2.90	30
Glutenin	3.40	100	2.32	15
Cyanoethyl.gluten	3.60	75	2.51	18
Wheat flour	3.58	68	1.70	0

Glue composition: 54 parts urea dry resin
5.4 parts hardener (60 % flour)
30.6 parts water
10 parts filler

Components of edible synthetic sausage skins

Finally, the inclusion of wheat gluten as a component in edible synthetic sausage skins was investigated in detail. At present, skins are manufactured exclusively from collagen by plasticization and extrusion. Without prior plasticization, vital gluten could not hitherto be processed with collagen (Mullen 1971, Turbak 1972). Using a new digestive technique involving intensive mechanical processing of the gluten after treatment with acetic acid, the resulting viscous mass could be mixed with collagen without any problem. Using mixtures with additions of 5-20 % wheat gluten, sausage skins of very high quality could be produced. They were assessed as equivalent in every respect in preparation and consumption tests.

TABLE V

Formaldehyde Release from Plywood Test Pieces

Composition	Formaldehyde release per unit area ($mg \cdot h^{-1} \cdot m^{-2}$)		Formaldehyde release per unit mass ($mg \cdot (100\ g)^{-1}$)	
	A	B	A	B
100 parts UF glue[a] (66 % dry matter)	30.5	9.3	15.1	4.6
90 parts UF glue 10 parts vital gluten (68.4 % dry matter)	18.2	12.7	10.5	7.3
90 parts UF glue 10 parts vital gluten 10 parts water (62.2 % dry matter)	22.3	14.5	14.3	9.3
90 parts UF glue 10 parts wheat flour 68.4 % dry matter)	22.6	21.7	13.0	12.6

A : Test pieces with narrow sides open
B : Test pieces with narrow sides sealed
a Urea-formaldehyde glue

CONCLUSIONS

Modern processes with high technological requirements for the chemical modification of wheat gluten for use as a co-binder in paper coating mixtures, as a filler in urea-formaldehyde resin glues, as a component in synthetic edible sausage skins, and, last but not least, as a possible constituent of plastics are setting new standards. They are in line with the generally recognized responsibility for maintaining the natural requirements for life, and the growing need for improvements in tapping renewable resources which have hitherto received little attention.

The successes justify further intensive endeavours to develop and optimize the chemical modification of wheat gluten to open new and novel applications in the chemical and industrial field. Further research projects are essential, not only to deepen theoretical knowledge but also to investigate how such knowledge gained in the laboratory can be applied on a semi-technical scale and finally in industrial practice.

The results described in detail here justify such investment because they have shown that it is possible to exploit agricultural by-products usefully. However, the chemical and industrial use of gluten which, after starch, is the largest component of wheat as an additive source material cannot significantly reduce the immense overproduction in agriculture and/or perceptibly counteract the increasing shortage of fossil resources. What it can certainly do is to make a reasonable contribution to solving these problems.

LITERATURE CITED

BARBER, K. H., and WARTHESEN, J. J. 1982. Some functional properties of acylated wheat gluten J. Agric. Food Chem. 30:930-934.

COCO, C. E. 1983. Modified Soy Protein - Tools fo Controlled Coating Properties. Pages 109-122 Coating Conf. Proc., Tappi Press, Atlanta, GA.

COCO, C. E. 1984. The Effect of Protein Latex Rati on Binder Migration, Runability and Resultan Coated Board Properties. Pages 131-141. Coatin Conf. Proc., Tappi Press, Atlanta, GA.

COFFMAN, J. R. 1951. Chemical modification o proteins. US Pat. 2 562 534, Cl. 260-8 31.7.1951, 3 p.

FRIEDMANN, M. 1978. Wheat gluten-alkali reactions Pages 81-100. Proc. 10th Natl. Conf. Whea Utilization Res.

FUJIMAKI, M., HARAGUCHI, T., ABE, K., HOMMA, S., a ARAI, S. 1980. Specific conditions th maximize formation of lysino-alanine in wheat gluten and fish protein concentrate. Agr. Biol. Chem. 44:1911-1916.

GAREY, C. L. 1987. Analyzing use of soy protein binders in coating formulation. Paper Trade J. 166, 17:41-44.

GRANT, D. R. 1973. The modification of wheat flour proteins with succinic anhydride. Cereal Chem. 50:417-428.

HESS, W. 1988. Personal Communication.

KEMPF, W., and PELECH, B. 1986. Erschliessung neuer Einsatzgebiete für die Nebenprodukte nachwachsender Rohstoffe. Institute of Starch and Potato Technology, Detmold, Federal Republic of Germany (Report on Research Project BML 81-NR-035).

KOTTE, H. 1978. Streichen und Beschichten von Papier und Karton. P. Keppler Publishing Company KG, Heusenstamm, Federal Republic of Germany.

MECHAM, D. K., and MOHAMMAD, A. 1955. Extraction of lipids from wheat products. Cereal Chem. 32:404-414.

MULLEN, J. D. 1971. Film formation from non-heat coaguable simple protein with filler and resulting products. US Pat. 3 615 171, Int. Cl A 22 c 13/00, 26.10.1971.

SCHWENKE, K. D. 1978. Beeinflussung funktioneller Eigenschaften von Proteinen durch chemische Modifizierung. Nahrung 22:101-102.

SILVERNAIL, L. H., and BAIN, W. M. 1961. Synthetic and Protein Adhesives for Paper Coating. Tappi Monograph Series No. 22, Tech. Assoc. Pulp Paper Ind., Atlanta, GA.

STRAUSS, R. 1975. Protein Binders in Paper and Paperboard Coating. Tappi Monograph Series No. 36. Tech. Assoc. Pulp Paper Ind., Atlanta, GA.

TURBAK, A. F. 1972. Edible vegetable protein casing. US Pat. 3 682 661, Int. Cl A 22 c 13/00, 8.8.1972.

WHALEN-SHAW, M. 1984. Protein-pigment interactions for controlled rotogravure printing properties. Mechanistic and structural considerations. Tappi J. 67:60-64.

33

THERMAL MODIFICATION OF GLUTEN AS RELATED TO END-USE PROPERTIES

J.C. Autran, O. Ait-Mouh and P. Feillet

Laboratoire de Technologie des Céréales
I.N.R.A., 2 place viala
34060 Montpellier Cedex, France

INTRODUCTION

Most of cereal food processes are characterized by hydrothermic treatments under a range of temperature (55°C to 220°C) and humidity (35 to 70 %), depending on the type of product (bread, cookie, pasta, vital gluten). These treatments are important and critical steps in cereal technology and they strongly affect the quality characteristics of end-products.

On the other hand, gluten proteins play a predominant role in most of cereal foods, contributing to structures that are desirable to the consumer or the industry. Their multifunctional characteristics enable them to take part in several stages occuring between raw materials and products. They are strongly modified by heat treatments and one of the first consequences of the temperature raise is a physical change known as protein denaturation.

THEORETICAL BASIS OF HEAT DENATURATION OF PROTEINS

As recently reviewed by Schofield and Booth (1983), Davies (1986) and Hoseney (1986), it is well known that exposing proteins to high temperatures for only short periods causes most of them to undergo conformational changes, of which the most visible effect is a decrease in solubility. Most globular proteins undergo denaturation when heated above 60 to 70°C, but considerable variations in the degree of change exist among proteins (Voutsinas et al. 1983). Formation of an insoluble white coagulum when egg white is boiled is a common example of protein denaturation but equally significant and profound is certainly the loss of biological activity of enzymes.

Since no covalent bonds of the polypeptide chain are broken during a relatively mild treatment, denaturation causes the native characteristic folded structure of the polypeptide chain to uncoil or unwind into a randomly looped chain. Then, thermal agitation contributes to new associations and sequentially alters polypeptides into aggregates and finally into insoluble components.

As unfolding occurs, functional groups previously associated within the molecule become available for external binding (increase of surface hydrophobicity), for interactions with other proteins or subunits, or other constituents (depending on heating temperature, heating time and humidity), giving rise to more highly aggregated complexes (Wall and Huebner 1981). If the denatured unfolded conformation has less free energy (i.e. is more stable) than the folded native conformation by only a very small margin, the change may be reversible. However, if the activation-energy barrier is high, the polypeptide may be irreversibly locked into the denatured (aggregated) conformation.

As the interactions of water molecules with ionic and polar groups strongly influence the folded conformation of proteins, the effect of heat on denaturation is markedly affected by the water content of protein : as the water content decreases, the amount of heat required to denature a protein significantly increases. Dry proteins are thus much more resistant to heat denaturation than proteins in solution (Neucere and Cherry 1982).

Heat disrupts the non-covalent forces, particularly H bonds and electrostatic interactions. All chemical bonds are weakened as the temperature increases. However, hydrophobic bond formation is an endothermic process which may be favored by increasing temperature up to about 60°C.

In gluten proteins, where the structure is mainly stabilized through S-S bonds and hydrophobic interactions, two major phenomena have been described. First, heat-induced unfolding causes the apolar residues to move towards the outside, to make more difficult the contact of the molecule with water (the protein becomes insoluble in aqueous media) and contributes to aggregates strengthened through hydrophobic interactions. Simultaneously, an SH/S-S interchange can be facilitated between exposed S-S and SH groups in adjacent molecules (Schofield 1986), giving rise to a more highly aggregated structural state having less free energy and therefore more stable than the native state.

IMPORTANCE OF THERMAL DENATURATION IN CEREAL PRODUCTS

The phenomenon of thermal denaturation is of primary importance in relation to food products. It is extremely important and affects their preparation, processing, nutritional value, quality and safety. Depending upon the particular application, it may be desirable or undesirable.

For many uses, undenatured proteins have superior functional properties (such as solubility, emulsifying, foaming and thermosetting properties) compared to denatured proteins. Therefore protein denaturation is usually viewed in a negative sense by food-protein chemists, particularly in relation to preparation of functional ingredients.

However, denaturation may have positive effects and, in cereal foods, heat treatments are even necessary for starch digestibility, texturization and determination of the technological and organoleptic qualities of the final product. For instance, in bread-making, the thermal denaturation of the gluten film around the gas droplet is an essential step for the formation of the unique foam texture of bread. It also allows selective thermal inactivation of certain undesirable components (amylases, yeast enzymes, lipases, lipoxygenases, peroxidases).

In contrast to many food systems (egg white, soy and milk proteins), relatively few reports have concerned denaturation of wheat gluten proteins. Since the pioneering studies of Mecham and Olcott (1947) and Pence et al (1953), few, if any, in-depth analyses of the alterations produced in flour or semolina proteins as a consequence of heat treatments were reported in the literature.

The objectives of the present paper are to review the major effects of thermal processing on wheat proteins and to develop an understanding of the complex physico-chemical changes they undergo, based on the study of four different examples of heat treatments :
-Model systems : hand-washed gluten or purified gliadins or glutenins.
-Bread-baking and cookie-making.
-High-temperature pasta drying and pasta cooking.
-Industrial vital gluten production and drying.

In these studies, protein denaturation has been monitored by the following methods : solubility in various solvents, electrophoresis in polyacrylamide gel in acid buffer (A-PAGE) or in the presence of sodium dodecyl sulfate (SDS-PAGE), conventional size-exclusion chromatography, size-exclusion (SE-HPLC) or reversed-phase (RP-HPLC) high-pressure liquid chromatography, circular dichroism spectroscopy (CD), prediction of secondary structures, viscoelastic measurements and differential scanning calorimetry (DSC) experiments.

MODEL SYSTEMS : HAND-WASHED GLUTEN OR PURIFIED GLIADINS OR GLUTENINS

Jeanjean et al (1980) extracted gluten om four common wheat and three durum wheat ltivars, cast it into a special cell of the scoelastograph and dipped it into boiling water r 0-7 min, after which viscoelasticity and otein solubility were examined. On heating, uten compressibility decreased, gluten firmness d elasticity increased (Fig. 1); some proteins, luble in 60 % ethanol were insolubilized. Results SDS-PAGE showed that some subunits of salt-solu- e and gliadin-like proteins participate in e formation of insoluble protein networks and oth-

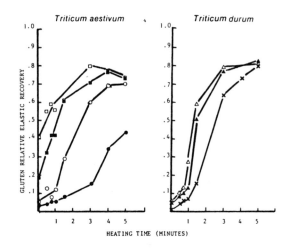

Fig. 1 Influence of heating time on gluten disk relative elastic recovery. Cultivars : ● = Clement, ○ = Maris Huntsman, ■ = Capitole, □ = Kolibri, ✗ = V39, ▲ = Lakota, △ = Agathe (Reprinted, by permission, from Jeanjean et al 1980).

ers do not. Moreover, differences between cultivars were significant : in common wheats with better baking quality and in durum wheats with better pasta-cooking quality, the tendency of ethanol-soluble proteins to aggregate during heating was greater. Since the aggregates could be disrupted further only with mercaptoethanol, it was postulated that protein insolubilization occured through the formation of new bonds, possibly disulfide bonds.

Complementary observations were reported on bread wheat glutens heated 0-90 min in boiling water, specifically an insolubilization of proteins associated with a gradual disappearance of streaking material and of bands in A-PAGE patterns or unreduced SDS-PAGE patterns (Autran and al 1982). Similar results were obtained by Autran and Berrier (1984) from durum wheat glutens heated 0-30 min. Standard patterns were essentially restored upon treatment of gluten by a reducing agent before SDS-PAGE confirming that disulfide bonds could be involved in protein aggregation by heat. Some protein fractions were more susceptible to insolubilization than others. Streaks and slot material, that correspond to extractable LMW and HMW glutenin fractions, disappeared after less than 5 min of heat treatment, and patterns from glutens heated from 30 to as long as 90 min consisted essentially of an enhanced group of ω-gliadin bands (Fig. 2).

When comparing different glutens extracted from flours that varied in breadmaking potential, Sadouki and Autran (1987) observed different behaviours for the various HMW glutenin subunits. In general, those (viz. n° 1 or n° 5-10) whose presence is positively correlated with high baking quality exhibited a higher tendency to form insoluble complexes upon heat treatment than those (viz. n° 2-12) whose presence is negatively correlated with baking quality.

Recently, more accurate studies were carried out from purified gliadin fractions. Menkovska et al (1987) submitted gliadin fractions to heat

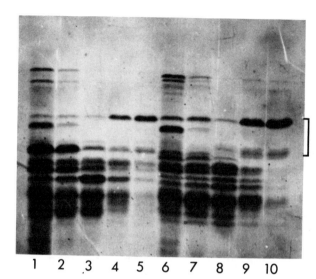

Fig. 2. SDS-PAGE patterns of ethanol-soluble fraction of gluten from two durum wheat cultivars: Montferrier (slots 1 to 5) and Calvinor (slots 6 to 10). Proteins were extracted from control semolina (1 and 6) and from gluten heated 0 min (2 and 7), 1.5 min (3 and 8), 10 min (4 and 9) and 30 min (5 and 10) (Reprinted, by permission, from Autran and Berrier 1984).

treatments (15 min at 100°C and 200°C) respectively, selected to represent the approximate temperatures that the crumb and crust approach during bread-baking. Subsequent analyses by A-PAGE and RP-HPLC showed that the highly hydrophobic fractions (which consist essentially of α-, β-, γ-gliadins, and possibly of extracted LMW-glutenin material) undergo major changes in the peak size of the RP-HPLC chromatograms and seem to be more heat labile than the less hydrophobic gliadins (which correspond to ω-gliadins). These results agree with SDS-PAGE data of Meier et al (1985) who showed an increase in molecular weights of the gliadin proteins on heating and also confirm those obtained from crude gliadins and glutenins by Anno (1981).

Using a different physico-chemical approach based on circular dichroism (CD) measurements and prediction of secondary structure, Tatham and Shewry (1985) studied the effect of temperature on the conformation of native and cysteine-modified sulfur-rich α-, β-, and γ-gliadins and of sulfur-poor ω-gliadins in ethanolic solutions. Increasing the temperature from 20° to 80°C resulted in an increase in aperiodic structure, with partial loss of the -helical content, confirming earlier reports of Kasarda et al (1968) on A-gliadin. A different result, however, was obtained on ω-gliadins in which the conformational change consisted of an increase of β-turns. It was concluded that whereas the ω-gliadins are stabilized by strong hydrophobic interactions, the main stabilizing forces in the α-, β-, and γ-gliadins are covalent disulfide bonds and non-covalent hydrogen bonds. Interestingly, the conformation of the modified gliadin was similar to that of the native protein, indicating that reduction and alkylation had little effect on the conformation. More recently, Tatham et al (1987) studied the effect of heating on the conformation of LMW subunits of glutenin (or aggregated gliadins) and found CD spectra more similar to those of α-, β-, and γ-gliadins than to those of ω-gliadins or HMW subunits of glutenins. No major differences in thermal stabilities of LMW fractions could be found to explain the differences in breadmaking quality between cultivars.

Therefore, irrespective of the experimental approach used for monitoring heat denaturation, all treatments clearly cause in gluten proteins a gradual loss in solubility associated with changes in functional properties (higher firmness and elastic recovery) that are likely to result from structural modifications. The different fractions do not have the same sensitivity to heat treatments : gliadins (especially fractions ω) are much less affected than glutenins whose thermal aggregation seems to involve the formation of new disulfide bonds.

BREAD-MAKING OR COOKIE-MAKING

In a regular bread-making process, after mixing and fermentation steps, the dough is placed into an oven at about 220°C. A steep temperature gradient, 200°C ⟶ 100°C, inward from the crust is established in the dough piece. Because bread is a moist product, its temperature cannot raise much above the boiling point of water unless it becomes completely dry. The only part of bread dried during baking is the crust, that can reach a temperature of 200°C, while the crumb does not reach a temperature greater than 100°C.

According to Hoseney (1986), the differences in characteristics and in size of the final baked product are determined by interactions that occur during heating. However, heat phenomena during the baking step are extremely complicated. The major change that takes place during the oven process is the redistribution of water from the gluten phase to the starch phase, thereby allowing the starch to undergo gelatinization. Gluten coagulation sets in at about 74°C and continues slowly until the end of baking. In the course of this process, the gluten matrix surrounding the individual gas cells is transformed into a semi-rigid film structure which becomes thinner as the gas cells expand and may rupture but not collapse.

In recent studies, attention has been focused on solubility and conformational changes in both gliadin and glutenin proteins of bread. Meier et al (1985) and Menkovska et al (1987), using RP-HPLC and A-PAGE demonstrated that, in contrast to the apparent absence of modification of the gliadin fraction during mixing of fermentation step (as evidenced by absence of changes in electrophoretic patterns), interactions or degradations of gliadin proteins essentially occurred during baking. For instance, when comparing an extract of the crumb to a flour extract, the A-PAGE gliadin bands with low relative mobility (RM $<$ 40) were more intensely stained and those with RM $>$ 40 were less intensely stained, while the intensity of all gliadin bands in the crust was dramatically reduced. Menkovska et al (1987) also demonstrated that the highly hydrophobic gliadins (RP-HPLC elution times $>$ 23 min) extracted from crumb were more heat labile (and probably interacted more with other flour components) than the less hydrophobic gliadins (elution times $<$ 23 min).

When comparing flours that varied in breadmaking potential, Menkovska et al (1988) found that the change from flour to bread crumb was more pronounced in good breadmaking flours than in poor breadmaking flours (i.e. the relative decrease from flour to crumb in gliadins of high mobility was much greater in the good to intermediate breadmaking flours than in the poor-breadmaking flours). From RP-HPLC elution curves, the change in intensity of the peaks was not equal for various groups of gliadins and for various flours and the extent of reduction of peak intensity was also much higher in good-than in poor-breadmaking quality flours (Fig. 3), thus indicating differences in interactions or heat lability. Menkovska et al (1988) postulated that heat-labile (and highly hydrophobic) α -, β -, and γ -gliadins were modified during baking and that the modification may be related, in part at least, to differences in breadmaking potential of flours.

All these results agree with those of Wrigley et al (1980) who studied the effect of baking on heat denaturation by gradient-PAGE of gliadins and those of several authors who devised methods for detecting gluten proteins in heated foods

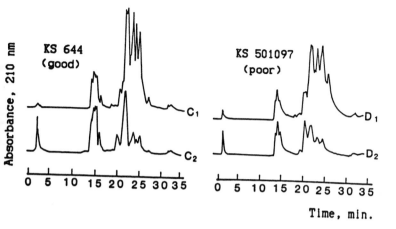

Fig. 3. Reversed-phase chromatography of gliadin proteins from flours (1) and corresponding bread crumbs (2) of lines KS 644 (C) and KS 501097 (D) (Reprinted, by permission, from Menkovska et al 1988).

nd baked goods (McCausland and Wrigley 1976; rigley 1977; Dysseler et al 1986). A complete liadin pattern was obtained for the crumb of xperimentally pan-baked bread, but the components f higher mobility did not stain as strongly s for flour. In contrast, only gliadins of low obility were extracted with 2M urea from the rumb of hearth baked bread. On the other hand, ess gliadin was extractable from bread crust, here protein denaturation due to heating was uch greater than in the crumb. After extreme reatments only a few zones of very low mobility ere noted. These probably represent polymerization roducts due to heating (McCausland and Wrigley 976). The addition to 2M urea or of a 0.2 % educing agent, β-mercaptoethanol, greatly increased the intensity of crumb protein staining in the gliadin mobility range. It was concluded that the reducing agent probably exerts this effect both by aiding the extraction of the gliadin proteins themselves and by reducing the S-S crosslinked glutenin polypeptides, thereby making them extractable also.

A comparison between the changes in gliadin proteins resulting from bread-making and cookie-making has been also carried out by Pomeranz et al (1987) using A-PAGE and RP-HPLC fractionations. It was observed that both in bread and in cookies, there was a reduction (from flour to baked product) in the intensity and resolution of highly hydrophobic (long elution time) HPLC peaks and that the decrease in resolution seemed to be more pronounced in good than in poor cookie-making flours. A possible involvement (probably through interaction with proteins or with other components) of gliadins in production of satisfactory products was proposed. A reduction in intensity and resolution of HPLC short elution time peaks (from flour to baked products) in cookies was also noted. These changes seemed to be primarily governed by the effect of high temperatures exerted during the baking of cookies and might be also related to functional properties of the flours.

HIGH-TEMPERATURE PASTA DRYING AND PASTA COOKING

Temperature is an important parameter in pasta technology. Firstly, because drying operations are more frequently performed above 70° or 90°C and, secondly, because during cooking, pasta is left in boiling water for about 10 minutes (depending on shape). Pasta making, however, differs from bread-making : if dough development is defined as formation of a continuous network of protein sheets and fibrils, then at the dough-water content of pasta goods, no full gluten development takes place (Dexter and Matsuo 1979).

Such an absence of full gluten development would explain why differences in cooking quality between samples of semolina may be essentially accounted for by the manner in which the proteins are modified during the drying or the cooking step.

High-temperature pasta drying

Recent pasta technologies using a drying step at 70° or 90°C may have a strong effect on quality of cooked pasta, which involves improvement of rheological characteristics and surface conditions (Dexter et al 1981; Feillet 1987).
While most of the information concerns the influence of temperature on lysine availability of pasta products (Cubadda et al 1968; Dexter et al 1984), only few reports relate to modifications of pasta proteins.
Drying spaghetti at 80°C extensively denaturates proteins, as demonstrated by solubility in dilute acetic acid (Dexter et al 1981; Ibrahim and McDonald 1981) and changes in gel filtration elution profiles. Increasing duration (from 1 to 9 hr) and temperature (from 70° to 90°C) of drying decreases proteins solubility in $MgSO_4$ and 0.1 % acetic acid (Resmini et al 1976). Additional information gained from microscopy studies showed that high temperatures contribute to the formation of a protein network and result in better pasta performance during cooking (Resmini et al 1976).
In other studies of heat treatment on pasta processed from both type γ-gliadin 45 and type ω-gliadin 42 durum varieties, Autran and Berrier (1984) confirmed the losses in ethanol or acetic acid solubility but also reported a weakening of gliadin bands in A-PAGE patterns compared to pasta dried at lower temperatures. However, this phenomenon essentially concerned the fast-moving components, while the slow-moving ω-gliadin bands remained present, what allowed to propose a method for detecting bread wheat flour in heat-treated pasta products based on quantitation

of ω-gliadin group from A-PAGE electrophoresis (Kobrehel et al 1985).

In a recent study, Feillet (1987) and Feillet et al (1987) investigated the effects of moisture content, and duration and temperature of pasta drying. For instance, submitting pasta (30 % moisture content) to heat treatment for 120 minutes led to a steep decrease of solubility in SDS, the initial solubility being restored only by further extraction with mercaptoethanol (Fig. 4); the changes could be explained by formation of disulfide bonds between pasta proteins during heat treatment. In another experiment, pasta samples with 24 %, 18 % and 12 % moisture contents were left for 2 hours at 90°C. The stronger the hydrothermic treatment (i.e. the pasta humidity), the larger was the loss of solubility

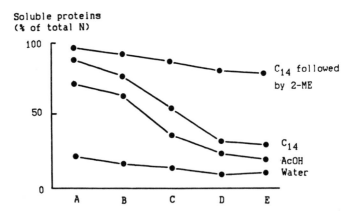

Fig. 4. Effect of heat treatment (2 hrs at 90°C) on pasta protein solubility. A-Semolina; B-Pasta dried at 55°C; Pasta left for 2 hrs at 90°C at 13 % (C), 18 % (D), and 24 % (E) moisture content. AcOH : Acetic acid; C14 : Sodium myristate; ME : 2-mercaptoethanol (Reprinted, by permission, from Feillet 1987).

in sodium myristate, which disrupts hydrophobic bonds. In addition, most residual proteins were soluble in β-mercaptoethanol.

A-PAGE fractionations of unreduced protein extracts revealed which proteins aggregate during heat treatment. ω-Gliadins, which have a very low sulfur content, are highly heat-resistant. Streaks and slot-proteins (which essentially consist of proteins with molecular weights from 35,000 to 50,000 daltons, i.e. in the range of major LMW subunits of glutenin) rapidly disappear from electrophoretic patterns upon gradually increased heat treatments (Feillet et al 1987).

The heat sensitivity of LMW glutenins was confirmed by SE-HPLC of SDS-phosphate (unreduced) extracts. All protein peaks decreased when the intensity of heat treatment was increased, but the phenomenon especially affected peaks 1 and 2 (which essentially consist of LMW glutenin aggregates), which rapidly disappeared from the elution curves (Fig. 5).

Since these phenomena occur in a similar way in both types of wheats, another question was why poor varieties can be improved through such treatments, especially as far as surface condition (absence of stickiness and surface deterioration) is concerned. It is now accepted Resmini and Pagani 1983; Feillet 1984) that during cooking in boiling water there is a competition between (1) protein coagulation into a continuous network and (2) starch swelling. If (1) prevails, starch particles are trapped in a continuous network, promoting high firmness and little stickiness of cooked pasta, while, if (2) prevails, proteins coagulate in discrete regions lacking a continuous network, giving soft and sticky pasta. High temperature pasta drying might partially overcome this competition by producing a coagulated protein network in dry pasta without starch swelling (Feillet 1984).

It is generally believed that gluten is not fully developed when pasta are prepared from doughs containing 25-35 % water. Still, in poten-

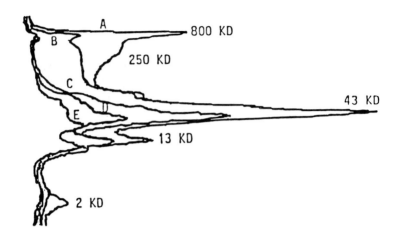

Fig. 5 Size exclusion HPLC chromatography of SDS-phosphate (unreduced extracts). A - Semolina; B - Pasta dried at 55°C; Pasta left for 2 hrs at 90°C at 13 % (C), 18 % (D), and 24 % (E) moisture content, respectively (Reprinted, by permission, from Feillet 1987).

tially strong varieties (type γ-45), a viscoelastic network might be already formed at the mixing or extrusion stage (before any heat treatment) due to pronounced functional properties (or more probably due to the very high proportion of LMW subunits of glutenin in the varieties belonging to type γ-45 (Autran et al 1987)). In contrast, poor varieties (type γ-42) have a lower quality potential and lack a viscoelastic network, because of the reduced tendency to aggregate (or the small amount) of their LMW proteins. To affect an improved firm structure, these varieties require a heat treatment, which may cause a rapid aggregation of their LMW (and HMW) glutenin subunits.

However, the proteins having a high tendency to coagulate upon heat treatment and to promote the retention of a good surface condition of pasta during cooking, may not be necessarily those that contribute to the formation of a viscoelastic network during mixing and extrusion. A new hypothesis has been proposed by Feillet et al (1988), following the discovery by Kobrehel and Alary (1988) of a new protein fraction called DSG (durum wheat sulfur-rich glutenin). SH + S-S content of DSG proteins and surface condition of cooked pasta were significantly correlated and DSG proteins were not extractable by sodium myristate from heat-treated pasta, but were extractable by mercaptoethanol. Consequently, Feillet et al (1988) proposed a functional role of DSG in preventing disaggregation of the surface of cooked pasta. They postulated that DSG proteins contribute to aggregation of LMW glutenins (and possibly of HMW glutenins) through hydrophobic and disulfide bonds and that these bonds are sufficiently strong to prevent starch leaching during pasta cooking and to maintain a satisfactory surface condition of cooked pasta.

Pasta cooking

The effect of cooking on protein solubility of pasta was thoroughly examined by Wasik (1978) who found that cooking decreased the amount of soluble proteins recovered by the Osborne procedure ; 10-15 % of the total recovered protein was soluble in water, saline, alcohol and acid solvents, in contrast to about 65-75 % from the uncooked product. Cooking reduced the solubility in AUC solvent from 95 % to 50 % and radically modified Sephadex G200 gel filtration profiles. Nevertheless, no change was observed in the SDS-PAGE patterns of proteins after reduction by mercaptoethanol.

According to Dexter and Matsuo (1979), protein extractability in dilute acetic acid rapidly decreased during spaghetti cooking up to about 2 min. At each cooking time examined, the poorer quality wheats exhibited significantly greater protein extractability than the better ones. This was explained by the greater proportion of extractable gluten protein in the poor-quality wheats. However, gel filtration elution profiles of acetic acid extracts at each cooking time revealed no significant quantitative differences in the pattern of protein denaturation.

In another study, spaghetti processed from several durum wheat varieties of a wide range of cooking quality were cooked in boiling water, frozen and freeze-dried, and examined for protein solubility and electrophoretic patterns by Autran and Berrier (1984). Upon cooking, a gradual loss of solubility in ethanol of several gliadin bands was observed. They included firstly some γ-gliadins (A-PAGE mobilities 40-51), then some β- and α-gliadins, whilst the whole slow moving group (ω-gliadins 20-23-26 and 33-35-38) remained obviously visible ; they dominate the patterns (and are even enhanced) after 10 min of pasta cooking. Also, streaks in the gliadin patterns disappear after only 1-2 min of cooking (Fig. 6).

A similar behaviour was observed for both good (Bidi 17) and poor (Tomclair) cooking quality cultivars. In particular, γ-gliadin 45 seemed to be insolubilized according to the same kinetics than γ-gliadin 42. It was especially interesting, however, to note that the partition between components that are resistant or susceptible to heat denaturation occurs within a genetic linkage group (where recombinations are usually not observed). The bands 40-42, or 45 are typical γ-gliadins and aggregate upon heat treatment; on the other hand the bands 33-35-38, or 35, are typical ω-gliadins and behave as highly resistant proteins. It is not clear how components coded by a cluster of closely linked genes, that presumably derive from an ancestor gene through point mutations, can differ so widely in functional properties and behaviour upon heat treatments.

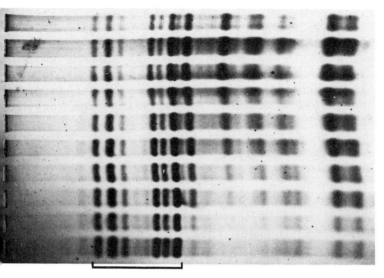

ω-gliadins

Fig. 6. A-PAGE of gliadins extracted from (S) semolina, (C) control uncooked spaghetti and spaghetti cooked 1, 2, 3, 4, 5, 6, 8 and 10 min, respectively (from Autran, unpublished results).

INDUSTRIAL VITAL GLUTEN PRODUCTION AND DRYING

The use of gluten by the Western European milling industry to replace strong imported wheats in the production of flours for conventional breadmaking and especially for other specialty breads has increased rapidly in recent years. It is essential that the gluten be vital, e.g., that it retains the desirable viscoelastic properties required for gas retention (Stenvert et al. 1981; McDermott 1985). However, gluten is obtained through wet processes and then dried in driers operating at elevated (50-75°C) temperatures. The commercial quality of these glutens is extremely variable. According to Wu and Inglett

(1974), Schofield et al (1983) and Schofield (1986), the degree of heat denaturation during drying is generally the main source of variation. Although little is known about the exact effect of drying conditions (freeze, drum, or spray drying) on the functional and biochemical properties of vital gluten, excessively high drying temperatures are thought to be a major factor in the reduced and variable baking performances of commercial glutens.

As laboratory heating of gluten may not exactly represent the commercial process, Booth et al (1980), Schofield and Booth (1983) and Schofield et al (1983) have studied the chemical and functional changes that gluten proteins undergo as a result of different conditions employed during industrial drying. They showed that extractibility in SDS buffer decreased and that glutenins of highest molecular weight (in SE-chromatography) were affected predominantly. Gliadins were also affected but at higher temperatures : α -, β -, and γ -gliadin fractions were made progressively inextractable, unlike ω -gliadins that were essentially unaffected.

Baking performance, assessed in reconstituted flours, declined progressively between 50 and 70°C and most of the functionality was lost by 75°C (Schofield et al 1983). When a gluten dough was heated to 80°C, a large increase in mixing time and a reduction in loaf volume resulted. Also, after steaming, the gluten was both more difficult to deform and relatively more elastic (Dreese et al 1988).

The fact that the functionality of heat-denaturated gluten can be restored, at least partially, by dough mixing in the presence of a reducing agent (Schofield et al 1984) tends to support the involvement of S-S bonds in the denaturation process. However, Schofield et al (1983) concluded that the level of total SH groups remained constant up to 100°C and that there was a transfer of SH groups from an SDS-extractable to an SDS-inextractable form (Fig. 7). Accordingly, gluten drying is causing the polymer system to become effectively more cross- linked, but without any decrease of the total SH groups. Thermal agitation, which allows to explore all possible conformations, may therefore promote the formation of new bonds through SH/S-S interchange and to form a more highly polymerized state with less free energy and more stability.

Similar conclusions were drawn by Hansen et al (1975) and, recently by Davies et al (1987). They suggested that heat setting of proteins results from an increase in branching from chain interactions through difulfide bond formation; they explained the increase in breaking stress by postulating the formation of new protein crosslinks between cystein residues exposed by temper-

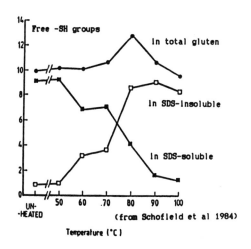

Fig. 7. Effect of heat on free sulfhydryl groups in SDS extractable (■——■) and SDS-inextractable (□——□) fractions of gluten and in the total gluten protein (●——●) (Reprinted, by permission, from Schofield et al 1984).

ature-induced unfolding of the proteins. The authors proposed, however that differences in tensile properties among varieties are determined primarily by the number and reactivity (accessibility) of cystine residues.

Additional, alternative, mechanisms have been reported in an attempt to describe the phenomena involved in heat modification of functional properties. For instance, Dreese et al (1988) compared rheological characteristics of laboratory hand-washed and lyophilized gluten and commercial gluten and concluded that the washing effect was more important than the drying effect. Also, Eliasson and Hegg (1980), studying heated gluten by differential scanning calorimetry (DSC), recorded a major thermal transition in the starch component, rather than in the protein fraction, probably because the method was not adapted to follow protein modifications. Further experiments of Schofield et al (1984), who demonstrated that inclusion of fungal α-amylase did not protect the gluten against loss of baking performance on heating up to 70°C, suggested that the effect on gluten functionality in this temperature range was due to changes in the protein component and not in the starch component.

In summary, as succinctly reviewed by Schofield (1986), all the available data suggest that the loss of functionality in heated gluten is likely to occur through heat-induced unfolding of glutenin polypeptides that facilitates SH/S-S interchange between exposed S-S and SH groups in adjacent molecules. This results in a tougher, harsher, more elastic gluten, whose baking performance is decreased. It was demonstrated that the glutenin fraction is more susceptible to heat denaturation than the gliadin fraction.

CONCLUDING REMARKS

The functional properties (elasticity, extensibility, viscosity) of most of the wheat products are largely determined by the gluten protein viscoelastic behaviour ; gliadins contribute to viscosity and extensibility of gluten, while glutenins impart elasticity. Although a number of models have been proposed to account for those properties, their precise molecular basis is still not known. Nevertheless, it is well established that, among gluten proteins, gliadins are monomers that aggregate by hydrogen bonding and hydrophobic interactions, while glutenins are part of very high molecular size complexes that are additionally stabilized by interpolypeptide disulfide bonds.

Obviously, the balance between fractions that aggregate by hydrogen bonding, hydrophobic interactions, or by formation of new disulfide bonds is critical in determining satisfactory dough strength and loaf volume and this balance is dramatically changed upon heat treatment. The changes that result from heat treatment may be much higher than the regular differences among varieties. A possible question, therefore, is why the higher levels of inextractable protein in heat treated gluten do not confer better quality on the samples since a higher content in insoluble or residual protein is generally associated with a higher baking quality. The obvious answer is that the physico-chemical basis of protein insolubility is different in both cases. Still, it is necessary to elucidate which bonds are involved in those differences. According to Booth et al (1980), either inextractable proteins differ in character from residue protein found naturally in wheat flours - a strong positive correlation between level of residue protein and baking quality has been reported by Orth and Bushuk (1972) - or that they are similar in character but present in excess of the level required for baking performance thus resulting in a tough gluten.

Regardless of the type of heat treatment and of the methods used for monitoring denaturation,

several general conclusions can be drawn.
1 - Thermal agitation causes a native folded structure to uncoil or unwind into a randomly looped chain causing different chemical groups to react or to interact (sulfhydryl into disulfide bonds, disulfide interchange reactions, non polar groups brought together giving rise to hydrophobic interactions) and to form aggregate or network structures that can be at least partly disrupted by reducing agents or detergents. The most visible effect of heat denaturation of gluten is therefore a decrease in protein solubility and a weakening or a disappearance of certain components from unreduced electrophoretic patterns, associated with changes in functional properties (higher firmness and elastic recovery) that are likely to result from structural modifications.
2 - The different gluten fractions or subunits do not have the same tendency to interact or to cross-link through thermal denaturation. Glutenins are extremely sensitive and strongly aggregate upon even mild heat treatments, while the most hydrophobic α -, β -, and γ -gliadins also become insoluble but in a more gradual way. The solubility of most fractions can be generally restored upon the use of reducing agents, indicating the occurence of new S-S bonds (possibly through SH/S-S interchange) in heat-treated pasta products.
3 - Sulfur-free (and less hydrophobic) ω -gliadins, the structure of which is more a random coil type, have a very low chemical reativity and therefore a very high resistance to heat denaturation. Extra uncoiling is however a possible reason why they can bind more dye and give more intense bands.
4 - Varietal differences have been observed in the overall decrease in solubility. Among bread wheats or durum wheats, the tendency of proteins to aggregate is greater in good- than in poor-breadmaking flours or -pastamaking semolinas. This indicates that the susceptibility of proteins

to heat modification may explain some differences in quality potential of flours. Among durum wheats, LMW glutenins (whose relative amount makes a major difference between the two quality classes of durum wheats) could be more directly involved in varietal differences in cooking quality than γ-gliadin markers 42 and 45, or HMW glutenins.
5 - Wheat proteins make an extremely complex model since proteins are synthesized in a wet environment that becomes drier upon grain development; they are hydrated again upon mixing and then modified upon drying, baking or cooking. The manner in which the structure can change during various heat denaturation treatments (flexibility, loosening of the overall conformation) could provide a powerful dynamic approach for investigating the physico-chemical basis of technological quality and for understanding the basis of functional properties in food processing.
6 - Many aspects are still elusive and more sophisticated physico-chemical methods are now required. Since most of the biochemical work in model systems has focussed on gliadin proteins, future investigations should concern glutenins which are the ones primarily involved in heat aggregation phenomena.

LITERATURE CITED

NNO, T. 1981. Studies on heat-induced aggregation of wheat gluten. J. Jpn Soc. Food Nutr. 34: 127-132.

UTRAN J.C., and BERRIER, R. 1984. Durum wheat functional subunits revealed through heat treatments. Biochemical and Genetic implications. In : Gluten Proteins : Proc. 2nd Int. Symp. Workshop on Gluten Proteins, Wageningen The Netherlands, A. Graveland A. and J.H.E. Moonen, eds. : 175-183.

UTRAN, J.C. HOULIAROPOULOS, E., and LAIGNELET, B., 1982. Viscoelastic properties of bread wheat gluten : measurements, use in breeding and biochemical basis. Cereal Foods World 27 :

469.
AUTRAN, J.C., LAIGNELET, B., and MOREL, M.H. 1987. Characterization and quantification of low-molecular-weight glutenins in durum wheats. Biochimie 69 : 699-711.
BOOTH, M.R., BOTTOMLEY, R.C., ELLIS, J.R.S., MALLOCH, G., SCHOFIELD, J.D., and TIMMS, M.F. 1980. The effect of heat on gluten physicochemical properties and baking quality. Ann. Technol. Agric. 29 : 399-408.
CUBADDA, R., FABRIANI, G., and RESMINI, P. 1968. Variazioni della lisina utilizzabile nelle paste alimentari indotte dai processi technologici d'essicamento. Quad. Nutr. 28 : 199-208.
DAVIES, A.P. 1986. Protein functionality in bakery products. In : Chemistry and Physics of Baking, Special Publication No. 56, The Royal Society of Chemistry, Burlington House, London; J.M.V. Blanshard, P.J. Frazier and T. Galliard, eds.: 89-104.
DAVIES, A.P., PATIENT, D.W., INGMAN, S.J., ABLETT, S., DRAGE, M., ASQUITH, M., and BARNES, D.J. 1987. Wheat protein properties and puff pastry structure. In : Proc. 3rd International Workshop on Gluten Proteins, Budapest, Hungary, R. Lasztity and F. Bekes, eds. World Scientific Singapore : 466-477.
DEXTER, J.E., and MATSUO, R.R. 1979. Changes in spaghetti protein solubility during cooking. Cereal Chem. 56 : 394-398.
DEXTER, J.E., MATSUO, R.R., and MORGAN, B.C. 1981. High temperature drying : effect on spaghetti properties. J. Food Sci. 46 : 1741-1756.
DEXTER, J.E., TKACHUK, R., and MATSUO, R.R. 1984. Amino acid composition of spaghetti : effect of drying conditions on total and available lysine. J. Food Sci. 49 : 225-228.
DREESE, P.C., FAUBION, J.M., and HOSENEY, R.C. 1988. The effect of different heating and washing procedures on the dynamic rheological properties of wheat gluten. Cereal Foods World, 33 : 225-228.

DYSSELER, P., JACQUAIN, D., and KARYDAS, J. 1986. Détermination électrophorétique des gliadines dans les produits alimentaires cuits. Belgian J. Food Chem. Biotechnol., 41 : 143-149.

ELIASSON, A.C., and HEGG, P. O. 1980. Thermal stability of wheat gluten. Cereal Chem. 57: 436-437.

FEILLET, P. 1984. The biochemical basis of pasta cooking quality. Its consequences for durum wheat breeders. Sciences des Aliments 4 : 551-566.

FEILLET, P. 1987. Séchage à haute et très haute température des pâtes alimentaires: rapport de synthèse. Institut National de la recherche Agronomique, Ministère de l'Agriculture, 269 p.

FEILLET, P., AUTRAN, J.C., and AIT MOUH, O. 1987. Role of low molecular weight glutenin in determining cooking quality of pasta products. Cereal Foods World 32 : 670.

FEILLET, P., AIT MOUH, O., KOBREHEL, K., and AUTRAN, J.C. 1988. Role of low molecular weight glutenins in determining cooking quality of pasta products : an overview. Cereal Chem. (in press).

HANSEN, L.P., JOHNSTON, P.H., and FERREL, R.E. 1975. Heat-moisture effect on wheat flour. I. Physical-chemical changes of flour proteins resulting from thermal processing. Cereal Chem. 52 : 459-472.

HOSENEY, R.C. 1986. Component interaction during heating and storage of baked products. In: Chemistry and Physics of Baking, Special Publication No. 56, The Royal Society of Chemistry, Burlington House, London. J.M.V. Blanshard, P.J. Frazier and T. Galliard, eds.) : 216-226.

IBRAHIM, R.H., and McDONALD, C.E. 1981. Experimental high-temperature drying of spaghetti Cereal Foods World 26 : 507.

JANJEAN, M.F., DAMIDAUX, R., and FEILLET, P. 1980. Effect of heat treatment on protein solubility and viscoelastic properties of wheat gluten. Cereal Chem. 57 : 325-331.

KASARDA, D.D., BERNARDIN, J.E., and GAFFIELD, W. 1968. Circular dichroism and optical rotatory dispersion of α-gliadin. Biochemistry 7 : 3950-3957.

KOBREHEL, K., AGAGA, D., and AUTRAN, J.C. 1985. Possibilité de détection de la présence de blé tendre dans les pâtes alimentaires ayant subi des traitements thermiques à haute température. An. Fals. Exp. Chim. 78 : 109-117.

KOBREHEL, K., and ALARY, R. 1988. The role of a low molecular weight glutenin fraction in the cooking quality of durum wheat pasta. J. Sci. Food Agric. (in press).

McCAUSLAND, J. and WRIGLEY, C.W. 1976. Electrophoretic analysis of wheat and rye mixtures in meal, flour and baked goods. J. Sci. Food Agric. 27 : 1197-1202.

McDERMOTT, E.E. 1985. The properties of commercial glutens. Cereal Foods World 30 : 169-171.

MECHAM, D.K., and OLCOTT, H.S. 1947. Effect of dry heat on proteins. Ind. Eng. Chem. 39: 1023-1027.

MEIER, P., WINDEMANN, H., and BAUMGARTNER, E. 1985. Auftrennung von Gesamtgliadin aus unterschiedlich hitzebelasteten Weizenmehlen mittels Phasenumkehr-Hochdruckflüssig-Chromatographie. Z. Lebensm. Unters. Forsch. 180 : 467-473.

MENKOVSKA, M., LOOKHART, G.L., and POMERANZ, Y. 1987. Changes in the gliadin fraction(s) during breadmaking : isolation and characterization by high-performance liquid chromatography and polyacrylamide gel electrophoresis. Cereal Chem. 64 : 311-314.

MENKOVSKA, M., POMERANZ, Y., LOOCKHART, G.L., and SHOGREN M.D. 1988. Gliadin in crumb of bread from high-protein wheat flours of varied breadmaking potential. Cereal Chem. 65 : 198-201.

NEUCERE, J.N., and CHERRY, J.P. 1982. Structural changes and metabolism of proteins following heat denaturation. In : Food Protein Deterioration, ACS Symp. Series No. 206, J.P. Cherry ed. American Chemical Society, Washington, 135-162.

ORTH, R.A., and BUSHUK, W. 1972. A comparative study of the proteins of wheats of diverse baking qualities. Cereal Chem. 49 : 268-275.

PENCE, J.W., MOHAMMAD, A., and MECHAM, D.K. 1953. Heat denaturation of gluten. Cereal Chem. 30 : 115-126.

POMERANZ, Y., LOOKHART, G.L., RUBENTHALER, G.L., and ALBERTS, L.A. 1987. Changes in gliadin proteins during cookie making. Cereal Foods World, 32 : 670.

RESMINI, P. and PAGANI, M.A. 1983. Ultrastructure of pasta. Food Microstructure, 2 : 1-12.

RESMINI, P., De BERNARDI, G., and MAZZOLINI, C. 1976. Influenza delle condizioni di essicazione su alcune caratteristiche della pasta alimentare. Bull. Lab. Chim. Prov. 27 : 283-293.

ADOUKI, H., and AUTRAN, J.C. 1987. Mise en évidence du rôle de certaines gluténines de haut poids moléculaire dans la qualité boulangère des blés tendres en Algérie. Lebensm.-Wiss. u.-Technol., 20 : 180-190.

SCHOFIELD, J.D. 1986. Flour proteins : structure and functionality in baked products. In : Chemistry and Physics of Baking, Special Publication No. 56, The Royal Society of Chemistry, Burlington House, London, J.M.V. Blanshard, P.J. Frazier and T. Galliard, eds.) 14-29.

SCHOFIELD, J.D., and BOOTH, M.R. 1983. Wheat proteins and their technological significance In : Developments in Food Proteins, Vol. 2, Applied Science Publishers, London, B.J.F. Hudson, ed. : 1-65.

SCHOFIELD, J.D., BOTTOMLEY, R.C., TIMMS, M.F., and BOOTH, M.R. 1983. The effect of heat on wheat gluten and the involvement of sulphydryl-disulfide interchange reactions. J. Cereal Sci. 1 : 241-253.

SCHOFIELD, J.D., BOTTOMLEY, R.C., LEGRYS, G.A. TIMMS, M.F., and BOOTH, M.R. 1984. Effect of heat on wheat gluten. In : Gluten Proteins, Proc. 2nd Int. Workshop on Gluten Proteins, Wageningen, The Netherlands. A. Graveland and J.H.E. Moonen, eds. TNO, The Netherlands: 81-90.

STENVERT, N.L., MOSS, R., and MURRAY, L. 1981. The role of dry vital gluten in breadmaking. Bakers Digest 55 (2) : 6-12.

TATHAM, A.S., and SHEWRY, P.R. 1985. The conformations of wheat gluten proteins. I. The secondary structures and thermal stabilities of α-, β-, γ- and ω-gliadins. J. Cereal Sci. 3 : 103-113.

TATHAM, A.S., FIELD, J.M., SMITH, S.J., and SHEWRY P.R. 1987. The conformations of wheat gluten proteins. II. Aggregated gliadins and low molecular weight subunits of glutenin. J. Cereal Sci. 5 : 203-214.

VOUTSINAS, L.P., NAKAI, S., and HARWALKAR, V.R. 1983. Relationships between protein hydrophobicity and thermal functional properties of food proteins. Can. Inst. Food Sci. Technol. J. 16 : 185-190.

WALL, J.S., and HUEBNER, F.R. 1981. Adhesion and cohesion. In : Protein Functionality in Foods, ACS Symp. Series No. 147, American Chemical Society, Washington. J.P. Cherry, ed. : 111-129.

WASIK, R.J. 1978. Relationship of protein composition of durum wheat with pasta quality and the effects of processing and cooking on these proteins. Can. Inst. Food Sci. Technol. J. 11 : 129-133.

WRIGLEY, C.W. 1977. Characterisation and analysis of cereal products in foods by protein electrophoresis. Food Technology in Australia 29: 17-20.

WRIGLEY, C.W., Du CROS, D.L., ARCHER, M.J., DOWNIE, P.G., and ROXBURGH, C.M. 1980. The sulfur content of wheat endosperm proteins and its relevance to grain quality. Aust. J. Plant Physiol. 7 : 755-766.

WU, Y.V., and INGLETT, G.E. 1974. Denaturation of plant proteins related to functionality and food applications. A review. J. Food Sci. 39 : 218-225.

34

THE INTERACTIONS THAT PRODUCE UNIQUE PRODUCTS FROM WHEAT FLOUR

R. Carl Hoseney

Department of Grain Science and Industry
Kansas State University
Manhattan, KS 66506

INTRODUCTION

Wheat flour is exceptional because of its ability to produce a large number of unique products. These would include such items as bread, cake, cookies, crackers, and many others. The other cereal grains do not produce acceptable versions of those products. The chemical composition of the various cereal grains is not that different. Certainly one can find wheat flours and corn flours that contain the same amounts of protein, starch, etc. However, the properties of their flours are quite different. For example, only wheat flour will produce a viscoelastic dough and only wheat flour dough will retain gas during fermentation and the early stages of baking.

In addition, the wheat flour dough will "set" during baking to give a baked product that will retain its light texture. To understand why wheat flour has those unusual properties, when compared to the other cereals, is a very formidable problem. It is one thing to state that the wheat protein, starch, or lipids are different than similar components found in other cereals; it is quite another problem to show how they are different and how this difference results in the unique products mentioned above.

Although wheat is unique in the production of several products we will restrict this discussion to its unique ability to produce bread. The major differences appear to be how the components of wheat flour interact during doughing and baking. It is the purpose of this paper to look at those interactions and discuss the approaches and techniques necessary to study them.

FORMATION OF DOUGH

The changes that flour goes through when it wetted with water and mixed are well known. The flo hydrates and then forms a cohesive mass that takes viscoelastic properties. One way to illustrate the changes is with a mixing curve such as the well kno mixogram.

Up to a point, as the dough is mixed resistance-to-extension (height of the curve) increase It is also obvious as mixing proceeds the amount of f water in the dough decreases. If we continue to mix dough beyond the point referred above the resistan to-extension decreases and the dough breaks d (Hoseney 1985).

To really understand what occurs during mix perhaps we should take a closer look. If we examine endosperm of hard wheat we find it to be vitreous glassy in appearance. When water is added, with without mixing, the endosperm adsorbs the water and protein undergoes a transition to a viscoelastic mas This was elegantly shown in the 1970's by Bernardin Kasarda (1973). Recently we were able to show that underlying transformation was the gluten protein go through a glass transition (Hoseney et al 1986). As water content of the polymer increases the gl transition temperature decreases (Fig. 1), until transition occurs at room temperature. Before viscoelastic dough can be formed, the gluten must through this transition turning the glassy endospe into the elastic dough.

Polymer science tells us that no particul structure of the protein is needed to give a dough gluten viscoelastic properties. Polymers when heat above their glass transition temperature (or plasticiz to the point that the transition occurs at ro temperature) will have rubbery (viscoelasti properties.

During the mixing of dough there appears to ta place an interchange reaction involving the disulfi bonds in the gluten protein. Mauritzen (196) Mauritzen and Stewart (1963), and Stewart and Maurit

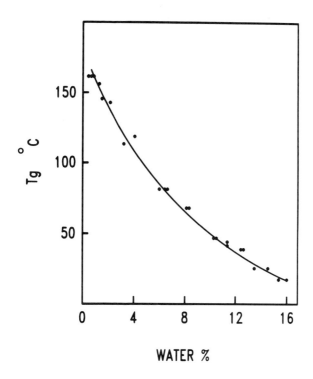

Figure 1. Change in T_g as a function of moisture content. Source: Hoseney et al (1986).

1966) showed that small molecular weight sulfhydryl compounds did interchange with the gluten proteins. In addition, a number of workers (Mecham, Sokol and Pence 1962, Mecham, Cole and Sokol 1963, Mecham, Cole and Pence 1965, Tsen 1967, and Danno and Hoseney 1982a) have shown that as a result of dough mixing the solubility of gluten proteins increases. This presumably is the result of the interchange between protein that leads to a lower molecular weight and therefore a larger solubility (Danno and Hoseney 1982b). The mechanism of the interaction is not completely clear, but some evidence suggest a free radical reaction.

As with all reactions involving dough and gluten,

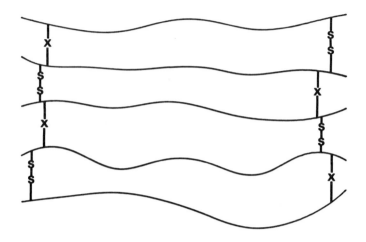

Figure 2. Illustration of difference between polymerized and crosslinked polymers.

it should be emphasized that both the gliadin and glutenin proteins are involved in the interactions. Because of their size and the fact that they have intermolecular disulfide bonds, the glutenin proteins receive the most attention. It should be emphasized that we are discussing the glutenin proteins and not the glutenin subunits. Another factor that appears to be confusing is that the glutenin proteins are polymerized but are not crosslinked to any extent (Ewart 1977 and 1979). The difference between crosslinking and polymerization is shown in Fig. 2. The difference responsible for large differences in the rheological properties of polymers. The -x- in Fig. 2 may represent additional disulfide bonds or other crosslinking bonds.

The viscoelastic properties of doughs are also affected to a great extent by oxidation. This can be the result of chemical oxidation by potassium bromate or similar chemicals being added to the dough or simply by the oxidation by oxygen in the air trapped during mixing. A third mechanism appears to be by the yeast used in doughs (Hoseney et al 1979). In all of these cases it appears that the oxidation does not occur to any extent on the protein but appears to be more an oxidation of low molecular weight entities. Removal of those entities will decrease the amount of interchange that occurs during mixing and thereby result in a stronger dough.

GAS RETENTION

One of the outstanding characteristics of wheat flour doughs is their ability to retain the gas produced during fermentation and the early stages of baking. Although the retention of gas is often viewed as trapping of the gas in a small balloon made of gluten, logic tells us that this is not true (Hoseney 1984). Because the yeast produces the CO_2 in the aqueous phase of the dough, there must be a mechanism for the gas to enter the cell. If the gas can diffuse into the cell, it must also be able to diffuse out. Therefore, there is no gas retaining membrane in dough. In fact we do not need a membrane, if the aqueous phase is saturated with CO_2 and the yeast continues to produce additional gas to keep it saturated, then the gas can only diffuse from the saturated aqueous phase to the unsaturated gases phase inside the cell. As the pressure increases inside the cell, the dough being viscoelastic will flow to equalize the pressure and as a result the dough expands.

Evidence to support the above scenario has been obtained by He and Hoseney (1988). Doughs made with flours other than wheat flour allow gas to diffuse at a much higher rate. This, of course, raises the question of how does the wheat flour dough slow the diffusion of gas to such a large degree. It is well known that the rate of diffusion of small molecules is affected to a large degree by the viscosity of the solution in which the molecules are diffusing. The continuous phase in wheat flour dough is the highly hydrated gluten matrix. The gluten forms an elastic mass with a high apparent viscosity. Thus, we would anticipate that the rate of diffusion of small molecules through this matrix would be slow.

With dough made from rice flour, corn flour, or starch alone the continuous phase is just water. The water in such doughs is relatively free and does not contain a large amount of dissolved material. In addition, the protein in these flours is glassy and does not go through a transition to become water compatible. Therefore, the viscosity of the continuous phase stays low and the diffusion in this phase is very rapid. This, of course, leads to the loss of large amounts of gas at the dough surface. Rye flours are known to retain gas better than any cereal flour, except for wheat. This is also shown clearly with the data of He and Hoseney (1988). The protein from rye is not like that of wheat, and does not result in the continuous phase. Instead the explanation for rye's ability to slow diffusion is because of the high level of water soluble pentosans that it contains (Weipert and Zwingelberg 1980). Casier et al (1973) have shown that the addition of pentosan material to a number of flour results in the dough retaining gas to a much better extent than in the doughs without the added pentosans

Another way of improving the gas retaining ability of starch or coarse flour doughs is with the addition of certain surfactants. Rotsch (1954) baked "bread" from starch with improved volume compared to just starch by the addition of monoglycerides. The monoglyceride were heated with water before they were added to the dough. This apparently formed a lipid phase so that diffusion was through a number of water:lipid interfaces. This would make diffusion very slow. similar approach was used by Rajapaksa et al (1983) using natural flour lipids as the surfactant. The results were just as spectacular. It had also been shown that bread with good volume can be produced from starch and certain gums (Kulp et al 1974). Xanthan gum appears to be particularly useful.

LOSS OF GAS

If gas diffusion is fast then the dough is losing a significant amount of gas throughout proof and the loaf never proofs. With wheat flour doughs, the amount of gas lost during proofing is very small compared to the amount of gas being produced. During the early stages of baking the amount of gas diffusing to and escaping at the surface remains relatively small. However, as baking proceeds the dough starts to lose much larger quantities of gas.

Two possible reasons for the sudden increase in the rate of gas loss come to mind. The dough could undergo some change that destroys its ability to retain gas. It is not clear what this change might be. The second possibility is that the dough undergoes a rheological change so that it no longer is viscous and now a much higher pressure is required to expand the dough. As the pressure in the gas cells increase, the driving force for diffusion also increases. We feel this is the reason for the increase in diffusion rate. The amount of gas being lost from the dough is still much less than what we would expect if the dough had become permeable to the gas.

The rheological change surmised above can be shown by probing the dough with an Instron (Moore and Hoseney 1986). The chemical changes that occur in the dough during heating and bring about the rheological change is yet to be determined.

EFFECT OF HEAT ON "SETTING" OF DOUGH

There are many references in the literature to the denaturation of the gluten proteins. However, if we consider denaturation to be the phase change from an ordered to a disordered structure this has not been shown for the gluten proteins. One of the best ways of following the change from an ordered to a disordered structure is with the differential scanning calorimeter (DSC). However, DSC of gluten proteins do not show a reasonable denaturation peak (Arntfield and Murry 1981, Eliasson and Hegg 1980, and Schofield et al 1984). Instead we find a small peak that is apparently gelatinization of the residual starch in the gluten, and two other very small peaks that may be denaturation of water soluble proteins associated with the gluten.

If the gluten shows no denaturation peak, and this

certainly appears to be the case, then what does this mean? It could well be that gluten is a random polymer. If gluten has no ordered structure then it cannot be disordered. Tatham et al (1985) have suggested that gluten has a beta-turn configuration. Their conclusion are based on CD measurements of dilute solutions. I is possible that the gluten has an ordered structure i solution with various solvents but not in th concentrated neutral system that we know as whea gluten.

Tatham et al (1985) have suggested that gluten i elastic because of being in the beta-turn configuration. Their conclusion is based on the elastic properties o elastin, the protein of connective tissue in animals However, if we look at the fundamentals of rheology w find that a specific structure is not necessary for polymer to have elastic properties (Graessley 1984). It is also true that the type of elastic character four in elastin is quite different than that of wheat gluter

If gluten is not denaturated by heating does thi mean that we can isolate vital gluten from bread? Th answer is no. While we feel that gluten is no denaturated, it is clear that it is changed during th heating process. Schofield et al (1983, 1984) hav shown that the solubility of gluten is decreased durin baking. This appears to be the result of disulfid sulphydryl interchange reactions. It is not that clea if this leads to just a polymerization of the prote or to a cross-linking, or to a combination of both.

Solubilization of the protein from bread with solvent containing sodium dodecyl sulfate a mercaptoethanol give nearly 100% solubilization (Hosen et al 1988). This shows that no new bonds other th disulfide bonds are formed during baking and the cau of the decrease in solubility is because of disulfi bonds. When bread is heated with steam the amount cross-linking appears to increase. This was shown extraction of protein with a sodium dodecyl sulfa solvent that contains various amounts mercaptoethanol. Bread that was steamed longer requir higher concentrations of mercaptoethanol to extract 10 of the protein. This infers that the cross-linki action requires not only temperature but also time. This may also infer that during baking a small number of cross-links are formed and it is this cross-linking that is responsible for the setting of the loaf of bread. This has been suggested by Bale and Muller (1970) based on their rheological work.

NONBREAD DOUGHS

Wheat gluten is an amorphous glassy polymer (Hoseney et al 1986). When dry, the glass transition (T_g) is at above 160°C. As the water content (plasticizer) is increased, the T_g decreases. The significance of this is that gluten must go through the glass transition to form an elastic material. This was graphically shown by Bernardin and Kasarda (1973) in their work with gluten fibrils. Thus, during the mixing of a bread dough sufficient water is added to decrease the glass transition temperature below room temperature. In many cookie doughs and perhaps in other dough systems where water is limiting and/or the sugar content is high the glass transition temperature may remain above room temperature and the gluten remains glassy. Thus, we do not have a developed dough in those cases. This describes a wire cut cookie dough. The flour particles remain intact and the dough does not become continuous until the dough is heated to a sufficient temperature. This temperature is the temperature at which the cookie sets. Therefore, the setting temperature is the temperature at which the gluten goes through the glass transition (Doescher et al 1987a and b). We think there are similar phenomena in cracker making and perhaps other systems.

LITERATURE CITED

ARNTFIELD, S. D. and MURRY, E. D. 1981. The influence of processing parameters on food protein functionality. I. Differential scanning calorimetry as an indicator of protein denaturation. J. Inst. Can. Sci. Tech. 14:436.

BALE, R. and MULLER, H. G. 1970. Application of statistical theory or rubber elasticity to the effect of heat on wheat gluten. J. Food Tech. 5:295.

BERNARDIN, J. E. and KASARDA, D. D. 1973. Hydrated protein fibrils from wheat endosperm. Cereal Chem. 50:529.

CASIER, J. P. J., DE PAEPE, G., and BRUMMER, J.-M. 1973. Einfluss der wasserunloslichen Weizen-und Roggen-Pentosane auf die Backeigenschaften von Weizenmehlen und anderen Rohstoffen. Getreide, Mehl, Brot. 27:36.

DANNO, G. and HOSENEY, R. C. 1982a. Effects of dough mixing and rheologically active compounds on relative viscosity of wheat proteins. Cereal Chem. 59:196.

DANNO, G. and HOSENEY, R. C. 1982b. Changes in flour proteins during dough mixing. Cereal Chem. 59:249.

DOESCHER, L. C., HOSENEY, R. C., MILLIKEN, G. A., and RUBENTHALER, G. L. 1987a. Effects of sugars and flours on cookie spread evaluated by time-lapse photography. Cereal Chem. 64:163.

DOESCHER, L. C., HOSENEY, R. C., and MILLIKEN, G. A. 1987b. A mechanism for cookie dough setting. Cereal Chem. 64:158.

ELIASSON, A.-C. and HEGG, P.-O. 1980. Thermal stability of wheat gluten. Cereal Chem. 57:436.

EWART, J. A. D. 1977. Re-examination of the linear glutenin hypothesis. J. Sci. Food Agric. 28:191.

EWART, J. A. D. 1979. Glutenin structure. J. Sci. Food Agric. 30:482.

GRAESSLEY, W. W. 1984. Viscoelasticity and flow in polymer melts and concentrated solutions. In: Physical Properties of Polymers. J.E. Mark, A. Eisenberg, W.W. Graessley, L. Mandelkern, and J.L. Koenig, eds. ACS, Washington, DC.

HE, HUIFEN and HOSENEY, R. C. 1988. Study of bread baking using the electric resistance oven system. Abstract 221. Cereal Foods World 33:694.

HOSENEY, R. C. 1984. Gas retention in bread doughs. Cereal Foods World 29:305.

HOSENEY, R. C. 1985. The mixing phenomenon. Cereal Foods World 30:453.

HOSENEY, R. C., HSU, K. H., and JUNGE, R. C. 1979. A simple spread test to measure the rheological properties of fermenting dough. Cereal Chem. 56:141.

HOSENEY, R. C., DREESE, P. C., DOESCHER, L. C., and FAUBION, J. M. 1988. Thermal properties of gluten. pp 518. Proc. 3rd Internat. Workshop on Gluten Proteins. R. Lasztity and F. Bekes, eds. World Scientific Publ., Teaneck, NJ.

HOSENEY, R. C., ZELEZNAK, K., and LAI, C. S. 1986. Wheat gluten: a glassy polymer. Cereal Chem. 63:285.

KULP, K., HEPBURN, F. N., and LEHMANN, T. A. 1974. Preparation of bread without gluten. Baker's Dig. 43(3):34.

LAURITZEN, C. A. M. and STEWART, P. 1963. Disulphide-sulphydryl exchange in dough. Nature 197:48.

LAURITZEN, C. A. M. 1967. The incorporation of cysteine-S and N-ethylmaleimide-C into doughs made from wheat flour. Cereal Chem. 44:170.

MECHAM, D. K., COLE, E. G., and SOKOL, H. A. 1963. Modification of flour proteins by dough mixing: Effects of sulfhydryl-blocking and oxidizing agents. Cereal Chem. 40:1.

MECHAM, D. K., SOKOL, H. A., and PENCE, J. W. 1962. Extractable protein and hydration characteristics of flours and doughs in dilute acid. Cereal Chem. 39:81.

MECHAM, D. K., COLE, E. G., and PENCE, J. W. 1965. Dough-mixing properties of crude and purified glutens. Cereal Chem. 42:409.

MOORE, W. R. and HOSENEY, R. C. 1986. Influence of shortening and surfactants on retention of carbon dioxide in bread dough. Cereal Chem. 63:67.

NAJAPAKSA, D., ELIASSON, A.-C., and LARSSON, K. 1983. Bread baked from wheat/rice mixed flours using liquid-crystalline lipid phases in order to improve bread volume. J. Cereal Sci. 1:53.

OTSCH, A. 1954. Chemische und backtechnische Untersuchungen an kunstlichen Teigen. Brot und Geback 8:129.

SCHOFIELD, J. D., BOTTOMLEY, R. C., TIMMS, M. F., and BOOTH, M. R. 1983. The effect of heat on wheat gluten and the involvement of sulphydryl-disulfide interchange reactions. J. Cereal Sci. 1:241-253.

SCHOFIELD, J. D., BOTTOMLEY, R. C., LEGRYS, G. A., TIMMS, M. F., and BOOTH, M. R. 1984. Effects of heat on wheat gluten. p. 81 In: Proc. 2nd Internat. Workshop on Gluten Proteins. A. Graveland and J.M.E. Moonen, eds. TNO, Wageningan, The Netherlands.

STEWART, P. R. and MAURITZEN, C. M. 1966. The incorporation of 35S cysteine into proteins of dough by disulfide-sulphydryl interchange. Austr. J. Biol. Sci. 19:1125.

TATHAM, A. S., MIFLIN, B. J., and SHEWRY, P. R. 1985. The beta-turn conformation in wheat gluten proteins: Relationship to gluten elasticity. Cereal Chem. 62:405.

TSEN, C. C. 1967. Changes in flour proteins during dough mixing. Cereal Chem. 44:308.

WEIPERT, D. and ZWINGELBERG, H. 1980. The pentosan-starch ratio in relation to quality of milled rye products. p. 495 In: Cereals for Foods and Beverages. G. Inglett and L. Munck, eds. Academic Press, NY.

35

ADVANCES IN WHEAT PROCESSING AND UTILIZATION IN JAPAN

Seiichi Nagao

Cereal and Food Research Laboratory
Research Center
Nisshin Flour Milling Co., Ltd.
Ohi-machi, Saitama 354, Japan

AUTOMATION

The third and newest unit added to the Chiba complex of Nisshin Flour Milling Co., Ltd. about two years ago is highly efficient and fully computerized. No-man automation is featured in the process of blending, tempering and milling. Every measure is taken to assure production efficiency, quality control, energy conservation, sanitation and prevention of contamination with foreign materials. The addition brought total milling capacity at the Chiba complex to about 1,000 tonnes of wheat per 24-hour day, producing about 17,000 cwts of flour.

The automatic warehousing process for palletized flour attached to the three mills is also unique to Japanese milling. About 40% of the mill's production is packed in 25-kilogram bags, with 60% moving by bulk truck. The flour warehouse at Chiba has a storage capacity of 3,160 tonnes of bagged flour. Remotely controlled from a delivery center, the carts, (each carrying one pallet-load of 50 flour bags), travel a circuitous route from the mill to designated places in the warehouse. The first pallet moves automatically into an elevator, where it is lifted to the top of one of columns. Leaving the pallet supported in place of the top, the elevator descends to receive the next pallet.

No-man automation has been tried by several flour millers in the different parts of the world, and some of them succeeded in operating just as they intended. Among them, the new unit of the Chiba complex is specially characterized by the high precision in quality control coping with diversities of material and product in high operation.

FLOUR TYPES

Over fifty types of flour which comprise flours for different kinds of bread, Chinese and Japanese type noodles, and confectionery, etc. are milled at the three mills of Chiba complex, offered on the market of Tokyo metropolitan area under brand names. It is usual to blend two to four kinds of wheat in order to get the flour quality as required as well as to assure its uniformity as shown in Table I. Several extraction patterns are followed in the milling of flour. Usually, first-grade, second-grade, and low-grade flours are obtained. The mill operates more than 22 days a month, and it is approaching its full operation.

Large flour milling concerns in Japan produce many types of flour for a variety of flour products having a good command of the latest milling technique in the field of wheat blending, grinding, stock selection, flour blending, air-classification, etc. Excluding third- and feed-grade flours, standard flours for food range from 0.3 to 0.8% in ash and 6. to 13.0% in protein content as shown in Table II. Most flours are designed to have characteristics suitable for processing to a certain type of flour products. In addition, some specific flours developed for special usage such as French bread, frozen dough, castilla, etc. have recently become popular among bakeries, and there will be a growing demand for them.

We have conducted research on flour products not only for developing specific flours but also for providing quality technical services to our customers in expectation of expanding flour sales. For example

Flour classification and types of wheat milled in Japan

Flour	Wheat
Strong	No.1 Canada Western Red Spring (Protein:13.5%)
	U.S. Dark Northern Spring (Protein:14.0%)
Semi-Strong	U.S. Hard Red Winter (Protein:13.0%)
	U.S. Hard Red Winter (Protein:11.5%)
Medium	Australian Standard White (Western Australia)
	Japanese Soft
Soft	U.S. Western White

TABLE II

Average Protein and Ash Contents of Flours in Japan

		Grade			
		First	Semi-First	Second	Third	Low
Ash (%)		0.3– 0.4	about 0.4	about 0.5	about 1.0	2–3
Protein (%)	Strong	11.5–12.5	11.7–12.7	12.0–13.0		
	Semi-Strong	10.4–12.0	10.5–12.2	11.5–12.5		
	Medium	7.5– 9.0	8.0– 9.5	9.0–10.5		
	Soft	6.5– 8.0	7.5– 8.5	8.0– 9.0		

the results of our study on frozen baking dough, frozen noodle, sour dough bread, rye bread, etc. were provided to our customers or were open to the public through the short courses held specially for flour processors. It seems to me that the efforts of this kind have partly contributed to the expansion of flour consumption as well as to the technological improvement of flour utilization in Japan.

Baking and noodle manufacturing industries have made remarkable progress in the processes for the last several decades. They are characterized by advanced automation and flexible manufacturing systems capable of producing many kinds of quality products. Above all, the introduction of different types of encrusting machines suitable for each purpose has contributed to the process automation not only in large bakeries but also in small ones. The continuous supply of dough and filling to the machine makes possible the production of a quality product, lowers unit cost, and meets sanitation requirements. Encrusting machines manufactured by Rheon Automatic Machinery Co., Ltd are now used worldwide especially for the production of small baked goods.

ENZYMIC GLUTEN MODIFICATION

Nisshin Flour Milling Co., Ltd. in cooperation with Oriental Yeast Co., Ltd. and Mitsubishi Heavy Industries has conducted research on developing continuous process for the enzymatic modification of wheat gluten by using a bioreactor (Motoi et al 1986, 1987, Fukudome et al 1987, and Motoi et al 1988). The purpose of this work is to produce a hydrolysate of wheat gluten, which is characteristic in its emulsifying and foaming properties, by means of limited hydrolysis with an immobilized protease.

In order to prepare an immobilized protease, chitin processed into the shape of porous beads and pepsin from pig's stomach were selected from several commercial products as a carrier and an enzyme respectively. After adsorbing pepsin to chitin, it was cross-linked with glutaraldehyde to prepare stable immobilized enzyme.

Wheat gluten extracted from hard wheat flour was
spersed and dissolved in a dilute acid solution
chanically. Prior to the treatment in a bioreactor,
was heated and reduced to enhance its affinity to
immobilized enzyme.

When wheat gluten was continuously treated in a
ni-column for a long time, a fall in enzyme activity
s observed. It was caused by the partial separation
enzyme from the carrier due to the interaction and
condary adsorption of charged wheat gluten with the
rrier. After a series of tests on the interaction
ong carrier, enzyme and substrate, we succeeded in a
able running system at 40-50°C and in the acid range
 pH 3.0. The conditions also helped to reduce the
ntamination by undesirable bacteria.

Making good use of our testing results on the
ni-column, we developed a bench scale reactor of 5L
lume and started a test of continuous hydrolysis
ig. 1). Quality hydrolysate of wheat gluten was

g. 1. Bench scale bioreactor system.

produced when a substrate stayed in the column for min. After 400ℓ of substrate solution was continuous treated for a week, 10 kg of hydrolysate was obtaine

Molecular weight distribution, amino acid co position and hydrophobic property of molecular surfa of gluten hydrolysate were measured, and their rel tion to functional properties were evaluated. It w found that the polypeptide, which comprised about 1 amino acids and was characterized by a relatively hi hydrophobic property of molecular surface, showed good foaming characteristic and a high stability the foams. The gluten hydrolysate is white in col and low in bitterness. It will be used for vario purposes including processing of wheat flour product Thus, for example, the addition of the glut hydrolysate to sponge-cake improved its volume a internal texture, even if no emulsifier was added.

WHEAT CLASSES

The U.S. produces all kinds of wheats, and c supply any class and grade of wheat complying with t requests of buyers. The diversity of U.S. whe classes enables overseas millers to meet the rath complicated requirements of their customers. Japanese millers are well aware of the quality fe tures of each class through their long experience U.S. wheats have been put to their proper use.

However, a totally unacceptable idea of elimina ing the classification in hard red wheat trading w suggested by some persons in the hard red whe producing areas. This is a very serious matter f oversears buyers. It is my firm belief that as ha red spring (HRS) and hard red winter (HRW) wheats a quite different in their milling and baking perfo mance, the U.S. wheat industry should make eve conceivable effort to maintain the current system classification, even if there are some difficulties the inspection process.

We have done a series of tests to determine t potential of reversed-phase high-performance liqu chromatography (RP-HPLC) as a means of classifying H and HRW wheats as well as to compare the milling a flour analytical properties between wheat classes. Some results of our analytical research work will be reported elsewhere.

The superiority in noodle processing characteristics of Australian standard white (ASW) wheat from Western Australia to that of other soft wheats was found, and this is thought to be related mainly to its starch characteristics. Amylograph gelatinization temperature for ASW flour is lower than that of other soft wheats. The relatively low gelatinization temperature was presumed to be a factor in soft and pliable noodles (Nagao et al 1977).

Gelatinization properties of ASW and wheats representing the three major varieties grown in Japan were compared with respect to their suitability for Japanese noodle in our laboratory (Endo et al 1988). Quality factors which affect the starch gelatinization properties were studied using the Brabender Amylograph. The addition of sodium cloride caused a pronounced difference in the area under the sticking-out portion of the amylogram during the cooling stage for flours milled from ASW and Japanese wheats. ASW,

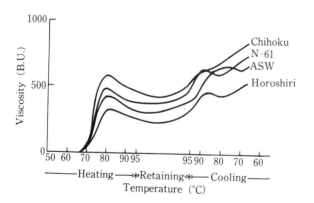

Fig. 2. Comparison of amylogram curves between ASW and Japanese wheats (N-61, Chihoku and Horoshiri). Tests were done for 60% extraction flours on the addition of 2% NaCl for each sample. (reproduced from Endo et al 1988)

which possessed favorable characteristics for noodle making, showed a large area, whereas no or only a small area was observed for Japanese wheats. This may relate to noodle quality, especially to noodle staling. We hope that the favorable noodle processing quality of ASW wheat will be maintained.

LITERATURE CITED

ENDO, S., KARIBE, S., OKADA, K., and NAGAO, S. 1988. Factors affecting gelatinzation properties of wheat starch. Nippon Shokuhin Kogyo Gakkaishi 35:7.

FUKUDOME, S., MOTOI, H., and OGAWA, G. 1987. Modification of wheat protein by using a bioreactor system (Part 3). 39th Ann. Mtg. Soc. Ferment. Technol. Japan.

MOTOI, H., FUKUDOME, S., and OGAWA, G. 1986. Modification of wheat protein by using a bioreactor system (Part 1). 30th Ann. Mtg. Agr. Chem. Soc. Japan.

MOTOI, H., FUKUDOME, S., and OGAWA, G. 1987. Modification of wheat protein by using a bioreactor system (Part 2). 31st Ann. Mtg. Agr. Chem. Soc. Japan.

MOTOI, H., FUKUDOME, S., and KITAMURA, I. 1988. Modification of wheat protein by using a bioreactor system (Part 4). 32nd Ann. Mtg. Agr. Chem. Soc. Japan.

NAGAO, S., ISHIBASHI, S., IMAI, S., SATO, T., KANBE, T., KANEKO, T., and OTSUBO, H. 1977. Quality characteristics of soft wheats and their utilization in Japan. I. Evaluation of wheats from the United States, Australia, France, and Japan. Cereal Chem. 54:198.

36

NEW AND POTENTIAL MARKETS FOR WHEAT STARCH

William M. Doane
Northern Regional Research Center
Agricultural Research Service
U.S. Department of Agriculture
1815 N. University Street
Peoria, IL 61604

ABSTRACT

Cereal starches have long enjoyed a wide diversity of markets in both the food and industrial areas. Facile conversion to glucose and oligosaccharides has established starch as a major source of sweeteners for the food and beverage industries. Intensified research efforts to improve bioconversion efficiencies of starch into low molecular weight acids, alcohols, polyols, aldehydes and ketones are beginning to show promise as routes to the multibillion pound markets for oxygenated compounds now derived from petroleum. More than three and one-half billion pounds of starch and modified starch are used yearly for their adhesive properties in the paper, corrugating board and textile industries, and these markets are expected to continue to expand. Starch in its polymeric form is finding new market opportunities as a component of plastics, especially where the introduction of a biodegradable component is important.

This chapter is in the public domain and not copyrightable. It may be freely reprinted with customary crediting of the source. American Association of Cereal Chemists, 1988.

INTRODUCTION

Starch is the major food reserve polysaccharide produced by cultivated crops. Although it has played an important role in the diet of human beings since before recorded history, it has also been significant for thousands of years in nonfood (industrial) uses. Egyptian papyrus dating to 4000 B.C. contained starch as an adhesive. Later, the use of starch as a size to produce ink-resistant surfaces and to improve paper strength properties emerged. Today, although scores of other industrial uses for starch have been developed, its use in paper is by far the largest, accounting for over 80% of the present industrial use of starch in the U.S.

Interest perhaps has never been higher than it is today in developing new and expanded markets for cereal starches. This is due largely to carry over of corn grain in excess of four billion bushels during each of the last three years. It is the starch from corn that enjoys more than 95% of the industrial starch market in the U.S., and starch from other grains such as wheat must compete in this market dominated by corn. Ideally, applications are desired that take advantage of some unique or unusual property of wheat starch. Some of these do exist, and we expect to learn more about these during this meeting. For many of the emerging opportunities for starch it appears that most starches would perform similarly. This paper will look briefly at the current industrial use for starch and to the future for new application opportunities for this important polysaccharide.

CURRENT MARKETS

The inherent adhesive and film-forming properties have earned starch a sizeable market in paper applications. Currently, about three-and-a-half billion pounds of starch are used in the paper, paperboard, and related industries where starch serves a variety of adhesive functions. A broad spectrum of chemically physically and biologically modified starches have been developed to build in the properties required for specific end-use applications. Largely through continuing efforts to determine the precise property required for a specific application, the researcher has come up with appropriately modified starches that have allowed starch to realize continued market growth in many adhesive applications. There is good reason to believe that starch will continue to enjoy expanding markets in the paper and paperboard industries. Recent articles by Kirby (1987) and Kennedy (1987) provide an overview of the use of starch in these applications and the projected opportunities for expanded markets in these industries.

The textile industry provides a market opportunity for several hundred million pounds of starch products. The principal use is as a warp size to strengthen warp yarns and improve their resistance to abrasion during weaving. They also are used in the finishing of fabrics and in printing. Unfortunately, starch has not retained its share of the market that it had at one time due to its replacement by synthetic polymers. However, in order to maintain a good share of the market, starch scientists have developed several modified starches that function alone or in combination with synthetic polymers to meet end-use requirements. Opportunity exists to capture a larger share of the market through the design and development of specifically tailored starch products. Meeting this challenge is certainly not beyond the capabilities of the starch scientist.

Beyond these major markets in the paper, paperboard and textile industries, starch, modified starch, starch-derived dextrins, oligosaccharides, and glucose are used in a myriad of ways to thicken, flocculate, stabilize, absorb, coat, adhere, dry, moisten, etc. Scores of starch-derived products are marketed to serve specialized needs of numerous industries. Again, these markets exist because of the recognized needs of a particular industry and the efforts of the starch scientist to develop a product to meet a particular requirement.

NEW MARKET OPPORTUNITIES

There should be little reason to doubt that a continued search for new markets, in concert with the commitment of resources to discover and develop technologies for new and improved starch-based products, will result in a healthy growth in new and expanded markets for starch. The commitment needs to be long term and not one that erratically swings back and forth due to a variety of reasons, mostly political. Certainly we have seen dramatic changes in support for such research by industry, academia and government during the last two decades. Support waned in the seventies and early eighties. Suddenly, in the mid-eighties, the country was faced with a carry over of more than four billion bushels of corn and the interest in support for research that might lead to new markets for corn reached a fever pitch. This level of interest has continued as the four-plus-billion-bushel carry over continued. It will be interesting to see what occurs in the level of support in 1989 as carry over of corn dips to a several-year low due to the 1988 drought. Hopefully, those with foresight will prevail and a high level of commitment will continue.

Interest in developing new starch-based products is not restricted to the U.S. Due to improved agricultural technologies, many countries, once so dependent on imported grain, now find themselves with an abundance beyond their domestic needs. Research is being conducted in several of these countries, looking at starch as a renewable raw material for a variety of industrial applications.

As most of the reports indicate, the major markets targeted for replacement by starch are those served by petrochemicals. It is understandable why this is so, starch being annually renewable produced in great abundance with opportunity to produce even more, reasonably inexpensive and convertible chemically, physically, and biologically to a broad array of useful low-molecular-weight compounds and polymerics. Although, technologically, it is possible to produce essentially all of the petrochemicals from starch, it is unreasonable, especially in the near term, to consider starch as a source for all petrochemicals. While we should not expect starch to become a source of all the chemicals now derived from petroleum, there are many opportunities to replace some of these chemicals either by direct substitution or by substitution of an alternate chemical. Advantage might accrue where the starch-derived alternate provides improved functionality and brings benefit to the end-use application.

We are now witnessing the entry of starch-derived products into market areas traditionally serviced by petrochemicals. Some of these products are alternate in structure to the traditional materials and do bring a beneficial property. We want to look at some of these materials and how they are being used as well as additional market opportunities for them and other starch-derived products. Let us classify and consider these products as either low-molecular-weight chemicals or as polymerics.

Chemicals

It has been stated that technologically starch could be used as a raw material to produce any chemical now derived from petroleum. Indeed, starch could be treated at high temperatures to yield synthesis gas (carbon monoxide and hydrogen, which can be further processed to a wide variety of chemicals). However, it is unlikely that starch will serve such a purpose to any considerable extent. Rather, the most likely consideration for chemicals from starch will derive from the glucose formed on depolymerization of the polysaccharide. Commercially, starch is efficiently converted by appropriate enzyme treatment to give nearly the theoretical yield of glucose.

Since starch and glucose contain approximately 50% oxygen, it is obvious that the greatest mass yield of conversion products will arise from those that retain considerable amounts of the oxygen. It is primarily for this reason that the greatest opportunity for replacing petroleum-derived chemicals will be those that contain oxygen. Of the more than 200 billion pounds of petroleum chemicals produced in this country each year, about 25% are oxygen-containing. It is this 50-billion-pound market that shows the most promise for replacement by products from glucose.

Fermentation Chemicals

Glucose can be converted to a variety of alcohols, cyclic and acyclic polyols, acids, aldehydes and ketones. Both chemical and biological processes are now employed commercially to provide selected low-molecular-weight chemicals from starch or starch-derived glucose.

Industrial fermentation is now used to produce the alcohols, ethanol, isopropanol and butanol; the polyols, glycerol and 1,4-butane diol; the ketone, acetone; and the acids, acetic, citric, gluconic, itaconic and lactic. Many of these oxygen-containing chemicals also are produced from petroleum, thus keeping the market share quite small for some of the fermentation products.

Ethanol is the chemical produced in greatest abundance from glucose. In 1988, the production of ethanol for the domestic fuel market will utilize nearly 400 million bushels of grain to produce the 950 million gallons to be blended with gasoline. Although the debate continues over the production and use of ethanol from grain as a gasoline additive, as oil prices rise and supplies diminish, and as technology improves for more efficient biological production and recovery of ethanol, the market for this alcohol will continue to expand.

It is possible to convert ethanol into ethylene, the major primary olefin produced from petroleum. U.S. production in 1987 was about 35 billion pounds. It remains unlikely, at least in the near term, a raw material such as starch will be considered as a source of ethylene or other olefins such as propylene and butadiene due to the larger mass loss due primarily to loss of oxygen from the glucose molecule. Although conversion efficiency of starch --> glucose --> ethanol --> ethylene is excellent at each step, overall mass yield is less than 30 percent.

Other oxychemicals such as acrylic acid, ethylene glycol, propylene glycol, methylethylketone, maleic anhydride and fumaric acid can be produced via fermentation routes. To capture a significant portion of the market for these and other oxychemicals, we must see further advances in bioconversion technologies. Organisms must be identified or developed that are more efficient in producing the desired product in higher yield, at a faster rate, and at higher glucose concentration. New bioreactors are needed to permit continuous processing, mixed-culture fermentations, and the use of solid substrates. Improved engineering (chemical and bio) technology for separation, recovery and purification of fermentation products will contribute significantly to the realization of expanded markets. As there will be byproducts in any fermentation process, efficient technology must be developed for their separation and recovery, and appropriate markets must be identified. Fortunately, we have witnessed during the last several years a research commitment addressing these various needs.

Admittedly, it is very enticing to look at the large multibillion-pound markets for chemicals derived from petroleum and to direct all of one's energies toward direct replacement of these chemicals. However, because this industry is, and has been, so highly competitive, companies have continued to improve their technology for conversion of their oil and natural gas feedstock such that, today, these chemicals are produced with great efficiency. Even with substantial improvement in fermentation technology, ever-increasing direct replacement of petrochemicals will depend on diminishing supplies and increasing costs of oil and natural gas feedstock. There is little doubt that starch-derived chemicals, due to their unique structure and/or properties, are promising as alternates to, or substitutes for, selected petroleum chemicals. Several such chemicals already enjoy sizeable markets and additional ones are expected to emerge.

Polyols
Currently produced from starch are polyols that compete well with synthetic polyols such as pentaerythritol. Sorbitol, now produced in excess of 200 million pounds, is used extensively for making surfactants and emulsifiers, especially for food applications. Smaller amounts of sorbitol are used in making specialty polyethers for urethane foam production.

Methyl alpha-D-glucoside is being produced commerically by the Horizon Chemical Division of A E. Staley. It is useful in applications such as surfactants, alkyds and urethane foams. Detergent applications for this glucoside and for higher alkyl glucosides made from it could continue to increase because of their functionality and biodegradability. Since petroleum-derived polyols enjoy a multibillion-pound market, ethylene glycol alone accounting for 5 billion pounds, research leading to new and improved technologies for starch-derived polyols could be most rewarding.

Organic Acids
New market opportunities are emerging for the carboxylic acids produced by fermentation of starch-derived carbohydrates. Acetic acid, now produced mostly from petroleum-derived ethylene, and in small amount from fermentation of sugars, ha

...en undergoing evaluation, after conversion to ...lcium magnesium acetate (CMA), as a deicer salt ...r highways. In 1982, the Federal Highway ...ministration initiated a program for complete ...aluation of CMA as an alternative deicer to the ...rrosive sodium chloride-calcium chloride salt now ...ed. Reports from studies conducted under this ...ogram and from various other projects and ...searchers suggest CMA is considerably less ...rrosive than the salt currently used and would ...ve little deleterious effect on plants or ...imals. One company marketed the deicer in the ...-87 winter. Although CMA can be derived from ...nthetic acetic acid or from biomass other than ...ain, it is hoped that a sizeable new market for ...ain will result from widespread use of CMA.

Citric acid, the largest volume acid derived ...om starch, shows good functionality for replacing ...osphates in detergents. Because of its excellent ...elating properties and its complete ...odegradability, new market opportunities for ...tric acid are being realized.

The current production of lactic acid from ...rn-based carbohydrates could increase by orders of ...gnitude if its potential could be realized as an ...termediate to polymerizable monomers such as ...rylic acid, acrylamide, alkylacrylates, and ...rylonitrile and other chemicals now made from ...troleum. Also, polymers and copolymers of lactic ...id exhibit exceptional properties for a number of ...d-use applications. Researchers at Battelle, ...lumbus, Ohio, have prepared and evaluated several ...ctic-acid-based polymers and have shown their good ...nctionality in several applications where ...nventional thermoplastics are now used (Sinclair ...87).

...icrobial Polysaccharides and Dextrins

Fermentation of starch-derived carbohydrates ...t only produces simple chemicals but also produces ...lysaccharides and unique dextrins. Several ...lysaccharide hydrocolloids arising from microbiol conversion of glucose have been isolated and characterized. Dextran and xanthan are two made available commercially. Pullulan, a biopolymer produced by a yeast-like fungus growing on starch or sugar, exhibits many interesting properties suggesting potential use as a biodegradable film, a coating to protect fruits and vegetables, a seed coating, a high tensile strength fiber and many other applications (Scott 1987). Improvement in production and recovery of this polysaccharide could enhance realization of the polymer's potential. As bioprocessing techniques continue to improve, markets for these and other microbial polysaccharides will undoubtedly expand.

Cyclodextrins, known for many years as unusual cyclic oligosaccharides produced during the growth of certain bacteria on a solution of starch, are now commercialized. These dextrins, which contain rings of 6, 7 or 8 glucose units, form molecular inclusion complexes with numerous low-molecular-weight chemicals. This property is useful for separating, purifying, protecting and delivering of chemicals for scores of applications. Already produced in Japan and Hungary, the American Maize-Products Company began production in the U.S. in 1988. Although not specifying the specific markets for their production, market opportunities exist in the pharmaceutical, agricultural chemical, household chemical, flavor and fragrance industries. Their use in processing streams to remove unwanted compounds appears most promising.

Polymers

The dramatic market growth of petroleum-derived polymers over the last several decades has attracted the attention of starch researchers for some time. Effort has been made to recapture some of the traditional markets in paper and textiles lost to the synthetics as well as to capture a share of new markets emerging for polymers. In order to enhance the functionality of starch, a broad range o chemical reactions has been conducted including crosslinking, oxidation, esterification and etherification. Several of these modifications have resulted in commercial products (Wurzburg 1986). Improved conversion technologies that result in more efficient reactions with better separation and recovery of product should allow the starch products to become more competitive.

Starch-Plastic Compositions

A large portion of the chemicals produced from petroleum is utilized to manufacture the nearly 60 billion pounds of synthetic polymers now used annually in the United States. Plastics, which now account for about 75% of the total synthetic polymer production, will double in usage volume during the next decade if raw materials are available. In recognition of this growth potential and the uncertainties in availability of sufficient petroleum feed stocks, interest has increased in the use of natural polymers as extenders for plastics or as total replacements for certain types of plastics. Not only is it the renewable aspect of raw materials such as starch that has piqued the interest of industry and the public, but also the potential of such natural polymers to impart degradability to plastic materials.

Because of concern over buildup in the environment of discarded plastic goods due to their resistance to microorganisms, increased attention is being given to this area. If plastics can be made readily degradable, new markets for such materials would materialize and the growth for plastics would likely exceed even the most liberal estimates.

Although plastic shopping bags containing low levels of starch were introduced in the U. K. in the mid-seventies (Maddever and Chapman 1987), and researchers in the U.S. were reporting on starch-based plastic films and molded articles also in the mid-seventies (Otey et al. 1974), it is only within the last two years that polyethylene bags containing starch have been produced in the U.S.

625

The St. Lawrence Starch Company, Limited, of Canada is marketing a specially treated cornstarch as an additive for making starch-containing plastic film and bottles (Maddever and Graham 1987). Their product, granular cornstarch treated with a silane coupling agent and dried to less than 1% moisture, currently is being used by some polyethylene film producers to manufacture starch-containing plastic bags.

Employing a different and as yet undisclosed technology, the Archer Daniels Midland Company of Decatur, IL has announced its entry into a new business which uses starch alone or in combination with other components to cause plastics to degrade. According to articles in the press, this company started marketing their product(s) in 1988.

Agri-Tech Industries, Incorporated, of Urbana, Illinois, during the last several months, has announced its plans to produce starch-plastic film for a variety of uses. Production is expected to begin in late 1988.

Research continues in this area, both to improve compatibility of starch-synthetic polymer systems and to better understand degradation of various products when buried in soil or placed in landfills.

Starch has been evaluated as an inert filler in poly(vinyl chloride) (PVC) plastics. Various techniques were investigated for incorporating large amounts of starch as a filler in PVC plastics by Westhoff et al. (Westhoff et al. 1974). Starch PVC films were prepared and their properties were measured in Weather-Ometer and outdoor exposure tests. By varying the composition, films were obtained that lasted from 40 to 900 hr. in the Weather-Ometer and from 30 to more than 120 days in the soil. All samples tested under standard conditions with common soil microorganisms showed mold growth, with the greatest amount of growth recorded for samples containing the highest amount of starch.

Starch-Urethane Composition

Not only can starch be mixed with synthetic polymers and exhibit utility as a filler, extender, or reinforcing agent; it can also become an integral part of such polymers through chemical bonds. Urethane is an example of a system where starch has been chemically bonded to a resin. To produce relatively low-cost rigid urethane plastics that might have application in solvent-resistant floor tile, a system was developed based on 10 to 60% starch, castor oil, the reactive products of castor oil, and starch-derived glycol glycosides, and polymeric diisocyanates. The addition of starch to the isocyanate resins substantially reduced chemical costs and improved solvent resistance and strength properties. Evidence showed that the starch chemically combined with the resin molecules.

The degree of reactivity between starch and isocyanates can be greatly enhanced by modifying starch with nonpolar groups, such as fatty esters, before the isocyanate reaction. Maximum reactivity of the modified starches was achieved when the degree of substitution was about 0.7. Elastomers have been prepared where starch was a filler and crosslinking agent for diisocyanate-modified polyesters. Also, starch can be incorporated into urethane systems to yield shock-resistant foams. Up to 40% starch or dextrin can be incorporated into rigid urethane foam, and such foams are more flame resistant and more readily attacked by microorganisms. By incorporating a modified starch into starch-rigid polyurethane foam, improved dimensional stability and flame retardancy of the foam have been realized (Koch and Roper 1988).

Starch Graft Copolymers

Another approach to chemically bonded natural polymer-synthetic polymer compositions is through graft polymerization. This technique has received considerable attention of scientists at the Northern Regional Research Center, especially for those systems where the natural polymer is starch.

Basically, the procedure used for synthesizing starch-graft polymer is to initiate a free radical on the starch backbone and then allow the radical to react with polymerizable vinyl or acrylic monomers. A number of free-radical initiating methods have been used to prepare graft copolymers, and these may be divided into two broad categories: initiation chemically and by irradiation. The choice depends in part on the particular monomer or combination of monomers to be polymerized. Both chemical and irradiation-initiating systems have been employed to graft-polymerize onto starch a wide variety of monomers, both alone and in selected combinations. Fanta and Doane (1986) have reviewed these systems.

For plastic or elastomeric copolymer compositions that can be extruded or milled monomers such as styrene, isoprene, acrylonitrile, and various alkyl acrylates and methacrylates were employed.

Starch graft-polystyrene, -poly (methyl methacrylate), -poly (methyl acrylate), and -poly (methyl acrylate-co-butyl acrylate) polymers have been prepared with approximately 50% add-on and evaluated for extrusion-processing characteristics.

Narayan and Potter (1987) have reported starch or cellulose graft copolymers where the side chains are synthetic polymers such as polyethylene or polystyrene. The process used by them allows considerable control over the number and molecular weight of the side chains. These workers propose that by more specific control of structures, graft copolymers can be tailored to meet specific end use requirements. The use of such copolymers as compatibilizing agents for blends of starch and synthetic polymers is worth further study.

Several other monomers have been graft polymerized onto granular and gelatinized starch, and several of the graft polymers show promise as thickeners for aqueous systems, flocculants, clarification aids for waste waters, retention aids in papermaking, and many other uses. The polymer that has received the most attention and is now being produced by three U.S. companies is made by graft polymerizing acrylonitrile onto gelatinized starch and subjecting the resulting starch graft-polyacrylonitrile copolymer to alkaline saponification to convert the nitrile functionalities to a mixture of carboxamide and alkali metal carboxylate groups. Removing the water from this polymer provides a solid that absorbs many hundreds of times its weight of water but does not dissolve. Fanta and Doane (1986) have reviewed the preparation, properties and commercial applications for this polymer. Applications include dewatering of fuels, treatment of skin ulcers, body powders, seed and root coatings, and disposable soft goods such as diapers, catamenials and bedpads.

The use of starch and starch-based polymers as a matrix for controlled delivery of active agents including insecticides, herbicides, nematicides fungicides, nutrients and medicaments has received the attention of researchers for some time. Sizeable new markets could develop for starch as a matrix polymer, if the projected benefits are fully realized. Field studies have shown that losses of pesticide due to volatility, leaching and decomposition by light are substantially reduced when the pesticide is applied in a starch matrix.

CONCLUSIONS

The potential for realization of new and expanded industrial markets for starch is excellent. Improved efficiencies of chemical and biological conversions of starch and glucose already have improved the competitive position of selected conversion products with those derived from petroleum. Continued improvement in conversion technologies will allow greater market capture for these products and new markets for a variety of other alcohols, polyols and acids. The large market enjoyed by synthetic polymers is a viable target for starch and starch-derived polymers. Incorporation of starch into synthetic polymer films and molded articles, just now surfacing on a commercial level, holds great promise for sizeable new markets in the near term. Copolymers where synthetic polymers are grafted onto starch have entered the commercial market as aqueous fluid absorbers for use in disposable goods, wound treatment, fuel filters and a variety of agricultural applications. Continued research on chemically bonded starch-synthetic polymer systems has pointed to several other market possibilities.

LITERATURE CITED

FANTA, G. F., and DOANE, W. M., 1986. Grafted starches. Pages 149-178 in: Modified Starches: Properties and Uses, Otto Wurzburg, Ed., CRC Press, Inc., Florida.

KENNEDY, H., 1987. Starch and dextrin based adhesive. Pages 267-281 in: Proceedings of the First Annual Corn Utilization Conference, St Louis, MO., June 11-12.

KIRBY, K. W., 1987. Starch in paper applications and coating materials. Pages 253-266 in: Proceedings of the First Annual Corn Utilization Conference, St. Louis, MO, June 11-12.

KOCH, H., and ROPER, H., 1988. New industrial products from starch. Starch/Starke, $\underline{40}$: 121.

MADDEVER, W. J., and CHAPMAN, G. M., 1987. Enhanced polymer properties and biodegrability through modified starch additions. Addendum to Proceedings of the First Annual Corn Utilization Conference, St. Louis, MO, June 11-12.

NARAYAN, R., and POTTER, A. A., 1987. Preparation of corn-based plastics for the materials applications. Pages 209-220 in: Proceedings o the First Annual Corn Utilization Conference St. Louis, MO, June 11-12.

OTEY, F. H., MARK, A., MEHLTRETTER, C. L., and RUSSELL, C. R., 1974. Starch-based film fo degradable agricultural mulch. Ind. Eng. Chem Prod. Res. Dev., $\underline{13}$: 90.

OTT, C. D., 1987. Corn products as chemical feedstocks: The corn refining industry. Pages 286-305 in: Proceedings of the First Annual Corn Utilization Conference, St. Louis, MO, June 11-12.

NCLAIR, R. G., 1987. Lactic acid polymers derived from corn in the controlled release of plant nutrients and other agricultural chemicals. Pages 221-236 in: Proceedings of the First Annual Corn Utilization Conference, St. Louis, MO, June 11-12.

STHOFF, R. P., OTEY, F. H., MEHLTRETTER, C. L., and RUSSELL, C. R., 1974. Starch-filled polyvinyl chloride plastics - preparation and evaluation. Ind. Eng. Chem. Prod. Res. Dev., 13: 123.

RZBURG, O., 1986. Modified Starches: Properties and Uses, CRC Press, Inc., Florida.

TRUDEX - A SPRAY DRIED TOTAL WHEAT SUGAR

Norman Wookey
Tenstar Products Ltd
Hythe Road, Ashford, Kent, TN 24 8BH

For many years glucose syrup produced by hydrolysis of various starches has been a valuable material for the food industry. The starches have mainly been derived from cereals or potato, the cereal source being mainly maize but with wheat becoming an important source in more recent times. To make glucose, the starch molecular chain is split by hydrolysis and the degree of splitting determines the type of syrup produced and in particular the amount of glucose and other sugars present. The hydrolysis can be effected either by treatment with hot acid or by enzyme action or by a combination of both. The degree of hydrolysis is usually denoted by the D.E. (dextrose equivalent) and a typical commercial glucose syrup might be of 42 D.E. and contain perhaps 20% of glucose.

Our interest here, however, is in the so-called high D.E. syrups, where the D.E. may be in the range 96-98 D.E. and the pure glucose content more than 90%, the residual sugars being mainly disaccharides with a small proportion of higher sugars.

Whilst glucose in liquid form is suitable for many applications, particularly confectionery, brewing and baking, there are other outlets where a solid form is desirable. To obtain solid material there are two possible methods, either slow crystallisation of pure glucose from the syrup leaving a residual mother liquor of impurities, or by solidifying the mix of glucose, higher sugars and impurities as a 'total sugar'.

The first method is, of course, the basis
dextrose monohydrate production in which glucose
the form of its monohydrate $C_6H_{12}O_6 \cdot H_2O$ is removed as
pure crystalline material and this is particular
suitable in pharmaceutical applications. This meth
is, however, quite expensive in capital equipment,
fairly slow in operation and part of the yield is lo
as a mother liquor which is sold at a low price.
is true that the final solid product is very pure, b
this is not a significant advantage in much of t
food industry where the material is mixed with oth
raw materials like cocoa, flour or milk products.

In view of the apparently better economics
making a total sugar from a high D.E. glucose syru
a number of attempts have been made in the pa
twenty years to achieve this on an industrial scal
The obvious method of using normal spray dryi
techniques is not practicable because the slow ra
of crystallisation of such glucose syrups mere
results in a sticky mass on the spray drier wall
Various other methods have been tried of which thr
have been reasonably successful from a technic
point of view.

a) The Kroyer granulator technique developed
Denmark, in which the syrup was sprayed into
heated rotating pan of previously crystallis
material in the form of small (1-2 mm) granules.

The main disadvantages of the method were t
requirement for a very high DE (greater than 98
preferably a deionised product, the high capital co
of the equipment and the very dense hard nature of t
finished glucose product, this leading to a sl
dispersing and dissolving rate when added to wate
The method was developed to a pilot plant stage b
was not adopted industrially.

b) The Pennick and Ford method, developed in t
USA and in Japan, of forming a slurry of gluco
syrup and previously made solid material and aft
standing for some hours, spray drying the partial
crystallised slurry in a traditional spray drie
This system worked on an industrial scale for so
years but had the same disadvantage of requiring a

very high D.E. and also careful control of the holding conditions prior to and after spray drying. The product however was a high quality free flowing product which found good application in the food industry.

) The method originally proposed by Niro of Denmark and later modified by Tenstar Products Ltd. of Ashford, Kent, U.K., of spraying droplets of glucose syrup on to a falling cloud of recycled powder, and of encouraging crystallisation by a holding period when the material clings to the walls of the spray drier, building up to a thickness of perhaps 5-10 cm before dropping off at intervals. The product is further dried and crystallised in a rotating horizontal drum situated under the spray drier and after grinding is a reasonably free flowing white powder. The basic final material is anhydrous dextrose, about 40% being in the α form and 60% in the β form. Due to its method of formation it has a very open structure which, coupled with the higher solubility of the β form, disperses and dissolves rapidly when added to water. The process has been operated industrially at Ashford in Kent for the past 15 years and the product which is made at about 1 tonne per hour finds a ready outlet in the food and confectionery industry.

The process was originally tried out in Denmark using potato starch as the starting material. It was not, however, developed to an industrial operation largely because of the practical difficulty of recycling a sticky product from the bottom to the top of a large spray drier. Partially crystallised dextrose is a very difficult material to handle because it readily forms a hard glass when compressed or moved by rotating or scraping machinery, so that rotary valves, elevators, screw feeds or similar equipment are soon brought to a halt by such a process. Similar difficulties were experienced when the plant was started at Ashford and the recycling operation was only achieved after much effort and by the introduction of a special valve which allowed a variable split in a recycled powder/air stream between the drier and the final product line.

The process is as shown below:

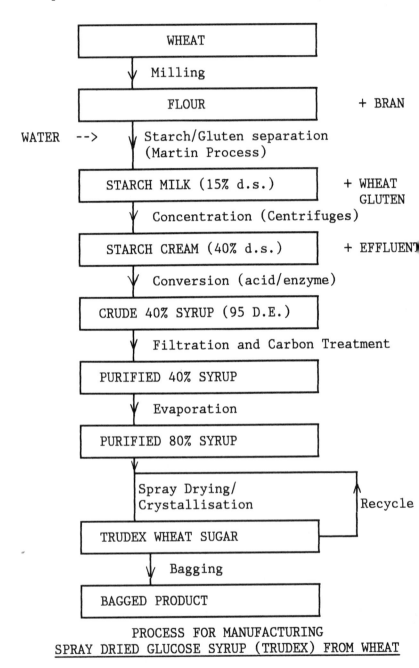

PROCESS FOR MANUFACTURING
SPRAY DRIED GLUCOSE SYRUP (TRUDEX) FROM WHEAT

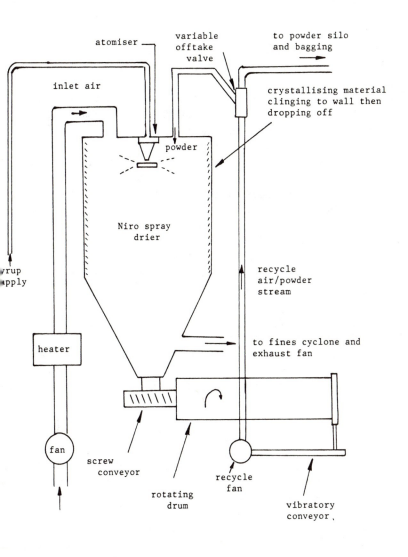

ARRANGEMENTS FOR SPRAY/CRYSTALLISING
OF HIGH D.E GLUCOSE SYRUP

The key points in the process are as follows:
1. The High D.E. Syrup, made either by acid-enzyme or all enzyme process, must be of least 93 D.E. although higher values make crystallisation process easier. The solids cont should be 78-80%. Spraying of the syrup at at 70°C is through a Niro Atomiser operating at 15, rpm.
2. The spray drier is of 5m diameter and cylindrical portion is about 5m high. The lc conical portion is inclined at 60° to horizontal. The horizontal drying drum below spray drier is 6m long and 1½ m diameter and rota at 4 rpm.
3. The weight ratio of recycle powder/syrup sol should be about 4 to 1. The air drying temperat is 130°C at inlet and 65-70°C at outlet.
4. Two items of plant are critical
(a) The Niro recycle fan transports all powder from the outlet of the drying drum to the of the spray drier using an air stream. Two f are used alternately as after about 12 hours use build-up of material occurs on the blades cleaning with hot water is necessary.
(b) A special valve at the top of the spray dr splits the recycled air/powder stream into two, as direct recycle powder to the drier and one a take off stream to a silo and bagging machine. T valve needs to be adjustable and not lead to build-up of glassy sugar on any of its surfaces. is essentially a scoop arrangement made from obliquely truncated cylinder partially rotata into the main recycle line, with all powder imp angles such that material build-up does not occur.

Inside the spray drier, droplets of sy impinge on to falling particles of recycle pow which seed the crystallisation process and slightly sticky particles stick on to the dr walls and build up to an average thickness of 5 cm. Periodically, this coating drops off the wa and falls into the bottom of the drier and the into the horizontal drying drum. The proc operates continuously and is only stopped cleaning every few months. It may, however, be

pped at any time and restarted without
ficulty.
The process and product has the following
antages:
Although the crystallisation process
ceeds more easily at high D.E.'s the process will
k satisfactorily with D.E.'s down to 93.
onisation of the syrup is not necessary.
) Again although the process is easier if the
rting material is all 'A' starch, it will work
isfactorily with a mix of 'A' and 'B' starches in
 normal proportions of the starch product made
m wheat flour. This "straight mix" starch may
e a 'B' starch content of about 15% by weight and
 mix may have a protein content of up to 0.6%.
re is an obvious economic advantage in using all
 'B' starch stream in this way compared with many
at starch plants where it is rejected or used for
 value outlets.
i) The final spray dried product has less than
 moisture compared with the 9% present in dextrose
ohydrate, this giving the customer a larger
unt of dextrose per unit weight purchased, and
owing direct weight for weight replacement of
rose in recipes.
) The end product (trade name "Trudex") has an
sual setting property due to its composition of
 α and 60% β crystalline anhydrous dextrose. The
orm has a higher initial solubility (about 68%)
n the α form (about 30%). When Trudex is mixed
h water a solution of about 68% solids first
ms. However, on standing for about one hour
arotation occurs with some of the β form changing
 the α form and when equilibrium is reached the
ubility has fallen by 10-15%. This less soluble
m then comes out of solution as the α monohydrate
 results in the material setting to a solid after
few hours. This property can be exploited by
king Trudex with fruit juice or even oils and fats
 obtaining a set hard product without use of
lling agents.
Other advantages of the product are that it
es not settle out of suspensions with water and
at it has only about 60% of the sweetness of

sucrose, this latter property improving t
perception of flavour in sugar/fruit mixes.
addition the equilibrium relative humidity of
70% solids solution of Trudex is about 77% compar
with 83% for dextrose monohydrate and 86% f
sucrose, leading to better shelf life properties
products in which it is incorporated. As t
fermentables content of Trudex is about 94% t
material is suitable for many brewing and oth
fermentation applications. Compared with sucrose
gives better browning in baked products due to t
Maillard reaction between glucose and protei
Trudex is particularly suitable for tabletti
purposes due to a tendency to cake slightly wh
compressed. This latter property, however, preven
the material being handled in bulk containers.

An interesting low calorie sweetener has al
been made by incorporating a little saccharin
Trudex and sieving the spray dried product to give
fraction with a density only half that of sucro
but having a similar sweetness to sucrose on
volume basis.

In summary, spray dried wheat glucose has be
developed as an interesting food product deriv
from wheat and its continued use for the la
fifteen years has established its significant val
to the baking and other industries.

38

THE GLUCOTECH PROCESS

Horst W. Doelle

Biotechnology Unit, Department of Microbiology
University of Queensland
St. Lucia-Brisbane, Australia 4067

William J. Wells III

BioCom USA Limited, Atlanta, Georgia, USA 40341

INTRODUCTION

There should be no doubt in anyone's mind that the agricultural industries in the world are in or going into their deepest crisis since the big depression in the 1930's. The only, but most significant, difference between the two depressions is that the depression in the 1930's was caused by drought and thus lack of crop availability, whereas now and in future we have and will always have a depression in the form of vast oversupply (Mildon 1986). Whereas the former depression was overcome with improving weather conditions, the latter cannot be overcome unless:
1. disaster strikes some areas of the globe;
2. farmland is reduced;
3. government subsidies are allocated;
4. alternative industries are found to convert the agricultural surplus into other products of stable, preferably domestic market demands (Sasson 1984, Raymond and Larvor 1986).

The most active introduction of alternative or biotechnological industries into agriculture started, of course, in Brazil and the U.S.A. with the production of ethanol in the former and High Fructo(se) Corn Syrup in the latter country.

AGRICULTURAL BIOTECHNOLOGY

In the development of a biotechnological indust(ry) in agriculture based on renewable substrates (Doel(le) 1986a; Cormack 1987), one has to realize that 60-7(0%) of the total costs involved are manifested in the r(aw) material. This means that the existing single produ(ct) processes, such as ethanol may always strugg(le) economically, because of the limited value which c(an) be added in such processes (Maiorella et al. 198(?); Kosaric et al. 1981; Rolz et al. 1983). This h(as) partly been overcome in the U.S.A. in the dry-milli(ng) industry in using the solid residue [Distillers Dri(ed) Grain] as an animal feed substitute and value add(ed) second product of the process.

It becomes apparent, therefore, that any furth(er) economic improvement is restricted to the 30-40% (of) the process, which includes fermentation technolog(y,) energy feasibility and costs, process cost(s,) by-product credit and capital investment. It furth(er) becomes apparent, particularly in the U.S. ethan(ol) industry, that ethanol yields of 8-10% (v/v(),) conversion efficiencies of 85-89% and fermentati(on) times of 50-70 hours, as they exist in the prese(nt) yeast technology, can not be sustained as economic(al) for such an industry.

In order to achieve an economic viability for t(he) industrial ethanol industry using fermentation, a n(ew) technology must be found (Doelle 1988), whi(ch) provides:
1. a high feed stock to ethanol conversi(on) efficiency (95-99%) with minimal to no carb(on) loss to biomass cell growth or non-usab(le) by-product formation;
2. a consistently high ethanol yield of 10-13% (v/(v)) or more, to reduce evaporation and wat(er) recirculation;
3. short fermentation times, to obtain fast(er) turn-around of fermenters and reduce downtime (of) the plant;
4. reduction of energy costs through elimination

sterilization, centrifugation of biomass, minimizing cooling costs etc.;
5. reduction in plant equipment costs;
6. low B.O.D. and C.O.D. levels in the effluent.

has become clear through extensive research up to ce that yeast technology can not provide these iditions owing to the thermodynamics, biochemistry, l enzymology of the yeast cell.

The potential for the use of Zymomonas mobilis in ice of the commonly used yeast in the ethanol lustry has been well documented (Lavers et al. 1981; gers et al. 1984; Doelle and Greenfield 1985a; vford 1986; Viikari 1988), although the initial tempt to introduce the bacterium into the starch ste industry (Bringer et al. 1984) have failed owing contamination or high glucose inhibition.

No attempt has been reported so far using the y-milling process feedstock.

ZYMOMONAS PROCESS DEVELOPMENT

Overwhelming evidence demonstrates that the netics of glucose batch fermentations of Zymomonas bilis allow significantly higher rates of ethanol oduction than yeast and that Zymomonas mobilis oduces less than one-third the cell mass that ccharomyces uvarum does (Cromie and Doelle 1980, 81, 1982; Lawford et al. 1982; Rogers et al. 1982; elle 1986b; Stevnsborg and Lawford 1986; Buchholz et . 1987). The latter means that Zymomonas mobilis n achieve greater volumetric productivities than uivalent yeast systems with the same amount of omass (Fig. 1 and 2). In addition, the different ll membrane structure allows for higher ethanol lerance (Barrow et al. 1983). Furthermore, momonas has a higher protein content with superior ino acid profile (Low and Rogers 1984).

A detailed investigation by our research group thin the Biotechnology Unit into the biochemical, zymological and regulatory mechanisms (Cromie and elle 1980, 1981, 1982; Lyness and Doelle 1980, 81a, b; Doelle 1982a, b; Doelle et al. 1982; ackbeard and Doelle 1983; Hoppner and Doelle 1983; rrill et al. 1983) revealed that:

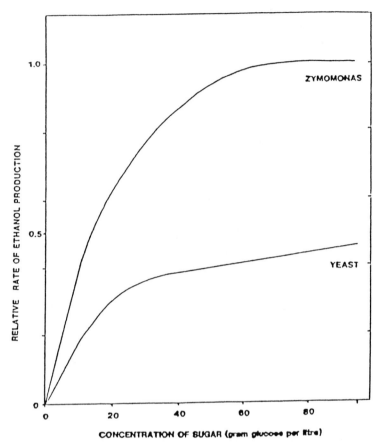

Fig. 1. Comparison of the relationship of glucose concentration to the relative rate of ethanol production between yeast and Zymomonas strains.

1. Zymomonas possesses a levansucrase instead of invertase with different control mechanisms;
2. Zymomonas does not possess a hexokinase stringently controlling glucose and fructose uptake into the cell, but possesses for each sugar a separate enzyme, glucokinase and fructokinase, allowing higher simultaneous monosaccharide uptake by the cell;
3. 'energetically uncoupled growth' can be induced by physical and chemical factors in the environment;

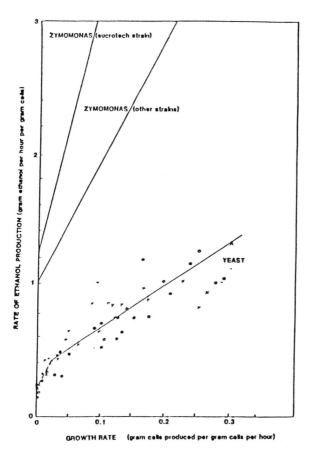

Fig. 2. Comparison of the relationship of growth rate and rate of ethanol production between yeast and Zymomonas strains.

4. the use of a different biochemical pathway together with 'uncoupled growth' conditions reduces heat evolution.

When yeasts are used, ethanol formation is a growth related process, which means that there exists a tight coupling between energy demanding and energy producing processes. As a result, very high concentrations of yeast are required to obtain an acceptable fermentation rate. In the case of Zymomonas, the energy supply can be diverted as it is not controlled by the energy demand of growth. The consequence of this phenomenon is that less biomass cells are required for the same amount of ethanol produced.

These findings and realizations were translat
into a fermentation technology (Doelle and Greenfie
1985a, b, c) and finally led to the development of t
two new Process Technologies, now referred to
Glucotech (U.S. Patent Appl. No. 106,744) a
Sucrotech (U.S. Appl. No. 594,580).

The Glucotech process has now obtained F
approval for the use of our process in the dry-milli
ethanol industry and the use of DDG from our proce
as animal feed.

GLUCOTECH PROCESS

Extensive industrial testing of this process h
substantiated the following advantages ov
traditional yeast fermentation:
1. increased ethanol production from feedstoc (12-13.5%, v/v);
2. reduced fermentation times;
3. foam reduction;
4. reduced heat generation;
5. practical elimination of infection;
6. elimination of centrifuge requirement for bioma cell recycle;
7. improved feed (DDG) quality;
8. no furfural, glycerol or fusel oils;
9. low cell concentration - low carbon loss;
10. high conversion efficiencies (95-99%).

The Glucotech process works extremely well in dry- a
wet-milling processes, using wheat, milo, cor
potatoes, etc. as feedstocks.

Potato and Wheat Starch Fermentation

According to our enzymatic analysis, one kg
potato mash contained 142.2 g of starch, which
equivalent to 155 g of glucose. Figure 3 exhibits
typical fermentation pattern of 478 g/L potato ma
containing 65.78 g/L starch.

Since it is common in the grain to ethan
industry to use higher starch concentrations a
recycle backset (= thin stillage) from the previc
fermentation, wheat starch in the form of maltrin w
added to the potato mash together with 20% backse

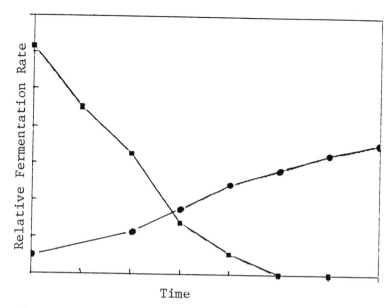

Fig. 3. Typical fermentation pattern of a small scale fermentation using potato mash as feedstock and Zymomonas mobilis [(■) glucose; (●) ethanol].

After 22 hours, 106.5 g/L ethanol or 13.5% (v/v) ethanol were obtained, which is equivalent to a 98% conversion efficiency.

Similar results were obtained using maltrin or wheat on its own.

Corn (maize) Starch Fermentation

Corn is at present either wet- or dry-milled for ethanol production, depending upon whether the manufacturer produces high fructose corn syrup or ethanol only.

Glucose syrup hydrolysate from the wet milling process containing corn steep liquor as nutrient was fermented within 24 hours with a 96-99% conversion efficiency calculated on the glucose available in the hydrolysate. No problems were encountered at concentrations up to 18% (w/v) glucose in the hydrolysate.

Figure 4 exhibits a typical medium scale (17,000 gallons) fermentation with and without backset addition resulting in 2.6-2.7 gallons/bushel ethanol in 35-42 hours using corn from a dry-milling process.

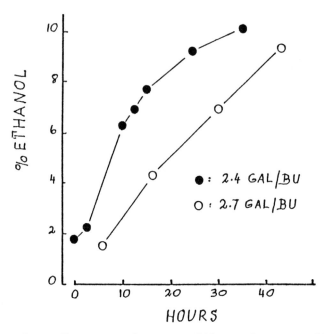

Fig. 4. Fermentation profile of a medium-sc fermentation using corn with (●) and without (O) addition of 30% backset and <u>Zymomonas</u> <u>mobilis</u>.

Using a mixture of corn, milo, and wheat flour an 1,200 gallon scale, the results exhibited in Fig 5 were obtained.

This dry-milling process fermentation has b carried out on an 1,200 gallon scale over a 4-w period without further biomass cell addition continuous inoculation from one fermenter to the ne without any deterioration of fermentation activity.

Milo Starch Fermentation

Large scale fermentations have been carried on milo flour. In Figure 6, a summary of such fermentation is presented. The whole fermentat took only 42 hours and yielded an approximately conversion efficiency.

The downstream processing by-product (Distill Dried Grain) has been collected and analyzed. Tabl shows that DDG from <u>Zymomonas</u> gives a higher prot

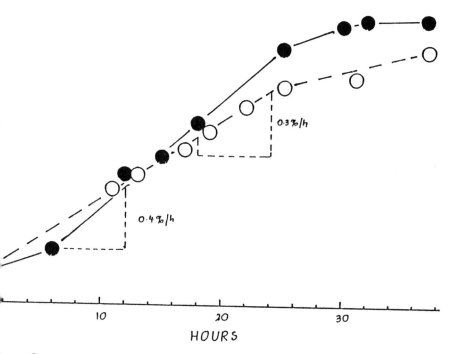

Fig. 5. Two successive medium-scale fermentations using a mixture of milo, corn, and wheat and <u>Zymomonas mobilis</u> [(O) fermentation run 1; (●) fermentation run 2 seeded from fermenter run 1].

level compared to yeast DDG. Preliminary feeding trials conducted over a 3 month period showed no preference by the animals for one or the other DDG.

SUMMARY

The Glucotech process is a new process technology for the ethanol industry based on the bacterium <u>Zymomonas mobilis</u>. This process, in our opinion, offers farmers and ethanol producers the opportunity to increase their economic viability.

Furthermore, shorter fermentation times mean more bushels of grain can be utilized without any additional capital expenditure provided the existing distillation capacity is not limiting. The ethanol producer, on the other hand, produces more ethanol owing to the higher conversion efficiency and at the

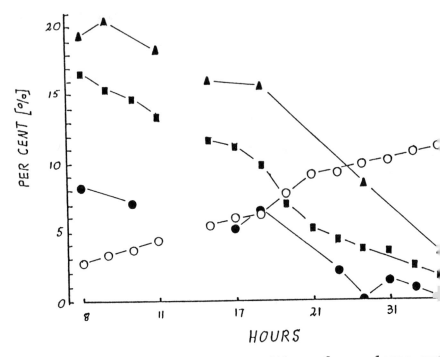

Fig. 6. Fermentation profile of a large-scale fermentation of milo using Zymomonas mobilis [ethanol; (●) residual sugar; (■) Brix; (▲) solids].

same time adds further value to the DDG by-product the lower B.O.D. and C.O.D. levels in the effluen Further savings come from the lack of fusel oils the ethanol and negligible, if any, glycerol in DDG product.

ACKNOWLEDGEMENTS

Our thanks go to Queensland Science Technology Ltd. in Australia and BioCom Internatio Ltd. and BioCom USA Ltd. for their financ assistance and generous support in our research development efforts.

Our success would, however, not be possi without the help received from the many etha producers in the USA, who so generously opened th doors and plants for our medium and large-sc

TABLE I

Laboratory Analysis of DDGS (Distillers Dried Grain plus syrup addition) and Syrup (concentrated thin stillage) Produced by the Glucotech Process (using milo as feedstock)

DDGS Yeast		Zymomonas
Moisture	10.7%	11.0%
Protein	30.9%	34.3%
Fiber	8.8%	7.5%
Fat	10.7%	9.5%
Ash	5.9%	4.6%
Starch	14.17%	12.1%
Total Digest.N	75.44%	76.9%
Total Sugars	0.36%	0.18%
Standard Plate Count	2,100/g	120/g
Bulk density	29.6 lbs/cu ft.	25.74 lbs/cu ft.
CDS : Liquid (Syrup)-		
Moisture	72.0%	60.6%
Protein	4.7%	12.4%
Fat	6.0%	7.3%
Fiber	<0.1%	<0.1%
Solids	27.6%	45%
Ash	3.57%	4.39%
Starch	0.07%	0.36%
Total Digest N.	24.35%	34.91%
Total Sugars	0.18%	0.15%
pH	4.24	4.93

periments. Our special thanks go to ADM (small ale), Shreveport Ethanol Company (medium scale) and ergy Fuel and Development Corporation (medium and rge scale) for their support. We are indebted to e Nebraska Gasohol Committee, Diversified Feed and el and Prof. T. Klopfenstein of the University of raska for carrying out our palatability and tensive feeding trials.

LITERATURE CITED

BARROW, K. D., COLLINS, J. G., ROGERS, P. L., and SMITH, G. M. 1983. Lipid composition of an ethanol-tolerant strain of Zymomonas mobilis. Biochim. Biophys. Acta 753:324-330.

BRINGER, S., SAHM, H., and SWYZEN, W. 1984. Ethanol production by Zymomonas mobilis and its application on an industrial scale. Biotech Bioeng. Symp. 14:311-319.

BLACKBEARD, J. R., and DOELLE, H. W. 1983. The effect of glucose on the sucrose hydrolyzing activity of Zymomonas mobilis. Eur. J. Appl Microbiol. Biotechnol. 17:261-263.

BUCHHOLZ, S. E., DOOLEY, M. M., and EVELEIGH, D. E. 1987. Zymomonas - an alcoholic enigma. Tibtec 5:199-204.

BURRILL, H., DOELLE, H. W., and GREENFIELD, P. F. 1983. The inhibitory effect of ethanol production by Zymomonas mobilis. Biotech Lettrs. 5:423-428.

CORMACK, C. A. 1987. Arable crops for the fermentation industry - a grower's viewpoint. Pages 13-27 In: Carbon Substrates in Biotechnology. J.D. Stowell, A.J. Beardmore, C.W. Keevil, and J.R. Woodward, eds. IRL Press Washington, DC.

CROMIE, S., and DOELLE, H. W. 1980. Relationship between maintenance energy requirements, mineral salts and efficiency of glucose to ethanol conversion by Zymomonas mobilis. Biotech Lettrs. 2:357-362.

CROMIE, S., and DOELLE, H. W. 1981. Nutritional effects on the kinetics of ethanol production from glucose by Zymomonas mobilis. Eur. J. Appl. Microbiol. Biotechnol. 11:116-119.

CROMIE, S., and DOELLE, H. W. 1982. Limitations substrate utilization efficiency by Zymomonas mobilis. Eur. J. Appl. Microbiol. Biotechnol. 14:69-73.

DOELLE, H. W. 1982a. The existence of two separate constitutive enzymes for glucose and fructose Zymomonas mobilis. Eur. J. Appl. Microbiol Biotechnol. 15:20-24.

ELLE, H. W. 1982b. Kinetic characteristics and regulatory mechanisms of glucokinase and fructokinase from Zymomonas mobilis. Eur. J. Appl. Microbiol. Biotechnol. 14:241-246.

ELLE, H. W. 1986 a. Biotechnology and the agricultural industry. 19th Conv. Australian Inst. Food Sci. Technol., Brisbane, May 1986.

ELLE, H. W. 1986b. Ethanol biosynthesis. XIIIth Intern. Congr. Microbiology, Manchester, U.K.

ELLE, H. W. 1988. Zymomonas - foe or friend of the alcohol industry. 20th Australian-New Zealand Brewing Conv., Brisbane, Australia.

ELLE, H. W., and GREENFIELD, P. F. 1985 a. Biotechnology and the sugar industry. Pages 23-30 In: Emerging Biotechnologies for Agriculture. R.N. Oram and B.G. Johnston, eds. Australia Inst. Agric. Sci., Melbourne.

ELLE, H. W., and GREENFIELD, P. F. 1985b. The production of ethanol from sucrose using Zymomonas mobilis. Appl. Microbiol. Biotechnol. 22:405-410.

ELLE, H. W., and GREENFIELD, P. F. 1985 c. Fermentation pattern of Zymomonas mobilis at high sucrose concentrations. Appl. Microbiol. Biotechnol. 22:411-415.

ELLE, H. W., PREUSSER, H. J., and ROSTEK, H. 1982. Electron microscopic investigations of Zymomonas cells grown in low and high glucose concentrations. Eur. J. Appl. Microbiol. Biotechnol. 16:136-141.

PPNER, T. C., and DOELLE, H. W. 1983. Purification and kinetic characteristics of pyruvate decarboxylase and ethanol dehydrogenase from Zymomonas mobilis in relation to ethanol production.

SARIC, N., DUVNJAK, Z., and STEWART, G. G. 1981. Fuel ethanol from biomass: production, economics and energy. Adv. Biochem. Eng. 20:119-151.

VERS, B. H., PANG, P., MACKENZIE, C. R., LAWFORD, G. R., PIK, J. P., and LAWFORD, H. G. 1981. Industrial alcohol production by high performance bacterial fermentation. Pages 195-200 In: Adv. Biotechnol. II, Pergamon Press.

LAWFORD, H. G. 1986. Zymomonas. An alternative yeast in alcohol production. ALLTECH's 6th A Nat. Course, Biotechnol. Centre, Nicholasvil Kentucky.

LAWFORD, G. R., LAVERS, B. H., GOOD, D., CHARLEY, FEIN, J., and LAWFORD, H. G. 1982. Zymomo ethanol fermentations: biochemistry bioengineering. Pages 482-506 In: Internatio Symposium on Ethanol from Biomass. H. Duckworth ed., Royal Soc. Canada, Winnipeg.

LOW, K. S., and ROGERS, P. L. 1984. macromolecular composition and essential am acid profiles of strains of Zymomonas mobil Appl. Microbiol. Biotechnol. 19:75-78.

LYNESS, E., and DOELLE, H. W. 1980. Effect temperature on sucrose to ethanol conversion Zymomonas mobilis strains. Biotech. Lett 2:549-554.

LYNESS, E., and DOELLE, H. W. 1981a. Fermentat pattern of sucrose to ethanol conversion Zymomonas mobilis. Biotech. Bioe 28:1449-1460.

LYNESS, E., and DOELLE, H. W. 1981 b. Etha production from cane juice by Zymomonas mobil Biotech. Lettrs. 3:257-260.

MAIORELLA, B., WILKE, C. R., and BLANCH, H. W. 198 Alcohol production and recovery. Adv. Bioch Eng. 20:43-92.

MILDON, R. 1986. Surpluses and the com agricultural policy. In: Alternative Uses Agricultural Surpluses. W.F. Raymon and Larvor, eds. Elsevier Appl. Sci., Barking, U.

RAYMOND, W. F., and LARVOR, P. (eds) 198 Alternative Uses for Agricultural Surplus Elsevier Appl. Sci., Barking, U.K.

ROGERS, P. L., LEE, K. J., SKOTNICKI, M. L., TRIBE, D. E. 1982. Ethanol production Zymomonas mobilis. Adv. Biochem. Eng. 23:37-8

ROGERS, P. L., SKOTNICKI, M. L., LEE, K. J., and L J. H. 1984. Recent developments in Zymomonas Process for ethanol production. Cr Rev. Biotechnol. 1:273-288.

OLZ, C., DECABRERA, S., CALZADA, F., GARCIA, R., DELEON, R., DEL CARMEN DE ARRIOLA, M., DEMICHEO, F., and MORALES, E. 1983. Concepts on the biotransformation of carbohydrates into fuel ethanol. Adv. Biotechnol. Processes 1:97-142.

ASSON, A. 1984. Biotechnologies: challenges and promises. Unesco, Paris.

EVNSBORG, N., and LAWFORD, H. G. 1986. Performance assessment of two patent strains of Zymomonas mobilis in batch and continuous fermentation. Appl. Microbiol. Biotechnol. 25:106-115.

IKARI, L. 1988. Carbohydrate metabolism in Zymomonas. Crit. Rev. Biotechnol. 7:237-261.

39

UTILIZATION OF BY-PRODUCTS OF WHEAT-BASED
ALCOHOL FERMENTATION

Y. V. Wu
Northern Regional Research Center
U.S. Department of Agriculture
Agricultural Research Service
1815 N. University St.
Peoria, IL 61604

ABSTRACT

Whole wheat and wheat flour were fermented to ke ethanol. The residue, after ethanol was stilled, was fractionated into distillers' grains, ntrifuged solids, and stillage solubles. trafiltration combined with reverse osmosis parated wheat stillage solubles into small volumes concentrated solutions and large volumes of rmeate suitable for reuse. Distillers' grains and ntrifuged solids had protein contents of 29 and %, respectively, and accounted for 36 and 21% of tal hard wheat nitrogen. Half of the nitrogen in our stillage solubles had molecular weights lower an 10,000. Wheat and wheat flour fermentation -products had lower protein solubility than the arting wheat or wheat flour. Lysine, expressed

---

This chapter is in the public domain and not pyrightable. It may be freely reprinted with istomary crediting of the source, American sociation of Cereal Chemists, 1988.

as g/16 g N, was considerably higher in distiller grains and centrifuged solids than in whea Freeze-dried wheat flour centrifuged solids retain some functional properties for breadmaking.

INTRODUCTION

Although corn is the most common substrate f commercial ethanol fermentation in the United State a small amount of wheat is also used for this purpo (Morris 1983). Fermentation of cereal grains to ma ethanol results in a protein-rich material (stillag after ethanol is distilled. The fermentation proce predominantly consumes the starch in cereal grain and other nutrients such as protein a concentrated. Satterlee et al (1976) prepar protein concentrates from fermented corn a fermented wheat by extraction with alkali. Tsen al (1982, 1983) incorporated corn distillers' dri grain flours in bread and cookies. Prentice (197 and Prentice et al (1978) blended brewers' spe grain with flour to make muffins and cookies. Finl and Hanamoto (1980) added various fractions brewers' spent grains to bread.

Optimum use of stillage is important f commercial success of the overall ethan fermentation process. Since fermentation of whe creates a higher percentage of stillage solubles th does fermentation of corn (Wu et al 1981, 1984), t economics of the wheat process depend even more recovery of stillage solubles at a minimal cos Stillage is usually screened or centrifuged to yie a solid fraction (distillers' grains) and a solub fraction (stillage solubles). For each gram ethanol produced from wheat or wheat flour, 10 ml stillage solubles results. Because stillage solubl from wheat and wheat flour contain about 5% d matter, evaporation of large amounts of water recover solids is expensive. Because ultrafiltrati (UF) and reverse osmosis (RO) do not evaporate wate large savings of energy and cost are possible. and RO can separate a large volume of dilute soluti into a small volume of concentrated solution and a large volume of permeate (solution that passes through the membrane) that can be reused as water or safely discarded. Preconcentration of apple juice by RO (Sheu and Wiley 1983), concentration of whole milk by RO for cheddar cheese production (Agbevari et al 1983), and UF and RO of oilseeds (Lawhon and Lusas 1984) are some examples of UF and RO. Since detailed knowledge of the composition and properties of by-products of wheat-based alcohol fermentation is essential for better utilization of these by-products, this review will cover fractionation and characterization of protein-rich residue from wheat alcohol distillation (Wu et al 1984), and recovery of stillage soluble solids from hard and soft wheat by RO and UF (Wu 1987).

MATERIALS AND METHODS

Fermentation and Fractionation of Stillage

A previous publication (Wu et al 1984) described experimental details. Newton, a hard red winter wheat, was grown in Kansas. Daws, a soft white winter wheat, was from Washington. Wheats were milled in a laboratory model Allis rolls until all passed through a No. 12 screen (1.65 mm aperture) for whole grain or in a Buhler mill with 70% of the wheat collected in the flour bins (70% extraction flour). Ground wheat or wheat flour was dispersed in tap water in a stainless steel, temperature-controlled, jacketed fermentor equipped with stirrers. Alpha-amylase was added to degrade starch to soluble dextrins, glucoamylase was used to convert dextrins to glucose, and yeast converted glucose to ethanol. Ethanol was distilled from the fermentor, and the residue (stillage) was filtered through cheesecloth to separate the distillers' grains. A continuous centrifuge separated thin stillage that passed through the cheesecloth into centrifuged solids and stillage solubles (Fig. 1).

Ultrafiltration and Reverse Osmosis

For analytical work, two kinds of membran (43 mm in diameter) with nominal molecular weig cutoffs of 500 (UM05) and 10,000 (PM 10) were us under 340 kPa (50 psi) nitrogen pressure in an Amic ultrafiltration cell to separate stillage solubl into permeate and concentrate fractions. For lar scale work, an OSMO Econo Pure RO unit equipped wi 1.0 m^2 membrane with a molecular weight cutoff 1,000 for organic compounds was used for UF at 6 kPa (100 psi). The solution that passed through t membrane was the permeate, and the solution retair by the membrane was the concentrate. The concentra stream was recirculated back to the initi solution. Samples of permeate and of concentra plus initial solution (later termed concentrate) we periodically removed for analysis. A Moc UHPROLA-100 RO system equipped with 1.1 m^2 membra with a molecular weight cutoff of 100 for orgar compounds was used at 5,440 kPa (800 psi) for RO the UF permeate. Details of UF and RO were describ elsewhere (Wu 1987).

Protein Extraction

Ground Newton wheat or flour was extract sequentially with 1% sodium chloride, water, 0.01 acetic acid (twice) and 0.03 $\underline{N}$ sodium hydroxide. I wheat distillers' grains and wheat centrifug solids, one more 0.03 $\underline{N}$ sodium hydroxide extracti and two more 0.1 $\underline{N}$ sodium hydroxide extractions we performed. For wheat flour centrifuged solids, c more 0.03 $\underline{N}$ sodium hydroxide extraction than whe flour was added. Wu et al (1984) gave additior details.

Analyses and Functional Properties

Protein, fat, crude fiber, and ash we determined by AACC approved methods (1983), a protein was calculated from Kjeldahl N X 5.7. Star was analyzed by a polarimetric method (Garcia a

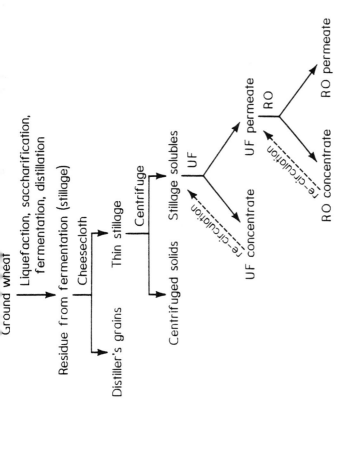

Fig. 1. Schematic diagram of wheat fermentation, fractionation of stillage, ultrafiltration, and reverse osmosis. From Wu 1987.

Wolf 1972), and moisture was determined by heati[ng] samples at 100°C to constant weight. Dieta[ry] fiber, the sum of lignin, water-insolub[le] hemicellulose, and cellulose, was measured by t[he] neutral detergent method (McQueen and Nichols 1979). Analyses for glucose, glycerol, and ethan[ol] were made by high-performance liquid chromatograp[hy] (Wu et al 1984). For amino acid analyses, ea[ch] sample was hydrolyzed in 6 $\underline{N}$ hydrochloric acid. T[he] hydrolyzate was analyzed in an amino acid analyze[r] and data were computed by the method of Cavins a[nd] Friedman (1968). More details were given elsewhe[re] (Wu et al 1984).

Gluten ball forming ability was determined [by] mixing 2 g of vital gluten, devitalized gluten, [or] wheat flour distillers' grains with 3 ml of 0.[5%] sodium chloride at room temperature. A 10.0 [g] Swanson-Working Mixograph described by Finney a[nd] Shogren (1972) was used for mixograms. For ha[rd] wheat flour and wheat flour centrifuged solid[s,] enough water and wheat starch were added to give 6[5%] water absorption and 11% protein. For soft whe[at] flour and wheat flour centrifuged solids, 58% wat[er] absorption and 8% protein were used. Mixograms we[re] stopped after 7 to 10 min of mixing.

RESULTS AND DISCUSSION

Composition and Yield of Fermentation Products

The yield of ethanol was higher for soft whe[at] and soft wheat flour than for the hard whe[at] counterparts, both on an absolute basis and as [a] percent of theoretical value. Table I lists t[he] protein, crude fiber, neutral detergent fiber, fa[t,] starch, and ash contents of hard wheat, hard whe[at] flour, and their fermentation by-products. For ha[rd] wheat grain, 35% of the original weight w[as] fermentation by-product, of which distillers' grai[n] accounted for 47%. The protein contents [of] fermentation by-products from hard wheat ranged fr[om] 29 to 57%, about 2 to 4 times that of the starti[ng] wheat. The fat content of distillers' grains w[as]

TABLE I

Composition and Yield of Fermentation By-Products[a]
from Hard Wheat and Hard Wheat Flour (Dry Basis)

	% of By-Product	Protein, N X 5.7	Fat	Crude Fiber	NDF	Ash	Starch
Hard Wheat		15.7	1.3	3.4	11.3	1.5	64.8
Distillers' grains	47	29.2	6.0	19.9	49.9	2.4	0
Centrifuged solids	15	57.2	1.8	6.8	23.5	2.2	0
Stillage solubles	38	37.4	nd	nd	nd	8.8	nd
Hard wheat flour		12.4	0.9	0.3	1.1	0.5	79.6
Centrifuged solids	36	49.0	6.1	0.5	30.4	1.7	0.7
Stillage solubles	64	51.0	nd	nd	nd	3.8	nd

[a]Fermentation by-products were 35% of wheat and 21% of wheat flour. nd = not determined. NDF = neutral detergent fiber. Adapted from Wu et al 1984.

several times that of the starting wheat, but the fa content of centrifuged solids was close to that c wheat. The stillage solubles had the highest as content of all fermentation by-products; part of th ash from stillage solubles was derived from sal formed as a result of neutralization of acid and ba: during pH adjustments. Centrifuged solids had twic the crude fiber and neutral detergent fiber content of starting wheat, but distillers' grains had th highest fiber value by far. Protein, fat, crud fiber, and ash contents of wheat distillers' grair and wheat stillage solubles in Table I are betwee the maximum and minimum values reported by Nation: Research Council (1956). However, our fractionation procedure included a centrifuged solic fraction not listed by National Research Counc (1956). Hard wheat distillers' grains had simil. protein and fat, higher crude fiber, and lower a contents than brewers' spent grains (Finley a Hanamoto 1980).

Hard wheat flour had lower protein, fat, cru fiber, neutral detergent fiber, and ash contents th. hard wheat (Table I). The lower yield of wheat flo fermentation by-product compared with whole wheat w due to the higher starch content of wheat flour. T smaller particle size of wheat flour yielded small particle size fermentation residue, which pass through cheesecloth completely. Only centrifug solids and stillage solubles were obtained aft fermentation of wheat flour. Stillage solubl accounted for two-thirds of the fermentati by-products of wheat flour in contrast to one-thi for wheat. Both centrifuged solids and stilla solubles had about 50% protein, four times that the flour. The fat and ash contents of centrifug solids and stillage solubles were higher than that flour. Crude fiber and neutral detergent fib contents of centrifuged solids were higher than tho of flour.

The protein, fat, crude fiber, neutral deterge fiber and ash contents of fermentation by-produc from soft wheat (Table II) were, in general, simil to those from hard wheat, except that prote

TABLE II

Composition and Yield of Fermentation By-Products[a] from Soft Wheat and Soft Wheat Flour (Dry Basis)

	% of By-Product	% Content					
		Protein, N X 5.7	Fat	Crude Fiber	NDF	Ash	Starch
Soft Wheat		10.5	1.5	2.6	11.3	1.6	68.5
Distillers' grains	48	23.0	6.7	18.9	54.2	2.2	0
Centrifuged solids	15	46.1	5.4	8.4	34.0	2.4	2.8
Stillage solubles	37	33.4	nd	nd	nd	9.1	nd
Soft wheat flour		8.8	0.9	0.2	2.3	0.5	81.5
Centrifuged solids	26	35.3	6.8	14.8	41.1	1.7	9.3
Stillage solubles	74	44.0	nd	nd	nd	4.0	nd

[a]Fermentation by-product was 35% of wheat and 21% of wheat flour. nd = not determined. NDF = neutral detergent fiber. Adapted from Wu et al 1984.

contents of soft wheat fermentation by-products w lower than those of the corresponding products f hard wheat, reflecting the lower protein content soft wheat. Also, the soft wheat flour fermentat by-products had lower protein contents t corresponding hard wheat flour by-products. addition, stillage solubles from soft wheat fl accounted for three-fourths of the fermentat by-products, a higher proportion than from hard wh flour.

The protein contents of soft wheat fermentat by-products were similar to the correspond fractions from corn, while fat contents of wh distillers' grains and centrifuged solids were lo than those of corresponding corn fractions (Wu et 1981). Also, wheat stillage solubles accounted fo larger fraction (38 vs. 20%) and higher percentage total nitrogen of fermentation by-products than corn stillage solubles (33 vs. 20).

UF and RO of Stillage Solubles

UF reduced the nitrogen and solids concentrati of hard wheat flour stillage solubles by three- ninefold, but only slightly reduced the concentration in the permeate fraction. nitrogen, solids, and ash concentrations of permeate at 1,360 kPa decreased 11- to 21-fold f those of UF permeate, whereas those of RO permeate 5,440 kPa decreased 278- to 978-fold from those of permeate. Almost 100% recovery of nitrogen, soli and ash was obtained at 5,440 kPa, compared with to 96% recovery at 1,360 kPa.

UF lowered the nitrogen and solids concentrati of hard wheat stillage solubles by two- to five-fo but reduced the ash concentration in the perme fraction by 20-27%. The nitrogen, solids, and concentrations of RO permeate at 1,360 kPa decrea nine- to 19-fold from those of UF permeate, but th of RO permeate at 5,440 kPa decreased 567- 784-fold from those of UF permeate. RO at 5,440 almost recovered 100% of nitrogen, solids, and compared with 90 to 95% at 1,360 kPa.

UF reduced the nitrogen and solids concentrations of soft wheat flour stillage solubles by three- to eightfold, but lowered the ash concentration in the permeate fraction by 21-24%. The nitrogen, solids, and ash concentrations of RO permeate at 1,360 kPa decreased 11- to 20-fold from those of UF permeate, whereas those of RO permeate at 5,440 kPa decreased 32- to 2640-fold from those of UF permeate.

Nitrogen Distribution and Content of Stillage Soluble Fractions

About 80% of the total nitrogen in stillage solubles from hard wheat, hard wheat flour and soft wheat flour was in compounds above 500 molecular weight. About 70% of the total nitrogen of soft wheat stillage solubles was from compounds above 500 molecular weight. Between 51 and 59% of the nitrogen in nitrogenous compounds of stillage solubles from wheat and wheat flour had molecular weights above 10,000. The nitrogen contents of the >10,000 molecular weight fractions were higher than those of the 500-10,000 molecular weight fractions, and the nitrogen contents of the 500-10,000 molecular weight fractions were higher than those of the <500 molecular weight fractions.

Protein Fractions of Wheat and Wheat Flour Fermentation By-Products

Solvents used to extract protein from wheat and its fermentation by-products are listed in Table I. Sodium chloride solution extracted albumin and globulin. A water wash was needed to remove most salt for subsequent extraction of gluten by acetic acid, since solubility of gluten in acetic acid is greatly reduced by salt. Residual salt caused less nitrogen in the first acetic acid extract than in the second extract. Acetic acid extracted only 5 to 8% nitrogen from distillers' grains and centrifuged solids, compared with 39% from wheat. Freeze-dried distillers' grains and centrifuged solids had lower soluble nitrogen than wheat, but higher soluble

TABLE III
Percent Nitrogen Distribution of Soluble Fractions of Hard Wheat and Its Fermentation Products[a]

Fraction	Hard Wheat	Distillers' Grains		Centrifuged Solids	
		Dried at 90°C	Freeze Dried	Dried at 90°C	Freeze Dried
1% NaCl Extract	18	15	24	7	8
Water Wash	5	6	7	6	7
1st 0.01 N HOAc Extract	12	3	5	3	6
2nd 0.01 N HOAc Extract	27	2	3	3	2
1st 0.03 N NaOH Extract	21	8	11	12	17
2nd 0.03 N NaOH Extract		11	10	13	29
1st 0.1 N NaOH Extract		11	7	21	8
2nd 0.1 N NaOH Extract		8	3	9	2
Residue	18	39	21	21	12

[a] From Wu et al 1984.

nitrogen than corresponding materials dried at 90°C. Lower solubility of proteins in wheat fermentation by-products resulted from drying at 90°C, as well as from fermentation and subsequent distillation of ethanol.

The nitrogen distribution of soluble fractions of hard wheat flour and its fermentation by-products is shown in Table IV. Solvents used were as in Table II, except the last two sodium hydroxide extractions were omitted. Acetic acid extracted 69% of the wheat flour nitrogen, but only 7 and 19% of the nitrogen from centrifuged solids dried at 90°C and freeze-dried, respectively. The protein in freeze-dried centrifuged solids was more soluble than that in 90°C-dried materials.

Amino Acid Composition

The amino acid analyses of wheat and its fermentation by-products revealed high levels of glutamic acid and proline (Table V). Wheat is deficient in lysine and low in isoleucine and threonine, compared with the amino acid pattern for high-quality protein for human consumption (National Academy of Science 1980). Distillers' grains and centrifuged solids had higher lysine, threonine, and isoleucine than wheat, and therefore, are expected to have better nutritional value. Stillage solubles had lower lysine and isoleucine than wheat, however, probably making them nutritionally inferior. Soft wheat and its fermentation by-products had higher lysine contents than corresponding materials from hard wheat.

Functional Properties

We evaluated wheat flour fermentation by-products for possible use in baked goods, because gluten accounts for most wheat protein, and is essential for that purpose. A vital wheat gluten, a devitalized wheat gluten, and the two wheat flour centrifuged solids (freeze-dried and dried at 90°C) were compared for their abilities to form gluten balls.

TABLE IV
Percent Nitrogen Distribution of Soluble Fractions of Hard Wheat
Flour and Its Fermentation Products[a]

Fraction	Hard Wheat Flour	Centrifuged Solids	
		Dried at 90°C	Freeze Dried
1% NaCl Extract	15	13	12
Water Wash	4	7	12
1st 0.01 N HOAc Extract	62	5	10
2nd 0.01 N HOAc Extract	7	2	9
1st 0.03 N NaOH Extract	9	9	19
2nd 0.03 N NaOH Extract		19	8
Residue	6	39	25

[a]From Wu et al 1984.

Amino Acid Compositions of Wheat, Distillers' Grains, Centrifuged Solids, and Stillage Solubles[a]

Amino Acid	Hard Wheat				Soft Wheat			
	Grain	Dist. Grain	Cent. Solids	Stillage Soluble	Grain	Dist. Grain	Cent. Solids	Stillage Soluble
Aspartic Acid	5.7	7.1	6.5	4.5	6.0	7.8	7.6	4.9
Threonine	3.2	4.3	4.2	3.2	3.3	4.5	4.5	5.5
Serine	5.0	5.7	5.4	5.4	5.1	5.4	5.7	5.5
Glutamic Acid	33.2	26.2	29.4	38.0	31.0	22.9	27.2	35.1
Proline	11.2	9.8	10.9	13.5	10.9	8.5	9.3	13.9
Glycine	4.4	5.4	4.4	5.0	4.8	5.8	4.8	5.3
Alanine	3.8	5.2	4.5	3.4	4.3	5.7	5.2	3.8
Valine	4.8	5.4	6.1	3.5	4.5	6.0	5.9	3.6
Methionine	1.2	1.5	1.8	0.9	1.4	1.6	1.6	0.9
Isoleucine	3.4	3.8	4.8	2.8	3.7	4.0	5.0	2.8
Leucine	7.1	8.0	8.5	6.0	7.4	8.1	8.9	6.0
Tyrosine	3.5	3.7	4.0	3.3	3.7	3.7	4.2	3.3
Phenylalanine	4.7	5.1	5.4	4.9	4.9	4.9	5.5	5.0
Lysine	2.8	3.9	3.7	2.6	3.2	4.4	4.1	2.8
Histidine	2.4	2.7	2.4	2.8	2.5	2.9	2.5	2.5
Arginine	5.7	7.6	5.8	4.6	6.0	7.8	6.0	4.6

[a] g Amino acid per 16 g nitrogen recovered. Tryptophan and half-cystine not determined. From Wu et al 1984.

Inability to form a gluten ball is a good indication of denaturation. Vital wheat gluten formed cohesive and elastic gluten ball, but the devitalized gluten could not form a gluten ball. The hard and soft wheat flour centrifuged solids that were freeze-dried formed moderately cohesive gluten balls. Drying these centrifuged solids at 90°C, however, led to weak gluten balls with little cohesiveness.

A Mixogram of hard wheat flour was compared with those of hard wheat flour centrifuged solids freeze-dried and dried at 90°C. The freeze-dried centrifuged solids showed some dough-forming properties, but centrifuged solids dried at 90° had little dough-forming properties. The same trend was observed when soft wheat flour was compared with soft wheat flour centrifuged solids (freeze-dried and dried at 90°C). Freeze-dried centrifuged solids retained some functional properties for breadmaking based on gluten ball forming and mixogram studies. However, centrifuged solids dried at 90°C had lost their breadmaking potential.

CONCLUSION

Fat contents of wheat fermentation by-products were lower than those of corresponding fractions from corn, but the protein contents of hard wheat fermentation by-products were higher than those of the corresponding fractions from corn (Wu et al, 1981). The lower fat contents of wheat fermentation by-products may enhance storage stability of wheat by-products, as compared with those from corn. The lysine contents of wheat distillers' grains and centrifuged solids were higher than those of whole wheat and corn. Since lysine is the first limiting amino acid for wheat and corn, the higher lysine contents of the wheat by-products may lead to more nutritious products than those from corn. Soft wheat and its fermentation by-products have better essential amino acid compositions than hard wheat and its by-products. The crude fiber content of centrifuged solids from hard wheat flour was nsiderably lower than that of hard wheat stillers' grains; use of hard wheat flour ntrifuged solids may thus be desirable when crude ber contents of products such as baby food may be miting. On the other hand, the high dietary fiber ntents of wheat distillers' grains may be useful in gh-fiber food products. Wheat centrifuged solids ve both high protein and high dietary fiber ntents, suggesting application as a high protein, gh dietary fiber source.

LITERATURE CITED

BEVARI, T., ROULEAU, D., and MAYER, R. 1983. Production and quality of cheddar cheese manufactured from whole milk concentrated by reverse osmosis. J. Food Sci. 48:642.

ERICAN ASSOCIATION of CEREAL CHEMISTS. 1983. Approved Methods of the AACC. 8th ed. Methods 08-03, 30-26, and 32-15, approved April 1961; Method 46-13, approved October 1976. The Association: St. Paul, MN.

VINS, J. F., and FRIEDMAN, M. 1968. Automatic integration and computation of amino acid analyses. Cereal Chem. 45:172.

NLEY, J. W., and HANAMOTO, M. M. 1980. Milling and baking properties of dried brewers' spent grains. Cereal Chem. 57:166.

NNEY, K. F., and SHOGREN, M. D. 1972. A ten-gram mixograph for determining and predicting functional properties of wheat flours. Baker's Dig. 46:32.

RCIA, W. J., and WOLF, M. J. 1972. Polarimetric determination of starch in corn with dimethyl sulfoxide as a solvent. Cereal Chem. 49:298.

WHON, J. T., and LUSAS, E. W. 1984. New techniques in membrane processing of oilseeds. Food Technol. 38(12):97.

QUEEN, R. E., and NICHOLSON, J. W. G. 1979. Modification of the neutral-detergent fiber procedure for cereals and vegetables by using alpha-amylase. J. Assoc. Off. Anal. Chem. 62:676.

MORRIS, C. E. 1983. Huge plant for ethanol and HFCS Food Eng. 55(6):107.
NATIONAL ACADEMY OF SCIENCES. 1980. Recommende Dietary Allowances, 9th ed. The Academy Washington, DC.
NATIONAL RESEARCH COUNCIL. 1956. Composition o Concentrate By-Product Feeding Stuffs. Publ 449. Natl. Acad. Sci., Washington, DC.
PRENTICE, N. 1978. Brewers' spent grain in high fibe muffins. Baker's Dig. 52(10):22.
PRENTICE, N., KISSELL, L. T., LINDSAY, R. C., an YAMAZAKI, W. T. 1978. High fiber cookie containing brewers' spent grain. Cereal Chem 55:712.
SATTERLEE L. D., VAVAK, D. M., ABDUL-KADIR, R., an KENDRICK. J. G. 1976. The chemical, functional and nutritional characterization of protei concentrates from distillers' grains. Cerea Chem. 53:739.
SHEU, M. J., and WILEY, R. C. 1983. Preconcentratio of apple juice by reverse osmosis. J. Food Sci 48:422.
TSEN, C. C., EYESTONE, W., and WEBER, J. L. 1982 Evaluation of the quality of cookies supplemente with distillers' dried grain flours. J. Foc Sci. 47:684.
TSEN, C. C., WEBER, J. L., and EYESTONE, W. 198: Evaluation of distillers' dried grain flour as bread ingredient. Cereal Chem. 60:295.
WU, Y. V. 1987. Recovery of stillage soluble solic from hard and soft wheat by reverse osmosis ar ultrafiltration. Cereal Chem. 64:260.
WU, Y. V., SEXSON, K. R., and LAGODA, A. A. 198 Protein-rich residue from wheat alcoho distillation: fractionation a characterization. Cereal Chem. 61:423.
WU, Y. V., SEXSON, K. R., and WALL, J. S. 198 Protein-rich residue from corn alcoho distillation: fractionation a characterization. Cereal Chem. 58:343.
WU, Y. V., SEXSON, K. R., and WALL, J. S. 198 Reverse osmosis of soluble fraction of co stillage. Cereal Chem. 60:248.

40

THE DEVELOPMENT AND UTILIZATION OF WHEAT DISTILLERS'
GRAINS WITH SOLUBLES (DDGS)
AS A FOOD INGREDIENT

B. A. Rasco

Institute for Food Science and Technology
University of Washington, HF-10
Seattle, WA 98195

INTRODUCTION

Wheat distillers' grains or distillers' grains with solubles (DDGS) are produced by the enzymic and yeast fermentation of whole grains and have potential for use as a high fiber food ingredient. Until recently, the use of these materials as food ingredients has received little attention due to the limited availability of food grade product and poor product quality. In this review, the chemical composition, nutritional quality, and baking properties of DDGS from wheat are compared with other fiber ingredients such as brewer's grains produced from barley and DDGS from corn. The nutritive and sensory properties of these materials vary depending on the feedstock grain and the manufacturing process employed. The protein quality, level of purines, content of certain B vitamins and mineral nutrients for distillers' grain materials produced from different cereals are presented. Distillers' grains from wheat can be successfully incorporated at replacement levels for flour of up to 25% (w/w) in breading and batter mixes, and 30% (w/w) in baked goods such as breads, sweet rolls, and cookies.

The production of ethanol from agricultural products using yeast is one of the most important and best known industrial fermentations. Production statistics for ethanol from agricultural products are available back at least as far as 1936 (Boruff and van Lanen 1947). Corn is the most popular grain used commercially to manufacture fuel alcohol. A common process utilizing dry milled corn involves a two-stage enzymic conversion of starch to glucose followed by fermentation of the glucose with yeast. At this stage, the ethanol may be recovered from the whole mash, or the solids and liquid fraction (solubles) may be separated from each other by centrifugation with the ethanol recovered from the liquid fraction only. Often times, carbon dioxide is recovered as well. Manufacturers may dry the distillers' grain (DDG) and sell this as an ingredient for animal feeds. The soluble fraction can be concentrated, blended with dry DDG, and co-dried yielding distillers' dried grains with solubles (DDGS). Many smaller manufacturers recover the distillers' grains and ship them in a wet form to local feedlots or dairy farms. In these cases, the distillers' solubles are generally discarded. Similar processes are used for the manufacture of ethanol from other cereal grains.

Surplus wheat was used as a carbohydrate source for the manufacture of 700 million gallons of fuel alcohol in the United States during World War I because molasses was in short supply. A discussion of the different types of wheat and their performance as substrates in this fermentation is presented by Star et al (1943). In general, soft wheats were preferred as fermentation substrates because of higher starch content. At the present time, wheat is not an economical choice as a fermentation substrate unless the DDG or DDGS can be sold at a higher price as a human food ingredient rather than as a component of animal feed (Rasco et al 1987a, b, c; Ranhotra et al 1982).

Research conducted over the past 10 years has concentrated on improving process technology and determining the chemical characteristics of distillers' grain materials, principally from corn or barley. Research into the nutritive properties of these materials has emphasized protein quality and food safety concerns with the intent of utilizing these ingredients in blended foods for export (Bookwalter et al 1988, 1984).

THE CHEMICAL COMPOSITION OF DISTILLERS' GRAIN MATERIALS

A major difficulty with evaluating much of the composition and nutritional data for distillers' grain materials is that information on the processing conditions used for manufacture and an adequate description of the fermentation feedstock are often not available. A compilation of compositional data for distillers' grains from wheat, corn and sorghum; protein concentrates and brewers' spent grains are presented in Table I.

Corn distillers' dried grains with solubles (C-DDGS) contains from 23 to 30% protein and 6 to 16% crude fiber (Wall et al 1984, Wu and Stringfellow, 1982, Wu et al 1981) and 32 to 36% dietary fiber (Dong and Rasco 1987, San Buenaventura et al 1987). Corn protein concentrate (CPC) is prepared by treating the residual solids remaining after the yeast fermentation of whole milled grain with sodium hydroxide and recovering the solids after alkaline treatment. An additional treatment involves treating the solids with alcohol following the alkaline treatment to increase the protein content (Satterlee et al 1976). Corn protein concentrate contains 50 to 81% protein. The protein level of milo distillers' grains range from 35 to 45%, and total dietary fiber from 29 to 70% (San Buenaventura et al 1987, Morad et al 1984, Wu and Jackson 1984).

Barley DDG and DDGS contain approximately 32% protein and from 53 to 85% total dietary fiber (San Buenaventura et al 1987, Weber and Chaudhary 1987, Wu 1985). Brewers' spent grains (BSG) contain up to 24% protein and approximately 54% total dietary fiber (Prentice and D'Appolonia 1977). The relative concentration of protein in the dried brewer's grains can be altered by dry milling and fractionation (Weber and Chaudhary 1987, Kissell and Prentice 1979). Fractionated BSG flours contain from 21 to 27% protein, and 26 to 76% dietary fiber.

TABLE I

Chemical Composition of Whole Grains, Distillers' Grains and Related Products
(% dry weight basis)

Product	Protein	Ash	Lipid	Crude Fiber	Dietary Fiber
White Wheat[abc]	6.8-14.3	1.5-1.8	1.5-1.7	2.1-2.9	10.5-1
White Wheat DDGS[abc]	19.6-38.4	2.2-8.4	3.7-6.7	7.6-18.9	33.9-3
Red Wheat[abc]	14.1-15.7	1.5-1.8	1.3-1.8	2.6-3.4	9.9
Red Wheat DDGS[abc]	29.2-33.9	2.4-6.8	2.0-6.0	6.9-19.9	34.8
Wheat Protein Concentrate[d]	59.8-87.9	0.8-1.0	2.2-10.3	ND	ND
Corn[ab]	7.4	1.3	3.7	2.1	12.1
Corn DDGS[abde]	23.0-30.0	2.4-10.1	9.0-12.6	6.3-16.1	32.0-3
Corn Protein Concentrate[de]	50.5-81.1	2.6-5.2	3.4-26.8	2.1-7.7	ND
Milo[bf]	10.7	1.5	3.2	2.2	10.1
Milo DDGS[bg]	34.9-45.3	2.1-4.6	6.3-12.3	8.5-11.6	29.1-6
Barley[bh]	13.4	2.6	2.2	3.9	21.7
Barley DDGS[bh]	32.6	4.4	6.0	16.6	84.7
Barley Bran Flour[i]	18.5	4.6	6.8	ND	70.0
Barley High Protein Flour[i]	35.5	4.6	8.7	ND	35.5
Barley Protein Concentrate[j]	54.3	4.1	8.4	3.5	ND
Brewer's Grains[ikm]	24.5-28.3	2.4-4.4	7.7-10.9	6.9-11.2	53.5
Brewer's Condensed Solubles[l]	8.9	2.5	1.4	2.1	ND

DDGS = Distillers' Dried Grains with Solubles
ND = not determined.

a Data from Rasco et al (1987a).
b Data from San Buenaventura et al (1987).
c Data from Wu et al (1984).
d Data from Satterlee et al (1976).
e Data from Wall et al (1984).
f Data from Wu and Sexson (1984).
g Data from Ranhotra et al (1982).
h Data from Wu (1985).
i Data from Weber and Chaudhary (1987).
j Data from Finley et al (1976).
k Data from Tsen et al (1982).
l Data from Sebree et al (1983).
m Data from Prentice and D'Appolonia (1977).

Wheat DDGS and products prepared by fractionation of dried wheat distillers' grains have protein contents ranging from 29 to 59% (Rasco et al 1987a, Wu et al 1984). The dietary fiber content of these materials ranges from 40 to 55% (Dong and Rasco 1987, San Buenaventura et al 1987). Wheat protein concentrate (WPC) contains approximately 88% protein (Satterlee et al 1976).

NUTRITIONAL QUALITY OF DISTILLERS' GRAINS FROM WHEAT

The utilization of the residue remaining after fermentation has received little attention until recently for uses other than animal feeds. Waelti and Ebeling (1982) determined the nutritional value of distillers' dried grains, distillers' wet grains and thin solubles from corn in rations for dairy and beef cattle, poultry and swine. Recent nutritional studies are concentrated on characterizing distillers' grain materials and related products made from a number of different feedstocks for their potential as ingredients in human food. Bookwalter et al (1983) predicted that the nutritional quality of distillers' grain from soft wheat may be slightly better than that of distillers' grain products made from hard wheat or corn.

The essential amino acid content of whole grains, DDGS and brewer's grain products, wheat and corn protein concentrates, and brewer's condensed solubles are given in Table II. The concentration of essential amino acids (g/100 g protein) are similar for whole wheat, wheat protein concentrate (WPC) and wheat DDGS. The level of phenylalanine was higher in the corn DDGS and corn protein concentrate (CPC) than whole corn. The content of essential amino acids in barley spent grains and the spent grains (distillery byproducts) were variable. These materials contained less valine, isoleucine and arginine; more methionine, leucine and phenylalanine than wheat DDGS; and comparable levels of lysine to wheat DDGS. Lysine was the limiting essential amino acid in the wheat distillers' grain products (Table II). Lysinoalanine levels were less than the detection limit (0.04% w/w) in distillers' grains made from white or red wheat (Dong et al 1987).

A unique feature of the DDGS and protein concentrates (CPC and WPC) are that the protein content is high enough to allow the material to be incorporated into a 10% protein diet for PER and NPR assays. Wheat flour has often been assayed at a 6% protein level (Satterlee et al 1976). Estimation of the protein quality of distillers' grain products and protein concentrates from wheat and corn as adjusted PER (protein

TABLE II

Essential Amino Acid Content of Whole Grains, Distillers' Grains and Related Products
(g/100 g protein)

Amino Acid	Ref. Casein[a]	White Wheat[a]	White Wheat DDGS[a,b]	Red Wheat[a]	Red Wheat DDGS[a,b]	WPC[c]	Corn[a]	Corn DDGS[a]	CPC[c]	Distillers' Spent Grains[d]	Barley Flour[e]	BCS[f]
Threonine	4.1	3.0	3.0-4.5	2.96	3.0-4.3	3.65	3.44	3.69	2.65	3.67- 4.13	2.96	3.34
Valine	6.4	4.5	5.0-6.0	4.25	4.9-5.4	5.14	3.10	5.04	5.58	2.41- 2.59	2.59	4.35
Methionine	2.4	1.3	1.5-1.6	1.26	1.5-2.0	2.15	1.85	2.00	1.74	2.87- 4.58	1.55	1.45
Isoleucine	4.9	3.6	3.9-4.0	3.60	3.7-3.8	3.23	1.92	3.85	3.94	1.76- 1.92	2.03	2.66
Leucine	9.0	6.9	7.6-8.1	6.93	7.1-8.1	9.64	11.89	13.50	12.03	10.22-11.47	7.73	5.93
Phenyl-alanine	5.0	4.6	4.9-5.2	4.78	4.8-5.1	4.14	4.35	5.21	5.23	5.13- 6.56	4.52	4.26
Histidine	2.9	2.5	2.7-2.9	2.45	2.4-2.7	2.68	3.98	2.89	2.13	3.30- 4.01	1.60	4.71
Lysine	7.9	3.0	2.8-4.4	2.79	2.5-3.9	3.09	2.32	2.26	3.47	2.50- 3.74	2.41	3.28
Arginine	3.9	5.0	5.6-7.8	4.96	4.8-7.6	4.92	4.78	4.36	4.61	2.67- 3.38	4.02	4.80
Tryptophan	1.1	1.0	1.07	0.90	0.9	1.59	0.97	0.49	0.65	ND	ND	1.50

[a] Data from Dong et al (1987). Refer to text for description of DDGS materials (distillers' dried grains with solubles).

[b] Data from Wu et al (1984). Distillers' grains from soft white winter wheat.

[c] Data from Satterlee et al (1976). Wheat protein concentrate (WPC) and corn protein concentrate (CPC). For more complete descriptions, refer to text.

[d] Data from Ranhotra et al (1982). Distillers' grains made from corn, barley, rye and/or milo.

[e] Data from Weber and Chaudhary (1987). Barley high protein flour, a byproduct of beer manufacture.

[f] Data from Sebree et al (1983). Solubles from distillers' grains made from barley (brewer's condensed solubles).

fficiency ratio) and in vitro protein digestibility are presented in Table III. The adjusted PER of wheat DGS was substantially lower (Dong et al 1987) than that reported for WPC by Satterlee et al (1976) even though the apparent protein digestibilities for the wheat DDGS and WPC in these two studies were comparable. The in vitro protein digestibility for the wheat DDGS is somewhat lower than for whole wheat suggesting that the fermentation or drying process makes the material less susceptible to proteolysis by the digestive enzymes used in the standard in vitro digestibility assay.

When male weanling rats were fed a diet containing white wheat DDGS plus supplemental essential amino acids, the animals exhibited significantly higher growth rates and food intake levels than a comparable group of animals fed an unfortified diet. The growth rate and feed intake for the animals fed the amino acid fortified diet approached that of the control group fed a casein diet (Dong et al 1987). Lack of essential amino acids, particularly lysine, rather than the presence of antinutritional factors, appeared to be the major reason for poor growth in animals fed DGS as a sole source of dietary protein.

The in vitro digestibility for corn DDGS was lower than for CPC and whole corn. The adjusted PER for the corn DDGS was lower than for corn. The PER values for CPC were highly variable with the lowest PER values for certain commercially manufactured products. The adjusted PER value for spent grain materials from other feedstocks were also variable.

Although whole grains contain significant levels of thiamin and riboflavin, little information is available on the vitamin content or effects of processing on the level of vitamins in distillers' grain products or similar materials. According to Keagy et al (1980), cereal products provide 41% of the thiamin and 21% of the riboflavin to the US diet, primarily through the enrichment of processed grain products. The content of B vitamins in wheat and the variability in vitamin content among several types and cultivars of wheat from different locales (Davis et al 1984, Keagy et al 1980) have been reported. The thiamin and

TABLE III
Protein Quality Assessment of Whole Grains and Distillers' Grain Products

Product	PER (adjusted)	Protein Digestibility (in vitro, %)
Wheat, Whole[a]	1.44	78.0
White Wheat, Whole[b]	ND	88.9
Red Wheat, Whole[b]	ND	86.6
Wheat Protein Concentrate[a]	1.26	82.8
Distillers' Grains w/ Solubles (DDGS) White Wheat[b]	0.2	79.9
DDGS, Red Wheat[b]	0.6	81.0
Corn[a]	1.40	78.3
Corn Protein Concentrate[a,c]	0.14–1.45	78.0–87.4
DDGS, Corn[b]	0.98–1.00	64.8–75.9[d]
Distillers' Grains[e]	0.50–1.66	ND

[a] Data from Satterlee et al (1976). For description of wheat and corn protein concentrates refer to text.

[b] Data from Dong et al (1987).

[c] Data from Wall et al (1984). Values are for commercial products.

[d] *In vivo* digestibility measurement using male Sprague Dawley weanling rats. (N intake-fecal N)/N intake x 100%.

[e] Data from Ranhotra et al (1982). Distillers' grains made from corn, barley, rye and/or milo.

ND = not determined.

riboflavin content of whole grains and distillers grain products, and the soluble solids recovered after centrifugation of the whole mash which had been concentrated by evaporation, are presented in Table IV.

There is a large loss of thiamin in wheat DDG compared to the level in whole wheat. If no losses resulted from processing, the level of vitamin should increase by roughly threefold, since approximately 3

TABLE IV

Thiamin and Riboflavin Content of Whole Grains and
Distillers' Grain Products (mg/100 g, dwb)

Product	Thiamin	Riboflavin
US, Hard Red Winter Wheat[a]	0.42-0.52	0.15-0.19
US, Soft White Winter Wheat[a]	0.46-0.50	0.11-0.12
Hard Wheat Flour[b]	0.162	0.047
Soft Wheat Flour[b]	0.168	0.043
Distillers' Dried Grains w/ Solubles (White Wheat)	0.09-0.19	0.17-0.50
Washed Distillers' Grains (White Wheat)	0.05-0.11	0.14-0.28
Distillers' Grains[c]	0.19-0.61	0.36-0.62
Brewer's Condensed Solubles[d]	0.58	1.28
Condensed Solubles from White Wheat Distillers' Grains	0.26	0.82

[a] Data from Davis et al (1984).

[b] Data from Keagy et al (1980).

[c] Data from Ranhotra et al (1982). Distillers' grains made from corn, barley, rye and/or milo.

[d] Data from Sebree et al (1983). Solubles from distillers' grains manufactured from barley.

of whole grain are required to produce 1 Kg of DDGS (...sco et al 1987a). The level of riboflavin in wheat ...S is similar to or greater than the level in the ...le grains. The thiamin and riboflavin contents of ...tillers' spent grains were variable, but tended to ... higher than for wheat DDGS. Wheat DDGS contained ...parable levels of riboflavin to enriched wheat ...ur (260 g/100 g) but less thiamin (440 g/100 g) ...nnington and Church 1985).

Removal of the soluble and suspended solids from the distillers' grain products led to a lower recovery of vitamin activity (Table IV). Removal of the soluble solids by washing or solvent extraction has been advocated as a means to improve the sensory qualities of distillers' grain products (Bookwalter et al 1988, 1984); however, removal of the soluble solids would reduce the thiamin and riboflavin content of these materials by an additional 25 to 50%.

Wheat distillers' grain products provide higher levels of calcium, iron, and zinc than whole wheat (Table V). Wheat protein concentrate had less of these three minerals than white wheat DDGS, but more iron than whole wheat flour or white wheat bran. Corn protein concentrate and WPC had comparable levels of these three mineral nutrients. The barley bran flour and barley high protein flour contained higher levels of calcium than wheat DDGS. The soluble mineral content for the distillers' grain materials is high as indicated by the data for the concentrated solubles. The availability of calcium, iron, and zinc should be somewhat higher than for wheat bran and possibly whole wheat flour. The level of phytic acid in DDGS would be substantially lower than in wheat bran and whole wheat flour due to enzymic degradation during fermentation.

Distillers' grain materials are a relatively high purine food, not only because of the presence of yeast, but also because the whole grain cereals from which they are made contain moderately high levels 50 to 150 mg purine/100 g (Pennington and Church 1985). A high intake of purines can lead to hyperurinemia, a condition which is associated with a number of different diseases. Values for nucleic acid nitrogen (NAN) for distillers' active dry yeast and other food yeasts range from 11 to 14% (Brule et al 1988). The purine content of distillers grain products from wheat ranges from 100 to 180 mg/100 g (dwb). The RNA content of wheat protein concentrate and corn protein concentrates are 90 mg/100 g and mg/100 g respectively (Satterlee et al 1976).

TABLE V

Calcium, Iron and Zinc Content of Whole Grains, Bran and Distillers' Grain Products (µg/g dwb)

Product	Calcium	Iron	Zinc
White Wheat, Whole	186	13.5	33
White Wheat Bran	42	46	127
Distillers' Grains w/ Solubles from Wheat (DDGS)	620	56	105
Concentrated Solubles from DDGS	2200	44	97
Wheat Protein Concentrate[a]	120	70	20
Barley, Whole	91	27.4	130
Barley, Bran Flour[b]	5648	164	90
Barley, High Protein Flour[b]	2690	171	69
Distillers' Grains[d]	490-830	7.7-29.9	59.4-69.8
Brewer's Concentrated Solubles[c]	1539	85	12.6
Corn Protein Concentrate[a]	110	70	30

Data from Weber and Chaudhary (1987). Bran and high protein flours recovered from beer manufacture.

Data from Sebree et al (1983). Solubles from distillers' grains and fractions from barley.

Data from Ranhotra et al (1982). Distillers' grains made from corn, barley, rye and/or milo.

SENSORY QUALITY OF DISTILLERS' GRAINS WITH SOLUBLES FROM WHEAT

Distillers' grain from a number of feedstock ains including wheat (Rasco et al 1987a, b, c), rghum (Morad et al 1984), barley (Dawson et al 1987, 35, 1984) and corn or corn based (Reddy et al 1986a, Wampler and Gould 1984, Tsen et al 1983, 1982, ssell and Prentice, 1979, Prentice and D'Appolonia 77) have been used as ingredients in baked goods and er foods to enhance the protein and dietary fiber tent. Sensory attributes of DDGS and DDG are strongly dependent upon the manufacturing process us (Rasco 1988). Tsen and coworkers (1982, 1983) not that the acceptability of distillers' grain produc from corn varied widely among commercial supplie depending on how the materials were processe Efforts have been made to increase the protein conte of distillers' grain materials and similar products fractionation without taking into consideration t effect these treatments have on the appearanc functionality, or flavor of the ingredient.

One readily quantifiable index that correlat well with sensory acceptability of these products color. For example, the color of sorghum DD products, DDGS from blended grains, and co (Bookwalter et al 1988, 1984, Rasco et al 1987b, vary significantly in color depending on the proces ing conditions. Drying is a primary factor determining the quality of the distillers' gra products and has been shown to have a marked effect product color as well as the sensory acceptability products made containing distillers' grains (Rasco al 1988, 1987b, c, Bookwalter et al 1984, Tsen et 1982, 1983). Distillers' grain products from barl and milo are darker than those from red or whi wheat. Corn DDGS has a lightness comparable to whe DDGS, but is more difficult to incorporate in formulations because of the distinctive yellow col and corn flavor which are difficult to mask. T color of blended flours containing white wheat DDGS 15 and 30% w/w replacement levels are similar to who wheat flour.

Partial replacement of all-purpose flour wi distillers' dried grain flour in light colored cooki was found to be unacceptable at 15 or 25% replaceme levels because of dark color (Tsen et al 1982 Distillers' dried grains from white sorghum have be successfully incorporated at flour replacement leve of 25% in molasses cookies (Morad et al 1984). The was limited success using darker varieties of sorgh in similar formulations due to dark color.

Wu et al (1984) and Bookwalter et al (198 1984) make reference to the distinct odor and taste DDG from corn and the detrimental effect of the flavor defects on the acceptability of formulated foods. Solvent extraction (Bookwalter et al 1988) and neutral water rinses were used (Bookwalter et al 1984) to remove components associated with the solubles fraction in dried distillers' grains from corn in an attempt to improve the sensory properties. Components thought to be responsible for the off flavors in these materials include oxidized or hydrolyzed lipid, and fermentation byproducts.

Changes in the lipid profiles of distillers' grains from barley during the manufacturing process have been reported (O'Palka et al 1987, Dawson et al 1987, 1984) with major changes in the lipid constituents resulting from lipid hydrolysis during liquefaction and distillation, and the loss of unsaturated fatty acids during fermentation via oxidative reactions. Lipid hydrolysis also occurs during drying of distillers' grain products made from wheat (Rasco 1988). The incorporation of antioxidants into the solids recovered from the whole mash following fermentation but prior to drying appears to do little to improve lipid stability during storage.

Removal of the lipid and bleaching with chemical agents have also been employed to improve either the color or flavor of distillers' grains produced from barley (O'Palka et al 1987, Dawson et al 1987, 1984). Dawson et al (1984) found that defatting distillers' dried grain produced from barley significantly improved the quality of DDG and bleached DDG. When delipidated materials were used at substitution levels of 15% in oatmeal cookies, the experimental products were as acceptable as the controls which contained none of the fiber ingredient.

White bread containing 6% brewer's grains flour had acceptable break and shred, grain and texture, but a slightly depressed loaf volume. A 12% substitution of brewers' spent grains for flour yielded an unacceptable product (Finley and Hanamoto 1980); decreased loaf volume and increased water absorption of brewers' grains or bran fractions from brewer's grains have been reported by others (Morad et al 1984, Reese and Hoseney 1982, Prentice and D'Appolonia 1977, Pomeranz et al 1977). Table VI provides data comparing the mixing properties of white wheat DDGS with other fiber ingredients at a 15% replacement level (w/w) in all-purpose flour doughs. Water absorption of DDGS doughs was less than in doughs containing wheat bran. Grinding distillers' grain products tended to improve crumb grain and loaf volume.

TABLE VI

Mixing Properties of Doughs Containing Wheat Distillers' Grains and Other Fiber Ingredients
(15% w/w replacement, as is basis)

Product	Absorption (%)	Development Time (min.)	Stability Time (min.)
Flour, Control	75	6	10.5
White Wheat DDGS	71	7.5	7.5
White Wheat Bran	79	7	10
Oat Bran	70	6	7
Barley Fiber (spent grains)	69	4	4.5

Ingredients:
All purpose flour (Gold Medal, Pillsbury, Co., Mnpls., MN); white wheat distillers' dried grains with solubles (DDGS) prepared according to procedure outlined in Rasco et al (1987a); white wheat bran (Fisher Mills, Harbor Island, WA); oat bran (Better Basics Oat Fiber, No. 757 P.D. Williamson and Co. Inc., Modesto, CA); barley fiber (Commercial Barley Fiber, MGF 200, Coors Food Products Co., Golden, CO).

The most complete evaluation of wheat distillers' grain products as food ingredients to date has been conducted in our laboratory. Chocolate chip cookies and banana bread containing 30% (w/w) substitution of white wheat DDGS for flour were rated as good to excellent by a consumer sensory panel; white and whole wheat breads containing 30% DDGS were rated as acceptable to good (Rasco et al 1987b). Deep-fried fish nuggets coated with a batter mix containing 25% DDGS from either corn, red or white wheat were rated favorably by a consumer panel (Rasco et al 1987c). For this particular application, th salty or yeast-like flavor of DDGS was a positive flavor attribute. In recent studies, we have found white wheat DDGS can be incorporated at a 25% substitution level in sweet roll or cookie dough, yielding products that were highly acceptable and which were rated as similar to the control (no DDGS) in forced preference tests.

In conclusion, distillers' grains from wheat contain relatively high levels of protein and dietary fiber. The protein quality is similar to that of other cereal products. Lysine is the limiting essential amino acid in wheat DDGS, and lack of essential amino acids rather than the presence of antinutritional factors appeared to be the reason for reduced growth in rats fed a diet containing DDGS as the sole source of dietary protein. The level of thiamin is lower in DDGS than in unenriched wheat flour, but the level of riboflavin is comparable. Wheat DDGS can be incorporated into baked goods formulations at higher replacement levels than previously described for most other distillers' grain products.

LITERATURE CITED

BOOKWALTER, G.N., KWOLEK, W.F., WALL, J.S., WARNER, K.A., WU, Y.V. and GUMBMANN, M.R. 1983. Investigation on the Use of Distiller's Grains or Fractions Thereof in Blended Foods for the Foods for Peace Program and Other Food Applications. Final Report. USDA. Agricultural Research Service, Northern Regional Research Center, Peoria, IL.

BOOKWALTER, G.N., WARNER, K., WALL, J.S. and WU, Y.V. 1984. Corn distillers' dried grains and other byproducts of alcohol production in blended foods. II. Sensory stability and processing studies. Cereal Chem. 61:509.

BOOKWALTER, G.N., WARNER, K. and WU, Y.V. 1988. Processing corn distillers' grains to improve flavor: Storage stability in corn-soy-milk blends. J. Food Sci. 53:623.

BORUFF, C.S. and VAN LANEN, J.M. 1947. Th fermentation industry during World War II. Ind Eng. Chem. 39:934.

BRULE, D., SAWAR, G. and SA VOIE, L. 1988. Purir content of selected Canadian food products. J Food Comp. and Anal. 1:130.

DAVIS, K.R., PETERS, L.J. and LE TOURNEAU, D. 1984 Variability of the vitamin content in wheat Cereal Foods World 29:364.

DAWSON, K.L., EIDET, I., O'PALKA, J. and JACKSON, L.L 1987. Barley neutral lipid changes during th fuel ethanol production process and produc acceptability from the dried distillers' grains J. Food Sci. 52:1348.

DAWSON, K.R., NEWMAN, R.K. and O'PALKA, J. 198! Effects of bleaching and defatting on barle distillers' grains used in muffins. Cerea Research Comm. 13:387.

DAWSON, K.R., O'PALKA, J., HETHER, N.W., JACKSON, L.I and GRAS, P.W. 1984. Taste panel preferen correlated with lipid composition of barl dried distillers' grains. J. Food Sci. 49:787

DONG, F.M. and RASCO, B.A. 1987. The neutra detergent, acid detergent, crude fiber a lignin content of distillers' dried grains wi solubles. J. Food Sci. 52:403.

DONG, F.M., RASCO, B.A. and GAZZAZ, S.S. 1987. protein quality assessment of wheat and co distillers' grains with solubles. Cereal Che 64:327.

DREESE, P.C. and HOSENEY, R.C. 1982. Baki properties of bran fraction from brewer's spe grains. Cereal Chem. 59:89.

FINLEY, J.W. and HANAMOTO, M.M. 1980. Milling a baking properties of dried brewers' spe grains. Cereal Chem. 57:166.

FINLEY, J.W., WALKER, C.E. and HAUTALA, E. 197 Utilization of press water from brewer's spe grains. J. Sci. Food. Agric. 27:655.

KEAGY, P.M., BORNSTEIN, H.B., RANUM, P., CONNER, M.A., LORENZ, L., HOBBS, W.E., HILL, G., BACHMAN, A.L., BOYD, W.A. and KULP, K. 1980. Natural levels of nutrients in commercially milled wheat flours. Vitamin analysis. Cereal Chem. 57:59.

KISSELL, L.T.I. and PRENTICE, N. 1979. Protein and fiber enrichment of cookie flour with brewer's spent grain. Cereal Chem. 56:261.

MORAD, M.M., DOHERTY, C.A. and ROONEY, L.W. 1984. Utilization of dried distillers' grains from sorghum in baked food systems. Cereal Chem. 61:409.

PALKA, J., EIDET, I. and JACKSON, L.L. 1987. Neutral lipids traced through the beverage alcohol production process. J. Food Sci. 52:515.

PENNINGTON, J.A.T. and CHURCH, H.N. 1985. Food Values of Portions Commonly Used. 14th Ed. J.B. Lippincott, Co. Philadelphia, PA.

POMERANZ, Y., SHOGREN, M.D., FINNEY, K.F. and BECHTEL, D.B. 1977. Fiber in breadmaking--Effects on functional properties. Cereal Chem. 54:25.

PRENTICE, N. and D'APPOLONIA, B.L. 1977. High fiber bread containing brewers' spent grains. Cereal Chem. 54:1084.

MINHOTRA, G.S., GELROTH, J.A., TORRENCE, F.A., BOCK, M.A., WINTERRINGER, G.L. and BATES, L.S. 1982. Nutritional characteristics of distillers' spent grain. J. Food Sci. 47:1184.

RASCO, B.A. 1988. Stability of lipids in distillers' dried grain products made from soft white winter wheat. Cereal Chem. 65:161.

RASCO, B.A., BORHAN, M. and OWUSU-ANSAH, Y. 1989. Effect of drying technique and incorporation of soluble solids on the chemical composition and color of distillers' grain products. Cereal Foods World. In press.

RASCO, B.A., DONG, F.M., HASHISAKA, A.E., GAZZAZ, S.S., DOWNEY, S.E. and SAN BUENAVENTURA, M.L. 1987a. Chemical composition of distillers' dried grains with solubles (DDGS) from soft white wheat, hard red wheat and corn. J. Food Sci. 52:236.

RASCO, B.A., DOWNEY, S.E., and DONG, F.M. 1987 Consumer acceptability of baked goods containi: distillers' dried grains with solubles from so white winter wheat. Cereal Chem. 64:139.

RASCO, B.A., DOWNEY, S.E., DONG, F.M. and OSTRANDE] J. 1987c. Consumer acceptability of color deep-fried fish batter made with distiller: dried grains with solubles (DDGS) from wheat a corn. J. Food Sci. 52:1506.

REDDY, N.R, COOLER, F.W. and PIERSON, M.D. 1986. Sensory evaluation of canned meat-based foo supplemented with dried distiller's grain flou J. Food Quality 9:233.

REDDY, N.R., PIERSON, M.D. and COOLER, F.W. 1986 Supplementation of wheat muffins with dri distillers' grain flour. J. Food Quality 9:24.

SAN BUENAVENTURA, M.L., DONG, F.M. and RASCO, B. 1987. The total dietary fiber content distillers' dried grains with solubles. Cere Chem. 64:135.

SATTERLEE, L.D., VAVAK, D.M., ABDUL-KADIR, R. a KENDRICK, J.G. 1976. The chemical, function and nutritional characterization of prote: concentrates from distillers' grains. Cere Chem. 53:739.

SEBREE, B.R., CHUNG, D.S. and SEIB, P.A. 198: Brewer's condensed solubles. I. Composition a physical properties. Cereal Chem. 60:147.

STARK, W.H., KOLACHOV, P. and WILKIE, J.F. 194: Wheat as a raw material for alcohol production Ind. Eng. Chem. 35:133.

TSEN, C.C., WEBER, J.L. and EYESTONE, W. 1983. Evalu ation of distillers' dried grain flour as bread ingredient. Cereal Chem. 60:295.

TSEN, C.C., EYESTONE, W. and WEBER, J.L. 1982. Evalu ation of the quality of cookies supplemente with distillers' dried grain flour. J. Fo Sci. 47:684.

WAELTI, H. and EBELING, J.N. 1982. Fuel alcohol distillers' dried grains--nutritional value Cooperative Extension Services, Washington Stat University, Pullman, WA.

L, J.S., WU, Y.V., KWOLEK, W.F., BOOKWALTER, G.N. and WARNER, K. 1984. Corn distillers' grains and other byproducts of alcohol production in blended foods. I. Compositional and nutritional studies. Cereal Chem. 61:504.

MPLER, D.J. and GOULD, W.A. 1984. Utilization of distillers' spent grain in extrusion processed doughs. J. Food Sci. 49:1321.

BER, F.E. and CHAUDHARY, V.K. 1987. Recovery and nutritional evaluation of dietary fiber ingredients from a barley by-product. Cereal Foods World 32:548.

, Y.V. 1985. Fractionation and characterization of protein-rich material from barley after alcohol distillation. Cereal Foods World 30:540.

, Y.V. and SEXSON, K.R. 1984. Fractionation and characterization of protein-rich material from sorghum alcohol distillation. Cereal Chem. 61:388.

, Y.V., SEXSON, K.R. and LAGODA, A.A. 1984. Protein-rich residue from wheat alcohol distillation. Fractionation and characterization. Cereal Chem. 61:423.

, Y.V., SEXSON, K.R. and WALL, J.S. 1981. Protein-rich residue from corn alcohol distillation: Fractionation and characterization. Cereal Chem. 50:343.

, Y.V. and STRINGFELLOW, A.C. 1982. Corn distillers' dried grains with solubles and corn distillers' dried grains: Dry fractionation and composition. J. Food Sci. 47:1155.

41

WHEAT STARCH IN THE FORMULATION OF DEGRADABLE PLASTICS

Gerald J.L.Griffin
Ecological Materials Research Institute
Epron Industries
Ketton Business Estate
Stamford, UK PE9 3SZ

INTRODUCTION

The public and political interest in degradable plastics is now so strong around the world that I will take their desirability as established for the purpose of this paper. Because the applications of plastics classified as transient, or "one trip", call for materials of the lowest achievable cost the many complex chemical approaches possible to producing degradable polymers are ruled out despite their potential suitability for medical and specialist applications. The Author's interest in the subject dates from his observation that common starches are compatible fillers for most thermoplastics provided that strict control is exercised over quality and moisture content (see Griffin, 1973, 1975, 1987) and a program of work on SFP (starch filled plastics) has continued since that date. Since 1973 there have been many developments in packaging technology most of which have resulted in a reduction in the thickness of the plastics films used in order to reduce costs. The original SFP work was commercialised using maize starch because its 15 μm particle size was acceptable in 50 μm plastics films but it is now necessary to produce films as thin as 12 μm and, although many vegetable starches have particle sizes small enough for this application, they are specialist products currently too costly to be considered. The cost per

tonne of the common commercial starches is much low
than the cost of the synthetic polymers most wide
used for packaging applications and the price gap
generally sufficient to cover most of the costs of t
drying, handling, and compounding operations needed
introduce the starch as a filler into the polyme
This has not been a constant ratio, as reference
Fig. 2   makes clear, because the cost of synthet
polymers is very closely linked to the cost of t
mineral oil from which they are derived and this
given to rather startling fluctuations over the year
         The unique geometry and granulometry
wheat   starch   means   that   it   deserves speci
examination as a potential filler for these th
films.

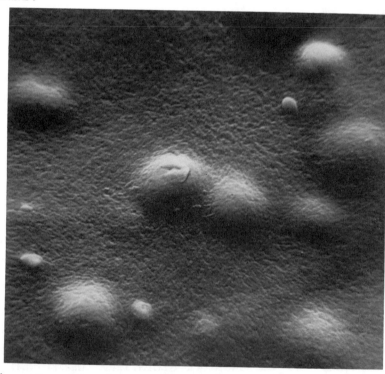

Fig. 1. SEM of the surface of 50µm polyethylene fi
containing 6% of maize starch. Each bump
attributable to a single 15µm starch particle benea
the surface

FILLERS IN POLYMER FILMS

The thin films of the plastics packaging industry are mostly made from polyolefines converted into film by the process of extrusion blowing. In this operation the polymer melt at a temperature of about 190° Centigrade is extruded by very high pressure, about 200 Bar, through an annular die into the form of a tube which is immediately expanded by inflation and drawn by traction so as to reduce its wall thickness to the desired value before it solidifies as the result of air cooling. Obviously any particulate material suspended in the melt will not deform along with the polymer and, if the particles are large by comparison with the final film thickness, then the surface of the product will be deformed. The usual particulates to be found in such films are mineral pigments of diameter less than 1 µm and concentration less than 2 weight percent. Because the SG (specific gravity) of these pigments is about 3, the effective volume fraction is extremely small and the effect on the extrusion process and the physical form of the product film is difficult to detect. The addition of maize starch, of SG between 1.25 and 1.5 and mean diameter 15 µm, at weight percentages between 5 and 6%, has a major effect on the surface texture of the film as is seen in Fig.1 which is a 50 µm film. If the film thickness is progressively reduced the manufacturing process becomes difficult to maintain at below 35 µm because only one oversize particle breaking through the film is needed to collapse the flimsy tube. More than 500 million shopping bags have been successfully made in the UK containing 6% maize starch in 50 µm film, but only trial amounts of thinner films have been achieved.

Fine Particle Starches

The Author has prepared extrusion blown polyolefine films of extreme thinness using cowcockle starch (Saponaria vaccaria), taro starch (Colocasia esculentia, various cultivars), rice starch (Oryza sativa), and amaranth (Amaranthus cruentus). O these only rice starch can be considered as a commo commercial material and is, even so, at least twice a expensive as maize starch. It is obviously possibl to promote the development of the exotic starches an there is considerable incentive to do so because th particle sizes available extend down to 1 μm and th product film is much stronger than can be produce even with rice starch at 5 μm. Such agricultura programs, however, have time scales measured in year and the need for thin degradable polymer films i urgent.

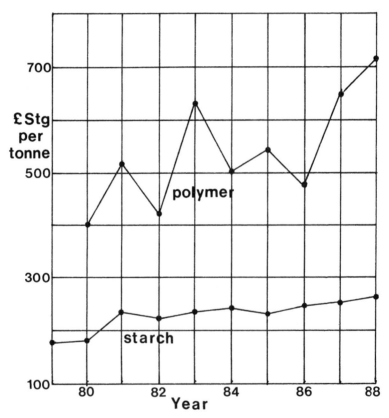

Fig. 2. Plot comparing the cost per tonne of lc density polyethylene with that of wheat starch ove the period 1979 to 1988. Prices, in pounds Sterlin per tonne, from various European commercial sources.

Bimodal Size Distribution Starches

The search for sub 10 μm starch particles raises the question of obtaining such material by classification of starches known to contain a wide size distribution. Curiously most starches have a particle size distribution that peaks narrowly around a single diameter value. Amongst the common exceptions, potato, barley, and wheat, we find that the size distribution is not simply broader but is truly bimodal. Potato starch (Fig. 3) is a very clear example and can be air classified with little difficulty. Unfortunately the small size fraction is still too large to be attractive for application in plastics films. Barley starch has a significant, if very variable, small particle fraction but the separation process is unexpectedly difficult and the starch itself is rather expensive. Wheat starch, on the other hand, is comparable in price with maize and has a further remarkable distinction in that the larger particles are not spherical or facetted but have a form approximating to the oblate ellipsoid which may offer a special benefit. Hoyer (1911) was

Fig. 3. SEM of potato starch showing large (approx. 30μm) and small (approx. 10μm) particles.

an early observer of the particle size of wheat starch whilst Buchanan and Naudain (1923) drew attention to the presence of the small particles in wheat starch and their effect on the 'strength' of flour. The bimodal particle size distribution was the subject of further studies by Klassen and Hill (1971). Later investigators concentrated on the variation in size distribution with wheat cultivar and maturity for example Brocklehurst and Evans (1977), Duffus and Murdoch (1979), Meredith (1981), Karlsson et al (1983),and Baruch et al (1983).

Size Classification of Starches

All starch manufacturing techniques discharge their product initially in the form of an aqueous slurry. It would seem appropriate to give serious consideration to using controlled sedimentation as potential method for separating the desired fine

Fig. 4. SEM of wheat starch showing large (approx. 15μm) ellipsoidal, and small (approx. 6μm) spherical particles.

particles. The wheat starch industry, however, is obliged to deal with the unique properties of wet gluten which, because of its unusual colloidal properties, complicates certain of the industrial processes used. As a result only a partial size classification is achieved in which some of the small particles collect with protein residues and other side products into the B grade material whilst the bulk of the small fraction appears with the purer A grade product. The B grade is not acceptable as a feedstock for SFP work because of its high protein content which can initiate Maillard reactions and consequential yellowing of the plastic products. The A grade material could, presumably, be further wet classified on a size basis alone or, if the economics proved favorable, be pneumatically dry classified using the Alpine zig-zag centrifuge or similar equipment. Because most starch markets are not granule size sensitive there should be no commercial difficulty caused by withdrawing the fine material.

ELLIPSOIDAL SHAPE, RHEOLOGICAL CONSEQUENCES

Behaviour of Particles in Melts

The flow of polymer melts through processing machinery is, in almost every case, strictly laminar, ie without turbulence. The melts also wet the walls of the dies through which they flow and therefore establish a velocity profile which, in the case of polyethylene, is almost Newtonian. Any solid particle of significant size suspended in such a flow will have various aspects of its surface in contact with viscous fluid moving at different velocities and, perhaps not surprisingly, it will adopt some motions other than simple translation in the general direction of flow. There is a substantial literature on the subject of flow orientation some noting the orientation of needle-like particles in the direction of flow, whilst others have considered the matter from a theoretical viewpoint. The practical aspects of flow orientation taking place during polymer melt processing of needle and plate-like particles have long been exploited in the manufacture of materials such as polarising filters and also paints and plastics having a simulated pearl appearance. A landmark in the development of our understanding of these particle movements is to be found in the remarkable theoretical paper by Jeffery (1922) in which he considered the specific case of the behaviour of oblate ellipsoidal bodies immersed in flowing viscous fluids and came to the conclusion that they would oscillate and finally adopt mean positions parallel to the flow plane, i.e. the energy associated with their interaction with the fluid would be minimised. The die of an extruder designed for making thin polymer films is a simple annulus with its major diameter very large by comparison with the width of the annular gap. If we consider the flow through a short peripheral segment of the annulus it will be very little different to flow between closely spaced parallel plates. Disc-like particles suspended in a viscous medium flowing through such a region will orientate until their axes are normal to the plane of the plates and this implies that the larger wheat starch particles frozen in position in a polymer film are likely to be found lying flat in the plane of the film.

Experimental Study

An A grade wheat starch of Australian origin was compounded as a 50% masterbatch in low density polyethylene of MFI (melt flow index) 2 and density 0.918 tonnes per metre cube. The resultant white product was cold blended with further polyethylene granules of the same type in a ratio to ensure a 5% concentration of starch in the final product. This blend was then blown into 50 um film using a laboratory film line made by Betol Ltd., UK. The blow-up ratio was about 2:1 and the film had the expected 'satin' surface instead of the glossy surface of an unfilled polymer. There was no evidence of yellowing due to pyrolytic degradation of the starch even though the melt temperature was 190 Centigrade.

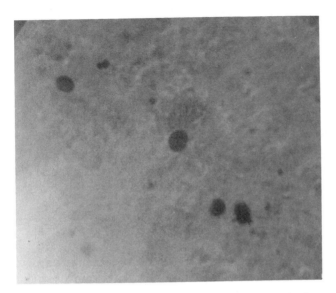

Fig. 5. Photomicrograph of 50μm polyethylene film containing 5% of wheat starch, iodine stained. Plan view and all particles appear circular.

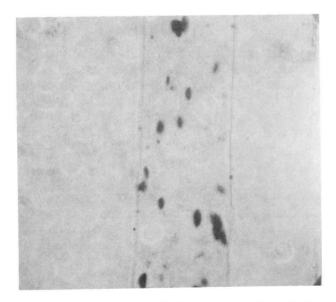

Fig. 6. Photomicrograph of the same polyethylene film as shown in Fig. 4 but taken of a cross section in which the particles, seen on edge, appear elliptical.

The starch particles in the product film were examined by optical microscopy after staining in iodine solution. For a plan view a simple square piece of the film was cut, immersed in the stain for 24 hours and then mounted for microscopy using Canada Balsam. For a section view it was necessary to cut a microtome section through the film, again immerse in iodine solution until all the starch particles were visible and then manouvre the slender strips of material into a position which presented the cut edge to the microscope objective by causing the sections to roll in the viscous mountant when the cover slip was displaced slightly sideways. This delicate motion was readily achieved by sliding the cover slip using rubber tipped eraser pencil. Simple optical transmission microscopy has been found best for such studies despite the need for the lengthy immersion in iodine solution because the alternatives of polarised light or phase contrast microscopy are greatly disturbed by the birefringence of the polymer matrix and its variable thickness. If Jeffery's theory holds for this system and the crystallisation of the polymer on cooling has not physically disturbed the orientation of the granules then we might expect to see disc like particles in the plan view of the extruded film and elongated lenticular shapes in cross section where the discs were only visible edge-on. Figs. 4 and 5 show that this was indeed the case and that true flow induced orientation of the larger wheat starch particles has taken place.

CONCLUSIONS

A special wheat starch fine fraction appears to be both economically and technically a potential candidate filler for degradable thermoplastic polymer films substantially thinner than those that are being made using maize starch. The demonstration of flow orientation of the flat larger particles might suggest that the classification process need not be over-critical. Technical collaboration with the wheat starch industry is already under way with a view to confirming these beliefs and to setting precise bounds on the separations that can be achieved and, consequently, the thinness of the plastics films that will result.

ACKNOWLEDGEMENT

The Author would like to record his gratitude to the Archer Daniels Midland Co. of Decatur, Illinois, current world licensees of his technology for their encouragement to pursue work on the utilisation of wheat starch in SFP.

LITERATURE CITED

BUCHANAN, J. H., and NAUDAIN, G. G. 1923. Influence of starch on the strength of wheat flour. Ind. Eng. Chem. 15:1050.

BROCKLEHURST, P. A., and EVANS, A. D. 1977. The size distribution of starch granules in endosperm of different size kernels of the wheat cultivar Maris Huntsman. J. Sci. Food Agric. 28(12):1084.

BARUCH, D. W., JENKINS, L. D., DENGATE, H. N., and MEREDITH, D. 1983. Nonlinear model of wheat starch granule distribution at several stages of development. Cereal Chem. 60(1):32.

DUFFUS, C. M., and MURDOCH, S. M. 1979. Variation in starch granule size and amylose content during development. Cereal Chem. 56(5):427.

GRIFFIN, G. J. L. 1972,73. UK Pat. 1485833, US Pat. 4021388, etc.

GRIFFIN, G. J. L. 1975. Biodegradable fillers in thermoplastics. ACS Advances Chem. Series No. 134:159.

GRIFFIN, G. J. L. 1987. Degradable plastics films. Proc. SPI Symp. Washington, 'Degradable Plastics':47.

MOYER, O. 1911. Die Korngrosse der Weizen und Kartoffel Starke. Eine vergleichende Untersuchung. Chem. Zentralblatt 2:305.

JEFFERY, G. B. 1922. The motion of ellipsoidal particles immersed in a viscous fluid. Proc. Roy. Soc. A102:161.

KLASSEN, A. J., and HILL, R. D. 1971. Comparison of starch from Triticale and its parental species. Cereal Chem. 48:647.

KARLSSON, R., OLERED, R., and ELIASSON, A. C. 1983. Changes in the starch granule size and starch gelatinization properties during development and maturation of wheat, barley, and rye. Staerke 35(10):335.

MEREDITH, P. 1981. Large and small starch granules in wheat: are they really different? Staerke 33(2):40.

WRAP-UP OF CONFERENCE

In preparation for this symposium, I had many conversations and brainstorm sessions with scientists from government, academia, and private institutes; representatives of the grain trade, the equipment-plant engineering companies, the wheat processing industry; and the wheat growers. Opinions about the advisability of organizing the symposium oscillated between those who kept saying there is nothing new under the sun in industrial processing of cereals and those who kept looking (hoping) for spectacular proposals and miraculous solutions that could be used to develop one-liners for use by the media.

What were the objectives of the conference? In the final analysis, this conference dealt with enhanced research for value-added food and non-food uses of wheat. Value-added research helps to maintain reasonable food costs and promotes processed agricultural products in world markets (Anon 1987).

In many of the presentations and discussions, the experience gained by the maize processing industry repeatedly came up. Consequently, a comparison of the products from wheat and maize processing is of particular interest; 100 kg grain yield:

Maize: 20 kg by-products, 3 kg oil, 4 kg germ, 5 kg protein, and 63 kg starch;

Wheat: 20-23 kg bran, 10 kg solubles, 8 kg gluten, and 45 to 47 kg starch A, plus 10 to 12 kg starch B.

In 1985, France used the following amounts of modified starches in food products (in tons): baked goods 2230, confectionery 920, dairy 135, ice cream 70, desserts 470, condiments, soups, and sauces 240, convenience foods 150, pet foods 85, diverse 250. Extensive research is conducted on the use of modified wheat starches in dietary foods (polydextrose, fat substitutes, etc.) and in the paper and textile industries. A large investment is made in research on utilization of starch (10% of the total) in industrial products. They include: biodegradable detergents, encapsulation of pesticides, biodegradable plastics, recovery of metals from dilute solutions, stabilization of emulsions, the building industry, pharmaceutical products, hygienic products (i.e. super absorbents for infants), miscellaneous adhesives and glues, and, of course, the huge paper and textile industry (Anon. 1988).

Finally, wheat flour can be used by the fermentation industry for conversion to organic acids, amino acids, ethanol, polysaccharides, and miscellaneous products (vitamin B12, antibiotics, enzymes, and nucleotides).

<u>What happened during the symposium?</u> We had the opportunity to review the new frontiers, learn about where we are at present, and hear some interesting and thought-provoking projections about what should, can, and will be done in the future. We reviewed not only the state-of-the-art, but also the state-of-the-possible and -impossible (or may be only the unlikely). We had all that review during a dialog and an exchange between cereal chemists and technologists, on one hand, and geneticists - plant breeders - wheat producers on the other hand. It was a true exchange of needs, information, and projections. In German, the two six letter words lehren (to teach) and lernen (to learn) are very similar because they compliment each other and make for an effective and meaningful two way communication.

This conference brought together a series of presentations, reviews, and discussions related to wheat industry utilization seen from different perspectives: of the biologist, plant breeder, and farmer who search for new techniques to produce more, better, and uniquely tailored crops; of the chemist, biochemist, engineer, and food technologist who study methods and processes to convert raw materials, by-products, and waste materials into chemicals, fermentation products, and fuel alternatives, and of the economist and systems expert who analyze the economical feasibility of exploiting, systematizing,

708

nd optimizing the conversion of wheat into food, feed, and miscellaneous new, novel, and needed products. We attempted to intergrate all those contributions into a unified picture so a multifaceted perspective of the complex issues may be obtained with due consideration to technological, environmental, and social impacts.

What have we learned? There are new processes, new technologies, and new products that were unknown a decade ago. Major progress has been made and is being made, and consequently things that were considered an impossible dream are in the realm of the attainable. We have identified opportunities for future developments and have realistically assessed impediments to those developments. The main question asked was how to translate the information gained during the conference into practice? Throughout the conference it has become abundantly clear that to increase the added value of wheat the plant breeder and farmer must recognize the inalienable facts about the marketing dimension of wheat production: a) that not all wheats are created equal from a standpoint of end-use properties, and b) we must produce wheats that can be marketed and not to market wheats merely because they were produced. Judicious use of improved technologies for the processing, marketing, storage, handling, and distribution of wheat no doubt will result in tangible benefits to the producer, the competitive marketing system, and the consumer. To what extent biotechnology, new advances in material science, and specifically tailored wheats for industrial uses will offer potential for new value-added agricultural products - remains to be determined.

Criticism (in part quite justified) has been voiced during the conference by wheat producers to the effect that:
a) some scientists have no well defined objectives, and
b) the invariable conclusion, (or so it seems, at least) of much theoretical research is....the need for more research.

Still, there is need for good theoretical information for very practical reasons. We need to define precisely what it is that we are after in the best wheat so that we can tell specifically the molecular biologist-geneticist-plant breeder what to put into the wheat plant or kernel to meet our objectives (Pomeranz 1981, 1987).

Postharvest technology research offers numerous benefits to producers of agricultural commodities, food processors, and members of the marketing system. It accomplishes those objectives through increased productivity in the postharvest processing and distribution system; added value; lower food costs; improved quality, safety, nutritional value, and convenience; or improved or novel products; and training personnel to accomplish the above. To accomplish those objectives, the research must:
a) define and characterize fundamental physical, chemical, and biological properties and structure of agricultural commodities;
b) understand the mechanisms to control or inhibit biological and chemical activity in agricultural commodities;
c) find innovative technologies for better utilization of agricultural products and development of new food and non-food products;
d) develop food safety methods to effectively control chemical and microbial agents; and
e) improve understanding of fundamental diet-health relationships (Anon 1987).

The final question that came up was <u>who should conduct the research?</u> To this end, we must identify and distinguish among the terms science, engineering and technology (Harrison 1984). Science involves investigation of physical, biological, social, and economic phenomena. In investigating those phenomena scientific knowledge consists of reliable and new data bases, sound methodologies, and confirmation or refutal of concepts. In engineering we investigate how to solve problems. Technology is concerned with the practical aspects of producing and delivering goods and services: more effectively, or with better characteristics, or of novel types. Science, engineering, and technology drive each other and in many ways are synergistic (Harrison 1984).

It is my belief that the scientific community should generate the know how in science and technology. Development of that know how in universities, government (state and federal), and private research institutions lacks, at present, adequate public and private support. The agricultural research system spends less that 15% of its research budget on research or development in the postharvest technology area. Over 75% of the cost to the consumer is added by processing, packaging, and distribution. Much of the current postharvest research program is of an applied nature. More fundamental research is needed to promote innovative and significant research and technological developments (Anon 1987). Production agriculture has seen significant increases in productivity over the past 20-40 years. However, the processing and distribution sectors of the agricultural system have not experienced similar productivity increases. Studies on postharvest technology and marketing emphasized the potential for increased productivity and reduced real cost of food from enhanced postharvest research.

I do believe, however, it would not be prudent or effective for the public institutions (including those of wheat growers) to engage in technology. Those activities are best implemented by the profit-oriented private sector. How far the research institutions should go in generating information in science and engineering and how it can be transferred best to the private sector, is a most difficult and vexing decision.

Conference guidelines were to have international contributors (and especially those from industry) who have something to say, have permission to say it (if proprietary information), know how to say (and write) it effectively, take the time to prepare well what they are about to say so it will have all the benefits of organized spontaneity, and are eager to say it while traveling a long way at their cost (Pomeranz 1981). Those guidelines are hard to meet. Proceedings of the conference document the extent to which we have succeeded. We have tried both avoiding generalities and everyone addressing the same question, while fully recognizing that most profound

questions have many facets that require interdisciplinary approaches.
It has been a challenging and professionally gratifying effort. I wish to thank the sponsors and authors for their fine contributions and for the opportunity to organize the conference.

November 1988
Pullman, WA

Y. Pomeranz
Technical Program Chair

LITERATURE CITED

ANON. 1987. Enhanced Research Agenda for Value-Added Food and Non-Food Uses of Agricultural Products. The Experiment Station Committee on Organization and Policy, CSRS-USDA, Washington, DC.

ANON. 1988. Les Cereales dans L'Agroindustrie Institut Technique des Cereales et des Fourages. Paris.

HARRISON, A. J. 1984. Science, engineering, and technology. Science 233:4636.

POMERANZ, Y. 1981. Cereals - a renewable resource. In: Cereals - A Renewable Resource; Theory and Practice. Y. Pomeranz and L. Munck, eds. pp. 709-728. Am. Assoc. Cereal Chem., St. Paul, MN.

POMERANZ, Y. 1987. Modern Cereal Science and Technology. VCH Publishers, Inc., New York.

SUBJECT INDEX

Agricultural
  biotechnology, 49-68, 642
  policies, 3-9, 49-68
  refinery, 63-67
Agri-commodity
  monetary considerations, 54-56
  present trends, 53-54
  production, 53-56
  versus agri-culture, 50-53
Air classification, 431-442
  baking quality, 441-442
  early studies, 431-433
  hard wheat fractions, 435-437
  particle size distribution, 437-438
  processing and utilization, 431, 441-442
  starch damage, 434-435
Alcohol fermentation by-products, 657-672, 675-689
Alpha-amylase, 103-112, 117-126, 203, 217
  inactivation, 119, 120, 123, 125
Amsace process, 529-530
Amsatin process, 524-526
Amylopectin, 218, 238
Amylose leaching, 222-223
Antibody
  applications of, 133
  identification tests, 135
  probes and tests, 132-143
  quality type identification, 138
Australian wheat
  exports, 22, 23, 25-26, 29
  domestic processing, 24-25
  production, 22, 23
  products, 26, 31-34
  uses of, 21-22, 25, 27
Automation in wheat processing, 607-608

B & M hydrocyclone process, 502
Bread, 89, 93, 94, 353, 571
Breakfast foods
  protein, 89

Carbohydrate digestion, 95-98
Caryopsis, 71, 72, 75, 76
Cereal
  breeding, antibody tests, 133
  mechanisms for development of, 371
  processing variables, 374
  quality diagnostics, 148
Cellulose yield, 62
Cholesterol
  effect of dietary fiber on, 92, 93, 94
Conference objectives I, 707-708
Conference program II, 708-712
Copolymers, graft, starch, 627
Corn (maize)
  amylopectin, 218
  amylose, 217, 222
  composition, 216
  crop disease diagnostics, 141-143
  gelatinization temperatures, 219
  glucose, 257
  lipids, 224, 259
  paste consistency, 222
  solubility and swelling power, 222
  starch, 216, 647
Corn wet milling industry, statistics, 7, 11

Dextrins, microbial, 623
Dietary guidelines, 87-88
Differential Scanning Calorimetry (DSC) 180, 182, 184, 237
Digestion, 95, 200, 201
Disc milling, 446-448
Distillers' grains, 657-672, 675-689
  chemical composition, 677-678
  development and utilization, 675-689
  nutritional quality, 679-684
  sensory quality, 685-689
Dough
  effect of heat on setting, 601-603
  formation, 596-599
  gas retention, 599-601
Dustiness, effects of lipids on, 346

EC policy and incentive prices, 5-6
Electrophoresis, 170, 305
Embryo recovery, 79
Endosperm, 76, 78, 79, 135, 194
Extrudates
  analytical characterization, 396-405, 407
  effect of lipids on, 358
Extrusion cooking, 369-378, 379-389, 395-427
  analytical model, 396-405
  biotechnology, 381-389
  enzymic modification, 382-384
  saccharification, 386-387
  system analytical model, 396-405
  parameters, 398-401, 405-416, 419-426
  product characteristics, 417-419, 420, 422-426
  wheat starch, 397-405

Far-Mar-Co process, 531-533
Fermentation
  by-products, 657-672, 675-689
Flour
  alpha-amylase content, 103
  characteristics of, 376
  costs, 42-43
  milling, 445-455
  protein, 45
  solubles, 47-48
  types in Japan, 608-610

Gelatinization of starch, 174-184, 202-204, 219-221, 464
Gel electrophoresis, 170, 305-306
Gel strength, 227
Glass-transition temperatures, 219, 220
Gliadins, 278-279
  ratio with glutenin, 292-293

Glucose
  synthesis, 286-289
  components, 257
  major uses, 251
  syrup production, 256-260, 386-387
  Trudex, spray dried total sugar, 633-640
  Zymomonas process, 643-651
Glucotech process, 641-651
Glues, urea-formaldehyde resin, 556
Gluten
  chemically modified, 541-560
  chemical properties, 266-270
  composition, 267-270
  developments in extraction, 479
  economics, 41
  enzymic modification, 610-612
  heat denaturation, 567-571
  physical properties, 266, 269
  production, 2, 15, 32-33, 41-42, 264-266, 509, 581
  properties of, 184-190, 266-271, 458-462, 471-476
  quality control tests, 140
  relations between properties and quality, 185
  separation, 1-20, 41-48, 467-477, 479-498, 501-508, 509-519, 521-539
  subsidies, 13
  supply/demand, 267
  thermal modification, 563-587
  utilization, 34, 271-274, 542-544, 545-559
  yield estimation, 44, 470-471
Glutenin, 279-297, 581
  drying, 581
  polymers, 280, 293, 294
  ratio with gliadin, 292-293
  structure, 277, 295-297
  subunits, 281-286
  synthesis, 286-289
Glycemic index, 96-97
Halle process, 527-529
Health-related
  limitations, 91-98
  potential, 91-98
Heat denaturation, of proteins, 563-584
  and breadmaking, 571-574
  and cookie making, 571-574
  and pasta making, 574-581
  and vital gluten, 581-584
High-Performance Liquid Chromatography (HPLC), 170-174, 303-315
Hormones diagnostics, 146-147
Immunoassay formats, 147-150
Immunodiagnostic
  approaches to process control, 131-150
  approaches to quality control, 131-150
Inhibition, sprout damaged wheat, 117, 119, 120

Lipid
  composition, 319-321, 325, 344-346
  content, 199, 224, 259
  different wheat classes and varieties, 348
  distribution, 321-324, 344-346
  dustiness, effect on, 346
  functionality in breadmaking, 327-333, 349-351, 353-355
  functional significance, 327-33 341, 355-360
  genetic control, 331, 333
  progress in methodology, 343-34
  redistribution on milling, 324-
  role in end-use properties, 353 355-360
  stored wheat and flour, 324, 34
  wet processing, 351-352
  wheat, 259, 348-349
Lipoxygenase, 328-329, 337
Longford-Slotter process, 530-531, 535-538

Maize (see corn)
Martin process, 2, 265, 523
Milling process, 445-455
  disc, 446-448
  short, for wheat, 445
Mixograph, 165-168
Mycotoxins
  identification, 143-144

Non-food
  major uses, 252-253, 462-466
  paper pulp, 37, 462-464
  utilization, 37, 59, 60-62
Nucleic acid probes, 132-133
Nutritional
  limitations, 85
  potential, 85-91

Organic acids, starch, 622-624
Osmosis, reverse, and ultrafiltra 660

Paper
  coating mixtures, 545
  making, starch in, 462-464
Pasta
  high temperature cooking, 574,
  high temperature drying, 574,
Paste consistency, 221-222
Pentosans, 258
Pesticides, immunological assay, 144
Pillsbury process, 533
Plant hormone diagnostics, 146-1
Plastic
  degradable, formulation, 695
  starch composition, 625
Polyethylene film, SEM, 696-697
Polymers, 624-627, 695-705
Polyols, starch, 622, 696
Pre-harvest sprouting of wheat, 117-126
Profitability of the EC wheat wa industry, 4-6

Quality of cereals and cereal pr control tests, 140-141, 148,

Raisio process, 265
Rapid Visco Analyser, 162-164

Refineries, agricultural, 63-66

Sausage, edible synthetic skins, 558
SDS-PAGE, glutenin, 281
Solubles, flour, 47-48
Spaghetti
  effect of lipids on, 358
Species identification tests, 135-138
Sprout
  damaged wheat, inhibition in, 104, 117, 123-126
  grading scale, 106-107
  visual assessment, 103, 106
Sprouting of wheat, 103-112, 117-126
Starch
  Alsace process, 529-530
  Alsatin process, 524-526
  B & M hydrocyclone process, 502-508
  chemical modification, 235-241, 616-630
  chemicals from, 619-624
  composition, 195-197, 207
  content, 194, 199, 206, 216
  current markets, 616
  damage, 434-435
  degradable plastics, 624-627, 695-705
  extrusion, 397-405
  Far-Mar-Co process, 531
  fine particle starches, 697
  gelatinization, 178-184, 202-204, 219-221, 464-466
  glucose, 257, 633-640, 641-651
  graft copolymers, 627-629
  granule composition, 195-200, 216-218
  granule development, 193-195
  granule digestibility, 200-202
  granule size, 178-181, 215, 216, 697-700
  Halle process, 527
  interactions with other components, 181-184
  lipids, 223-227
  Longford-Slotter process, 530-531, 535-538
  major uses, 252, 253, 462-466
  microbial polysaccharides and dextrins, 623-624
  minor components of, 253-260
  modification, chemical, 235-241, 616-630
  modification, physical, 241-247
  new and potential markets for, 615, 618
  non-food uses, 59-60
  organic acids, 622-623
  physical modification, 241-247
  Pillsbury process, 533
  plastic compositions, 625, 697-700
  olymers, 624-629, 697-700
  olyols, 622
  roduction, 15, 33-35, 57-59, 509, 521
  operties, 204-209, 215, 227-229, 237, 242, 701-704
  trogradation, 178, 180
  paration, 1-20, 41-48, 467-477, 479-498, 501-508, 509-519, 521-539
  ze classification, 700

solubility, 222
structure, 216-218
swelling, 202-204
urethane composition, 627
viscosity, 162-165
wheat, fermentation, 646, 657
yield, 44, 45, 62
Stillage
  fermentation and fractionation of, 659-662, 664-665
  nitrogen distribution and content, 667, 672
  functional properties, 669, 672
  solubles, UF and RO, 666
Subsidies for gluten prices, 13-14
Sugar, Trudex, 633-640
System analytical model, extrusion cooking, 396-405

Thermodification, gluten, 563, 567-571
Total sugar spray dried, Trudex, 633-640

Ultrafiltration and reverse osmosis, 660
Urea-formaldehyde resin glues, 556
Urethane starch composition, 627

Variety identification tests, 135-138, 168-174, 305-310
Viscosity of starch, 162-165

Wet milling, corn, 7, 11, 481-482
Wet milling system for gluten and starch, 483-498, 509-519, 521-539
Wheat
  aleurone, 75-76
  breeding, 310-311
  caryopsis alteration, 62-81
  classification, 311-313, 612-614
  crease, 72-75
  endosperm, 76-79
  germ, 79-81
  marketing, 313-314
  modification during sprouting, 106-112
  non-food utilization, 35-36
  production and future uses in Australia, 22-24
  utilization and quality, 314-315
Wheat fractionation, 1-15, 457-466, 467-477, 479-498, 501-508
  economic constraints, 6
  economic developments, 6, 10
  future trade impacts, 14, 15
  growth, 1
  past profitability, 4
  revenues, 8, 12
  technical characteristics, 2, 12
  technical constraints, 6, 9
Wheat washing industry
  future developments, 6-10
  future trade impacts, 14-17
  statistics, 8, 12
  technical characteristics, 2-3
Whole-crop harvesting, 65-66

Zymomonas process, 388, 643-651

715

Erschienen
aus Anlaß des
XXI. Deutschen Orientalistentages
in Berlin